Ten Lectures on Wavelets

CBMS-NSF REGIONAL CONFERENCE SERIES IN APPLIED MATHEMATICS

A series of lectures on topics of current research interest in applied mathematics under the direction of the Conference Board of the Mathematical Sciences, supported by the National Science Foundation and published by SIAM.

GARRETT BIRKHOFF, *The Numerical Solution of Elliptic Equations*
D. V. LINDLEY, *Bayesian Statistics, A Review*
R. S. VARGA, *Functional Analysis and Approximation Theory in Numerical Analysis*
R. R. BAHADUR, *Some Limit Theorems in Statistics*
PATRICK BILLINGSLEY, *Weak Convergence of Measures: Applications in Probability*
J. L. LIONS, *Some Aspects of the Optimal Control of Distributed Parameter Systems*
ROGER PENROSE, *Techniques of Differential Topology in Relativity*
HERMAN CHERNOFF, *Sequential Analysis and Optimal Design*
J. DURBIN, *Distribution Theory for Tests Based on the Sample Distribution Function*
SOL I. RUBINOW, *Mathematical Problems in the Biological Sciences*
P. D. LAX, *Hyperbolic Systems of Conservation Laws and the Mathematical Theory of Shock Waves*
I. J. SCHOENBERG, *Cardinal Spline Interpolation*
IVAN SINGER, *The Theory of Best Approximation and Functional Analysis*
WERNER C. RHEINBOLDT, *Methods of Solving Systems of Nonlinear Equations*
HANS F. WEINBERGER, *Variational Methods for Eigenvalue Approximation*
R. TYRRELL ROCKAFELLAR, *Conjugate Duality and Optimization*
SIR JAMES LIGHTHILL, *Mathematical Biofluiddynamics*
GERARD SALTON, *Theory of Indexing*
CATHLEEN S. MORAWETZ, *Notes on Time Decay and Scattering for Some Hyperbolic Problems*
F. HOPPENSTEADT, *Mathematical Theories of Populations: Demographics, Genetics and Epidemics*
RICHARD ASKEY, *Orthogonal Polynomials and Special Functions*
L. E. PAYNE, *Improperly Posed Problems in Partial Differential Equations*
S. ROSEN, *Lectures on the Measurement and Evaluation of the Performance of Computing Systems*
HERBERT B. KELLER, *Numerical Solution of Two Point Boundary Value Problems*
J. P. LASALLE, *The Stability of Dynamical Systems*
D. GOTTLIEB AND S. A. ORSZAG, *Numerical Analysis of Spectral Methods: Theory and Applications*
PETER J. HUBER, *Robust Statistical Procedures*
HERBERT SOLOMON, *Geometric Probability*
FRED S. ROBERTS, *Graph Theory and Its Applications to Problems of Society*
JURIS HARTMANIS, *Feasible Computations and Provable Complexity Properties*
ZOHAR MANNA, *Lectures on the Logic of Computer Programming*
ELLIS L. JOHNSON, *Integer Programming: Facets, Subadditivity, and Duality for Group and Semi-Group Problems*
SHMUEL WINOGRAD, *Arithmetic Complexity of Computations*
J. F. C. KINGMAN, *Mathematics of Genetic Diversity*
MORTON E. GURTIN, *Topics in Finite Elasticity*
THOMAS G. KURTZ, *Approximation of Population Processes*
JERROLD E. MARSDEN, *Lectures on Geometric Methods in Mathematical Physics*
BRADLEY EFRON, *The Jackknife, the Bootstrap, and Other Resampling Plans*
M. WOODROOFE, *Nonlinear Renewal Theory in Sequential Analysis*
D. H. SATTINGER, *Branching in the Presence of Symmetry*
R. TEMAM, *Navier–Stokes Equations and Nonlinear Functional Analysis*

MIKLÓS CSÖRGO, *Quantile Processes with Statistical Applications*
J. D. BUCKMASTER AND G. S. S. LUDFORD, *Lectures on Mathematical Combustion*
R. E. TARJAN, *Data Structures and Network Algorithms*
PAUL WALTMAN, *Competition Models in Population Biology*
S. R. S. VARADHAN, *Large Deviations and Applications*
KIYOSI ITÔ, *Foundations of Stochastic Differential Equations in Infinite Dimensional Spaces*
ALAN C. NEWELL, *Solitons in Mathematics and Physics*
PRANAB KUMAR SEN, *Theory and Applications of Sequential Nonparametrics*
LÁSZLÓ LOVÁSZ, *An Algorithmic Theory of Numbers, Graphs and Convexity*
E. W. CHENEY, *Multivariate Approximation Theory: Selected Topics*
JOEL SPENCER, *Ten Lectures on the Probabilistic Method*
PAUL C. FIFE, *Dynamics of Internal Layers and Diffusive Interfaces*
CHARLES K. CHUI, *Multivariate Splines*
HERBERT S. WILF, *Combinatorial Algorithms: An Update*
HENRY C. TUCKWELL, *Stochastic Processes in the Neurosciences*
FRANK H. CLARKE, *Methods of Dynamic and Nonsmooth Optimization*
ROBERT B. GARDNER, *The Method of Equivalence and Its Applications*
GRACE WAHBA, *Spline Models for Observational Data*
RICHARD S. VARGA, *Scientific Computation on Mathematical Problems and Conjectures*
INGRID DAUBECHIES, *Ten Lectures on Wavelets*
STEPHEN F. MCCORMICK, *Multilevel Projection Methods for Partial Differential Equations*
HARALD NIEDERREITER, *Random Number Generation and Quasi-Monte Carlo Methods*
JOEL SPENCER, *Ten Lectures on the Probabilistic Method, Second Edition*
CHARLES A. MICCHELLI, *Mathematical Aspects of Geometric Modeling*
ROGER TEMAM, *Navier–Stokes Equations and Nonlinear Functional Analysis, Second Edition*
GLENN SHAFER, *Probabilistic Expert Systems*
PETER J. HUBER, *Robust Statistical Procedures, Second Edition*
J. MICHAEL STEELE, *Probability Theory and Combinatorial Optimization*
WERNER C. RHEINBOLDT, *Methods for Solving Systems of Nonlinear Equations, Second Edition*
J. M. CUSHING, *An Introduction to Structured Population Dynamics*
TAI-PING LIU, *Hyperbolic and Viscous Conservation Laws*
MICHAEL RENARDY, *Mathematical Analysis of Viscoelastic Flows*
GÉRARD CORNUÉJOLS, *Combinatorial Optimization: Packing and Covering*
IRENA LASIECKA, *Mathematical Control Theory of Coupled PDEs*
J. K. SHAW, *Mathematical Principles of Optical Fiber Communications*
ZHANGXIN CHEN, *Reservoir Simulation: Mathematical Techniques in Oil Recovery*
ATHANASSIOS S. FOKAS, *A Unified Approach to Boundary Value Problems*
MARGARET CHENEY AND BRETT BORDEN, *Fundamentals of Radar Imaging*
FIORALBA CAKONI, DAVID COLTON, AND PETER MONK, *The Linear Sampling Method in Inverse Electromagnetic Scattering*
ADRIAN CONSTANTIN, *Nonlinear Water Waves with Applications to Wave-Current Interactions and Tsunamis*
WEI-MING NI, *The Mathematics of Diffusion*
ARNULF JENTZEN AND PETER E. KLOEDEN, *Taylor Approximations for Stochastic Partial Differential Equations*
FRED BRAUER AND CARLOS CASTILLO-CHAVEZ, *Mathematical Models for Communicable Diseases*

Ingrid Daubechies
Rutgers University and
AT&T Bell Laboratories

Ten Lectures on Wavelets

SIAM® SOCIETY FOR INDUSTRIAL AND APPLIED MATHEMATICS
PHILADELPHIA

15 14 13 12 11 10

Library of Congress Cataloging-in-Publication Data

Daubechies, Ingrid.
Ten lectures on wavelets / Ingrid Daubechies.
p. cm. — (CBMS-NSF regional conference series in applied mathematics ; 61)
Includes bibliographical references and index.
ISBN 978-0-898712-74-2
1. Wavelets (Mathematics)—Congresses. I. Title. II. Series.
QA403.3.D38 1992
515'.2433—dc20

92-13201

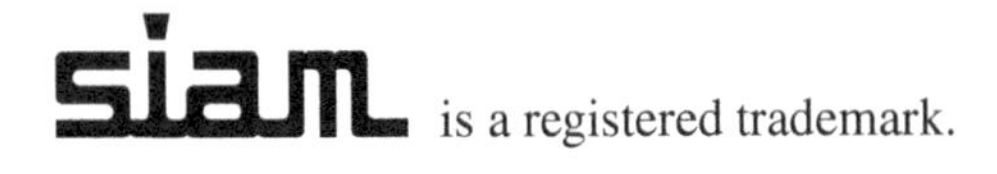

To my mother, who gave me the will to be independent.
To my father, who stimulated my interest in science.

Contents

Introduction

Wavelets are a relatively recent development in applied mathematics. Their name itself was coined approximately a decade ago (Morlet, Arens, Fourgeau, and Giard (1982), Morlet (1983), Grossmann and Morlet (1984)); in the last ten years interest in them has grown at an explosive rate. There are several reasons for their present success. On the one hand, the concept of wavelets can be viewed as a synthesis of ideas which originated during the last twenty or thirty years in engineering (subband coding), physics (coherent states, renormalization group), and pure mathematics (study of Calderón–Zygmund operators). As a consequence of these interdisciplinary origins, wavelets appeal to scientists and engineers of many different backgrounds. On the other hand, wavelets are a fairly simple mathematical tool with a great variety of possible applications. Already they have led to exciting applications in signal analysis (sound, images) (some early references are Kronland-Martinet, Morlet and Grossmann (1987), Mallat (1989b), (1989c); more recent references are given later) and numerical analysis (fast algorithms for integral transforms in Beylkin, Coifman, and Rokhlin (1991)); many other applications are being studied. This wide applicability also contributes to the interest they generate.

This book contains ten lectures I delivered as the principal speaker at the CBMS conference on wavelets organized in June 1990 by the Mathematics Department at the University of Lowell, Massachusetts. According to the usual format of the CBMS conferences, other speakers (G. Battle, G. Beylkin, C. Chui, A. Cohen, R. Coifman, K. Gröchenig, J. Liandrat, S. Mallat, B. Torrésani, and A. Willsky) provided lectures on their work related to wavelets. Moreover, three workshops were organized, on applications to physics and inverse problems (chaired by B. DeFacio), group theory and harmonic analysis (H. Feichtinger), and signal analysis (M. Vetterli). The audience consisted of researchers active in the field of wavelets as well as of mathematicians and other scientists and engineers who knew little about wavelets and hoped to learn more. This second group constituted the largest part of the audience. I saw it as my task to provide a tutorial on wavelets to this part of the audience, which would then be a solid grounding for more recent work exposed by the other lecturers and myself. Consequently, about two thirds of my lectures consisted of "basic wavelet theory," the other third being devoted to more recent and unpublished work. This division is reflected in the present write-up as well. As a result, I believe that this book will be useful as an introduction to the subject, to be used either for individual reading, or for a seminar or graduate course. None of the other lectures or workshop papers presented at the CBMS conference have been incorporated here. As a result, this presentation is biased more toward my own work than the CBMS conference was. In many instances I have included pointers to references for further reading or a detailed exposition of particular applications, complementing the present text. Other books on wavelets published include *Wavelets and Time Frequency Methods* (Combes, Grossmann, and Tchamitchian (1987)), which contains the proceedings of the International Wavelet Conference held in Marseille, France, in December 1987, *Ondelettes*, by Y. Meyer (1990) (in French; English translation expected soon), which contains a mathematically more expanded treatment than the present lectures, with fewer forays into other fields however, *Les Ondelettes en* 1989, edited by P. G. Lemarié (1990), a collection of talks given at the Université Paris XI in the spring of 1989, and *An Introduction to Wavelets*, by C. K. Chui (1992b), an introduction from the approximation theory viewpoint. The proceedings of the International Wavelet Conference in May 1989, held again in Marseille, are due to come out soon (Meyer (1992)). Moreover, many of the other contributors to the CBMS conference, as well as some wavelet researchers who could not attend, were invited to write an essay on their wavelet work; the result is the essay collection *Wavelets and their Applications* (Ruskai et al. (1992)), which can be considered a companion book to this one. Another wavelet essay book is *Wavelets: A Tutorial in Theory and Applications*, edited by C. K. Chui (1992c); in addition, I know of several other wavelet essay books in preparation (edited by J. Benedetto and M. Frazier, another by M. Barlaud), as well as a monograph by M. Holschneider; there was a special wavelet issue of IEEE *Trans. Inform. Theory* in March of 1992; there will be another one, later in 1992, of *Constructive Approximation Theory*, and one in 1993, of IEEE *Trans. Sign. Proc.* In addition, several recent books include chapters on wavelets. Examples are *Multirate Systems and Filter Banks* by P. P. Vaidyanathan (1992) and *Quantum Physics, Relativity and Complex Spacetime: Towards a New Synthesis* by G. Kaiser (1990). Readers interested in the present lectures will find these books and special issues useful for many details and other aspects not fully presented here. It is moreover clear that the subject is still developing rapidly.

This book more or less follows the path of my lectures: each of the ten chapters stands for one of the ten lectures, presented in the order in which they were delivered. The first chapter presents a quick overview of different aspects of the wavelet transform. It sketches the outlines of a big fresco; subsequent chapters then fill in more detail. From there on, we proceed to the continuous wavelet transform (Chapter 2; with a short review of bandlimited functions and Shannon's theorem), to discrete but redundant wavelet transforms (frames; Chapter 3) and to a general discussion of time-frequency density and the possible existence of orthonormal bases (Chapter 4). Many of the results in Chapters 2–4 can be formulated for the windowed Fourier transform as well as the wavelet transform, and the two cases are presented in parallel, with analogies and differences pointed out as we go along. The remaining chapters all focus on orthonormal bases of wavelets: multiresolution analysis and a first general strategy for the construction of orthonormal wavelet bases (Chapter 5), orthonormal bases of compactly supported wavelets and their link to subband coding (Chapter 6), sharp regularity estimates for these wavelet bases (Chapter 7), symmetry for compactly supported wavelet bases (Chapter 8). Chapter 9 shows that orthonormal bases are "good" bases for many functional spaces where Fourier methods are not well adapted. This chapter is the most mathematical of the whole book; most of its material is not connected to the applications discussed in other chapters, so that it can be skipped by readers uninterested in this aspect of wavelet theory. I included it for several reasons: the kind of estimates used in the proof are very important for harmonic analysis, and similar (but more complicated) estimates in the proof of the "T(1)"-theorem of David and Journé have turned out to be the groundwork for the applications to numerical analysis in the work of Beylkin, Coifman, and Rokhlin (1991). Moreover, the Calderón–Zygmund theorem, explained in this chapter, illustrates how techniques using different scales, one of the forerunners of wavelets, were used in harmonic analysis long before the advent of wavelets. Finally, Chapter 10 sketches several extensions of the constructions of orthonormal wavelet bases: to more than one dimension, to dilation factors different from two (even noninteger), with the possibility of better frequency localization, and to wavelet bases on a finite interval instead of the whole line. Every chapter concludes with a section of numbered "Notes," referred to in the text of the chapter by superscript numbers. These contain additional references, extra proofs excised to keep the text flowing, remarks, etc.

This book is a mathematics book: it states and proves many theorems. It also presupposes some mathematical background. In particular, I assume that the reader is familiar with the basic properties of the Fourier transform and Fourier series. I also use some basic theorems of measure and integration theory (Fatou's lemma, dominated convergence theorem, Fubini's theorem; these can be found in any good book on real analysis). In some chapters, familiarity with basic Hilbert space techniques is useful. A list of the basic notions and theorems used in the book is given in the Preliminaries.

The reader who finds that he or she does not know all of these prerequisites should not be dismayed, however; most of the book can be followed with just the basic notions of Fourier analysis. Moreover, I have tried to keep a very pedestrian pace in almost all the proofs, at the risk of boring some mathematically sophisticated readers. I hope therefore that these lecture notes will interest people other than mathematicians. For this reason I have often shied away from the "Definition–Lemma–Proposition–Theorem–Corollary" sequence, and I have tried to be intuitive in many places, even if this meant that the exposition became less succinct. I hope to succeed in sharing with my readers some of the excitement that this interdisciplinary subject has brought into my scientific life.

I want to take this opportunity to express my gratitude to the many people who made the Lowell conference happen: the CBMS board, and the Mathematics Department of the University of Lowell, in particular Professors G. Kaiser and

M. B. Ruskai. The success of the conference, which unexpectedly turned out to have many more participants than customary for CBMS conferences, was due in large part to its very efficient organization. As experienced conference organizer I. M. James (1991) says, "every conference is mainly due to the efforts of a single individual who does almost all the work"; for the 1990 Wavelet CBMS conference, this individual was Mary Beth Ruskai. I am especially grateful to her for proposing the conference in the first place, for organizing it in such a way that I had a minimal paperwork load, while keeping me posted about all the developments, and for generally being the organizational backbone, no small task. Prior to the conference I had the opportunity to teach much of this material as a graduate course in the Mathematics Department of the University of Michigan, in Ann Arbor. My one-term visit there was supported jointly by a Visiting Professorship for Women from the National Science Foundation, and by the University of Michigan. I would like to thank both institutions for their support. I would also like to thank all the faculty and students who sat in on the course, and who provided feedback and useful suggestions. The manuscript was typeset by Martina Sharp, who I thank for her patience and diligence, and for doing a wonderful job. I wouldn't even have attempted to write this book without her. I am grateful to Jeff Lagarias for editorial comments. Several people helped me spot typos in the galley proofs, and I am grateful to all of them; I would like to thank especially Pascal Auscher, Gerry Kaiser, Ming-Jun Lai, and Martin Vetterli. All remaining mistakes are of course my responsiblity. I also would like to thank Jim Driscoll and Sharon Murrel for helping me prepare the author index. Finally, I want to thank my husband Robert Calderbank for being extremely supportive and committed to our two-career-track with family, even though it occasionally means that he as well as I prove a few theorems less.

Ingrid Daubechies
Rutgers University
and
AT&T Bell Laboratories

In subsequent printings minor mistakes and many typographical errors have been corrected. I am grateful to everybody who helped me to spot them. I have also updated a few things: some of the previously unpublished references have appeared and some of the problems that were listed as open have been solved. I have made no attempt to include the many other interesting papers on wavelets that have appeared since the first printing; in any case, the list of references was not and is still not meant as a complete bibliography of the subject.

Preliminaries and Notation

This preliminary chapter fixes notation conventions and normalizations. It also states some basic theorems that will be used later in the book. For those less familiar with Hilbert and Banach spaces, it contains a very brief primer. (This primer should be used mainly as a reference, to come back to in those instances when the reader comes across some Hilbert or Banach space language that she or he is unfamiliar with. For most chapters, these concepts are not used.)

Let us start by some notation conventions. For $x \in \mathbb{R}$, we write $\lfloor x \rfloor$ for the largest integer not exceeding x,

$$\lfloor x \rfloor = \max \{n \in \mathbb{Z};\quad n \leq x\} .$$

For example, $\lfloor 3/2 \rfloor = 1$, $\lfloor -3/2 \rfloor = -2$, $\lfloor -2 \rfloor = -2$. Similarly, $\lceil x \rceil$ is the smallest integer which is larger than or equal to x.

If $a \to 0$ (or ∞), then we denote by $O(a)$ any quantity that is bounded by a constant times a, by $o(a)$ any quantity that tends to 0 (or ∞) when a does.

The end of a proof is always marked with a $\blacksquare$; for clarity, many remarks or examples are ended with a $\square$.

In many proofs, C denotes a "generic" constant, which need not have the same value throughout the proof. In chains of inequalities, I often use $C, C', C'', \cdots$ or $C_1, C_2, C_3, \cdots$ to avoid confusion.

We use the following convention for the Fourier transform (in one dimension):

$$(\mathcal{F}f)(\xi) = \hat{f}(\xi) = \frac{1}{\sqrt{2\pi}} \int_{-\infty}^{\infty} dx\; e^{-ix\xi} f(x) . \tag{0.0.1}$$

With this normalization, one has

$$\begin{aligned} \|\hat{f}\|_{L^2} &= \|f\|_{L^2} , \\ |\hat{f}(\xi)| &\leq (2\pi)^{-1/2}\, \|f\|_{L^1} , \end{aligned}$$

where

$$\|f\|_{L^p} = \left[\int dx\; |f(x)|^p \right]^{1/p} . \tag{0.0.2}$$

Inversion of the Fourier transform is then given by

$$\begin{aligned} f(x) &= \frac{1}{\sqrt{2\pi}} \int_{-\infty}^{\infty} d\xi \; e^{i\xi x} (\mathcal{F}f)(\xi) = (\mathcal{F}f)^{\vee}(x) \;, \\ \check{g}(x) &= \hat{g}(-x) \;. \end{aligned} \tag{0.0.3}$$

Strictly speaking, (0.0.1), (0.0.3) are well defined only if f, respectively $\mathcal{F}f$, are absolutely integrable; for general L^2-functions f, e.g., we should define $\mathcal{F}f$ via a limiting process (see also below). We will implicitly assume that the adequate limiting process is used in all cases, and write, with a convenient abuse of notation, formulas similar to (0.0.1) and (0.0.3) even when a limiting process is understood.

A standard property of the Fourier transform is:

$$\mathcal{F}\left(\frac{d^\ell}{dx^\ell} f\right) = (i\xi)^\ell \; (\mathcal{F}f)(\xi) \;,$$

hence

$$\int dx \; |f^{(\ell)}(x)|^2 < \infty \leftrightarrow \int d\xi \; |\xi|^{2\ell} \; |\hat{f}(\xi)|^2 < \infty \;,$$

with the notation $f^{(\ell)} = \frac{d^\ell}{dx^\ell} f$.

If a function f is compactly supported, i.e., $f(x) = 0$ if $x < a$ or $x > b$, where $-\infty < a < b < \infty$, then its Fourier transform $\hat{f}(\xi)$ is well defined also for complex ξ, and

$$\begin{aligned} |\hat{f}(\xi)| &\le (2\pi)^{-1/2} \int_a^b dx \; e^{(\mathrm{Im}\,\xi)x} \; |f(x)| \\ &\le (2\pi)^{-1/2} \; \|f\|_{L^1} \begin{cases} e^{b\;(\mathrm{Im}\,\xi)} & \text{if} \;\; \mathrm{Im}\,\xi \ge 0 \\ e^{a\;(\mathrm{Im}\,\xi)} & \text{if} \;\; \mathrm{Im}\,\xi \le 0 \;. \end{cases} \end{aligned}$$

If f is moreover infinitely differentiable, then the same argument can be applied to $f^{(\ell)}$, leading to bounds on $|\xi|^\ell \; |\hat{f}(\xi)|$. For a C^∞ function f with support $[a, b]$ there exist therefore constants C_N so that the analytic extension of the Fourier transform of f satisfies

$$|\hat{f}(\xi)| \le C_N(1 + |\xi|)^{-N} \begin{cases} e^{b \; \mathrm{Im}\,\xi} & \text{if} \;\; \mathrm{Im}\,\xi \ge 0 \\ e^{a \; \mathrm{Im}\,\xi} & \text{if} \;\; \mathrm{Im}\,\xi \le 0 \;. \end{cases} \tag{0.0.4}$$

Conversely, any entire function which satisfies bounds of the type (0.0.4) for all $N \in \mathbb{N}$ is the analytic extension of the Fourier transform of a C^∞ function with support in $[a, b]$. This is the Paley–Wiener theorem.

We will occasionally encounter (tempered) distributions. These are linear maps T from the set $\mathcal{S}(\mathbb{R})$ (consisting of all C^∞ functions that decay faster than any negative power $(1 + |x|)^{-N}$) to $\mathbb{C}$, such that for all $m, n \in \mathbb{N}$, there exists $C_{n,m}$ for which

$$|T(f)| \le C_{n,m} \sup_{x\in\mathbb{R}} \; |(1 + |x|)^n \; f^{(m)}(x)|$$

holds, for all $f \in \mathcal{S}(\mathbb{R})$. The set of all such distributions is called $\mathcal{S}'(\mathbb{R})$. Any polynomially bounded function F can be interpreted as a distribution, with $F(f) = \int dx\, \overline{F(x)}\, f(x)$. Another example is the so-called "δ-function" of Dirac, $\delta(f) = f(0)$. A distribution T is said to be supported in $[a,b]$ if $T(f) = 0$ for all functions f the support of which has empty intersection with $[a,b]$. One can define the Fourier transform $\mathcal{F}T$ or $\hat{T}$ of a distribution T by $\hat{T}(f) = T(\check{f})$ (if T is a function, then this coincides with our earlier definition). There exists a version of the Paley–Wiener theorem for distributions: an entire function $\hat{T}(\xi)$ is the analytic extension of the Fourier transform of a distribution T in $\mathcal{S}'(\mathbb{R})$ supported in $[a,b]$ if and only if, for some $N \in \mathbb{N}$, $C_N > 0$,

$$|\hat{T}(\xi)| \le C_N (1 + |\xi|)^N \begin{cases} e^{b\ \mathrm{Im}\,\xi} & \mathrm{Im}\ \xi \ge 0 \\ e^{a\ \mathrm{Im}\,\xi} & \mathrm{Im}\ \xi \le 0\ . \end{cases}$$

The only measure we will use is Lebesgue measure, on $\mathbb{R}$ and $\mathbb{R}^n$. We will often denote the (Lebesgue) measure of S by $|S|$; in particular, $|[a,b]| = b - a$ (where $b > a$).

Well-known theorems from measure and integration theory which we will use include

Fatou's lemma. *If $f_n \ge 0$, $f_n(x) \to f(x)$ almost everywhere (i.e., the set of points where pointwise convergence fails has zero measure with respect to Lebesgue measure), then*

$$\int dx\ f(x) \le \limsup_{n\to\infty} \int dx\ f_n(x)\ .$$

In particular, if this lim sup *is finite, then f is integrable.*

(The lim sup of a sequence is defined by

$$\limsup_{n\to\infty}\ \alpha_n = \lim_{n\to\infty}\ [\sup\ \{\alpha_k;\ k \ge n\}]\ ;$$

every sequence, even if it does not have a limit (such as $\alpha_n = (-1)^n$), has a lim sup (which may be ∞); for sequences that converge to a limit, the lim sup coincides with the limit.)

Dominated convergence theorem. *Suppose $f_n(x) \to f(x)$ almost everywhere. If $|f_n(x)| \le g(x)$ for all n, and $\int dx\ g(x) < \infty$, then f is integrable, and*

$$\int dx\ f(x) = \lim_{n\to\infty} \int dx\ f_n(x)\ .$$

Fubini's theorem. *If $\int dx[\int dy\ |f(x,y)|] < \infty$, then*

$$\begin{aligned} \int dx \int dy\ f(x,y) &= \int dx \left[\int dy\ f(x,y)\right] \\ &= \int dy \left[\int dx\ f(x,y)\right], \end{aligned}$$

i.e., the order of the integrations can be permuted.

In these three theorems the domain of integration can be any measurable subset of $\mathbb{R}$ (or $\mathbb{R}^2$ for Fubini).

When Hilbert spaces are used, they are usually denoted by $\mathcal{H}$, unless they already have a name. We will follow the mathematician's convention and use scalar products which are linear in the *first* argument:

$$\langle \lambda_1 u_1 + \lambda_2 u_2,\ v\rangle = \lambda_1 \langle u_1,\ v\rangle + \lambda_2 \langle u_2,\ v\rangle .$$

As usual, we have

$$\langle v, u\rangle = \overline{\langle u, v\rangle} ,$$

where $\bar{\alpha}$ denotes the complex conjugate of α, and $\langle u, u\rangle \geq 0$ for all $u \in \mathcal{H}$. We define the norm $\|u\|$ of u by

$$\|u\|^2 = \langle u, u\rangle . \tag{0.0.5}$$

In a Hilbert space, $\|u\| = 0$ implies $u = 0$, and all Cauchy sequences (with respect to $\|\ \ \|$) have limits within the space. (More explicitly, if $u_n \in \mathcal{H}$ and if $\|u_n - u_m\|$ becomes arbitrarily small if n, m are large enough—i.e., for all $\epsilon > 0$, there exists n_0, depending on ϵ, so that $\|u_n - u_m\| \leq \epsilon$ if $n, m \geq n_0$—, then there exists $u \in \mathcal{H}$ so that the u_n tend to u for $n \to \infty$, i.e., $\lim_{n\to\infty} \|u - u_n\| = 0$.)

A standard example of such a Hilbert space is $L^2(\mathbb{R})$, with

$$\langle f, g\rangle = \int dx\ f(x)\ \overline{g(x)} .$$

Here the integration runs from $-\infty$ to ∞; we will often drop the integration bounds when the integral runs over the whole real line.

Another example is $\ell^2(\mathbb{Z})$, the set of all square summable sequences of complex numbers indexed by integers, with

$$\langle c, d\rangle = \sum_{n=-\infty}^{\infty} c_n\ \overline{d_n} .$$

Again, we will often drop the limits on the summation index when we sum over all integers. Both $L^2(\mathbb{R})$ and $\ell^2(\mathbb{Z})$ are infinite-dimensional Hilbert spaces. Even simpler are finite-dimensional Hilbert spaces, of which $\mathbb{C}^k$ is the standard example, with the scalar product

$$\langle u, v\rangle = \sum_{j=1}^{k} u_j\ \bar{v}_j ,$$

for $u = (u_1, \cdots, u_k)$, $v = (v_1, \cdots, v_k) \in \mathbb{C}^k$.

Hilbert spaces always have orthonormal bases, i.e., there exist families of vectors e_n in $\mathcal{H}$

$$\langle e_n, e_m\rangle = \delta_{n,m}$$

and

$$\|u\|^2 = \sum_n |\langle u, e_n\rangle|^2$$

for all $u \in \mathcal{H}$. (We only consider separable Hilbert spaces, i.e., spaces in which orthonormal bases are countable.) Examples of orthonormal bases are the Hermite functions in $L^2(\mathbb{R})$, the sequences e_n defined by $(e_n)_j = \delta_{n,j}$, with $n, j \in \mathbb{Z}$ in $\ell^2(\mathbb{Z})$ (i.e., all entries but the nth vanish), or the k vectors $e_1, \cdots, e_k$ in $\mathbb{C}^k$ defined by $(e_\ell)_m = \delta_{\ell,m}$, with $1 \leq \ell,\ m \leq k$. (We use Kronecker's symbol δ with the usual meaning: $\delta_{i,j} = 1$ if $i = j$, 0 if $i \neq j$.)

A standard inequality in a Hilbert space is the Cauchy–Schwarz inequality,

$$|\langle v, w\rangle| \leq \|v\|\ \|w\|\ , \tag{0.0.6}$$

easily proved by writing (0.0.5) for appropriate linear combinations of v and w. In particular, for $f, g \in L^2(\mathbb{R})$, we have

$$\left|\int dx\ f(x)\ \overline{g(x)}\right| \leq \left(\int dx\ |f(x)|^2\right)^{1/2} \left(\int dx\ |g(x)|^2\right)^{1/2} ,$$

and for $c = (c_n)_{n\in\mathbb{Z}}$, $d = (d_n)_{n\in\mathbb{Z}} \in \ell^2(\mathbb{Z})$,

$$\sum_n c_n \overline{d_n} \leq \left(\sum_n |c_n|^2\right)^{1/2} \left(\sum_n |d_n|^2\right)^{1/2} .$$

A consequence of (0.0.6) is

$$\|u\| = \sup_{v,\ \|v\|\leq 1} |\langle u, v\rangle| = \sup_{v,\ \|v\|=1} |\langle u, v\rangle|\ . \tag{0.0.7}$$

"Operators" on $\mathcal{H}$ are linear maps from $\mathcal{H}$ to another Hilbert space, often $\mathcal{H}$ itself. Explicitly, if A is an operator on $\mathcal{H}$, then

$$A(\lambda_1 u_1 + \lambda_2 u_2) = \lambda_1 A u_1 + \lambda_2 A u_2\ .$$

An operator is continuous if $Au - Av$ can be made arbitrarily small by making $u - v$ small. Explicitly, for all $\epsilon > 0$ there should exist δ (depending on ϵ) so that $\|u - v\| \leq \delta$ implies $\|Au - Av\| \leq \epsilon$. If we take $v = 0$, $\epsilon = 1$, then we find that, for some $b > 0$, $\|Au\| \leq 1$ if $\|u\| \leq b$. For any $w \in \mathcal{H}$ we can define $w' = \frac{b}{\|w\|} w$; clearly $\|w'\| \leq b$ and therefore $\|Aw\| = \frac{\|w\|}{b}\ \|Aw'\| \leq b^{-1}\|w\|$. If $\|Aw\|/\|w\|$ $(w \neq 0)$ is bounded, then the operator A is called bounded. We have just seen that any continuous operator is bounded; the reverse is also true. The norm $\|A\|$ of A is defined by

$$\|A\| = \sup_{u\in\mathcal{H},\ \|u\|\neq 0} \|Au\|/\|u\| = \sup_{\|u\|=1} \|Au\|\ . \tag{0.0.8}$$

It immediately follows that, for all $u \in \mathcal{H}$,

$$\|Au\| \leq \|A\|\ \|u\|\ .$$

Operators from $\mathcal{H}$ to $\mathbb{C}$ are called "linear functionals." For bounded linear functionals one has Riesz' representation theorem: for any ℓ: $\mathcal{H} \rightarrow \mathbb{C}$, linear and bounded, i.e., $|\ell(u)| \leq C\|u\|$ for all $u \in \mathcal{H}$, there exists a unique $v_\ell \in \mathcal{H}$ so that $\ell(u) = \langle u, v_\ell \rangle$.

An operator U from $\mathcal{H}_1$ to $\mathcal{H}_2$ is an isometry if $\langle Uv, Uw \rangle = \langle v, w \rangle$ for all $v, w \in \mathcal{H}_1$; U is unitary if moreover $U\mathcal{H}_1 = \mathcal{H}_2$, i.e., every element $v_2 \in \mathcal{H}_2$ can be written as $v_2 = Uv_1$ for some $v_1 \in \mathcal{H}_1$. If the e_n constitute an orthonormal basis in $\mathcal{H}_1$, and U is unitary, then the Ue_n constitute an orthonormal basis in $\mathcal{H}_2$. The reverse is also true: any operator that maps an orthonormal basis to another orthonormal basis is unitary.

A set D is called dense in $\mathcal{H}$ if every $u \in \mathcal{H}$ can be written as the limit of some sequence of u_n in D. (One then says that the closure of D is all of $\mathcal{H}$. The closure of a set S is obtained by adding to it all the v that can be obtained as limits of sequences in S.) If Av is only defined for $v \in D$, but we know that

$$\|Av\| \leq C\|v\| \quad \text{for all } v \in D , \tag{0.0.9}$$

then we can extend A to all of $\mathcal{H}$ "by continuity." Explicitly: if $u \in \mathcal{H}$, find $u_n \in D$ so that $\lim_{n\to\infty} u_n = u$. Then the u_n are necessarily a Cauchy sequence, and because of (0.0.9), so are the Au_n; the Au_n have therefore a limit, which we call Au (it does not depend on the particular sequence u_n that was chosen).

One can also deal with unbounded operators, i.e., A for which there exists no finite C such that $\|Au\| \leq C\|u\|$ holds for all $u \in \mathcal{H}$. It is a fact of life that these can usually only be defined on a dense set D in $\mathcal{H}$, and cannot be extended by the above trick (since they are not continuous). An example is $\frac{d}{dx}$ in $L^2(\mathbb{R})$, where we can take $D = C_0^\infty(\mathbb{R})$, the set of all infinitely differentiable functions with compact support, for D. The dense set on which the operator is defined is called its domain.

The adjoint A^* of a bounded operator A from a Hilbert space $\mathcal{H}_1$ to a Hilbert space $\mathcal{H}_2$ (which may be $\mathcal{H}_1$ itself) is the operator from $\mathcal{H}_2$ to $\mathcal{H}_1$ defined by

$$\langle u_1, A^*u_2 \rangle = \langle Au_1, u_2 \rangle ,$$

which should hold for all $u_1 \in \mathcal{H}_1$, $u_2 \in \mathcal{H}_2$. (The existence of A^* is guaranteed by Riesz' representation theorem: for fixed u_2, we can define a linear functional ℓ on $\mathcal{H}_1$ by $\ell(u_1) = \langle Au_1, u_2 \rangle$. It is clearly bounded, and corresponds therefore to a vector v so that $\langle u_1, v \rangle = \ell(u_1)$. It is easy to check that the correspondence $u_2 \to v$ is linear; this defines the operator A^*.) One has

$$\|A^*\| = \|A\|, \qquad \|A^*A\| = \|A\|^2 .$$

If $A^* = A$ (only possible if A maps $\mathcal{H}$ to itself), then A is called self-adjoint. If a self-adjoint operator A satisfies $\langle Au, u \rangle \geq 0$ for all $u \in \mathcal{H}$, then it is called a positive operator; this is often denoted $A \geq 0$. We will write $A \geq B$ if $A - B$ is a positive operator.

Trace-class operators are special operators such that $\sum_n |\langle Ae_n, e_n \rangle|$ is finite for *all* orthonormal bases in $\mathcal{H}$. For such a trace-class operator, $\sum_n \langle Ae_n, e_n \rangle$ is independent of the chosen orthonormal basis; we call this sum the trace of A,

$$\text{tr } A = \sum_n \langle Ae_n, e_n \rangle .$$

If A is positive, then it is sufficient to check whether $\sum_n \langle Ae_n, e_n \rangle$ is finite for only one orthonormal basis; if it is, then A is trace-class. (This is not true for non-positive operators!)

The spectrum $\sigma(A)$ of an operator A from $\mathcal{H}$ to itself consists of all the $\lambda \in \mathbb{C}$ such that $A - \lambda \,\mathrm{Id}$ (Id stands for the identity operator, $\mathrm{Id}\, u = u$) does not have a bounded inverse. In a finite-dimensional Hilbert space, $\sigma(A)$ consists of the eigenvalues of A; in the infinite-dimensional case, $\sigma(A)$ contains all the eigenvalues (constituting the point spectrum) but often contains other λ as well, constituting the continuous spectrum. (For instance, in $L^2(\mathbb{R})$, multiplication of $f(x)$ with $\sin \pi x$ has no point spectrum, but its continuous spectrum is $[-1, 1]$.) The spectrum of a self-adjoint operator consists of only real numbers; the spectrum of a positive operator contains only non-negative numbers. The spectral radius $\rho(A)$ is defined by

$$\rho(A) = \sup \{|\lambda|;\ \lambda \in \sigma(A)\} .$$

It has the properties

$$\rho(A) \le \|A\| \quad \text{and} \quad \rho(A) = \lim_{n \to \infty} \|A^n\|^{1/n} .$$

Self-adjoint operators can be diagonalized. This is easiest to understand if their spectrum consists only of eigenvalues (as is the case in finite dimensions). One then has

$$\sigma(A) = \{\lambda_n;\ n \in \mathbb{N}\} ,$$

with a corresponding orthonormal family of eigenvectors,

$$Ae_n = \lambda_n e_n .$$

It then follows that, for all $u \in \mathcal{H}$,

$$Au = \sum_n \langle Au, e_n \rangle e_n = \sum_n \langle u, Ae_n \rangle e_n = \sum_n \lambda_n \langle u, e_n \rangle e_n ,$$

which is the "diagonalization" of A. (The spectral theorem permits us to generalize this if part (or all) of the spectrum is continuous, but we will not need it in this book.) If two operators commute, i.e., $ABu = BAu$ for all $u \in \mathcal{H}$, then they can be diagonalized simultaneously: there exists an orthonormal basis such that

$$Ae_n = \alpha_n e_n \quad \text{and} \quad Be_n = \beta_n e_n .$$

Many of these properties for bounded operators can also be formulated for unbounded operators: adjoints, spectrum, diagonalization all exist for unbounded operators as well. One has to be very careful with domains, however. For instance, generalizing the simultaneous diagonalization of commuting operators requires a careful definition of commuting operators: there exist pathological examples where A, B are both defined on a domain D, where AB and BA both make sense on D and are equal on D, but where A and B nevertheless are not simultaneously diagonalizable (because D was chosen "too small"; see, e.g., Reed and Simon (1971) for an example). The proper definition of commuting for unbounded self-adjoint operators uses associated bounded operators: H_1 and H_2 commute if their associated unitary evolution operators commute. For a self-adjoint operator H, the associated unitary evolution operators U_t are defined as follows: for any $v \in D$, the domain of H (beware: the domain of a self-adjoint operator is not just any dense set on which H is well defined), $U_T v$ is the solution $v(t)$ at time $t = T$ of the differential equation

$$i \frac{d}{dt} v(t) = H v(t) ,$$

with initial condition $v(0) = v$.

Banach spaces share many properties with but are more general than Hilbert spaces. They are linear spaces equipped with a norm (which need not be and generally is not derived from a scalar product), complete with respect to that norm (i.e., all Cauchy sequences converge; see above). Some of the concepts we reviewed above for Hilbert spaces also exist in Banach spaces; e.g., bounded operators, linear functionals, spectrum and spectral radius. An example of a Banach space that is not a Hilbert space is $L^p(\mathbb{R})$, the set of all functions f on $\mathbb{R}$ such that $\|f\|_{L^p}$ (see (0.0.2)) is finite, with $1 \le p < \infty$, $p \ne 2$. Another example is $L^\infty(\mathbb{R})$, the set of all bounded functions on $\mathbb{R}$, with $\|f\|_{L^\infty} = \sup_{x \in \mathbb{R}} |f(x)|$. The dual E^* of a Banach space E is the set of all bounded linear functionals on E; it is also a linear space, which comes with a natural norm (defined as in (0.0.8)), with respect to which it is complete: E^* is a Banach space itself. In the case of the L^p-spaces, $1 \le p < \infty$, it turns out that elements of L^q, where p and q are related by $p^{-1} + q^{-1} = 1$, define bounded linear functionals on L^p. Indeed, one has Hölder's inequality,

$$\left| \int dx\ f(x)\ \overline{g(x)} \right| \le \|f\|_{L^p}\ \|g\|_{L^q} .$$

It turns out that all bounded linear functionals on L^p are of this type, i.e., $(L^p)^* = L^q$. In particular, L^2 is its own dual; by Riesz' representation theorem (see above), every Hilbert space is its own dual. The adjoint A^* of an operator A from E_1 to E_2 is now an operator from E_2^* to E_1^*, defined by

$$(A^* \ell_2)(v_1) = \ell_2(A v_1) .$$

There exist different types of bases in Banach spaces. (We will again only consider separable spaces, in which bases are countable.) The e_n constitute a Schauder basis if, for all $v \in E$, there exist unique $\mu_n \in \mathbb{C}$ so that $v = \lim_{N \to \infty} \sum_{n=1}^N \mu_n e_n$ (i.e., $\|v - \sum_{n=1}^N \mu_n e_n\| \to 0$ as $N \to \infty$). The uniqueness requirement of the μ_n forces the e_n to be linearly independent, in the sense that no e_n can be in the closure of the linear span of all the others, i.e., there exist no γ_m so that $e_n = \lim_{N \to \infty} \sum_{m=1,\ m \ne n}^N \gamma_m e_m$. In a Schauder basis, the ordering of the e_n may be important. A basis is called unconditional if in addition it satisfies one of the following two equivalent properties:

- whenever $\sum_n \mu_n e_n \in E$, it follows that $\sum_n |\mu_n| e_n \in E$;
- if $\sum_n \mu_n e_n \in E$, and $\epsilon_n = \pm 1$, randomly chosen for every n, then $\sum_n \mu_n \epsilon_n e_n \in E$.

For an unconditional basis, the order in which the basis vectors are taken does not matter. Not all Banach spaces have unconditional bases: $L^1(\mathbb{R})$ and $L^\infty(\mathbb{R})$ do not.

In a Hilbert space $\mathcal{H}$, an unconditional basis is also called a Riesz basis. A Riesz basis can also be characterized by the following equivalent requirement: there exist $\alpha > 0$, $\beta < \infty$ so that

$$\alpha \|u\|^2 \leq \sum_n |\langle u, e_n \rangle|^2 \leq \beta \|u\|^2 , \tag{0.0.10}$$

for all $u \in \mathcal{H}$. If A is a bounded operator with a bounded inverse, then A maps any orthonormal basis to a Riesz basis. Moreover, all Riesz bases can be obtained as such images of an orthonormal basis. In a way, Riesz bases are the next best thing to an orthonormal basis. Note that the inequalities in (0.0.10) are not sufficient to guarantee that the e_n constitute a Riesz basis: the e_n also need to be linearly independent!

CHAPTER 1

The What, Why, and How of Wavelets

The wavelet transform is a tool that cuts up data or functions or operators into different frequency components, and then studies each component with a resolution matched to its scale. Forerunners of this technique were invented independently in pure mathematics (Calderón's resolution of the identity in harmonic analysis—see e.g., Calderón (1964)), physics (coherent states for the $(ax + b)$-group in quantum mechanics, first constructed by Aslaksen and Klauder (1968), and linked to the hydrogen atom Hamiltonian by Paul (1985)) and engineering (QMF filters by Esteban and Galland (1977), and later QMF filters with exact reconstruction property by Smith and Barnwell (1986), Vetterli (1986) in electrical engineering; wavelets were proposed for the analysis of seismic data by J. Morlet (1983)). The last five years have seen a synthesis between all these different approaches, which has been very fertile for all the fields concerned.

Let us stay for a moment within the signal analysis framework. (The discussion can easily be translated to other fields.) The wavelet transform of a signal evolving in time (e.g., the amplitude of the pressure on an eardrum, for acoustical applications) depends on two variables: scale (or frequency) and time; wavelets provide a tool for time-frequency localization. The first section tells us what time-frequency localization means and why it is of interest. The remaining sections describe different types of wavelets.

1.1. Time-frequency localization.

In many applications, given a signal $f(t)$ (for the moment, we assume that t is a continuous variable), one is interested in its frequency content *locally in time*. This is similar to music notation, for example, which tells the player which notes (= frequency information) to play at any given moment. The standard Fourier transform,

$$(\mathcal{F}f)(\omega) = \frac{1}{\sqrt{2\pi}} \int dt\; e^{-i\omega t} f(t) \ ,$$

also gives a representation of the frequency content of f, but information concerning time-localization of, e.g., high frequency bursts cannot be read off easily from $\mathcal{F}f$. Time-localization can be achieved by first windowing the signal f, so as to cut off only a well-localized slice of f, and then taking its Fourier transform:

$$(T^{\text{win}} f)(\omega, t) = \int ds \; f(s) \; g(s-t) e^{-i\omega s} \; . \tag{1.1.1}$$

This is the windowed Fourier transform, which is a standard technique for time-frequency localization.[1] It is even more familiar to signal analysts in its discrete version, where t and ω are assigned regularly spaced values: $t = nt_0$, $\omega = m\omega_0$, where m, n range over $\mathbb{Z}$, and $\omega_0, t_0 > 0$ are fixed. Then (1.1.1) becomes

$$T^{\text{win}}_{m,n}(f) = \int ds \; f(s) \; g(s - nt_0) \; e^{-im\omega_0 s} \; . \tag{1.1.2}$$

This procedure is schematically represented in Figure 1.1: for fixed n, the $T^{\text{win}}_{m,n}(f)$ correspond to the Fourier coefficients of $f(\cdot)g(\cdot - nt_0)$. If, for instance, g is compactly supported, then it is clear that, with appropriately chosen ω_0, the Fourier coefficients $T^{\text{win}}_{\cdot,n}(f)$ are sufficient to characterize and, if need be, to reconstruct $f(\cdot)g(\cdot - nt_0)$. Changing n amounts to shifting the "slices" by steps of t_0 and its multiples, allowing the recovery of all of f from the $T^{\text{win}}_{m,n}(f)$. (We will discuss this in more mathematical detail in Chapter 3.) Many possible choices have been proposed for the window function g in signal analysis, most of which have compact support and reasonable smoothness. In physics, (1.1.1) is related to coherent state representations; the $g^{\omega,t}(s) = e^{i\omega s} g(s-t)$ are the coherent states associated to the Weyl–Heisenberg group (see, e.g., Klauder and Skagerstam (1985)). In this context, a very popular choice is a Gaussian g. In all applications, g is supposed to be well concentrated in both time and frequency; if g and $\hat{g}$ are both concentrated around zero, then $(T^{\text{win}} f)(\omega, t)$ can be interpreted loosely as the "content" of f near time t and near frequency ω. The windowed Fourier transform provides thus a description of f in the time-frequency plane.

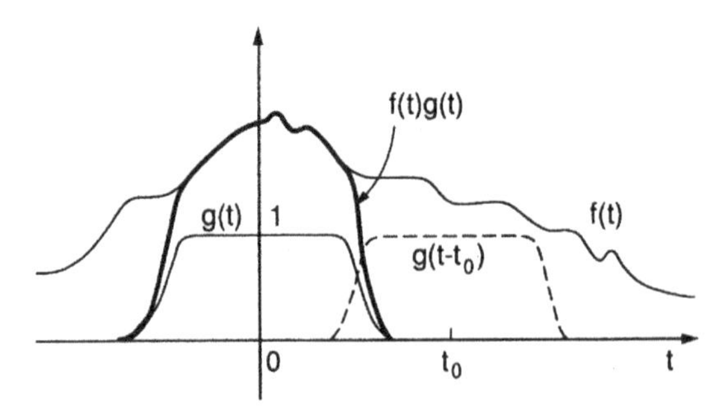

FIG. 1.1. *The windowed Fourier transform: the function $f(t)$ is multiplied with the window function $g(t)$, and the Fourier coefficients of the product $f(t)g(t)$ are computed; the procedure is then repeated for translated versions of the window, $g(t - t_0)$, $g(t - 2t_0)$, $\cdots$.*

1.2. The wavelet transform: Analogies and differences with the windowed Fourier transform.

The wavelet transform provides a similar time-frequency description, with a few important differences. The wavelet transform formulas analogous to (1.1.1) and (1.1.2) are

$$(T^{\text{wav}} f)(a,b) = |a|^{-1/2} \int dt\ f(t)\, \psi\left(\frac{t-b}{a}\right) \tag{1.2.1}$$

and

$$T^{\text{wav}}_{m,n}(f) = a_0^{-m/2} \int dt\ f(t)\, \psi(a_0^{-m}t - nb_0)\ . \tag{1.2.2}$$

In both cases we assume that ψ satisfies

$$\int dt\ \psi(t) = 0 \tag{1.2.3}$$

(for reasons explained in Chapters 2 and 3).

Formula (1.2.2) is again obtained from (1.2.1) by restricting a, b to only discrete values: $a = a_0^m$, $b = nb_0a_0^m$ in this case, with m, n ranging over $\mathbb{Z}$, and $a_0 > 1$, $b_0 > 0$ fixed. One similarity between the wavelet and windowed Fourier transforms is clear: both (1.1.1) and (1.2.1) take the inner products of f with a family of functions indexed by two labels, $g^{\omega,t}(s) = e^{i\omega s}g(s-t)$ in (1.1.1), and $\psi^{a,b}(s) = |a|^{-1/2}\ \psi\left(\frac{s-b}{a}\right)$ in (1.2.1). The functions $\psi^{a,b}$ are called "wavelets"; the function ψ is sometimes called "mother wavelet." (Note that ψ and g are implicitly assumed to be real, even though this is by no means essential; if they are not, then complex conjugates have to be introduced in (1.1.1), (1.2.1).) A typical choice for ψ is $\psi(t) = (1-t^2)\exp(-t^2/2)$, the second derivative of the Gaussian, sometimes called the mexican hat function because it resembles a cross section of a Mexican hat. The mexican hat function is well localized in both time and frequency, and satisfies (1.2.3). As a changes, the $\psi^{a,0}(s) = |a|^{-1/2}\psi(s/a)$ cover different frequency ranges (large values of the scaling parameter $|a|$ correspond to small frequencies, or large scale $\psi^{a,0}$; small values of $|a|$ correspond to high frequencies or very fine scale $\psi^{a,0}$). Changing the parameter b as well allows us to move the time localization center: each $\psi^{a,b}(s)$ is localized around $s = b$. It follows that (1.2.1), like (1.1.1), provides a time-frequency description of f. The difference between the wavelet and windowed Fourier transforms lies in the shapes of the analyzing functions $g^{\omega,t}$ and $\psi^{a,b}$, as shown in Figure 1.2. The functions $g^{\omega,t}$ all consist of the same envelope function g, translated to the proper time location, and "filled in" with higher frequency oscillations. All the $g^{\omega,t}$, regardless of the value of ω, have the same width. In contrast, the $\psi^{a,b}$ have time-widths adapted to their frequency: high frequency $\psi^{a,b}$ are very narrow, while low frequency $\psi^{a,b}$ are much broader. As a result, the wavelet transform is better able than the windowed Fourier transform to "zoom in" on very short-lived high frequency phenomena, such as transients in signals (or singularities

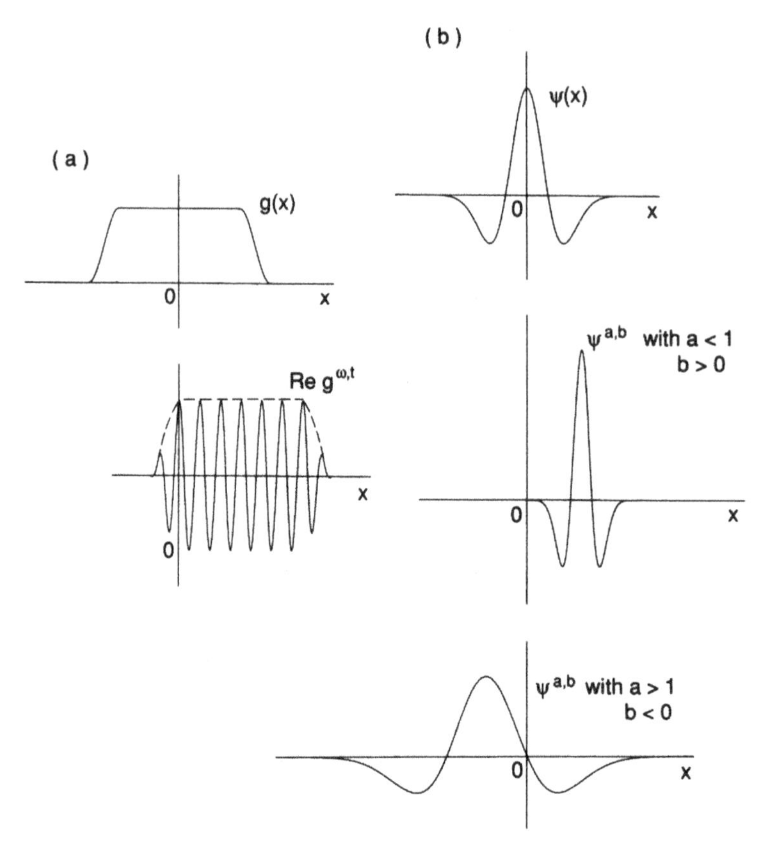

FIG. 1.2. *Typical shapes of* (a) *windowed Fourier transform functions* $g^{\omega,t}$, *and* (b) *wavelets* $\psi^{a,b}$. *The* $g^{\omega,t}(x) = e^{i\omega x}g(x-t)$ *can be viewed as translated envelopes* g, *"filled in" with higher frequencies; the* $\psi^{a,b}$ *are all copies of the same functions, translated and compressed or stretched.*

in functions or integral kernels). This is illustrated by Figure 1.3, which shows windowed Fourier transforms and the wavelet transform of the same signal f defined by

$$f(t) = \sin(2\pi\nu_1 t) \ + \ \sin(2\pi\nu_2 t) \ + \ \gamma[\delta(t-t_1) + \delta(t-t_2)] \ .$$

In practice, this signal is not given by this continuous expression, but by samples, and adding a δ-function is then approximated by adding a constant to one sample only. In sampled version, we have then

$$f(n\tau) = \sin(2\pi\nu_1 n\tau) \ + \ \sin(2\pi\nu_2 n\tau) \ + \ \alpha[\delta_{n,n_1} + \delta_{n,n_2}] \ .$$

For the example in Figure 1.3a, $\nu_1 = 500$ Hz, $\nu_2 = 1$ kHz, $\tau = 1/8,000$ sec (i.e., we have 8,000 samples per second), $\alpha = 1.5$, and $n_2 - n_1 = 32$ (corresponding to 4 milliseconds between the two pulses). The three spectrograms (graphs of

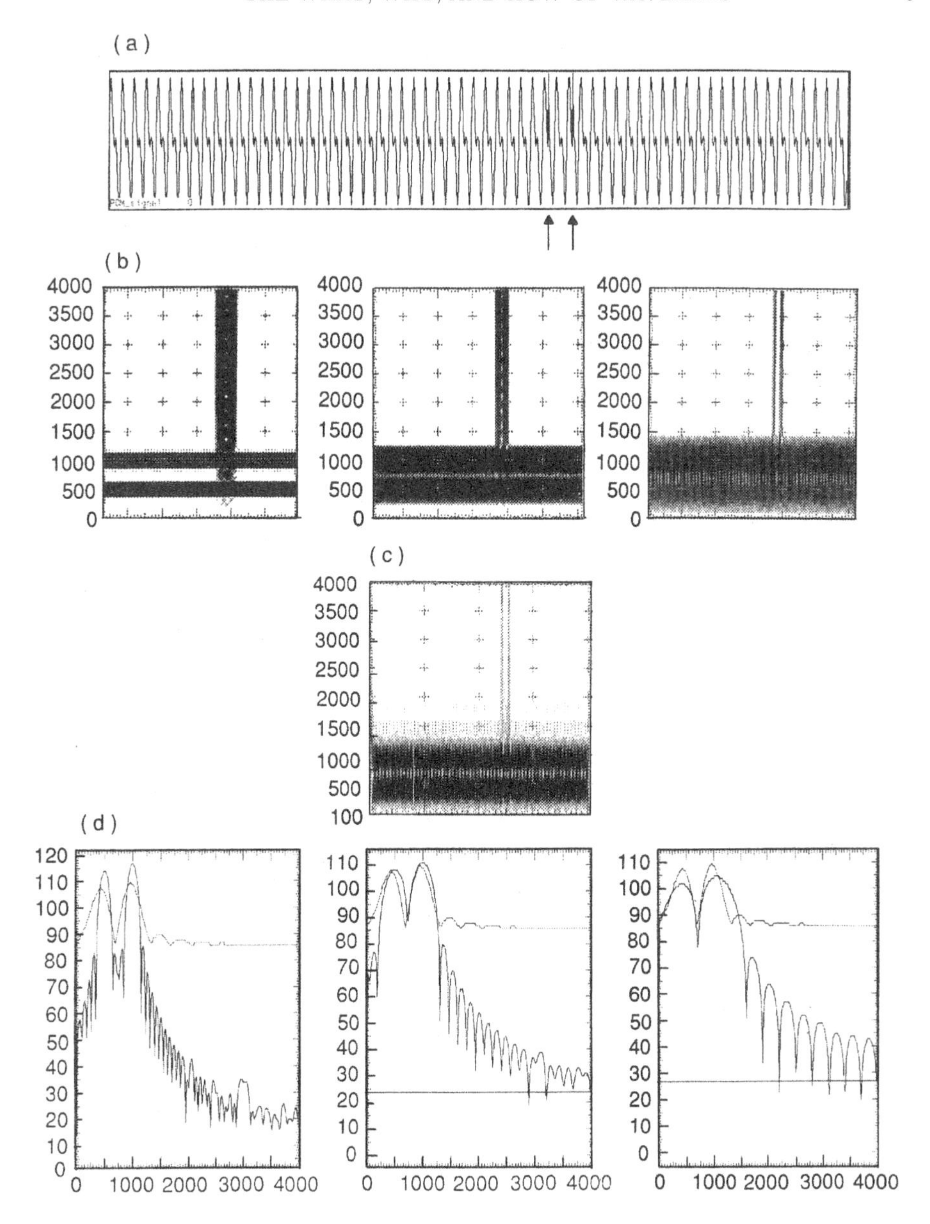

FIG. 1.3. (a) *The signal* $f(t)$. (b) *Windowed Fourier transforms of* f *with three different window widths. These are so-called spectrograms: only* $|T^{\text{win}}(f)|$ *is plotted (the phase is not rendered on the graph), zero = white, intermediate grey levels are assigned proportional to* $\log|T^{\text{win}}(f)|$) *in the* t(*abscissa*), ω(*ordinate*) *plane.* (c) *Wavelet transform of* f. *To make the comparison with* (b) *we have also plotted* $|T^{\text{wav}}(f)|$, *with the same grey level method, and a linear frequency axis (i.e., the ordinate corresponds to* a^{-1}). (d) *Comparison of the frequency resolution between the three spectrograms and the wavelet transform. I would like to thank Oded Ghitza for generating this figure.*

the modulus of the windowed Fourier transform) in Figure 1.3b use standard Hamming windows, with widths 12.8, 6.4, and 3.2 milliseconds, respectively. (Time t varies horizontally, frequency ω vertically, on these plots; the grey levels indicate the value of $|T^{\rm win}(f)|$, with black standing for the highest value.) As the window width increases, the resolution of the two pure tones gets better, but it becomes harder or even impossible to resolve the two pulses. Figure 1.3c shows the modulus of the wavelet transform of f computed by means of the (complex) Morlet wavelet $\psi(t) = C\, e^{-t^2/\alpha^2}(e^{i\pi t} - e^{-\pi^2\alpha^2/4})$, with $\alpha = 4$. (To make comparison with the spectrograms easier, a linear frequency axis has been used here; for wavelet transforms, a logarithmic frequency axis is more usual.) One already sees that the two impulses are resolved even better than with the 3.2 msec Hamming window (right in Figure 1.3b), while the frequency resolution for the two pure tones is comparable with that obtained with the 6.4 msec Hamming window (middle in Figure 1.3b). This comparison of frequency resolutions is illustrated more clearly by Figure 1.3d: here sections of the spectrograms (i.e., plots of $|(T^{\rm win} f)(\cdot, t)|$ with fixed t) and of the wavelet transform modulus ($|(T^{\rm wav} f)(\cdot, b)|$ with fixed b) are compared. The dynamic range (ratio between the maxima and the "dip" between the two peaks) of the wavelet transform is comparable to that of the 6.4 msec spectrogram. (Note that the flat horizontal "tail" for the wavelet transform in the graphs in Figure 1.3d is an artifact of the plotting package used, which set a rather high cut-off, as compared with the spectrogram plots; anyway, this cut-off is already at -24 dB.)

In fact, our ear uses a wavelet transform when analyzing sound, at least in the very first stage. The pressure amplitude oscillations are transmitted from the eardrum to the basilar membrane, which extends over the whole length of the cochlea. The cochlea is rolled up as a spiral inside our inner ear; imagine it unrolled to a straight segment, so that the basilar membrane is also stretched out. We can then introduce a coordinate y along this segment. Experiment and numerical simulation show that a pressure wave which is a pure tone, $f_\omega(t) = e^{i\omega t}$, leads to a response excitation along the basilar membrane which has the same frequency in time, but with an envelope in y, $F_\omega(t, y) = e^{i\omega t}\ \phi_\omega(y)$. In a first approximation, which turns out to be pretty good for frequencies ω above 500 Hz, the dependence on ω of $\phi_\omega(y)$ corresponds to a shift by $\log\,\omega$: there exists one function ϕ so that $\phi_\omega(y)$ is very close to $\phi(y - \log\,\omega)$. For a general excitation function f, $f(t) = \frac{1}{\sqrt{2\pi}} \int d\omega\ \hat{f}(\omega) e^{i\omega t}$, it follows that the response function $F(t, y)$ is given by the corresponding superposition of "elementary response functions,"

$$\begin{aligned} F(t,y) &= \frac{1}{\sqrt{2\pi}} \int d\omega\ \hat{f}(\omega)\ F_\omega(t,y) \\ &= \frac{1}{\sqrt{2\pi}} \int d\omega\ \hat{f}(\omega)\ e^{i\omega t} \phi(y - \log\,\omega)\ . \end{aligned}$$

If we now introduce a change of parameterization, by defining

$$\hat{\psi}(e^{-x}) = (2\pi)^{-1/2}\ \phi(x), \qquad G(a,t) = F(t, \log a)\ ,$$

then it follows that

$$G(a,t) = \int dt'\ f(t')\ \psi(a(t-t'))\ ,$$

which (up to normalization) is exactly a wavelet transform. The dilation parameter comes in, of course, because of the logarithmic shifts in frequency in the ϕ_ω. The occurrence of the wavelet transform in the first stage of our own biological acoustical analysis suggests that wavelet-based methods for acoustical analysis have a better chance than other methods to lead, e.g., to compression schemes undetectable by our ear.

1.3. Different types of wavelet transform.

There exist many different types of wavelet transform, all starting from the basic formulas (1.2.1), (1.2.2). In these notes we will distinguish between

A. The continuous wavelet transform (1.2.1), and

B. The discrete wavelet transform (1.2.2).

Within the discrete wavelet transform we distinguish further between

B1. Redundant discrete systems (frames) and

B2. Orthonormal (and other) bases of wavelets.

1.3.1. The continuous wavelet transform. Here the dilation and translation parameters a, b vary continuously over $\mathbb{R}$ (with the constraint $a \neq 0$). The wavelet transform is given by formula (1.2.1); a function can be reconstructed from its wavelet transform by means of the "resolution of identity" formula

$$f = C_\psi^{-1} \int_{-\infty}^{\infty} \int_{-\infty}^{\infty} \frac{da\ db}{a^2}\ \langle f, \psi^{a,b} \rangle\ \psi^{a,b}\ , \tag{1.3.1}$$

where $\psi^{a,b}(x) = |a|^{-1/2}\ \psi\left(\frac{x-b}{a}\right)$, and $\langle\ ,\ \rangle$ denotes the L^2-inner product. The constant C_ψ depends only on ψ and is given by

$$C_\psi = 2\pi \int_{-\infty}^{\infty} d\xi\ |\hat{\psi}(\xi)|^2\ |\xi|^{-1}\ ; \tag{1.3.2}$$

we assume $C_\psi < \infty$ (otherwise (1.3.1) does not make sense). If ψ is in $L^1(\mathbb{R})$ (this is the case in all examples of practical interest), then $\hat{\psi}$ is continuous, so that C_ψ can be finite only if $\hat{\psi}(0) = 0$, i.e., $\int dx\ \psi(x) = 0$. A proof for (1.3.1) will be given in Chapter 2. (Note that we have implicitly assumed that ψ is real; for complex ψ, we should use $\bar{\psi}$ instead of ψ in (1.2.1). In some applications, such complex ψ are useful.)

Formula (1.3.1) can be viewed in two different ways: (1) as a way of reconstructing f once its wavelet transform $T^{\text{wav}} f$ is known, or (2) as a way to write f as a superposition of wavelets $\psi^{a,b}$; the coefficients in this superposition are exactly given by the wavelet transform of f. Both points of view lead to interesting applications.

The correspondence $f(x) \rightarrow (T^{\text{wav}} f)(a, b)$ represents a one-variable function by a function of two variables, into which lots of correlations are built in (see Chapter 2). This redundancy of the representation can be exploited; a beautiful application is the concept of the "skeleton" of a signal, extracted from the continuous wavelet transform, which can be used for nonlinear filtering (see, e.g., Torrésani (1991), Delprat et al. (1992)).

1.3.2. The discrete but redundant wavelet transform-frames. In this case the dilation parameter a and the translation parameter both take only discrete values. For a we choose the integer (positive and negative) powers of one fixed dilation parameter $a_0 > 1$, i.e., $a = a_0^m$. As already illustrated by Figure 1.2, different values of m correspond to wavelets of different widths. It follows that the discretization of the translation parameter b should depend on m: narrow (high frequency) wavelets are translated by small steps in order to cover the whole time range, while wider (lower frequency) wavelets are translated by larger steps. Since the width of $\psi(a_0^{-m}x)$ is proportional to a_0^m, we choose therefore to discretize b by $b = nb_0 a_0^m$, where $b_0 > 0$ is fixed, and $n \in \mathbb{Z}$. The corresponding discretely labelled wavelets are therefore

$$\begin{aligned} \psi_{m,n}(x) &= a_0^{-m/2}\ \psi(a_0^{-m}(x - nb_0 a_0^m)) \\ &= a_0^{-m/2}\ \psi(a_0^{-m}x - nb_0)\ . \end{aligned} \tag{1.3.3}$$

Figure 1.4a shows schematically the lattice of time-frequency localization centers corresponding to the $\psi_{m,n}$. For a given function f, the inner products $\langle f, \psi_{m,n} \rangle$ then give exactly the discrete wavelet transform $T_{m,n}^{\text{wav}}(f)$ as defined in (1.2.2) (we assume again that ψ is real).

In the discrete case, there does not exist, in general, a "resolution of the identity" formula analogous to (1.3.1) for the continuous case. Reconstruction of f from $T^{\text{wav}}(f)$, if at all possible, must therefore be done by some other means. The following questions naturally arise:

(1) Is it possible to characterize f completely by knowing $T^{\text{wav}}(f)$?

(2) Is it possible to reconstruct f in a numerically stable way from $T^{\text{wav}}(f)$?

These questions concern the recovery of f from its wavelet transform. We can also consider the dual problem (see §1.3.1), the possibility of expanding f into wavelets, which then leads to the dual questions:

(1′) Can any function be written as a superposition of $\psi_{m,n}$?

(2′) Is there a numerically stable algorithm to compute the coefficients for such an expansion?

Chapter 3 addresses these questions. As in the continuous case, these discrete wavelet transforms often provide a very redundant description of the original

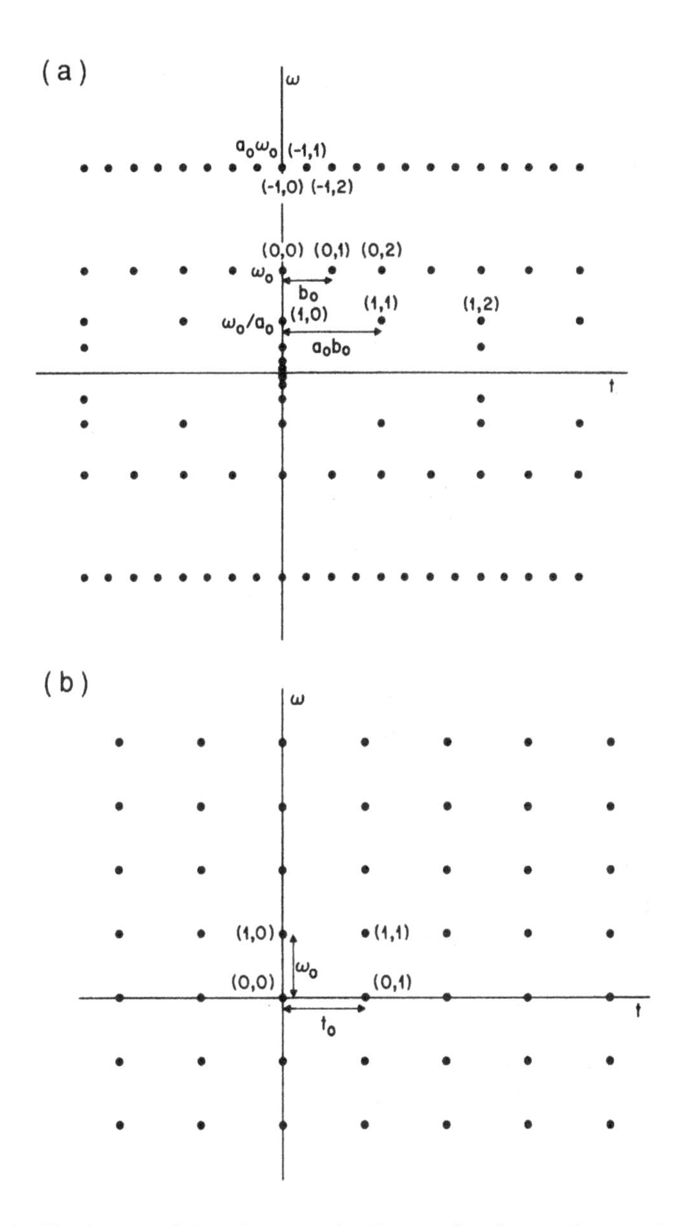

FIG. 1.4. *The lattices of time-frequency localization for the wavelet transform and windowed Fourier transform.* (a) *The wavelet transform: $\psi_{m,n}$ is localized around $a_0^m nb_0$ in time. We assume here that $|\hat{\psi}|$ has two peaks in frequency, at $\pm\xi_0$ (this is the case, e.g., for the Mexican hat wavelet $\psi(t) = (1-t^2)e^{-t^2/2}$); $|\hat{\psi}_{m,n}(\xi)|$ then peaks at $\pm a_0^m \xi_0$, which are the two localization centers of $\psi_{m,n}$ in frequency.* (b) *The windowed Fourier transform: $g_{m,n}$ is localized around nt_0 in time, around $m\omega_0$ in frequency.*

function. This redundancy can be exploited (it is, for instance, possible to compute the wavelet transform only approximately, while still obtaining reconstruction of f with good precision), or eliminated to reduce the transform to its bare essentials (such as in the image compression work of Mallat and Zhong (1992)). It is in this discrete form that the wavelet transform is closest to the "ϕ-transform" of Frazier and Jawerth (1988).

The choice of the wavelet ψ used in the continuous wavelet transform or in frames of discretely labelled families of wavelets is essentially only restricted by the requirement that C_ψ, as defined by (1.3.2), is finite. For practical reasons, one usually chooses ψ so that it is well concentrated in both the time and the frequency domain, but this still leaves a lot of freedom. In the next section we will see how giving up most of this freedom allows us to build orthonormal bases of wavelets.

1.3.3. Orthonormal wavelet bases: Multiresolution analysis. For some very special choices of ψ and a_0, b_0, the $\psi_{m,n}$ constitute an orthonormal basis for $L^2(\mathbb{R})$. In particular, if we choose $a_0 = 2$, $b_0 = 1$,[2] then there exist ψ, with good time-frequency localization properties, such that the

$$\psi_{m,n}(x) = 2^{-m/2}\, \psi(2^{-m}x - n) \tag{1.3.4}$$

constitute an orthonormal basis for $L^2(\mathbb{R})$. (For the time being, and until Chapter 10, we restrict ourselves to $a_0 = 2$.) The oldest example of a function ψ for which the $\psi_{m,n}$ defined by (1.3.4) constitute an orthonormal basis for $L^2(\mathbb{R})$ is the Haar function,

$$\psi(x) = \begin{cases} 1 & 0 \le x < \frac{1}{2} \\ -1 & \frac{1}{2} \le x < 1 \\ 0 & \text{otherwise}\ . \end{cases}$$

The Haar basis has been known since Haar (1910). Note that the Haar function does not have good time-frequency localization: its Fourier transform $\hat{\psi}(\xi)$ decays like $|\xi|^{-1}$ for $\xi \to \infty$. Nevertheless we will use it here for illustration purposes. What follows is a proof that the Haar family does indeed constitute an orthonormal basis. This proof is different from the one in most textbooks; in fact, it will use multiresolution analysis as a tool.

In order to prove that the $\psi_{m,n}(x)$ constitute an orthonormal basis, we need to establish that

(1) the $\psi_{m,n}$ are orthonormal;

(2) any L^2-function f can be approximated, up to arbitrarily small precision, by a finite linear combination of the $\psi_{m,n}$.

Orthonormality is easy to establish. Since support $(\psi_{m,n}) = [2^m n, 2^m(n+1)]$, it follows that two Haar wavelets of the same scale (same value of m) never overlap, so that $\langle \psi_{m,n}, \psi_{m,n'} \rangle = \delta_{n,n'}$. Overlapping supports are possible if the two wavelets have different sizes, as in Figure 1.5. It is easy to check, however, that if $m < m'$, then support $(\psi_{m,n})$ lies wholly within a region where $\psi_{m',n'}$ is constant (as on the figure). It follows that the inner product of $\psi_{m,n}$ and $\psi_{m',n'}$ is then proportional to the integral of ψ itself, which is zero.

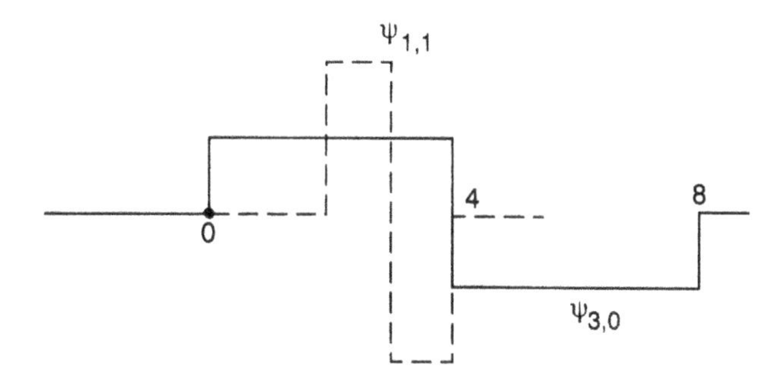

FIG. 1.5. *Two Haar wavelets; the support of the "narrower" wavelet is completely contained in an interval where the "wider" wavelet is constant.*

We concentrate now on how well an arbitrary function f can be approximated by linear combinations of Haar wavelets. Any f in $L^2(\mathbb{R})$ can be arbitrarily well approximated by a function with compact support which is piecewise constant on the $[\ell 2^{-j}, (\ell+1)2^{-j}[$ (it suffices to take the support and j large enough). We can therefore restrict ourselves to such piecewise constant functions only: assume f to be supported on $[-2^{J_1}, 2^{J_1}]$, and to be piecewise constant on the $[\ell 2^{-J_0}, (\ell+1)2^{-J_0}[$, where J_1 and J_0 can both be arbitrarily large (see Figure 1.6). Let us denote the constant value of $f^0 = f$ on $[\ell 2^{-J_0}, (\ell+1)2^{-J_0}[$ by f_ℓ^0. We now represent f^0 as a sum of two pieces, $f^0 = f^1 + \delta^1$, where f^1 is an approximation to f^0 which is piecewise constant over intervals twice as large as originally, i.e., $f^1|_{[k2^{-J_0+1},(k+1)2^{-J_0+1}[} \equiv \mathit{constant} = f_k^1$. The values f_k^1 are given by the averages of the two corresponding constant values for f^0, $f_k^1 = \frac{1}{2}(f_{2k}^0 + f_{2k+1}^0)$ (see Figure 1.6). The function δ^1 is piecewise constant with the same stepwidth as f^0; one immediately has

$$\delta_{2\ell}^1 = f_{2\ell}^0 - f_\ell^1 = \tfrac{1}{2}(f_{2\ell}^0 - f_{2\ell+1}^0)$$

and

$$\delta_{2\ell+1}^1 = f_{2\ell+1}^0 - f_\ell^1 = \tfrac{1}{2}(f_{2\ell+1}^0 - f_{2\ell}^0) = -\delta_{2\ell}^1 \ .$$

It follows that δ^1 is a linear combination of scaled and translated Haar functions:

$$\delta^1 = \sum_{\ell=-2^{J_1+J_0-1}+1}^{2^{J_1+J_0-1}} \delta_{2\ell}^1 \psi(2^{J_0-1}x - \ell) \ .$$

We have therefore written f as

$$f = f^0 = f^1 + \sum_\ell c_{-J_0+1,\ell}\ \psi_{-J_0+1,\ell} \ ,$$

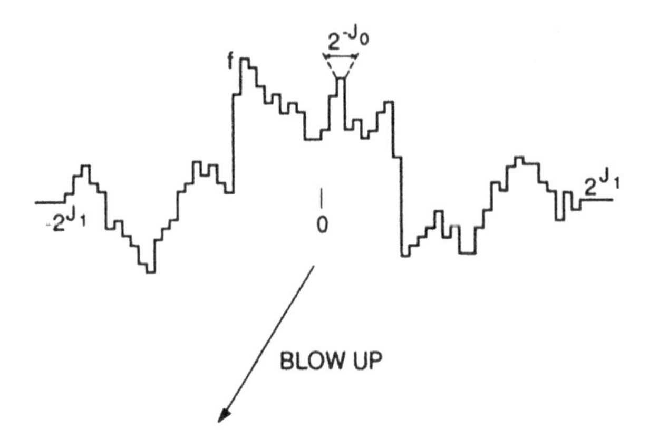

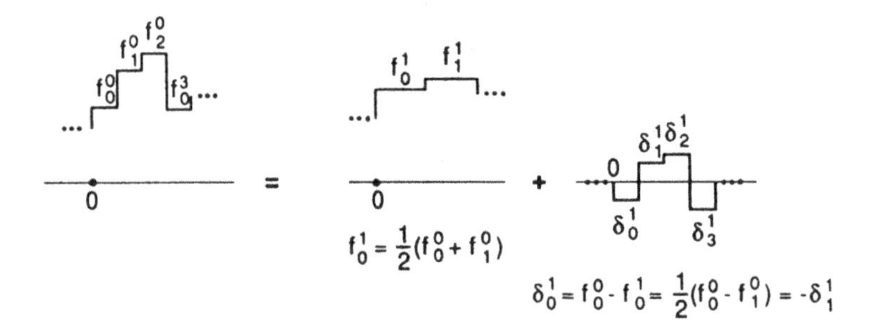

FIG. 1.6. (a) *A function f with support $[-2^{J_1}, 2^{J_1}]$, piecewise constant on the $[k2^{-J_0}, (k+1)2^{-J_0}[$.* (b) *A blowup of a portion of f. On every pair of intervals, f is replaced by its average ($\longrightarrow f^1$); the difference between f and f^1 is δ^1, a linear combination of Haar wavelets.*

where f^1 is of the same type as f^0, but with stepwidth twice as large. We can apply the same trick to f^1, so that

$$f^1 = f^2 + \sum_{\ell} c_{-J_0+2,\ell}\ \psi_{-J_0+2,\ell}\ ,$$

with f^2 still supported on $[-2^{J_1}, 2^{J_1}]$, but piecewise constant on the even larger intervals $[k2^{-J_0+2}, (k+1)2^{-J_0+2}[$. We can keep going like this, until we have

$$f = f^{J_0+J_1} + \sum_{m=-J_0+1}^{J_1} \sum_{\ell} c_{m,\ell}\ \psi_{m,\ell}\ .$$

Here $f^{J_0+J_1}$ consists of two constant pieces (see Figure 1.7), with $f^{J_0+J_1}|_{[0,2^{J_1}[} \equiv f_0^{J_0+J_1}$ equal to the average of f over $[0, 2^{J_1}[$, and $f^{J_0+J_1}|_{[-2^{J_1},0[} \equiv f_{-1}^{J_0+J_1}$ the average of f over $[-2^{J_1}, 0[$.

Even though we have "filled out" the whole support of f, we can still keep going with our averaging trick: nothing stops us from widening our horizon from

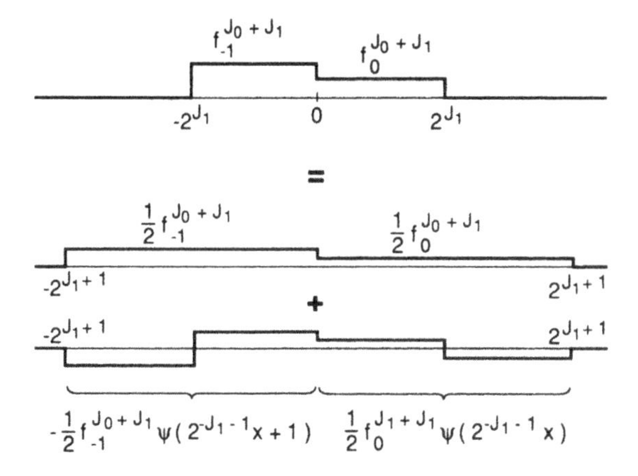

FIG. 1.7. *The averages of f on $[0,\ 2^{J_1}]$ and $[-2^{J_1},\ 0]$ can be "smeared" out over the bigger intervals $[0,\ 2^{J_1+1}]$, $[-2^{J_1+1},\ 0]$; the difference is a linear combination of very stretched out Haar functions.*

2^{J_1} to 2^{J_1+1}, and writing $f^{J_1+J_2} = f^{J_1+J_2+1} + \delta^{J_1+J_2+1}$, where

$$f^{J_1+J_2+1}|_{[0,2^{J_1+1}[} \equiv \tfrac{1}{2} f_0^{J_1+J_2}, \quad f^{J_1+J_2+1}|_{[-2^{J_1+1},0[} \equiv \tfrac{1}{2} f_{-1}^{J_1+J_2}$$

and

$$\delta^{J_1+J_2} = \tfrac{1}{2} f_0^{J_1+J_2} \psi(2^{-J_1-1}x) - \tfrac{1}{2} f_{-1}^{J_1+J_2} \psi(2^{-J_1-1}x+1)$$

(see Figure 1.7). This can again be repeated, leading to

$$f = f^{J_0+J_1+K} + \sum_{m=-J_0+1}^{J_1+K} \sum_{\ell} c_{m,\ell}\, \psi_{m,\ell}\ ,$$

where support $(f^{J_0+J_1+K}) = [-2^{J_1+K}, 2^{J_1+K}]$, and

$$f^{J_0+J_1+K}|_{[0,2^{J_1+K}[} = 2^{-K} f_0^{J_0+J_1},\ f^{J_0+J_1+K}|_{[-2^{J_1+K},0[} = 2^{-K} f_{-1}^{J_0+J_1}\ .$$

It follows immediately that

$$\begin{aligned} \left\| f - \sum_{m=-J_0+1}^{J_1+K} \sum_{\ell} c_{m,\ell}\, \psi_{m,\ell} \right\|_{L^2}^2 &= \|f^{J_0+J_1+K}\|_{L^2}^2 \\ &= 2^{-K/2} \cdot 2^{J_1/2}\, [|f_0^{J_0+J_1}|^2 + |f_{-1}^{J_0+J_1}|^2]^{1/2}\ , \end{aligned}$$

which can be made arbitrarily small by taking sufficiently large K. As claimed, f can therefore be approximated to arbitrary precision by a finite linear combination of Haar wavelets!

The argument we just saw has implicitly used a "multiresolution" approach: we have written successive coarser and coarser approximations to f (the f^j, averaging f over larger and larger intervals), and at every step we have written the difference between the approximation with resolution 2^{j-1}, and the next coarser level, with resolution 2^j, as a linear combination of the $\psi_{j,k}$. In fact, we have introduced a ladder of spaces $(V_j)_{j\in\mathbb{Z}}$ representing the successive resolution levels: in this particular case, $V_j = \{f \in L^2(\mathbb{R});\ f$ piecewise constant on the $[2^j k, 2^j(k+1)[,\ k \in \mathbb{Z}\}$. These spaces have the following properties:

(1) $\cdots \subset V_2 \subset V_1 \subset V_0 \subset V_{-1} \subset V_{-2} \subset \cdots$;

(2) $\bigcap_{j\in\mathbb{Z}} V_j = \{0\}, \quad \overline{\bigcup_{j\in\mathbb{Z}} V_j} = L^2(\mathbb{R})$;

(3) $f \in V_j \ \leftrightarrow\ f(2^j\cdot) \in V_0$;

(4) $f \in V_0 \ \rightarrow\ f(\cdot - n) \in V_0$ for all $n \in \mathbb{Z}$.

Property 3 expresses that all the spaces are scaled versions of one space (the "multiresolution" aspect). In the Haar example we found then that there exists a function ψ so that

$$\mathrm{Proj}_{V_{j-1}} f = \mathrm{Proj}_{V_j} f + \sum_{k\in\mathbb{Z}} \langle f, \psi_{j,k}\rangle\ \psi_{j,k}\ . \tag{1.3.5}$$

The beauty of the multiresolution approach is that whenever a ladder of spaces V_j satisfies the four properties above, together with

(5) $\exists \phi \in V_0$ so that the $\phi_{0,n}(x) = \phi(x-n)$ constitute an orthonormal basis for V_0,

then there exists ψ so that (1.3.5) holds. (In the Haar example above, we can take $\phi(x) = 1$ if $0 \le x < 1$, $\phi(x) = 0$ otherwise.) The $\psi_{j,k}$ constitute automatically an orthonormal basis. It turns out that there are many examples of such "multiresolution analysis ladders," corresponding to many examples of orthonormal wavelet bases. There exists an explicit recipe for the construction of ψ: since $\phi \in V_0 \subset V_{-1}$, and the $\phi_{-1,n}(x) = \sqrt{2}\ \phi(2x-n)$ constitute an orthonormal basis for V_{-1} (by (3) and (5) above), there exist $\alpha_n = \sqrt{2}\ \langle \phi, \phi_{-1,n}\rangle$ so that $\phi(x) = \sum_n \alpha_n\ \phi(2x-n)$. It then suffices to take $\psi(x) = \sum_n (-1)^n \alpha_{-n+1}\ \phi(2x-n)$. The function ϕ is called a *scaling function* of the multiresolution analysis. The correspondence multiresolution analysis $\rightarrow$ orthonormal basis of wavelets will be explained in detail in Chapter 5, and further explored in subsequent chapters. This multiresolution approach is also linked with subband filtering, as explained in §5.6 (Chapter 5).

Figure 1.8 shows some examples of pairs of functions ϕ, ψ corresponding to different multiresolution analyses which we will encounter in later chapters. The Meyer wavelets (Chapters 4 and 5) have compactly supported Fourier transform; ϕ and ψ themselves are infinitely supported; they are shown in Figure 1.8a. The Battle–Lemarié wavelets (Chapter 5) are spline functions (linear in Figure 1.8b, cubic in Figure 1.8c), with knots at $\mathbb{Z}$ for ϕ, at $\frac{1}{2}\mathbb{Z}$ for ψ. Both ϕ and ψ have infinite support, and exponential decay; their numerical decay is faster than for the Meyer wavelets (for comparison, the horizontal scale is the same in (a),

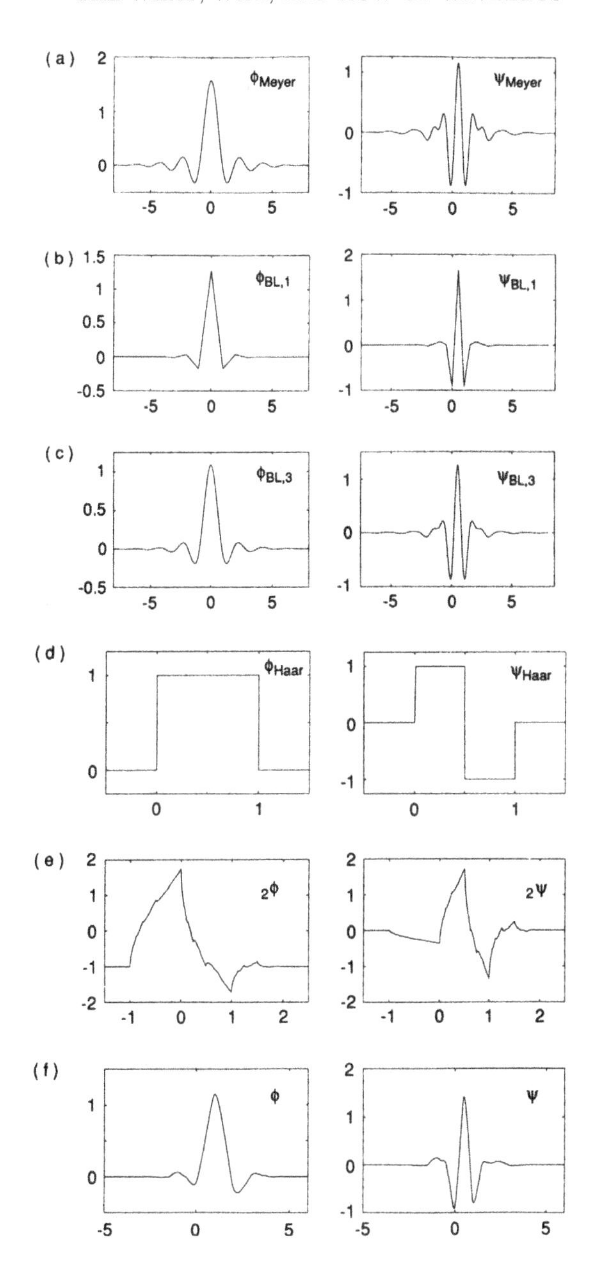

FIG. 1.8. *Some examples of orthonormal wavelet bases. For every* ψ *in this figure, the family* $\psi_{j,k}(x) = 2^{-j/2}\psi(2^{-j}x - k)$, $j, k \in \mathbb{Z}$, *constitutes an orthonormal basis of* $L^2(\mathbb{R})$. *The figure plots* ϕ *(the associated scaling function) and* ψ *for different constructions which we will encounter in later chapters.* (a) *The Meyer wavelets;* (b) *and* (c) *Battle–Lemarié wavelets;* (d) *the Haar wavelet;* (e) *the next member of the family of compactly supported wavelets,* $_2\psi$; (f) *another compactly supported wavelet, with less asymmetry.*

(b), and (c) of Figure 1.8). The Haar wavelet, in Figure 1.8d, has been known since 1910. It can be viewed as the smallest degree Battle–Lemarié wavelet ($\psi_{\text{Haar}} = \psi_{BL,0}$) or also as the first of a family of compactly supported wavelets constructed in Chapter 6, $\psi_{\text{Haar}} = {}_1\psi$. Figure 1.8e plots the next member of the family of compactly supported wavelets ${}_N\psi$; ${}_2\phi$ and ${}_2\psi$ both have support width 3, and are continuous. In this family of ${}_N\psi$ (constructed in §6.4), the regularity increases linearly with the support width (Chapter 7). Finally, Figure 1.8f shows another compactly supported wavelet, with support width 11, and less asymmetry (see Chapter 8).

Notes.

1. There exist other techniques for time-frequency localization than the windowed Fourier transform. A well-known example is the Wigner distribution. (See, e.g., Boashash (1990) for a good review on the use of the Wigner distribution for signal analysis.) The advantage of the Wigner distribution is that, unlike the windowed Fourier transform or the wavelet transform, it does not introduce a reference function (such as the window function, or the wavelet) against which the signal has to be integrated. The disadvantage is that the signal enters in the Wigner distribution in a quadratic rather than linear way, which is the cause of many interference phenomena. These may be useful in some applications, especially for, e.g., signals which have a very short time duration (an example is Janse and Kaiser (1983); Boashash (1990) contains references to many more examples); for signals which last for a longer time, they make the Wigner distribution less attractive. Flandrin (1989) shows how the absolute values of both the windowed Fourier transform and the wavelet transform of a function can also be obtained by "smoothing" its Wigner distribution in an appropriate way; the phase information is lost in this process however, and reconstruction is not possible any more.

2. The restriction $b_0 = 1$, corresponding to (1.3.4), is not very serious: if (1.3.4) provides an orthonormal basis, then so do the $\tilde{\psi}_{m,n}(x) = 2^{-m/2}$ $\tilde{\psi}(2^{-m}x - nb_0)$, with $\tilde{\psi}(x) = |b_0|^{-1/2}\psi(b_0^{-1}x)$, where $b_0 \neq 0$ is arbitrary. The choice $a_0 = 2$ cannot be modified by scaling, and in fact a_0 cannot be chosen arbitrarily. The general construction of orthonormal bases we will expose here can be made to work for all rational choices for $a_0 > 1$, as shown in Auscher (1989), but the choice $a_0 = 2$ is the simplest. Different choices for a_0 correspond of course to different ψ. Although the constructive method for orthonormal wavelet bases, called multiresolution analysis, can work only if a_0 is rational, it is an open question whether there exist orthonormal wavelet bases (necessarily not associated with a multiresolution analysis), with good time-frequency localization, and with irrational a_0.

CHAPTER 2

The Continuous Wavelet Transform

The images of L^2-functions under the continuous wavelet transform constitute a reproducing kernel Hilbert space (r.k.H.s.). Such r.k.H.s.'s occur and are useful in many different contexts. One of the simplest examples is the space of all bandlimited functions, discussed in §§2.1 and 2.2. In §2.3 we introduce the concept of band and time limiting; of course no nonzero function can be strictly time-limited (i.e., $f(t) \equiv 0$ for t outside $[-T, T]$) and band-limited ($\hat{f}(\xi) \equiv 0$ for $\xi \notin [-\Omega, \Omega]$), but one can still introduce time-and-band-limiting operators. We present a short review of the beautiful work of Landau, Pollak, and Slepian on this subject. We then switch to the continuous wavelet transform: the resolution of the identity in §2.4 (with a proof of (1.3.1)), the corresponding r.k.H.s. in §2.5. In §2.6 we briefly show how the one-dimensional results of the earlier sections can be extended to higher dimensions. In §2.7 we draw a parallel with the continuous windowed Fourier transform. In §2.8 we show how a different kind of time-and-band-limiting operator can be built from the continuous windowed Fourier transform or from the wavelet transform. Finally, we comment in §2.9 on the "zoom-in" property of the wavelet transform.

2.1. Bandlimited functions and Shannon's theorem.

A function f in $L^2(\mathbb{R})$ is called *bandlimited* if its Fourier transform $\mathcal{F}f$ has compact support, i.e., $\hat{f}(\xi) \equiv 0$ for $|\xi| > \Omega$. Let us suppose, for simplicity, that $\Omega = \pi$. Then $\hat{f}$ can be represented by its Fourier series (see Preliminaries),

$$\hat{f}(\xi) = \sum_{n \in \mathbb{Z}} c_n \, e^{-in\xi} ,$$

where

$$\begin{aligned} c_n &= \frac{1}{2\pi} \int_{-\pi}^{\pi} d\xi \; e^{in\xi} \hat{f}(\xi) \\ &= \frac{1}{2\pi} \int_{-\infty}^{\infty} d\xi \; e^{in\xi} \hat{f}(\xi) = \frac{1}{\sqrt{2\pi}} \, f(n) \, . \end{aligned}$$

It follows that

$$\begin{aligned}
f(x) &= \frac{1}{\sqrt{2\pi}} \int_{-\infty}^{\infty} d\xi \; e^{ix\xi} \hat{f}(\xi) \\
&= \frac{1}{\sqrt{2\pi}} \int_{-\pi}^{\pi} d\xi \; e^{ix\xi} \sum_n c_n \; e^{-in\xi} \\
&= \frac{1}{\sqrt{2\pi}} \sum_n c_n \int_{-\pi}^{\pi} d\xi \; e^{i(x-n)\xi} \\
&= \sum_n f(n) \; \frac{\sin\, \pi(x-n)}{\pi(x-n)} \; , \qquad (2.1.1)
\end{aligned}$$

where we have interchanged integral and summation in the third step, which is only a priori justifiable if $\sum |c_n| < \infty$ (e.g., if only finitely many c_n are nonzero). By a standard continuity argument, the final result holds for all bandlimited f (for every x, the series is absolutely summable because $\sum_n |f(n)|^2 = 2\pi \sum_n |c_n|^2 < \infty$). Formula (2.1.1) tells us that f is completely determined by its "sampled" values $f(n)$. If we lift the restriction $\Omega = \pi$ and assume support $\hat{f} \subset [-\Omega, \Omega]$, with Ω arbitrary, then (2.1.1) becomes

$$f(x) = \sum_n f\left(n\frac{\pi}{\Omega}\right) \; \frac{\sin\, (\Omega x - n\pi)}{\Omega x - n\pi} \; ; \qquad (2.1.2)$$

the function is now determined by its samples $f(n\frac{\pi}{\Omega})$, corresponding to a "sampling density" of $\Omega/\pi = \frac{|\text{support } \hat{f}|}{2\pi}$. (We use the notation $|A|$ for the "size" of a set $A \subset \mathbb{R}$, as measured by the Lebesgue measure; in this case $|\text{support } \hat{f}| = |[-\Omega, \Omega]| = 2\Omega$.) This sampling density is usually called the Nyquist density. The expansion (2.1.2) goes by the name of Shannon's theorem.

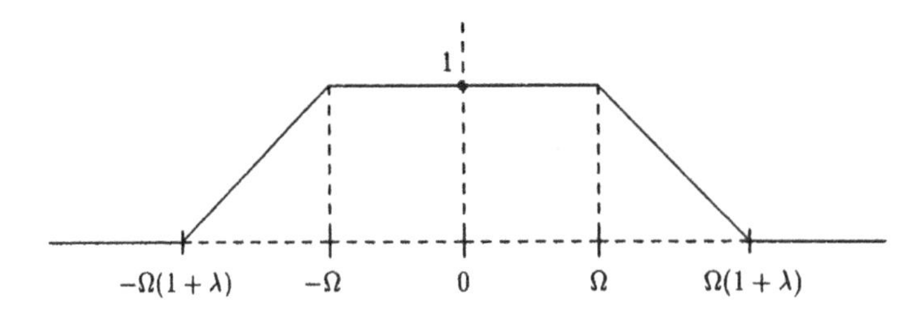

FIG. 2.1. *Graph of* $\hat{g}_\lambda$.

The "elementary building blocks" $\frac{\sin \Omega x}{\Omega x}$ in (2.1.2) decay very slowly (they are not even absolutely integrable). "Oversampling" makes it possible to write f as a superposition of functions with faster decay. Suppose that f is still bandlimited in $[-\Omega, \Omega]$ (i.e., support $\hat{f} \subset [-\Omega, \Omega]$), but that f is sampled at a rate $(1+\lambda)$ faster than the Nyquist rate, with $\lambda > 0$. Then f can be recovered from the $f(n\pi/[\Omega(1+\lambda)])$ in the following way. Define g_λ by

$$\hat{g}_\lambda(\xi) = \begin{cases} 1, & |\xi| \le \Omega \\ 1 - \frac{|\xi|-\Omega}{\lambda\Omega}, & \Omega \le |\xi| \le (1+\lambda)\Omega, \\ 0, & |\xi| \ge (1+\lambda)\Omega \end{cases}$$

(see Figure 2.1). Because $\hat{g}_\lambda \equiv 1$ on support $\hat{f}$, we have $\hat{f}(\xi) = \hat{f}(\xi)\hat{g}_\lambda(\xi)$. We can now repeat the same construction as before.

$$\begin{aligned} \hat{f}(\xi) &= \sum_n c_n \; e^{-in\xi\pi/[\Omega(1+\lambda)]} \\ \text{with } c_n &= \frac{\sqrt{2\pi}}{2\Omega(1+\lambda)} \, f\left(\frac{n\pi}{\Omega(1+\lambda)}\right); \end{aligned}$$

hence

$$\begin{aligned} f(x) &= \frac{1}{\sqrt{2\pi}} \int_{-\Omega(1+\lambda)}^{\Omega(1+\lambda)} d\xi \; e^{ix\xi} \; \hat{g}_\lambda(\xi) \sum_n \; c_n \; e^{-in\xi\pi/[\Omega(1+\lambda)]} \\ &= \sum_n f\left(\frac{n\pi}{\Omega(1+\lambda)}\right) \; G_\lambda\left(x - \frac{n\pi}{\Omega(1+\lambda)}\right), \end{aligned}$$

where

$$G_\lambda(x) = \frac{\sqrt{2\pi}}{2\Omega(1+\lambda)} \; g_\lambda(x) = \frac{2\sin[x\Omega(1+\lambda/2)] \; \sin(x\Omega\lambda/2)}{\lambda\Omega^2(1+\lambda)x^2} \; .$$

These G_λ have faster decay than $\frac{\sin \Omega x}{\Omega x}$; note that if $\lambda \to 0$, then $G_\lambda \to \frac{\sin \Omega x}{\Omega x}$, as expected. One can obtain even faster decay by choosing $\hat{g}_\lambda$ smoother, but it does not pay to put too much effort into making $\hat{g}_\lambda$ very smooth: true, G_λ will have very fast decay for asymptotically large x, but the size of λ imposes some restrictions on the *numerical* decay of G_λ. In other words, a C^∞ choice of $\hat{g}_\lambda$ leads to G_λ decaying faster than any inverse polynomial,

$$|G_\lambda(x)| \le C_N(\lambda)(1+|x|)^{-(N+1)},$$

but the constant $C_N(\lambda)$ can be very large: it is related to the range of values of the Nth derivative of $\hat{g}_\lambda$ on $[\Omega, \Omega(1+\lambda)]$, so that it is roughly proportional to λ^{-N}.

What happens if f is "undersampled," i.e., if support $\hat{f} = [-\Omega, \Omega]$, but only the $f(n\pi/[\Omega(1-\lambda)])$ are known, where $1 > \lambda > 0$? We have

$$\begin{aligned} f\left(n\frac{\pi}{\Omega(1-\lambda)}\right) &= \frac{1}{\sqrt{2\pi}} \int_{-\Omega}^{\Omega} d\xi \; \hat{f}(\xi) \; e^{in\pi\xi/[\Omega(1-\lambda)]} \\ &= \frac{1}{\sqrt{2\pi}} \int_{-\Omega(1-\lambda)}^{\Omega(1-\lambda)} d\xi \; e^{in\pi\xi/[\Omega(1-\lambda)]} \\ &\qquad [\hat{f}(\xi) + \hat{f}(\xi + 2\Omega(1-\lambda)) + \hat{f}(\xi - 2\Omega(1-\lambda))] \; , \end{aligned}$$

where we have used that the $e^{in\pi\xi/\alpha}$ have period 2α, and where we have assumed $\lambda \le \frac{2}{3}$ (otherwise more terms would intervene in the sum in the last integrand). This means that the undersampled $f(n\frac{\pi}{\Omega(1-\lambda)})$ behave exactly as if they were the Nyquist-spaced samples of a function of narrower bandwidth, the Fourier transform of which is obtained by "folding over" $\hat{f}$ (see Figure 2.2). In the "folded" version of $\hat{f}$, some of the high frequency content of f is found back in lower frequency regions; only the $|\xi| \le \Omega(1-2\lambda)$ are unaffected. This phenomenon is called aliasing; for undersampled acoustic signals, for instance, it is very clearly audible as a metallic clipping of the sound.

2.2. Bandlimited functions as a special case of a reproducing kernel Hilbert space.

For any α, β, $-\infty \le \alpha < \beta \le \infty$, the set of functions

$$\{f \in L^2(\mathbb{R});\ \ \text{support } f \subset [\alpha, \beta]\}$$

constitutes a closed subspace of $L^2(\mathbb{R})$, i.e., it is a subspace, and all Cauchy sequences composed of elements of the subspace converge to an element of the subspace. By the unitarity of the Fourier transform on $L^2(\mathbb{R})$, it follows that the set of all bandlimited functions

$$\mathcal{B}_\Omega = \{f \in L^2(\mathbb{R});\ \ \text{support } \hat{f} \subset [-\Omega, \Omega]\}$$

is a closed subspace of $L^2(\mathbb{R})$. By the Paley–Wiener theorem (see Preliminaries), any function f in $\mathcal{B}_\Omega$ has an analytic extension to an entire function on $\mathbb{C}$, which we also denote by f, and which is of exponential type. More precisely,

$$|f(z)| \le \frac{1}{\sqrt{2\pi}} \|\hat{f}\|_{L^1}\, e^{|\text{Im }z|\Omega}\ .$$

In fact, $\mathcal{B}_\Omega$ consists of exactly those L^2-functions for which there exists an analytic extension to an entire function satisfying a bound of this type. We can

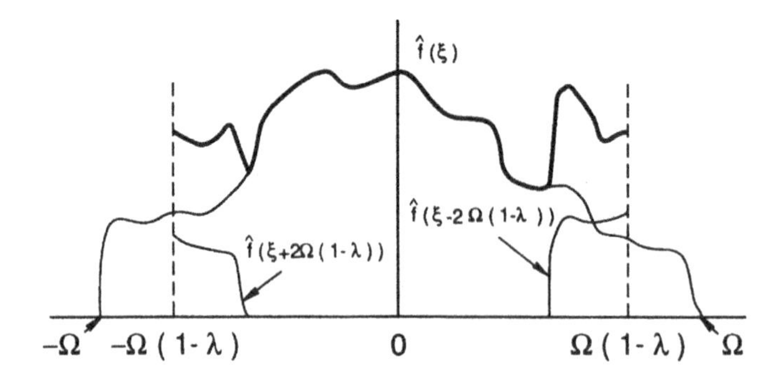

FIG. 2.2. *The three terms $\hat{f}(\xi)$, $\hat{f}(\xi+2\Omega(1-\lambda))$, and $\hat{f}(\xi-2\Omega(1-\lambda))$ for $|\xi| \le \Omega(1-\lambda)$, and their sum (thick line).*

therefore consider $\mathcal{B}_\Omega$ to be a Hilbert space of entire functions. For f in $\mathcal{B}_\Omega$ we have

$$\begin{aligned} f(x) &= \frac{1}{\sqrt{2\pi}} \int_{-\Omega}^{\Omega} d\xi \; e^{ix\xi} \; \hat{f}(\xi) \\ &= \frac{1}{2\pi} \int_{-\Omega}^{\Omega} d\xi \; e^{ix\xi} \int dy \; f(y) \; e^{-i\xi y} \\ &= \int dy \; f(y) \; \frac{\sin \Omega(x-y)}{\pi(x-y)} \; . \end{aligned} \tag{2.2.1}$$

(The interchange of integrals in the last step is permissible if $f \in L^1$, i.e., if $\hat{f}$ is sufficiently smooth. Since, for all x, $[\pi(x-.)]^{-1} \sin \Omega(x-.)$ is in $L^2(\mathbb{R})$, the conclusion then extends to all f in $\mathcal{B}_\Omega$ by the standard trick explained in Preliminaries.) Introducing the notation $e_x(y) = \frac{\sin \Omega(x-y)}{\pi(x-y)}$, we can rewrite (2.2.1) as

$$f(x) = \langle f, e_x \rangle \; . \tag{2.2.2}$$

Note that $e_x \in \mathcal{B}_\Omega$, since $\hat{e}_x(\xi) = (2\pi)^{-1/2} \; e^{-ix\xi}$ for $|\xi| < \Omega$, $\hat{e}_x(\xi) = 0$ for $|\xi| > \Omega$.

Formula (2.2.2) is typical for a reproducing kernel Hilbert space (r.k.H.s). In an r.k.H.s. $\mathcal{H}$ of functions, the map associating to a function f its value $f(x)$ at a point x is a continuous map (this does not hold in most Hilbert spaces of functions, in particular not in $L^2(\mathbb{R})$ itself), so that there necessarily exists $e_x \in \mathcal{H}$ such that $f(x) = \langle f, e_x \rangle$ for all $f \in \mathcal{H}$ (by Riesz' representation lemma; see Preliminaries). One also writes

$$f(x) = \int dy \; K(x\; , y) \; f(y),$$

where $K(x, y) = \overline{e_x(y)}$ is the *reproducing kernel*. In the particular case of $\mathcal{B}_\Omega$, there even exist special $x_n = \frac{n\pi}{\Omega}$ so that the e_{x_n} constitute an orthogonal basis for $\mathcal{B}_\Omega$, leading to Shannon's formula (2.1.2). Such special x_n need not exist in a general r.k.H.s. We will meet several examples of other r.k.H.s.'s in what follows.

2.3. Band- and timelimiting.

Functions cannot be both band- and timelimited: if f is bandlimited (with arbitrary finite bandwidth), then f is the restriction to $\mathbb{R}$ of an entire analytic function; if f were timelimited as well, support $f \subset [-T, T]$ with $T < \infty$, then $f \equiv 0$ would follow (nontrivial analytic functions can only have isolated zeros). Nevertheless, many practical situations correspond to an *effective* band- and timelimiting: imagine, for instance, that a signal gets transmitted (e.g. over a telephone line) in such a way that frequencies above Ω are lost (most realistic transmission means suffer from this kind of bandlimiting); imagine, also, that the signal (such as a telephone conversation) has a finite time duration. The transmitted signal is then, for all practical purposes, effectively band- and timelimited. How can this be? And how well can a function be represented by such a time- and bandlimited representation? Many researchers worked on these problems, until they were elegantly solved by the work of H. Landau, H. Pollack, and D. Slepian, in their series of papers Slepian and Pollak (1961) and Landau and Pollak (1961, 1962). An excellent review, with many more details than are given here, is Slepian (1976).

The example mentioned above (signal with finite time duration transmitted over a bandlimiting channel) can be modeled as follows: let Q_T, P_Ω be the orthonormal projection operators in $L^2(\mathbb{R})$ defined by

$$(Q_T f)(x) = f(x) \quad \text{for } |x| < T, \qquad (Q_T f)(x) = 0 \quad \text{for } |x| > T,$$

and

$$(P_\Omega f)^\wedge(\xi) = \hat{f}(\xi) \quad \text{for } |\xi| < \Omega, \qquad (P_\Omega f)^\wedge(\xi) = 0 \quad \text{for } |\xi| > \Omega \, .$$

Then a signal which is timelimited to $[-T, T]$ satisfies $f = Q_T f$, and transmitting it over a channel with bandwidth Ω gives as end product $P_\Omega f = P_\Omega Q_T f$ (provided there is no other distortion). The operator $P_\Omega Q_T$ represents the total time + band limiting process. How well the transmitted $P_\Omega Q_T f$ approaches the original f is measured by $\|P_\Omega Q_T f\|^2/\|f\|^2 = \langle Q_T\, P_\Omega\, Q_T\, f, f\rangle/\|f\|^2$.

The maximum value of this ratio is the largest eigenvalue of the symmetric operator $Q_T\, P_\Omega\, Q_T$, given explicitly by

$$(Q_T\, P_\Omega\, Q_T\, f)(x) = \begin{cases} \displaystyle\int_{-T}^{T} dy\, \frac{\sin \Omega(x-y)}{\pi(x-y)}\, f(y) & \text{if } |x| < T, \\ 0 & \text{if } |x| > T. \end{cases} \tag{2.3.1}$$

The eigenvalues and eigenfunctions of this operator are now known explicitly because of a fortunate accident: $Q_T\, P_\Omega\, Q_T$ commutes with the second order differential operator A,

$$(Af)(x) = \frac{d}{dx}(T^2 - x^2)\,\frac{df}{dx} - \frac{\Omega^2}{\pi^2}\, x^2 f(x) \, .$$

The eigenfunctions of this operator, which had been studied for different reasons long before their connection with band- and timelimiting was discovered, are called the prolate spheroidal wave functions, and many of their properties are known. Because A commutes with $Q_T\, P_\Omega\, Q_T$ (and because the eigenvalues of A are all simple), the prolate spheroidal wave functions are also the eigenfunctions of $Q_T\, P_\Omega\, Q_T$ (with different eigenvalues, of course). More specifically, if we denote the prolate spheroidal wave functions by ψ_n, $n \in \mathbb{N}$, ordered so that the corresponding eigenvalues α_n of A increase as n increases, then

$$\begin{aligned} &Q_T\, P_\Omega\, Q_T\, \psi_n = \lambda_n \psi_n, \\ &Q_T\, P_\Omega\, Q_T\, f = 0 \Leftrightarrow f \perp \psi_n \ \text{ for all } n \\ &\qquad\qquad\qquad\quad\ \Leftrightarrow f \text{ is supported on } \{x; |x| \geq T\}, \\ &\lambda_n \text{ decreases as } n \text{ increases, and } \lim_{n\to\infty} \lambda_n = 0 \, . \end{aligned}$$

The eigenvalues λ_n depend on T and Ω, of course; an easy scaling argument (substitute $x = Tx'$, $y = Ty'$ in the expression for $(Q_T\, P_\Omega\, Q_T\, f)(x))$ shows that the λ_n depend only on the product $T\Omega$. For fixed $T\Omega$, the behavior of λ_n as n increases is schematically represented in Figure 2.3. Typically, the λ_n stay close to 1 for small n, plunge to zero near the threshold value $2T\Omega/\pi$, and stay close

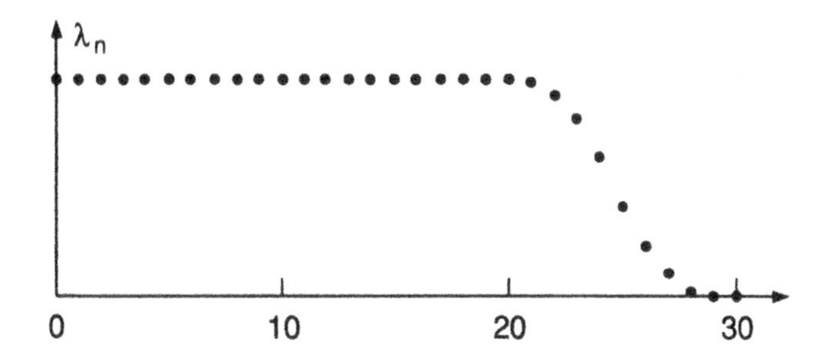

FIG. 2.3. *The eigenvalues λ_n for $Q_T P_\Omega Q_T$ for $2T\Omega/\pi = 25$.*

to zero afterwards. More precisely, for any (arbitrarily small) $\epsilon > 0$, there exists a constant C_ϵ so that

$$\begin{aligned} \# \quad & \{n; \lambda_n \geq 1 - \epsilon\} \leq \tfrac{2T\Omega}{\pi} - C_\epsilon \log(T\Omega), \\ \# \quad & \{n; 1 - \epsilon \geq \lambda_n \geq \epsilon\} \leq 2C_\epsilon \log(T\Omega), \end{aligned} \tag{2.3.2}$$

which means that the "plunge region" has width proportional to $\log(T\Omega)$. Since $\lim_{x\to\infty} x^{-1} \log x = 0$, the width of the plunge region becomes negligibly small, when compared to the threshold value $2T\Omega/\pi$, as $T, \Omega \to \infty$. In fact, (2.3.2) is a rigorous version of the fact that a time- and bandlimited region $[-T, T] \times [-\Omega, \Omega]$ corresponds to $2T\Omega/\pi$ "degrees of freedom," i.e., there exist (up to an error, small compared to $T\Omega$) $2T\Omega/\pi$ independent functions (and not more) that are essentially timelimited to $[-T, T]$ and bandlimited to $[-\Omega, \Omega]$. Note that $2T\Omega/\pi$ is exactly the area of $[-T, T] \times [-\Omega, \Omega]$, divided by 2π. This number is therefore equal to the number of sampling times within $[-T, T]$ specified by Shannon's theorem for a function with bandwidth Ω; this heuristic way of counting the "independent degrees of freedom" was part of the folklore of communication theory long before it was justified by Landau, Pollak, and Slepian. Independently, it was also known to physicists that a region in phase space (= space-momentum, or time-frequency such as here) with area S corresponds to $S/2\pi$ "independent states" in the semiclassical limit (i.e., when S is much larger than $\hbar$; the expression $S/2\pi$ corresponds to units such that $\hbar = 1$). We will extend the definition of Nyquist density from its original sampling background, and use it for the critical time-frequency density $(2\pi)^{-1}$ present in all these examples.

It is time to return to the wavelet transform. In what follows we will develop the continuous versions of both the wavelet transform and the windowed Fourier transform.

2.4. The continuous wavelet transform.

We restrict ourselves, for the time being, to one-dimensional wavelets. We always suppose that $\psi \in L^2(\mathbb{R})$; the analyzing wavelet should moreover satisfy the admissibility condition already mentioned in §1.3,

$$C_\psi = 2\pi \int d\xi \; |\xi|^{-1} \; |\hat{\psi}(\xi)|^2 < \infty \; . \tag{2.4.1}$$

The role of this condition will soon become clear. If $\psi \in L^1(\mathbb{R})$, then $\hat{\psi}$ is continuous and (2.4.1) can only be satisfied if $\hat{\psi}(0) = 0$, or $\int dx \; \psi(x) = 0$. On the other hand, if $\int dx \; \psi(x) = 0$ and we impose a slightly stronger condition than integrability on ψ, namely $\int dx \; (1 + |x|)^\alpha \; |\psi(x)| < \infty$ for some $\alpha > 0$, then $|\hat{\psi}(\xi)| \le C|\xi|^\beta$, with $\beta = \min\,(\alpha, 1)$, and (2.4.1) is satisfied. It follows that, for all practical purposes, (2.4.1) is equivalent to the requirement that $\int dx \; \psi(x) = 0$. (In practice, we will impose far more stringent decay conditions on ψ than those needed in this argument.)

We generate a doubly-indexed family of wavelets from ψ by dilating and translating,

$$\psi^{a,b}(x) = |a|^{-1/2} \; \psi\left(\frac{x-b}{a}\right) \; ,$$

where $a, b \in \mathbb{R}$, $a \neq 0$ (we use negative as well as positive a at this point). The normalization has been chosen so that $\|\psi^{a,b}\| = \|\psi\|$ for all a, b. We will assume that $\|\psi\| = 1$. The continuous wavelet transform with respect to this wavelet family is then

$$\begin{aligned} (T^{\text{wav}} f)(a,b) &= \langle f, \psi^{a,b} \rangle \\ &= \int dx \; f(x) \; |a|^{-1/2} \; \overline{\psi\left(\frac{x-b}{a}\right)} \; . \end{aligned}$$

Note that $|(T^{\text{wav}} f)(a,b)| \le \|f\|$.

A function f can be recovered from its wavelet transform via the resolution of the identity, as follows.

PROPOSITION 2.4.1. *For all* $f, g \in L^2(\mathbb{R})$,

$$\int_{-\infty}^{\infty} \int_{-\infty}^{\infty} \frac{da\,db}{a^2} \; (T^{\text{wav}} f)(a,b) \; \overline{(T^{\text{wav}} g)(a,b)} = C_\psi \langle f, g \rangle . \tag{2.4.2}$$

Proof.

$$\begin{aligned} &\int_{-\infty}^{\infty} \int_{-\infty}^{\infty} \frac{da\,db}{a^2} \; (T^{\text{wav}} f)(a,b) \; \overline{(T^{\text{wav}} g)(a,b)} \\ &\quad = \int \int \frac{da\,db}{a^2} \left[\int d\xi \; \hat{f}(\xi) \; |a|^{1/2} e^{-ib\xi} \; \overline{\hat{\psi}(a\xi)} \right] \\ &\qquad\quad \left[\int d\xi' \; \overline{\hat{g}(\xi')} \; |a|^{1/2} \; e^{ib\xi'} \; \hat{\psi}(a\xi') \right] . \end{aligned} \tag{2.4.3}$$

The expression between the first pair of brackets can be viewed as $(2\pi)^{1/2}$ times the Fourier transform of $F_a(\xi) = |a|^{1/2}\ \hat{f}(\xi)\ \overline{\hat{\psi}(a\xi)}$; the second has a similar interpretation as $(2\pi)^{1/2}$ times the complex conjugate of the Fourier transform of $G_a(\xi) = |a|^{1/2}\ \hat{g}(\xi)\ \overline{\hat{\psi}(a\xi)}$. By the unitarity of the Fourier transform it follows that

$$
\begin{aligned}
(2.4.3) \quad &= \quad 2\pi \int \frac{da}{a^2} \int d\xi\ F_a(\xi)\ \overline{G_a(\xi)} \\
&= \quad 2\pi \int \frac{da}{|a|} \int d\xi\ \hat{f}(\xi)\ \overline{\hat{g}(\xi)}\ |\hat{\psi}(a\xi)|^2 \\
&= \quad 2\pi \int d\xi\ \hat{f}(\xi)\ \overline{\hat{g}(\xi)} \int \frac{da}{|a|}\ |\hat{\psi}(a\xi)|^2 \\
&\qquad \text{(interchange is allowed by Fubini's theorem)} \\
&= \quad C_\psi\ \langle f, g \rangle \\
&\qquad \text{(make a change of variables } \zeta = a\xi \text{ in the second integral).} \quad \blacksquare
\end{aligned}
$$

It is now clear why we imposed (2.4.1): if C_ψ were infinite, then the resolution of the identity (2.4.2) would not hold.

Formula (2.4.2) can be read as

$$
f = C_\psi^{-1} \int_{-\infty}^{\infty} \int_{-\infty}^{\infty} \frac{da\,db}{a^2}\ (T^{\mathrm{wav}} f)(a, b)\ \psi^{a,b}\ , \tag{2.4.4}
$$

with convergence of the integral "in the weak sense," i.e., taking the inner product of both sides of (2.4.4) with any $g \in L^2(\mathbb{R})$, and commuting the inner product with the integral over a, b in the right-hand side, leads to a true formula. The convergence also holds in the following, slightly stronger sense:

$$
\lim_{\substack{A_1 \to 0 \\ A_2, B \to \infty}} \left\| f - C_\psi^{-1} \iint_{\substack{A_1 \le |a| \le A_2 \\ |b| \le B}} \frac{da\,db}{a^2}\ (T^{\mathrm{wav}} f)(a, b)\ \psi^{a,b} \right\| = 0\ . \tag{2.4.5}
$$

Here the integral stands for the unique element in $L^2(\mathbb{R})$ that has inner products with $g \in L^2(\mathbb{R})$ given by

$$
\iint_{\substack{A_1 \le |a| \le A_2 \\ |b| \le B}} \frac{da\,db}{a^2}\ (T^{\mathrm{wav}} f)(a, b)\ \langle \psi^{a,b},\ g \rangle\ ;
$$

since the absolute value of this is bounded by

$$
\iint_{\substack{A_1 \le |a| \le A_2 \\ |b| \le B}} \frac{da\,db}{a^2}\ \|f\|\ \|\psi^{a,b}\|\ \|g\| = 4B \left(\frac{1}{A_1} - \frac{1}{A_2} \right)\ \|f\|\ \|g\|\ ,
$$

we can give a sense to the integral in (2.4.5) by Riesz' lemma. The proof of (2.4.5) is then simple:

$$\begin{aligned}
&\left\| f - C_\psi^{-1} \iint\limits_{\substack{A_1 \le |a| \le A_2 \\ |b| \le B}} \tfrac{da\,db}{a^2}\, (T^{\text{wav}} f)(a,b)\, \psi^{a,b} \right\| \\
&\quad = \sup_{\|g\|=1} \left| \left\langle f - C_\psi^{-1} \iint\limits_{\substack{A_1 \le |a| \le A_2 \\ |b| \le B}} \frac{da\,db}{a^2}\, (T^{\text{wav}} f)(a,b)\, \psi^{a,b},\ g \right\rangle \right| \\
&\quad \le \sup_{\|g\|=1} \left| C_\psi^{-1} \iint\limits_{\substack{|a| \ge A_2 \\ \text{or } |a| \le A_1 \\ \text{or } |b| \ge B}} \frac{da\,db}{a^2}\, (T^{\text{wav}} f)(a,b)\, \overline{(T^{\text{wav}} g)(a,b)} \right| \\
&\quad \le \sup_{\|g\|=1} \left[C_\psi^{-1} \iint\limits_{\substack{|a| \ge A_2 \\ \text{or } |a| \le A_1 \\ \text{or } |b| \ge B}} \frac{da\,db}{a^2}\, |(T^{\text{wav}} f)(a,b)|^2 \right]^{1/2} \\
&\qquad\qquad \left[C_\psi^{-1} \iint \frac{da\,db}{a^2}\, |(T^{\text{wav}} g)(a,b)|^2 \right]^{1/2} .
\end{aligned}$$

By Proposition 2.4.1, the expression between the second pair of brackets is $\|g\|^2 = 1$, and the expression between the first pair of brackets converges to zero as $A_1 \to 0$, $A_2, B \to \infty$, because the infinite integral converges. This establishes (2.4.5).

Formula (2.4.5), which shows that any f in $L^2(\mathbb{R})$ can be arbitrarily well approximated by a superposition of wavelets, may seem paradoxical: after all, wavelets have integral zero, so how can any superposition of them (which necessarily still has integral zero) then be a good approximation to f if f itself happens to have nonzero integral? The solution to this paradox does not lie (as solutions to paradoxes so often do) in the mathematical sloppiness of the question. We can easily make it all rigorous: if we take $f \in L^1(\mathbb{R}) \cap L^2(\mathbb{R})$, and if ψ itself is in $L^1(\mathbb{R})$, then one easily checks that the

$$C_\psi^{-1} \iint\limits_{\substack{A_1 \le |a| \le A_2 \\ |b| \le B}} \frac{da\,db}{a^2}\, (T^{\text{wav}} f)(a,b)\, \psi^{a,b}$$

are indeed all in $L^1(\mathbb{R})$ (with norm bounded by $8C_\psi^{-1}\ \|f\|_{L^2}\ \|\psi\|_{L^2}\ \|\psi\|_{L^1}\ B(A_1^{-1/2} - A_2^{-1/2})$), and that they have integral zero, whereas f itself, the function they are approaching as $A_1 \to 0$, $A_2, B \to \infty$, may well have nonzero integral.

The explanation to this apparent paradox is that the limit (2.4.5) holds in L^2-sense, but not in L^1-sense. As $A_1 \to 0$, $A_2, B \to \infty$,

$$f(x) - C_\psi^{-1} \iint_{\substack{A_1 \le |a| \le A_2 \\ |b| \le B}} \frac{da\,db}{a^2} W f(a,b) \psi^{a,b}$$

becomes a very flat, very stretched-out function, which still has the same integral as f itself, but vanishingly small L^2-norm. (This is similar to the observation that the functions $g_n(x) = (2n)^{-1}$ for $|x| \le n$, 0 otherwise, satisfy $\int g_n = 1$ for all n, even though $g_n(x) \to 0$ for all x, and $\|g_n\|_{L^2} = (2n)^{-1/2} \to 0$ for $n \to \infty$; the g_n do not converge in $L^1(\mathbb{R})$.)

Several variations on (2.4.4) are possible, in which we restrict ourselves to positive a only (as opposed to the use of both positive and negative a in (2.4.4)). One possibility is to require that ψ satisfy an admissibility condition slightly more stringent than (2.4.1), namely

$$C_\psi = 2\pi \int_0^\infty d\xi\ |\xi|^{-1}\ |\hat{\psi}(\xi)|^2 = 2\pi \int_{-\infty}^0 d\xi\ |\xi|^{-1}\ |\hat{\psi}(\xi)|^2 < \infty\,. \tag{2.4.6}$$

Equality of these two integrals follows immediately if, e.g., ψ is a real function, because then $\hat{\psi}(-\xi) = \overline{\hat{\psi}(\xi)}$. The resolution of the identity then becomes, with this new C_ψ,

$$f = C_\psi^{-1} \int_0^\infty \frac{da}{a^2} \int_{-\infty}^\infty db\ T^{\text{wav}} f(a,b)\ \psi^{a,b}\ , \tag{2.4.7}$$

to be understood in the same weak or slightly stronger senses as (2.4.4). (The proof of (2.4.7) is entirely analogous to that of (2.4.4).)

Another variation occurs if f is a real function, and if support $\hat{\psi} \subset [0, \infty)$. In this case, one easily proves that

$$f = 2C_\psi^{-1} \int_0^\infty \frac{da}{a^2} \int_{-\infty}^\infty db\ \text{Re}\ [T^{\text{wav}} f(a,b)\ \psi^{a,b}]\ , \tag{2.4.8}$$

with C_ψ as defined by (2.4.1). (To prove (2.4.8), use that $f(x) = (2\pi)^{-1/2}$ $2\text{Re} \int_0^\infty d\xi\ e^{ix\xi} \hat{f}(\xi)$, because $\hat{f}(-\xi) = \overline{\hat{f}(\xi)}$.) Formula (2.4.8) can of course be rewritten in terms of $\psi_1 = \text{Re}\ \psi$ and $\psi_2 = \text{Im}\ \psi$, two wavelets which are each other's Hilbert transform. Using a complex wavelet, even for the analysis of real functions, may have its advantages. In Kronland-Martinet, Morlet, and Grossmann (1987), for example, a complex wavelet ψ with support $\hat{\psi} \subset [0, \infty)$ is used, and the wavelet transform $T^{\text{wav}} f$ is represented by graphs of its modulus and its phase.

If both f and ψ are so-called "analytical signals," i.e., if support $\hat{f}$ and support $\hat{\psi} \subset [0, \infty)$, then $T^{\text{wav}} f(a,b) = 0$ if $a < 0$, so that (2.4.4) immediately simplifies to

$$f = C_\psi^{-1} \int_0^\infty \frac{da}{a^2} \int_{-\infty}^\infty db\ T^{\text{wav}} f(a,b) \psi^{a,b}\ , \tag{2.4.9}$$

with C_ψ again as defined by (2.4.1). Finally, we can adapt (2.4.9) to the case where support $\hat{\psi} \subset [0, \infty)$, but support $\hat{f} \not\subset [0, \infty)$. We write $f = f_+ + f_-$, with support $\hat{f}_+ \subset [0, \infty)$, support $\hat{f}_- \subset (-\infty, 0]$, $\psi_+ = \psi$, and we introduce $\hat{\psi}_-(\xi) = \hat{\psi}(-\xi)$; clearly support $\hat{\psi}_- \subset (-\infty, 0]$. Then $\langle f_+, \psi_-^{a,b} \rangle = 0$ and $\langle f_-, \psi_+^{a,b} \rangle = 0$ for $a > 0$, so that, by straightforward application of (2.4.9),

$$f = C_\psi^{-1} \int_0^\infty \frac{da}{a^2} \int_{-\infty}^\infty db \left[(T_+^{\rm wav} f)(a, b) \psi_+^{a,b} + (T_-^{\rm wav} f)(a, b) \psi_-^{a,b} \right] , \qquad (2.4.10)$$

where $(T_+^{\rm wav} f)(a, b) = \langle f_+, \psi_+^{a,b} \rangle = \langle f, \psi_+^{a,b} \rangle$, $(T_-^{\rm wav} f)$ is defined analogously, and C_ψ is as in (2.4.1).

Another important variation consists of introducing a different function for the reconstruction than for the decomposition. More explicitly, if ψ_1, ψ_2 satisfy that

$$\int d\xi \; |\xi|^{-1} \; |\hat{\psi}_1(\xi)| \; |\hat{\psi}_2(\xi)| < \infty , \qquad (2.4.11)$$

then the same argument as in the proof of Proposition 2.4.1 shows that

$$\int \frac{da}{a^2} \int db \; \langle f, \psi_1^{a,b} \rangle \langle \psi_2^{a,b}, g \rangle = C_{\psi_1, \psi_2} \langle f, g \rangle , \qquad (2.4.12)$$

with $C_{\psi_1,\psi_2} = 2\pi \int d\xi \; |\xi|^{-1} \; \overline{\hat{\psi}_1(\xi)} \; \hat{\psi}_2(\xi)$. If $C_{\psi_1,\psi_2} \neq 0$, then we can rewrite (2.4.12) as

$$f = C_{\psi_1,\psi_2}^{-1} \int \frac{da}{a^2} \int db \; \langle f, \psi_1^{a,b} \rangle \; \psi_2^{a,b} . \qquad (2.4.13)$$

Note that ψ_1 and ψ_2 may have very different properties! One may be irregular, the other smooth; both need not even be admissible: if $\hat{\psi}_1(\xi) = O(\xi)$ for $\xi \to 0$, then $\hat{\psi}_2(0) \neq 0$ is allowed. We will not use this extra freedom here. In Holschneider and Tchamitchian (1990), the freedom in the choices of ψ_1, ψ_2 is exploited to prove some very interesting results (see also §2.9). One can, for instance, choose ψ_2 to be compactly supported, support $\psi_2 \subset [-R, R]$, so that, for any x, only the $\langle f, \psi_1^{a,b} \rangle$ with $|b - x| \leq |a| R$ will contribute to $f(x)$ in the reconstruction formula (2.4.13); the set $\{(a, b);\; |b - x| \leq |a| R\}$ is then called the "cone of influence" of ψ_2 on x. Holschneider and Tchamitchian (1990) also prove that, with mild conditions on f, (2.4.13) is true pointwise as well as in the L^2-sense.

PROPOSITION 2.4.2. *Suppose that* $\psi_1, \psi_2 \in L^1(\mathbb{R})$, *that* ψ_2 *is differentiable with* $\psi_2' \in L^2(\mathbb{R})$, *that* $x\psi_2 \in L^1(\mathbb{R})$, *and that* $\hat{\psi}_1(0) = 0 = \hat{\psi}_2(0)$. *If* $f \in L^2(\mathbb{R})$ *is bounded, then* (2.4.13) *holds pointwise in every point* x *where* f *is continuous, i.e.,*

$$f(x) = C_{\psi_1,\psi_2}^{-1} \lim_{\substack{A_1 \to 0 \\ A_2 \to \infty}} \int_{A_1 \leq |a| \leq A_2} \frac{da}{a^2} \int_{-\infty}^\infty db \; \langle f, \psi_1^{a,b} \rangle \; \psi_2^{a,b}(x) . \qquad (2.4.14)$$

Proof.

1. We can rewrite the right-hand member of (2.4.14) (before taking the limit) as

$$\begin{aligned} f_{A_1,A_2}(x) &= C_{\psi_1,\psi_2}^{-1} \int_{A_1\le|a|\le A_2} \frac{da}{a^2} \int_{-\infty}^{\infty} dy \\ &\quad \cdot \int_{-\infty}^{\infty} db\ f(y)|a|^{-1}\ \overline{\psi_1\left(\frac{y-b}{a}\right)}\ \psi_2\left(\frac{x-b}{a}\right) \\ &= \int_{-\infty}^{\infty} dy\ M_{A_1,A_2}(x-y)f(y)\ , \end{aligned} \tag{2.4.15}$$

where all the changes of order of integration are permitted by Fubini's theorem (the integral converges absolutely). Here M_{A_1,A_2} is defined by

$$M_{A_1,A_2}(x) = C_{\psi_1,\psi_2}^{-1} \int_{A_1\le|a|\le A_2} \frac{da}{|a|^3} \int_{-\infty}^{\infty} db\ \overline{\psi_1\left(-\frac{b}{a}\right)}\ \psi_2\left(\frac{x-b}{a}\right)\ .$$

2. One easily computes that the Fourier transform of M_{A_1,A_2} is

$$\hat{M}_{A_1,A_2}(\xi) = (2\pi)^{1/2}\ C_{\psi_1,\psi_2}^{-1} \int_{A_1\le|a|\le A_2} \frac{da}{|a|}\ \hat{\psi}_2(a\xi)\ \overline{\hat{\psi}_1(a\xi)} \tag{2.4.16}$$

$$= \hat{M}(A_1\xi)\ -\ \hat{M}(A_2\xi)\ , \tag{2.4.17}$$

where $\hat{M}(\xi) = (2\pi)^{1/2}\ C_{\psi_1,\psi_2}^{-1} \int_{|a|\ge|\xi|} \frac{da}{|a|}\ \hat{\psi}_2(a)\ \overline{\hat{\psi}_1(a)}$, as follows from a change of variables $a{\to}a\xi$ in (2.4.16). Since $a\hat{\psi}_2(a) \in L^2(\mathbb{R})$ and $\hat{\psi}_1(a)$ is bounded, we have

$$\begin{aligned} |\hat{M}(\xi)| &\le C \left(\int_{|a|\ge|\xi|} \frac{da}{|a|^4}\ |\hat{\psi}_1(a)|^2\right)^{1/2} \left(\int da\ |a|^2\ |\hat{\psi}_2(a)|^2\right)^{1/2} \\ &\le C'\ |\xi|^{-3/2}\ . \end{aligned}$$

By (2.4.11), $\hat{M}$ is also bounded, so that

$$|\hat{M}(\xi)| \le C(1+|\xi|)^{-3/2}\ , \tag{2.4.18}$$

implying that M, the inverse Fourier transform of $\hat{M}$, is well defined, bounded, and continuous.

3. The decay of M is governed by the regularity of $\hat{M}$. For $\xi \neq 0$, one easily checks that $\hat{M}$ is differentiable in ξ, with

$$\frac{d}{d\xi}\hat{M}(\xi) = (2\pi)^{1/2}\ C_{\psi_1,\psi_2}^{-1} \frac{-1}{\xi}\ \left[\hat{\psi}_2(\xi)\ \overline{\hat{\psi}_1(\xi)} + \hat{\psi}_2(-\xi)\ \overline{\hat{\psi}_1(-\xi)}\right]\ .$$

Because $x\psi_2 \in L^1$, $\hat{\psi}_2$ is differentiable, so that, for $\xi = 0$,

$$\frac{d}{d\xi}\hat{M}\bigg|_{\xi=0} = -(2\pi)^{1/2}\, C^{-1}_{\psi_1,\psi_2}\, 2\, \overline{\hat{\psi}_1(0)}\, \hat{\psi}_2'(0) = 0\ .$$

It follows that $\hat{M}$ is differentiable. Moreover, since $x\psi_2 \in L^1$, we have

$$\begin{aligned} |\hat{\psi}_2(\xi)| &= |\hat{\psi}_2(\xi) - \hat{\psi}_2(0)| \\ &\le C \int dx\ |e^{-i\xi x} - 1|\ |\psi_2(x)| \\ &\le C\,|\xi| \int dx\ |x\psi_2(x)| \le C'\ |\xi|\ , \end{aligned}$$

which implies $\left|\frac{d}{d\xi}\hat{M}(\xi)\right| \le C''\left[|\hat{\psi}_1(\xi)| + |\hat{\psi}_1(-\xi)|\right]$, so that $\frac{d}{d\xi}\hat{M} \in L^2$. It then follows from

$$\begin{aligned} \int dx\ |M(x)| &\le \left[\int dx\ (1+x^2)^{-1}\right]^{1/2} \left[\int dx\ (1+x^2)\ |M(x)|^2\right]^{1/2} \\ &\le C \left[\int d\xi\ \left(|\hat{M}(\xi)|^2 + \left|\frac{d}{d\xi}\hat{M}(\xi)\right|^2\right)\right]^{1/2} < \infty \end{aligned}$$

that $M \in L^1(\mathbb{R})$. Moreover, $\hat{M}(0) = (2\pi)^{1/2}\ C^{-1}_{\psi_1,\psi_2} \int \frac{da}{|a|}\hat{\psi}_2(a)\ \overline{\hat{\psi}_1(a)} = (2\pi)^{-1/2}$, or $\int dx\ M(x) = 1$.

4. Using (2.4.17) we can rewrite (2.4.15) as

$$f_{A_1,A_2}(x) = \int_{-\infty}^{\infty} dy\ \frac{1}{A_1} M\left(\frac{x-y}{A_1}\right) f(y) - \int_{-\infty}^{\infty} dy\ \frac{1}{A_2} M\left(\frac{x-y}{A_2}\right) f(y)\ .$$

Because M is continuous, integrable, and of integral 1, the first term tends to $f(x)$ for $A_1 \to 0$ if f is bounded, and continuous in x. (This follows from a simple application of the dominated convergence theorem.) The second term is bounded by

$$\begin{aligned} &\left|\int_{-\infty}^{\infty} dy\ \frac{1}{A_2} M\left(\frac{x-y}{A_2}\right) f(y)\right| \\ &\quad\le \left[\int_{-\infty}^{\infty} dy\ \frac{1}{A_2^2} \left|M\left(\frac{x-y}{A_2}\right)\right|^2\right]^{1/2} \left[\int dy\ |f(y)|^2\right]^{1/2} \\ &\quad\le A_2^{-1/2}\ \|M\|_{L^2}\ \|f\|_{L^2} \le C\ A_2^{-1/2}\ , \end{aligned}$$

because $M \in L^2(\mathbb{R})$ by (2.4.18). This term therefore tends to zero if $A_2 \to \infty$. ■

REMARK. In Holschneider and Tchamitchian (1990), this theorem is proved under slightly more general conditions on f as well as on ψ_1, ψ_2. □

2.5. The reproducing kernel Hilbert space underlying the continuous wavelet transform.

As a special case of (2.4.2), we have, for $f \in L^2(\mathbb{R})$,

$$C_\psi^{-1} \int\int \frac{da\, db}{a^2}\, |(T^{\rm wav} f)(a,b)|^2 = \int dx |f(x)|^2 .$$

In other words, $T^{\rm wav}$ maps $L^2(\mathbb{R})$ isometrically into $L^2(\mathbb{R}^2;\, C_\psi^{-1} a^{-2}\, da\, db)$, the space of all complex valued functions F on $\mathbb{R}^2$ for which $|||F|||^2 = C_\psi^{-1} \int\int \frac{da\, db}{a^2}$ $|F(a,b)|^2$ converges; equipped with the norm $|||\ \ |||$, this is a Hilbert space. The image $T^{\rm wav} L^2(\mathbb{R})$ constitutes only a closed subspace, not all of $L^2(\mathbb{R}^2;$ $C_\psi^{-1} a^{-2}\, da\, db)$; we will denote this subspace $\mathcal{H}$.

The following argument shows that $\mathcal{H}$ is a r.k.H.s. For any $F \in \mathcal{H}$, we can find $f \in L^2(\mathbb{R})$ so that $F = T^{\rm wav} f$. It follows then from (2.4.2) that

$$\begin{aligned} F(a,b) &= \langle f,\, \psi^{a,b} \rangle \\ &= C_\psi^{-1} \int\int \frac{da'\, db'}{a'^2}\, (T^{\rm wav} f)(a',b')\, \overline{(T^{\rm wav} \psi^{a,b})(a',b')} \\ &= C_\psi^{-1} \int\int \frac{da'\, db'}{a'^2}\, K(a,b;\, a',b')\, F(a',b') \end{aligned} \tag{2.5.1}$$

with

$$\begin{aligned} K(a,b;\, a',b') &= \overline{(T^{\rm wav} \psi^{a,b})(a',b')} \\ &= \langle \psi^{a',b'}, \psi^{a,b} \rangle . \end{aligned}$$

Formula (2.5.1) shows that $\mathcal{H}$ is indeed an r.k.H.s. embedded as a subspace in $L^2(\mathbb{R}^2;\, C_\psi^{-1} a^{-2}\, da\, db)$. (It also immediately shows that $\mathcal{H}$ is not all of $L^2(\mathbb{R}^2;$ $C_\psi^{-1} a^{-2}\, da\, db)$, since such a reproducing kernel formula could not hold for the whole space $L^2(\mathbb{R}^2;\, C_\psi^{-1} a^{-2}\, da\, db)$.)

In particular cases, $\mathcal{H}$ becomes a Hilbert space of analytic functions. Let us restrict ourselves again to functions f such that support $\hat{f} \subset [0,\infty)$; these functions form a closed subspace of $L^2(\mathbb{R})$ which we denote by H^2 (it is one of a family of Hardy spaces). For ψ we choose, for instance, $\hat{\psi}(\xi) = 2\xi e^{-\xi}$ for $\xi \geq 0$, $\hat{\psi}(\xi) = 0$ for $\xi \leq 0$ (ψ is also in H^2). Then the functions in $T^{\rm wav}\ H^2$ can be written as (consider only $a \geq 0$; see (2.4.9))

$$\begin{aligned} F(a,-b) = \langle f, \psi^{a,-b} \rangle &= 2a^{1/2} \int_0^\infty d\xi\ \hat{f}(\xi)\ a\xi\ e^{-i(b+ia)\xi} \\ &= (2\pi)^{1/2}\ a^{3/2}\ G(b+ia), \end{aligned}$$

where G is analytic on the upper half plane (Im $z > 0$). Moreover, one easily checks that

$$\int_0^\infty da \int_{-\infty}^\infty db\ a|G(b+ia)|^2 = \int dx\ |f(x)|^2 ,$$

so that T^{wav} can be interpreted as an isometry from H^2 to the Bergman space of all analytic function on the upper half plane, square integrable with respect to the measure $\text{Im}\, z\ d(\text{Im}\, z)\ d(\text{Re}\, z)$. On the other hand, one can prove that any function in this Bergman space is associated, via the wavelet transform with this particular ψ, to a function in H^2: the isometry is onto, and is therefore a unitary map. For other choices of ψ, such as $\psi \in H^2$ with $\hat{\psi}(\xi) = N_\beta\, \xi^\beta\, e^{-\xi}$ for $\xi \geq 0$, the image $T^{\text{wav}} H^2$ can be identified with other Bergman spaces of analytic functions on the upper half plane. [1]

Since $T^{\text{wav}} L^2$ or $T^{\text{wav}} H^2$ can be identified with a reproducing kernel Hilbert space, it should be no surprise that there exist discrete families of points $(a_\alpha,\ b_\alpha)$ such that f is completely determined by, and can be reconstructed from, $(T^{\text{wav}} f)\ (a_\alpha,\ b_\alpha)$. In particular, if $T^{\text{wav}} f$ can be identified with a function in a Bergman space, then it is obvious that its values at certain discrete families of points completely determine the function, since it is, after all, an analytic function. Reconstructing it in a numerically stable way may be another matter: the situation is not as simple as in the bandlimited case, where there exists a special family of points x_n such that the e_{x_n} constitute an orthonormal basis for $\mathcal{B}_\Omega$. There is no such convenient orthonormal basis e_{a_α, b_α} in our $T^{\text{wav}} L^2$ or $T^{\text{wav}} H^2$. We will see in the next chapter how this problem can be tackled.

Finally, before we leave this section, it should be remarked that (2.4.4), or the equivalent r.k.H.s. formulation, can be viewed as a consequence of the theory of square integrable group representations. I do not wish to dwell on this in detail here; readers who are interested in learning more about them should consult the references in the notes. [2] The $\psi^{a,b}$ are in fact the result of the action of the operators $U(a, b)$, defined by

$$[U(a,b)f](x) = |a|^{-1/2}\, f\left(\frac{x-b}{a}\right) ,$$

on the function ψ. The operators $U(a, b)$ are all unitary on $L^2(\mathbb{R})$, and constitute a representation of the $ax + b$-group:

$$U(a,b)\ U(a',b') = U(aa',\ b + ab') .$$

This group representation is *irreducible* (i.e., for any $f \neq 0$, there exists no nontrivial g orthogonal to all the $U(a, b)f$, which is equivalent to saying that the $U(a, b)f$ span the entire space). The following result is true: if U is an irreducible unitary representation in $\mathcal{H}$ of a Lie-group G with left invariant measure $d\mu$, and if for some f in $\mathcal{H}$,

$$\int_G d\mu(g)\ |\langle f,\ U(g)f\rangle|^2 < \infty , \tag{2.5.2}$$

then there exists a dense set D in $\mathcal{H}$ so that property (2.5.2) holds for any element $\tilde{f}$ of D. Moreover, there exists a (possibly unbounded) operator A, well defined in D, so that, for all $\tilde{f} \in D$ and all $h_1, h_2 \in \mathcal{H}$,

$$\int_G d\mu(g)\ \langle h_1, U(g)\tilde{f}\rangle\ \overline{\langle h_2, U(g)\tilde{f}\rangle} = C_{\tilde{f}} \langle h_1, h_2\rangle , \tag{2.5.3}$$

with $C_{\tilde{f}} = \langle A\tilde{f}, \tilde{f} \rangle$. In the wavelet case, the left invariant measure is $a^{-2}\, da\, db$, and A is the operator

$$(Af)^\wedge(\xi) = |\xi|^{-1}\, \hat{f}(\xi)\ .$$

Note that (2.5.3) is a general resolution of the identity!

In what follows, we will not exploit this group structure underlying the wavelet transform, mainly because we will soon go to discretely labelled wavelet families, and these do not correspond to subgroups of the $ax+b$-group.

In quantum physics, resolutions of the identity (2.5.3) have been studied and used for many different groups G. The associated families $U(g)f$ are there called coherent states, a name that was first used in connection with the Weyl–Heisenberg group (see also next section), but later spilled over to all the other groups as well (and even to some related constructions which were not generated by a group). An excellent review and a collection of important papers on this subject can be found in Klauder and Skägerstam (1985). The coherent states associated with the $ax+b$-group, which are now called wavelets, were first constructed by Aslaksen and Klauder (1968, 1969).

2.6. The continuous wavelet transform in higher dimensions.

There exist several possible extensions of (2.4.4) to $L^2(\mathbb{R}^n)$ with $n > 1$. One possibility is to choose the wavelet $\psi \in L^2(\mathbb{R}^n)$ so that it is spherically symmetric. Its Fourier transform is then spherically symmetric as well,

$$\hat{\psi}(\xi) = \eta(|\xi|)\ ,$$

and the admissibility condition becomes

$$C_\psi = (2\pi)^n \int_0^\infty \frac{dt}{t}\ |\eta(t)|^2 < \infty\ .$$

Along the same lines as in the proof of Proposition 2.1 one can then prove that, for all $f, g \in L^2(\mathbb{R}^n)$,

$$\int_0^\infty \frac{da}{a^{n+1}} \int_{-\infty}^\infty db\ (T^{\text{wav}} f)(a,b)\ \overline{(T^{\text{wav}} g)(a,b)} = C_\psi \langle f, g\rangle\ , \tag{2.6.1}$$

where $(T^{\text{wav}} f)(a,b) = \langle f, \psi^{a,b}\rangle$, as before, and $\psi^{a,b}(x) = a^{-n/2}\ \psi(\frac{x-b}{a})$, with $a \in \mathbb{R}_+$, $a \neq 0$, and $b \in \mathbb{R}^n$. Formula (2.6.1) can again be rewritten as

$$f = C_\psi^{-1} \int_0^\infty \frac{da}{a^{n+1}} \int_{\mathbb{R}^n} db\ (T^{\text{wav}} f)(a,b)\ \psi^{a,b}\ . \tag{2.6.2}$$

It is also possible to choose a ψ that is not spherically symmetric, and to introduce rotations as well as dilations and translations. In two dimensions, for instance, we then define

$$\psi^{a,b,\theta}(x) = a^{-1}\psi\left(R_\theta^{-1}\left(\frac{x-b}{a}\right)\right)\ ,$$

where $a > 0$, $b \in \mathbb{R}^2$, and where R_θ is the matrix

$$\begin{pmatrix} \cos\theta & -\sin\theta \\ \sin\theta & \cos\theta \end{pmatrix} .$$

The admissibility condition then becomes

$$C_\psi = (2\pi)^2 \int_0^\infty \frac{dr}{r} \int_0^{2\pi} d\theta \, |\hat{\psi}(r\cos\theta, \ r\sin\theta)|^2 < \infty ,$$

and the corresponding resolution of the identity is

$$f = C_\psi^{-1} \int_0^\infty \frac{da}{a^3} \int_{\mathbb{R}^2} db \int_0^{2\pi} d\theta \, (T^{\rm wav} f)(a, b, \theta) \psi^{a,b,\theta} .$$

A similar construction can be made in dimensions larger than 2. These wavelets with rotation angles were studied by Murenzi (1989), and applied by Argoul et al. (1989) in a study of DLA (diffusion-limited aggregates) and other two-dimensional fractals.

2.7. Parallels with the continuous windowed Fourier transform.

The windowed Fourier transform of a function f is given by

$$(T^{\rm win} f)(\omega, t) = \langle f, \ g^{\omega,t} \rangle , \qquad (2.7.1)$$

where $g^{\omega,t}(x) = e^{i\omega x} g(x-t)$. Arguments completely similar to those in the proof of Proposition 2.4.1 show that, for all $f_1, f_2 \in L^2(\mathbb{R})$,

$$\int\int d\omega \, dt \, (T^{\rm win} f_1)(\omega, t) \overline{(T^{\rm win} f_2)(\omega, t)} = 2\pi \, \|g\|^2 \langle f_1, f_2 \rangle ,$$

which can be rewritten as

$$f = (2\pi \|g\|^2)^{-1} \int\int d\omega \, dt \, (T^{\rm win} f)(\omega, t) g^{\omega,t} . \qquad (2.7.2)$$

There is no admissibility condition in this case: any window function g in L^2 will do. A convenient normalization for g is $\|g\|_{L^2} = 1$. (The absence of an admissibility condition is due to the unimodularity of the Weyl–Heisenberg group—see Grossmann, Morlet, and Paul (1985).)

The continuous windowed Fourier transform can again be viewed as a map from $L^2(\mathbb{R})$ to an r.k.H.s.; the functions $F \in T^{\rm win} L^2(\mathbb{R})$ are all in $L^2(\mathbb{R}^2)$ and moreover satisfy

$$F(\omega, t) = \frac{1}{2\pi} \int\int d\omega \, dt \, K(\omega, t; \ \omega', t') \, F(\omega', t') ,$$

where $K(\omega, t; \ \omega', t') = \langle g^{\omega' t'}, \ g^{\omega,t} \rangle$. (We assume $\|g\| = 1$ here.) Again there exist very special choices for g which reduce this r.k.H.s. to a Hilbert space of analytic functions: for $g(x) = \pi^{-1/4} \exp(-x^2/2)$, one finds

$$(T^{\rm win} f)(\omega, t) = \exp\left[-\frac{1}{4}(\omega^2 + t^2) - \frac{i}{2}\omega t \right] \, \phi(\omega + it) , \qquad (2.7.3)$$

where ϕ is an entire function. The set of all entire functions ϕ which can be obtained in this way constitutes the Bargmann Hilbert space (Bargmann (1961)).

The $g^{\omega,t}$ obtained from $g(x) = g_0(x) = \pi^{-1/4}\ \exp(-x^2/2)$ are often called the canonical coherent states (see the primer in Klauder and Skägerstam (1985)); the associate continuous windowed Fourier transform is the canonical coherent state representation. It has many beautiful and useful properties, of which we will explain one that will be used in the next section. Applying the differential operator $H = -\frac{d^2}{dx^2} + x^2 - 1$ to $g_0(x)$ leads to

$$\left(-\frac{d^2}{dx^2} + x^2 - 1\right)\ \pi^{-1/4}\ \exp(-x^2/2) = 0\ ,$$

i.e., g_0 is an eigenfunction of H with eigenvalue 0. In quantum mechanics language, H is the harmonic oscillator Hamiltonian operator, and g_0 is its ground state. (Strictly speaking, H is really *twice* the standard harmonic oscillator Hamiltonian.) The other eigenfunctions of H are given by higher order Hermite functions,

$$\phi_n(x) = \pi^{-1/4}\ 2^{-n/2}(n!)^{-1/2}\ \left(x - \frac{d}{dx}\right)^n\ \exp(-x^2/2)\ ,$$

which satisfy

$$H\phi_n = 2n\ \phi_n\ . \tag{2.7.4}$$

(The standard and easiest way to derive (2.7.4) is to write $H = A^*A$, where $A = x + \frac{d}{dx}$, and A^* is its adjoint $A^* = x - \frac{d}{dx}$, and to show that $Ag_0 = 0$, $A(A^*)^n = (A^*)^n A + 2n(A^*)^{n-1}$, so that $H\phi_n = \alpha_n\ A^*A(A^*)^n\ g_0 = \alpha_n\ A^*\ 2n(A^*)^{n-1}\ g_0 = 2n\ \phi_n$; the normalization α_n can be computed easily as well.) It is well known that the $\{\phi_n;\ n \in \mathbb{N}\}$ form an orthonormal basis for $L^2(\mathbb{R})$; they constitute therefore a "complete set of eigenfunctions" for H.[3]

Let us now consider the one-parameter families $\psi_s = \exp(-iHs)\psi$. These are the solutions to the equation

$$i\partial_s\psi_s = H\psi_s\ , \tag{2.7.5}$$

with initial condition $\psi_0 = \psi$. In the very special case where $\psi_0(x) = g_0^{\omega,t}(x) = \pi^{-1/4}\ e^{i\omega x}\ \exp[-(x-t)^2/2]$, we find $\psi_s = e^{i\alpha_s}\ g_0^{\omega_s,t_s}$, where $\omega_s = \omega\ \cos 2s - t\ \sin 2s$, $t_s = \omega\ \sin 2s + t\ \cos 2s$, and $\alpha_s = \frac{1}{2}(\omega t - \omega_s t_s)$ (as can easily be verified by explicit computation). That is, a canonical coherent state, when "evolved" under (2.7.5), remains a canonical coherent state (up to a phase factor which will be unimportant to us); the label (ω_s, t_s) of the new coherent state is obtained from the initial (ω, t) by a simple rotation in the time-frequency plane.

2.8. The continuous transforms as tools to build useful operators.

The resolutions of the identity (2.4.4), (2.7.2) can be rewritten in yet another way:

$$C_\psi^{-1}\ \int\int \frac{da\ db}{a^2}\ \langle\cdot,\ \psi^{a,b}\rangle\ \psi^{a,b} = \mathrm{Id}\ , \tag{2.8.1a}$$

$$\frac{1}{2\pi}\int\int d\omega\, dt\ \langle\cdot,\ g^{\omega,t}\rangle\ g^{\omega,t} = \text{Id}\ , \tag{2.8.1b}$$

where $\langle\cdot,\ \phi\rangle\phi$ stands for the operator on $L^2(\mathbb{R})$ that sends f to $\langle f,\ \phi\rangle\phi$; this is a rank one projection operator (i.e., its square and its adjoint are both identical to the operator itself, and its range is one-dimensional). Formulas (2.8.1) state that a "superposition," with equal weights, of the rank one projection operators corresponding to a family of wavelets (or a family of windowed Fourier functions) is exactly the identity operator. (As before, the integrals in (2.8.1) have to be taken in the weak sense.) What happens if we take similar superpositions, but give different weights to the different rank one projection operators? If the weight function is at all reasonable, we end up with a well-defined operator, different from the identity operator. If the weight function is bounded, then the corresponding operator is as well, but in many examples it is advantageous to consider even unbounded weight functions, which may give rise to unbounded operators. [4] We will review a few interesting examples (bounded and unbounded) in this section.

We start with the windowed Fourier case. Let us rewrite (2.8.1b) in the p, q (momentum, position) notation customary in quantum mechanics (rather than the ω, t notation for the frequency-time plane), and insert a weight function $w(p, q)$:

$$W = \frac{1}{2\pi}\int\int dp\, dq\ w(p,q)\ \langle\cdot,\ g^{p,q}\rangle\ g^{p,q}\ . \tag{2.8.2}$$

If $w \notin L^\infty(\mathbb{R}^2)$, then W may be unbounded and hence not everywhere defined; as a domain for W we can then take $\{f;\ \int\int dp\ dq\ |w(p,q)|^2\ |\langle f,\ g^{p,q}\rangle|^2 < \infty\}$, which is dense for reasonable w and g. [5] Two useful examples in quantum mechanics are (1) $w(p,q) = p^2$, which leads to $W = -\frac{d^2}{dx^2} + C_g\ \text{Id}$, where $C_g = \int d\xi\ \xi^2|\hat{g}(\xi)|^2$, and (2) $w(p,q) = v(q)$, for which W is a multiplicative potential operator: $(Wf)(x) = V_g(x)\ f(x)$, with $V_g(x) = \int dq\ v(q)|g(x-q)|^2$. Readers familiar with the basics of quantum mechanics will notice that in both cases the operator W corresponds pretty well to the "quantized version" of the phase space function $w(p,q)$ (in units such that $\hbar = 1$), with a slight twist: the extra constant C_g in the first case, the substitution of $v * |g|^2$ for the potential function v in the second case. In fact both formulas were used in Lieb (1981) to prove that Thomas–Fermi theory, a semiclassical theory for atoms and molecules, is "asymptotically" correct (for $Z \to \infty$, i.e., for very heavy atoms); it gives the leading order term of the much more complicated quantum mechanical model. Lieb's proof used the two examples above in three dimensions (rather than one); the operators he really wanted to consider were, of course, $-\Delta = -\partial^2_{x_1} - \partial^2_{x_2} - \partial^2_{x_3}$ and $V(x) = [x_1^2 + x_2^2 + x_3^2]^{-1/2}$, so that he had to choose an appropriate g and deal with the extra constant C_g and the difference between V and $V * |g|^2$ by some other means. Note that choosing $v(q)$ with an integrable singularity (such as the three-dimensional Coulomb potential) always leads to a nonsingular V_g: operators of type (2.8.2) cannot represent such singularities.

Many other applications of operators of type (2.8.2) exist. In pure mathematics they are sometimes called Toeplitz operators, and whole books have been written on them. In quantum optics they are also called "operators of type P," and again there exists an extensive literature on the subject (see Klauder and Skägerstam (1981)).[6] But let us go back to signal analysis, and see how (2.8.2) can be used to build time-frequency localization operators.

Let S be any measurable subset of $\mathbb{R}^2$. Let us return to time-frequency notation, and define, via (2.8.2), the operator L_S corresponding to the indicator function of S, $a(\omega,t) = 1$ if $(\omega,t) \in S$, 0 if $(\omega,t) \notin S$,

$$L_S = \frac{1}{2\pi} \iint_{(\omega,t)\in S} d\omega\, dt\, \langle \cdot,\, g^{\omega,t}\rangle\, g^{\omega,t} .$$

It follows immediately from the resolution of the identity that

$$\begin{aligned} \langle L_S f,\, f\rangle &= \frac{1}{2\pi} \iint_{(\omega,t)\in S} d\omega\, dt\, |\langle f, g^{\omega,t}\rangle|^2 \\ &\leq \frac{1}{2\pi} \iint d\omega\, dt\, |\langle f, g^{\omega,t}\rangle|^2 = \|f\|^2 ; \end{aligned}$$

on the other hand, obviously $\langle L_S f,\, f\rangle \geq 0$. In other words,

$$0 \leq L_S \leq \mathrm{Id} .$$

If S is a bounded set, then the operator L_S is trace-class (see Preliminaries), since, for any orthonormal basis $(u_n)_{n\in\mathbb{N}}$ in $L^2(\mathbb{R})$,

$$\sum_n \langle L_S\, u_n,\, u_n\rangle = \frac{1}{2\pi} \iint_{(\omega,t)\in S} d\omega\, dt\, \sum_n |\langle u_n, g^{\omega,t}\rangle|^2$$

(order of integral and summation may be inverted by Lebesgue's dominated convergence theorem)

$$= \frac{1}{2\pi} \iint_{(\omega,t)\in S} d\omega\, dt\, \|g^{\omega,t}\|^2 = |S| ,$$

where $|S|$ is the measure of S. It follows that there exists then a complete set of eigenvectors for L_S, with eigenvalues decreasing to zero,

$$\begin{aligned} &L_S \phi_n = \lambda_n \phi_n , \\ &\lambda_n \geq \lambda_{n+1} \geq 0, \quad \lim_{n\to\infty} \lambda_n = 0 , \\ &\{\phi_n;\ n \in \mathbb{N}\} \text{ orthonormal basis for } L^2(\mathbb{R}) . \end{aligned}$$

Such an operator L_S has a very natural interpretation. If the window function g is reasonably well localized and centered around zero in both time and frequency, then $\langle f, g^{\omega,t}\rangle\, g^{\omega,t}$ can be viewed as an "elementary component" of f, localized in the time-frequency plane around (ω, t). Summing all these components gives f again; $L_S f$ is the sum of only those components for which $(\omega, t) \in S$. Therefore, $L_S f$ corresponds to the extraction from f of only that information that pertains to the region S in the time-frequency plane, and the construction from that localized information of a function that "lives" on S only (or very nearby). This is the essence of a time-frequency localization operator such as we saw in §2.3! We can now, moreover, study L_S for sets S much more general than rectangles $[-\Omega, \Omega] \times [-T, T]$. (Note, however, that even for $S = [-\Omega, \Omega] \times [-T, T]$, our operators L_S are different from the $Q_T\, P_\Omega\, Q_T$ considered in §2.3.) Unfortunately, for most choices of S and g, the eigenfunctions and eigenvalues of L_S are hard to characterize, and this construction is of limited usefulness. However, there is one choice of g and one particular family of sets S for which everything is transparent. Take $g(x) = g_o(x) = \pi^{-1/4}\, \exp(-x^2/2)$, and $S_R = \{(\omega, t);\ \omega^2 + t^2 \le R^2\}$. Let us denote the corresponding localization operator by L_R,

$$L_R = \frac{1}{2\pi} \iint\limits_{\omega^2+t^2 \le R^2} d\omega\, dt\, \langle \cdot,\, g_o^{\omega,t}\rangle\, g_o^{\omega,t}\ .$$

These operators L_R commute with the harmonic oscillator Hamiltonian $H = -\frac{d^2}{dx^2} + x^2 - 1$ from §2.7, as can be seen by the following argument. Since

$$e^{-iHs}\ g_o^{\omega,t} = e^{i\alpha_s}\ g_o^{\omega_s, t_s}\ ,$$

with $\alpha_s = (\omega t - \omega_s t_s)/2 \in \mathbb{R}$, we have

$$\langle e^{-iHs} f,\ g_o^{\omega,t}\rangle\, g_o^{\omega,t} = \langle f,\ e^{iHs} g_o^{\omega,t}\rangle\, g_o^{\omega,t} = e^{-i\alpha_{-s}}\ \langle f,\ g_o^{\omega_{-s},t_{-s}}\rangle\, g_o^{\omega,t}\ ;$$

hence

$$L_R\, e^{-iHs} = \frac{1}{2\pi} \iint\limits_{\omega^2+t^2 \le R^2} d\omega\, dt\, \langle \cdot,\, g_o^{\omega_{-s},t_{-s}}\rangle\, e^{-i(\omega t - \omega_{-s} t_{-s})/2}\, g_o^{\omega,t}\ .$$

If we substitute $\omega' = \omega_{-s}$, $t' = t_{-s}$, then one easily checks (use the explicit formulas for α_s, ω_s, t_s at the end of §2.7) that $g_o^{\omega,t} = g_o^{\omega'_s,t'_s} = \exp\left[-\frac{i}{2}(\omega' t' - \omega t)\right]$ $e^{-iHs}\, g_o^{\omega',t'}$. On the other hand, the domain of integration is invariant under the transformation $(\omega, t) \rightarrow (\omega', t')$ (because this transformation is simply a rotation in time-frequency space!), so that

$$\begin{aligned} L_R\, e^{-iHs} &= \frac{1}{2\pi} \iint\limits_{\omega'^2+t'^2 \le R^2} d\omega'\, dt'\, \langle \cdot,\, g_o^{\omega',t'}\rangle\, e^{-iHs}\, g_o^{\omega' t'} \\ &= e^{-iHs}\, L_R\ , \end{aligned}$$

and L_R commutes with H, as announced. It follows that there exists an orthonormal basis in which both L_R and H are diagonal (see Preliminaries). But, since the eigenvalues of H are all nondegenerate, there exists only one basis that diagonalizes H, namely the Hermite functions (see §2.7). It follows that the Hermite functions ϕ_n are necessarily the eigenfunctions of L_R. The eigenvalues of L_R can be computed from

$$\langle \phi_n,\, g_o^{\omega,t} \rangle = (n!\, 2^n)^{-1/2} (-i)^n (\omega + it)^n \,\exp\left[-\frac{1}{4}(\omega^2 + t^2) - \frac{i}{2}\omega t \right] .$$

(There are many ways to compute this expression. One way, via the Bargmann Hilbert space, is explained in Note 3 at the end of this chapter.) We then have

$$L_R\, \phi_n = \lambda_n(R) \phi_n,$$

with

$$\begin{aligned} \lambda_n(R) &= \langle L_R\, \phi_n,\, \phi_n \rangle \\ &= \frac{1}{2\pi} \iint\limits_{\omega^2+t^2 \le R^2} d\omega\, dt\, |\langle \phi_n,\, g_o^{\omega,t} \rangle|^2 \\ &= \frac{1}{2\pi} \iint\limits_{\omega^2+t^2 \le R^2} d\omega\, dt\, \frac{1}{n!\, 2^n}\, (\omega^2 + t^2)^n \,\exp\left[-\frac{1}{2}(\omega^2 + t^2) \right] \\ &= \int_0^R dr\, r\, \frac{1}{n!\, 2^n}\, r^{2n} \,\exp\left(-\frac{1}{2}\, r^2 \right) \\ &= \frac{1}{n!} \int_0^{R^2/2} ds\, s^n\, e^{-s} , \end{aligned}$$

which is a so-called incomplete Γ-function. From this explicit formula for $\lambda_n(R)$, it is now possible to study its behavior as a function of n and R. I will summarize only the results here (details can be found in Daubechies (1988)); Figure 2.4 also shows a plot of $\lambda_n(R)$ for three different values of R. For every R, the $\lambda_n(R)$ decrease monotonically as n increases; for small n they are close to 1, for large n close to zero. The threshold value around which they make this "plunge," as defined, for example, by $n_{\text{thr}} = \max\{n;\ \lambda_n \ge 1/2\}$, is $n_{\text{thr}} \simeq R^2/2$. Note that this is again equal to $\pi R^2/2\pi$, i.e., the area of the time-frequency localization region S_R multiplied by the Nyquist density, just as in §2.3. The width of the plunge region is wider than in §2.3; however,

$$\#\ \{n;\ 1 - \epsilon \ge \lambda_n \ge \epsilon\} \le C_\epsilon R ,$$

(as compared to the logarithmic width in (2.3.2)), but it is still negligible, for large R, when compared with n_{thr}. Another striking difference with §2.3 is that the eigenfunctions ϕ_n in this case are independent of the size of the region S_R (unlike the prolate spheroidal wave functions): the R-dependence is completely concentrated in the $\lambda_n(R)$.[7]

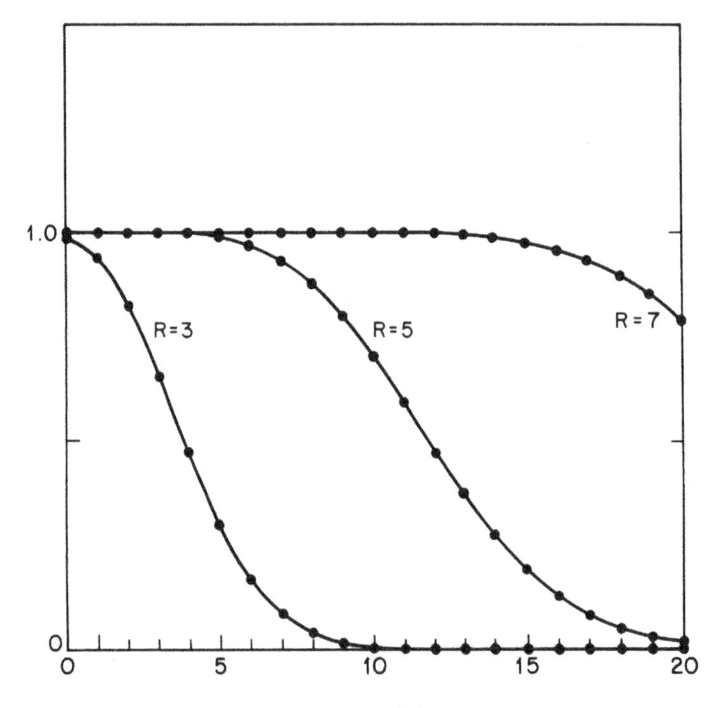

FIG. 2.4. *The eigenvalues $\lambda_n(R)$ for $R = 3,\ 5,$ and 7.*

Examples similar to all of the above exist for the continuous wavelet transform. We can again insert a non-constant function $w(a, b)$ in the integral in (2.8.1a), and construct operators W different from the identity operator. An example is $w(a, b) \sim a^2$ in three dimensions, with a spherically symmetric ψ (where the resolution of the identity is given by (2.6.2)), i.e.,

$$(Wf)(x) = C_\psi^{-1} \int_0^\infty \frac{da}{a^4} \int_{\mathbb{R}^3} db\ \frac{C_\psi}{\tilde{C}_\psi}\ a^2 (T^{\text{wav}} f)(a, b) \psi^{a,b}(x)\ , \tag{2.8.3}$$

where $\hat{\psi}(\xi) = \phi(|\xi|)$ and $\tilde{C}_\psi = (2\pi)^3 \int_0^\infty ds\ s\phi(s)$.

Because the three-dimensional Fourier transform of $g(x) = |x|^{-2}$ is $\hat{g}(\xi) = \sqrt{2}/(\sqrt{\pi}\,|\xi|)$ (in the sense of distributions), one easily checks that Wf can also be written as

$$(Wf)(x) = \frac{1}{4\pi} \int dy\ \frac{1}{|x - y|}\ f(y)\ , \tag{2.8.4}$$

so that $\langle Wf, g\rangle$ represents the interaction Coulomb potential energy for two charge distributions f and g. This formula was used in, e.g., the relativistic stability of matter paper by Fefferman and de la Llave (1986). Note that $\langle Wf, g\rangle$ becomes "diagonal" in the representation (2.8.3) (which, incidentally, is why it turned out to be useful in Fefferman and de la Llave (1986)). Note also that this diagonal wavelet representation completely captures the singularity of the kernel in (2.8.4): no "clipping off" of the singularity as in the windowed Fourier case. This is due to the fact that wavelets can zoom in on singularities (an extreme version of very short-lived high frequency features!), whereas the windowed Fourier functions cannot (see §1.2 or §2.9). [8]

We can also, as in the windowed Fourier case, choose to restrict the integral in (2.8.1a) to a subset S of (a, b)-space, thus defining time-frequency localization operators L_S. These are well defined for measurable S, and $0 \leq L_S \leq 1$. For compact S not containing any points with $a = 0$, L_S is a trace-class operator. For general S, the eigenfunctions and eigenvalues may again be hard to characterize, but there exist again special choices of ψ and S so that the eigenfunctions and eigenvalues of L_S are known explicitly. Their analysis is similar to the windowed Fourier case, but a bit more tricky. We will only sketch the results here; for full details the reader should consult Paul (1985) or Daubechies and Paul (1988). One such special ψ is $\hat{\psi}(\xi) = 2\xi\ e^{-\xi}$ for $\xi \geq 0$, 0 for $\xi \leq 0$; the associated resolution of the identity from which we start is (see (2.4.9))

$$C_\psi^{-1} \int_0^\infty \frac{da}{a^2} \int_{-\infty}^\infty db\ [\langle \cdot,\ \psi_+^{a,b} \rangle \psi_+^{a,b} + \langle \cdot,\ \psi_-^{a,b} \rangle \psi_-^{a,b}] = 1\ ,$$

where $\psi_+ = \psi$, $\hat{\psi}_-(\xi) = \hat{\psi}(-\xi)$. The operators $L_C = L_{S_C}$ we consider are given by

$$L_C = C_\psi^{-1} \iint_{(a,b) \in S_C} \frac{da\ db}{a^2}\ [\langle \cdot,\ \psi_+^{a,b} \rangle \psi_+^{a,b} + \langle \cdot,\ \psi_-^{a,b} \rangle \psi_-^{a,b}]\ ,$$

with $S_C = \{(a, b) \in \mathbb{R}_+ \times \mathbb{R};\ a^2 + b^2 + 1 \leq 2aC\}$, and $C \geq 1$. In the representation of (a, b)-space as the upper half complex plane ($z = b + ia$), the S_C correspond to the disks $|z - iC|^2 \leq C^2 - 1$. The role of the harmonic oscillator Hamiltonian is now played by the operator H defined by

$$H(f)^\wedge(\xi) = \left[-\xi \frac{d^2}{d\xi^2} - \frac{d}{d\xi} + \xi + \frac{1}{\xi} \right] \hat{f}(\xi)\ .$$

For this H we have then

$$\exp(-i\ Ht)\psi_+^{a,b} = e^{i\alpha_t(a,b)}\ \psi_+^{a(t),b(t)}\ ,$$

where

$$b(t) + ia(t) = z(t) = \frac{z\ \cos t + \sin t}{\cos t - z \sin t}\ ,$$

with $z = b + ia$. One easily checks that the flow $z \rightarrow z(t)$ preserves all the circles $|z - iC|^2 = C^2 - 1$, as illustrated by Figure 2.5. It follows that H and the L_C commute, so that they can be diagonalized simultaneously. [9] The eigenvalues of H all have degeneracy 2; for every eigenvalue $E_n = 3 + 2n$ we can find two eigenfunctions,

$$(\psi_n^+)^\wedge(\xi) = \begin{cases} 2\sqrt{2}\ [(n+2)(n+1)]^{-1/2}\ \xi\ L_n^2(2\xi) e^{-\xi} & \text{for}\, \xi \geq 0\ , \\ 0 & \text{for}\, \xi \leq 0\ , \end{cases}$$

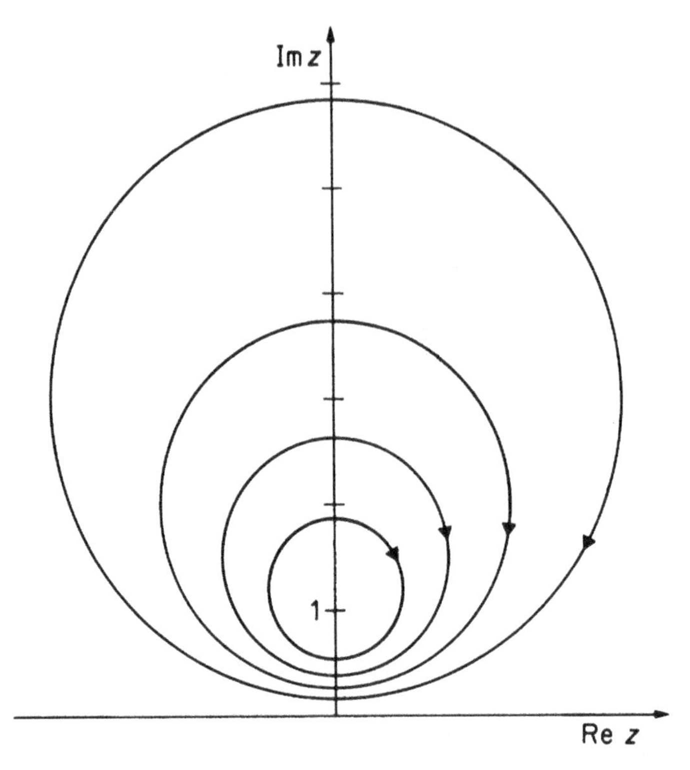

FIG. 2.5. *Flow lines for* $z(t) = \frac{z\cos t + \sin t}{\cos t - z\sin t}$.

and $(\psi_n^-)^\wedge(\xi) = (\psi_n^+)^\wedge(-\xi)$. Here L_n^2 is the Laguerre polynomial (2 is a superscript, not a power) as given by the general formula

$$\begin{aligned} L_n^\alpha(x) &= \frac{1}{n!}\, e^x\, x^{-\alpha}\, \frac{d^n}{dx^n}\, (e^{-x} x^{n+\alpha}) \\ &= \sum_{m=0}^{n} (-1)^m \frac{\Gamma(n+\alpha+1)}{\Gamma(n-m+1)\,\Gamma(\alpha+m+1)}\, \frac{1}{m!}\, x^m \,. \end{aligned}$$

Since the operators L_C commute with the parity operation $(\Pi f)^\wedge(\xi) = \hat{f}(-\xi)$, it follows that ψ_n^+, ψ_n^- are eigenfunctions for the L_C as well (because of the degeneracy of H, not every eigenfunction of H is a priori an eigenfunction of L_C!). The corresponding eigenvalues of L_C are

$$\lambda_n^+ = \lambda_n^- = (n+1)\left(1 - \frac{2}{C+1}\right)^{n+1} \left(\frac{2}{C+1} + \frac{1}{n+1}\right) .$$

(This means that L_C has the *same* degeneracy as H, so that in this case every eigenfunction of H is in fact an eigenfunction of L_C as well.) We can therefore also use $\psi_n^e = \frac{1}{\sqrt{2}}(\psi_n^+ + \psi_n^-)$ and $\psi_n^o = -\frac{i}{\sqrt{2}}\,(\psi_n^+ - \psi_n^-)$ as eigenfunctions; these have the advantage that they are real. Figure 2.6 shows the plots of the first few ψ_n^e, ψ_n^o (e for even, o for odd). A graph of $\lambda_n(C)$, for various values of C, is given in Figure 2.7. For reasonably large C, the $\lambda_n(C)$ behave as we by now expect of the eigenvalues of a time-frequency localization operator: they are close to 1 for n small, with $\lambda_o(C) = 1 - \frac{4}{(C+1)^2}$, and they plunge to 0 for a larger, C-dependent value of n. More precisely, for any $\gamma \in (0,1)$, the value of n for which $\lambda_n(C)$ crosses γ is equal to $n = \eta C + O(1)$ (C large), with $\eta(2+\eta^{-1})(1-2C^{-1})^{\eta C} = \gamma$ or $2\eta - \ln(1+2\eta) = -\ln\gamma + O(C^{-1})$. This implies that

$$\begin{aligned} \#\,\{\text{eigenfunctions; } \lambda_n(C) \geq \gamma\} &= 2\#\,\{n;\ \lambda_n(C) \geq \gamma\} \\ &= 2C\, F^{-1}\,(-\ln\gamma) + O(1)\ , \end{aligned}$$

where $F(t) = 2t - \ln(1+2t)$. In particular,

$$\#\,\{\text{eigenfunctions; } \lambda_n(C) \geq 1/2\} = 2C\, F^{-1}(\ln 2) + O(1)\ . \tag{2.8.5}$$

In order to compare this with the Nyquist density, we first need to find the area in time-frequency space corresponding to L_C. To do this, we go back to the $\psi_\pm^{a,b}$. We have

$$\begin{aligned} \int dx\ |\psi_\pm^{a,b}(x)|^2 &= b\ , \\ \int d\xi\ |(\psi_\pm^{a,b})^\wedge(\xi)|^2\xi &= \pm\frac{3}{2a}\ . \end{aligned}$$

Hence $S_C = \{(a,b) \in \mathbb{R}_+ \times \mathbb{R};\ a^2+b^2+1 \leq 2aC\}$ corresponds to the time-frequency set

$$\tilde{S}_C = \left\{(\omega,t) \in \mathbb{R}^2;\ t^2 + \frac{9}{4\omega^2} + 1 \leq \frac{3C}{|\omega|}\right\}\ .$$

This corresponds to a low frequency as well as a high frequency cut-off; see Figure 2.8 for a comparison between this time-frequency localization set and the disks of the windowed Fourier case.[10] The area of $\tilde{S}_C$ is $|\tilde{S}_C| = 6\pi(C-1)$. Combining this with (2.8.4) gives

$$\begin{aligned} \frac{\#\,\{\text{eigenfunctions; } \lambda_n(C) \geq 1/2\}}{|\tilde{S}_C|} &= \frac{1}{3\pi}\ F^{-1}(\ln 2) + O(C^{-1}) \\ &\simeq \frac{.646}{2\pi} + O(C^{-1})\ , \end{aligned}$$

which is different from the Nyquist density! This contradiction is only apparent, and is due to the fact that the width of the "plunge region" of the $\lambda_n(C)$ is proportional to C, and thus to $|\tilde{S}_C|$. We have indeed, for $\epsilon > 0$,

$$\begin{aligned} &\#\,\{\text{eigenfunctions; } \epsilon \leq \lambda \leq 1-\epsilon\} \\ &= \frac{1}{2\pi}\ |\tilde{S}_C|\ \left\{\frac{2}{3}\ [F^{-1}(|\ln(1-\epsilon)|)\ -\ F^{-1}(|\ln\ \epsilon|)]\ +\ O(|\tilde{S}_C|^{-1})\right\}\ , \end{aligned}$$

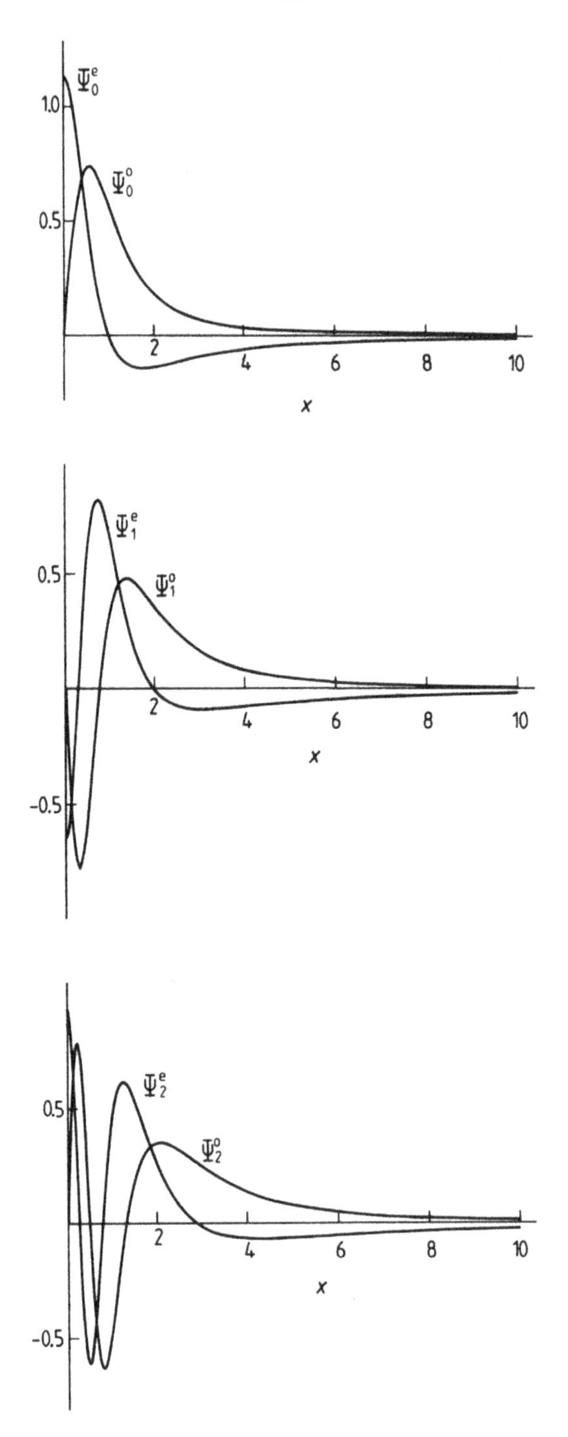

FIG. 2.6. *Plots of the eigenfunctions* ψ_n^e, ψ_n^o *for* $n = 0,\ 1,$ *and* 2.

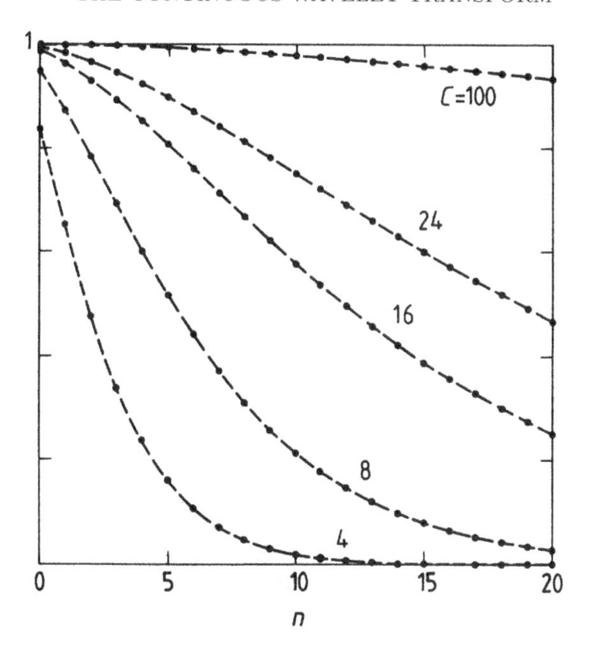

FIG. 2.7. *The eigenvalues $\lambda_n(C)$ for different values of C.*

in contrast to the prolate spheroidal wave case, where the analog of the expression between curly brackets tends to zero as $|S|^{-1}\log|S|$ for $|S|\to\infty$, and with the windowed Fourier case, where it behaves like $|S|^{-1/2}$ for $|S|\to\infty$. The fact that in the present case the width of the plunge region is of the same order as $|\tilde{S}_C|$ itself results from the non-uniform time-frequency localization of the $\psi_{\pm}^{a,b}$; it is an indication that we have to be careful with time-frequency-density-based intuition when dealing with wavelets. We will come back to this in Chapter 4.

2.9. The continuous wavelet transform as a mathematical zoom: The characterization of local regularity.

This section is entirely borrowed from Holschneider and Tchamitchian (1990), who developed these techniques in part to study local regularity properties of a nondifferentiable function proposed by Riemann.

THEOREM 2.9.1. *Suppose that $\int dx(1+|x|)\ |\psi(x)| < \infty$, and $\hat{\psi}(0) = 0$. If a bounded function f is Hölder continuous with exponent α, $0 < \alpha \leq 1$, i.e.,*

$$|f(x) - f(y)| \leq C|x-y|^\alpha ,$$

then its wavelet transform satisfies

$$|T^{\text{wav}}(a,b)| = |\langle f, \psi^{a,b}\rangle| \leq C'\ |a|^{\alpha+1/2} .$$

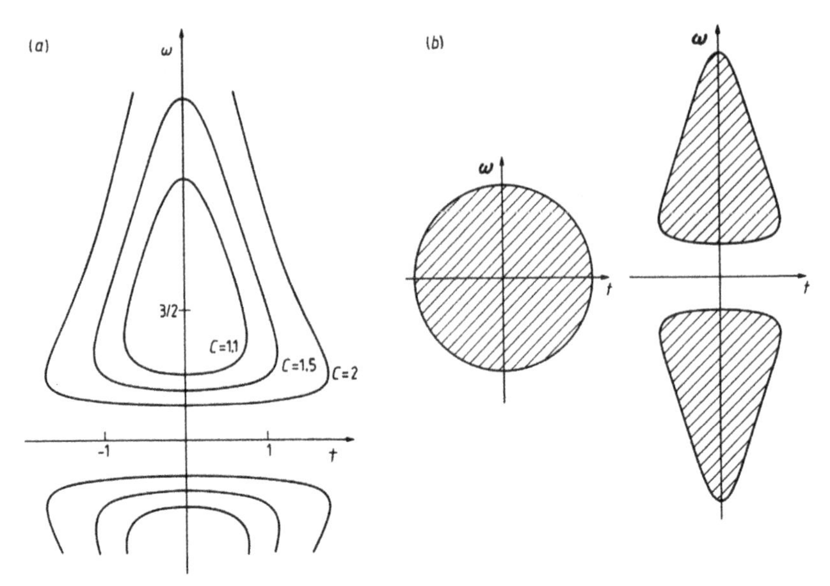

FIG. 2.8. (a) *The set* $\tilde{S}_C = \left\{(t,\omega);\ t^2 + \frac{g}{\omega^2} + 1 \le \frac{3C}{|\omega|}\right\}$ *for different values of* C. (b) *A comparison between the time-frequency localization sets for the windowed Fourier transform (the disk* $S_R = \{(t,\omega);\ t^2+\omega^2 \le R^2\}$ *at left) and for the wavelet transform (at right).*

Proof. Since $\int dx\ \psi(x) = 0$ we have

$$\langle \psi^{a,b}, f\rangle = \int dx\ |a|^{-1/2}\ \psi\left(\frac{x-b}{a}\right)[f(x)-f(b)]\ ;$$

hence

$$\begin{aligned} |\langle \psi^{a,b}, f\rangle| &\le \int dx\ |a|^{-1/2}\left|\psi\left(\frac{x-b}{a}\right)\right|\ C|x-b|^\alpha \\ &\le C\ |a|^{\alpha+1/2}\int dy\ |\psi(y)|\ |y|^\alpha \\ &\le C'\ |a|^{\alpha+1/2}\ . \end{aligned}$$ ∎

The following is a converse theorem.

THEOREM 2.9.2. *Suppose that* ψ *is compactly supported. Suppose also that* $f \in L^2(\mathbb{R})$ *is bounded and continuous. If, for some* $\alpha \in\ ]0,1[$, *the wavelet transform of* f *satisfies*

$$|\langle f, \psi^{a,b}\rangle| \le C|a|^{\alpha+1/2}\ , \tag{2.9.1}$$

then f *is Hölder continuous with exponent* α.

Proof.

1. Choose ψ_2 compactly supported and continuously differentiable, with $\int dx\ \psi_2(x) = 0$. Normalize ψ_2 so that $C_{\psi,\psi_2} = 1$. Then, by Proposition 2.4.2,

$$f(x) = \int_{-\infty}^{\infty} \frac{da}{a^2} \int_{-\infty}^{\infty} db\ \langle f, \psi^{a,b}\rangle\ \psi_2^{a,b}(x)\ .$$

We will split the integral over a into two parts, $|a| \le 1$ and $|a| \ge 1$, and call the two terms $f_{SS}(x)$ (small scales) and $f_{LS}(x)$ (large scales).

2. First of all, note that f_{LS} is bounded uniformly in x:

$$\begin{aligned} |f_{LS}(x)| &\le \int_{|a|\ge 1} \frac{da}{a^2} \int_{-\infty}^{\infty} db\ |\psi_2^{a,b}(x)|\ \|f\|_{L^2}\ \|\psi\|_{L^2} \\ &\le C \int_{|a|\ge 1} \frac{da}{a^2} \int_{-\infty}^{\infty} db\ |a|^{-1/2} \left|\psi_2\left(\frac{x-b}{a}\right)\right| \\ &\le C\ \|\psi_2\|_{L^1} \int_{|a|\ge 1} da\ |a|^{-3/2} = C' < \infty\ . \end{aligned} \tag{2.9.2}$$

Next, we look at $|f_{LS}(x+h) - f_{LS}(x)|$ for $|h| \le 1$:

$$\begin{aligned} |f_{LS}(x+h) - f_{LS}(x)| \le \int_{|a|\ge 1} \frac{da}{|a|^3} \int_{-\infty}^{\infty} db \int_{-\infty}^{\infty} dy\ |f(y)| \\ \left|\psi\left(\frac{y-b}{a}\right)\right| \left|\psi_2\left(\frac{x+h-b}{a}\right) - \psi_2\left(\frac{x-b}{a}\right)\right| \end{aligned} \tag{2.9.3}$$

Since $|\psi_2(z+t) - \psi_2(z)| \le C|t|$, and since support ψ, support $\psi_2 \subset [-R, R]$ for some $R < \infty$, we can bound this by

$$\begin{aligned} (2.9.3) &\le C'\ |h| \int_{|a|\ge 1} da\ a^{-4} \int_{\substack{|x-b|\le |a|R+1 \\ |y-b|\le |a|R}} db \int dy\ |f(y)| \\ &\le C''\ |h| \int_{|a|\ge 1} da\ |a|^{-3} \int_{|y-x|\le 2|a|R+1} dy\quad |f(y)| \\ &\le C''\ |h|\ \|f\|_{L^2} \int_{|a|\ge 1} da\ |a|^{-3} (4|a|R+2)^{1/2} \le C'''\ |h|\ . \end{aligned}$$

This holds for all $|h| \le 1$; together with the bound (2.9.2), we conclude that $|f_{LS}(x+h) - f_{LS}(x)| \le C|h|$ for all h, uniformly in x. Note that we did not even use (2.9.1) in this estimate: f_{LS} is always regular.

3. The small scale part f_{SS} is also uniformly bounded:

$$|f_{SS}(x)| \le C \int_{|a|\le 1} \frac{da}{a^2} \int_{-\infty}^{\infty} db\ |a|^{\alpha+1/2}\ |a|^{-1/2} \left|\psi_2\left(\frac{x-b}{a}\right)\right|$$

$$\leq \quad C\|\psi_2\|_{L^1} \int_{|a|\leq 1} da \; |a|^{-1+\alpha} = C' < \infty \, .$$

4. We therefore again only have to check $|f_{SS}(x+h) - f_{SS}(x)|$ for small h, such as $|h| \leq 1$. Using again $|\psi_2(z+t) - \psi_2(z)| \leq C|t|$, we have

$$\begin{aligned}
&|f_{SS}(x+h) - f_{SS}(x)| \\
&\leq \int_{|a|\leq |h|} \frac{da}{a^2} \int_{-\infty}^{\infty} db \; |a|^{\alpha} \left(\left|\psi_2\left(\frac{x-b}{a}\right)\right| + \left|\psi_2\left(\frac{x+h-b}{a}\right)\right| \right) \\
&\quad + \int_{|h|\leq |a|\leq 1} \frac{da}{a^2} \int_{|x-b|\leq |a|R+|h|} db \; |a|^{\alpha} \; C \left|\frac{h}{a}\right| \\
&\leq C' \left[\|\psi_2\|_{L^2} \int_{|a|\leq |h|} da \; |a|^{-1+\alpha} + |h| \int_{|h|\leq |a|\leq 1} da \; |a|^{-3+\alpha}(|a|R + |h|) \right] \\
&= C'' \; |h|^{\alpha} \, .
\end{aligned}$$

It follows that f is Hölder continuous with exponent α. ∎

Together, Theorems 2.9.1 and 2.9.2 show that the Hölder continuity of a function can be characterized by the decay in a of the absolute value of its wavelet transform. (Except for $\alpha = 1$, where we do not have complete equivalence.) Note that we did not assume any regularity for ψ itself: apart from decay conditions on ψ, we only exploited that $\int dx \; \psi(x) = 0$. (Although this condition is not stated explicitly in Theorem 2.9.2, ψ nevertheless satisfies it: the bound (2.9.1) cannot hold otherwise.) Higher order differentiability of f and Hölder continuity of its highest order well-defined derivative can be characterized similarly by means of the decay of the wavelet coefficients if ψ has more moments zero: in order to characterize $f \in C^n$ and Hölder continuity with exponent α of $f^{(n)}$ we will need a wavelet ψ so that $\int dx \; x^m \psi(x) = 0$ for $m = 0, 1, \cdots, n$. For such a wavelet we have, for $\alpha \in \,]0, 1[$,

> $f \in C^n$, with all the $f^{(m)}$, $m = 0, \cdots, n$ bounded and square integrable, and $f^{(n)}$ Hölder continuous with exponent α

$$\Longleftrightarrow$$

$$|\langle f, \psi^{a,b} \rangle| \leq C|a|^{n+1/2+\alpha}, \quad \text{uniformly in } a \, .$$

Again, we require no regularity for ψ.

What is most striking about all these characterizations is that they only involve the absolute value of the wavelet transform. Note that one can also derive regularity of f from the decay in ω of the absolute value of its windowed Fourier transform $T^{\text{win}}(\omega, t)$, if the window g is chosen sufficiently smooth. In most cases however the value for the Hölder exponent computed from $|T^{\text{win}}(\omega, t)|$ will not be optimal. To obtain a true characterization, the phase of $T^{\text{win}}(\omega, t)$ should also be taken into consideration, for instance via Littlewood–Paley type estimates (see, e.g., Frazier, Jawerth, and Weiss (1991)).

The wavelet transform can also be used to characterize *local* regularity, something that cannot be achieved, even when phase information is taken into account, by the windowed Fourier transform. The following two theorems are again borrowed from Holschneider and Tchamitchian (1990).

THEOREM 2.9.3. *Suppose that $\int dx\,(1+|x|)\,|\psi(x)| < \infty$ and $\int dx\,\psi(x) = 0$. If a bounded function f is Hölder continuous in x_0, with exponent $\alpha \in\,]0,1]$, i.e.,*

$$|f(x_0+h) - f(x_0)| \le C|h|^\alpha ,$$

then

$$|\langle f, \psi^{a,x_0+b}\rangle| \le C|a|^{1/2}\,(|a|^\alpha + |b|^\alpha) .$$

Proof. By translating everything we can assume that $x_0 = 0$. Because $\int dx\,\psi(x) = 0$, we again have

$$\begin{aligned} |\langle f, \psi^{a,b}\rangle| &\le \int dx\,|f(x) - f(0)|\,|a|^{-1/2}\left|\psi\left(\frac{x-b}{a}\right)\right| \\ &\le C\int dx\,|x|^\alpha\,|a|^{-1/2}\left|\psi\left(\frac{x-b}{a}\right)\right| \\ &\le C|a|^{\alpha+1/2}\int dy\,\left|y+\frac{b}{a}\right|^\alpha\,|\psi(y)| \\ &\le C'\,|a|^{1/2}\,(|a|^\alpha + |b|^\alpha) . \quad \blacksquare \end{aligned}$$

THEOREM 2.9.4. *Suppose that ψ is compactly supported. Suppose also that $f \in L^2(\mathbb{R})$ is bounded and continuous. If, for some $\gamma > 0$ and $\alpha \in\,]0,1[$,*

$$|\langle f, \psi^{a,b}\rangle| \le C|a|^{\gamma+1/2} \quad \textit{uniformly in } b,$$

and

$$|\langle f, \psi^{a,b+x_0}\rangle| \le C|a|^{1/2}\left(|a|^\alpha + \frac{|b|^\alpha}{|\log|b||}\right) ,$$

then f is Hölder continuous in x_0 with exponent α.

Proof.

1. The proof starts exactly like the proof of Theorem 2.9.2, of which the first three points carry over without change, with γ taking over the role of α in point 3.

2. We therefore only have to check $|f_{SS}(x_0+h) - f_{SS}(x_0)|$ for small h. By translating everything, we can assume $x_0 = 0$, and we obtain

$$\begin{aligned} &|f_{SS}(h) - f_{SS}(0)| \\ &\le \int_{|a|\le|h|^{\alpha/\gamma}} \frac{da}{a^2}\int_{-\infty}^{\infty} db\,|a|^\gamma\left|\psi_2\left(\frac{h-b}{a}\right)\right| \\ &+ \int_{|h|^{\alpha/\gamma}\le|a|\le|h|} \frac{da}{a^2}\int_{-\infty}^{\infty} db\left(|a|^\alpha + \frac{|b|^\alpha}{|\log|b||}\right)\left|\psi_2\left(\frac{h-b}{a}\right)\right| \end{aligned}$$

$$+ \int_{|a|\le|h|} \frac{da}{a^2} \int_{-\infty}^{\infty} db \left(|a|^\alpha + \frac{|b|^\alpha}{|\log|b||}\right) \left|\psi_2\left(-\frac{b}{a}\right)\right|$$

$$+ \int_{|h|\le|a|\le 1} \frac{da}{a^2} \int_{-\infty}^{\infty} db \left(|a|^\alpha + \frac{|b|^\alpha}{|\log|b||}\right) \left|\psi_2\left(\frac{h-b}{a}\right) - \psi_2\left(-\frac{b}{a}\right)\right| , \tag{2.9.4}$$

where we have assumed $\alpha > \gamma$. (If $\alpha \le \gamma$, things become simpler.) Let us denote the four terms in the right-hand side of (2.9.4) by T_1, T_2, T_3, and T_4.

3. $T_1 \le \int_{|a|\le|h|^{\alpha/\gamma}} da\ |a|^{-1+\gamma}\ \|\psi_2\|_{L^1} \le C|h|^\alpha$.

4. In the second term, we use support $\psi_2 \subset [-R, R]$ to derive

$$\begin{aligned} T_2 &\le \int_{|a|\le|h|} da\ |a|^{-1+\alpha}\|\psi_2\|_{L^1} \\ &\quad + \int_{|h|^{\alpha/\gamma}\le|a|\le|h|} da\ |a|^{-1}\|\psi_2\|_{L^1}\ \frac{(|a|R+|h|)^\alpha}{|\log(|a|R+|h|)|} \\ &\le C|h|^\alpha \left[1 + \frac{1}{|\log|h||} \int_{|h|^{\alpha/\gamma}\le|a|\le|h|} da\ |a|^{-1}\right] \\ &\le C'\ |h|^\alpha\ . \end{aligned}$$

5. Similarly, for sufficiently small $|h|$,

$$\begin{aligned} T_3 &\le \int_{|a|\le h} da\ |a|^{-1+\alpha}\ \|\psi_2\|_{L^1} + \int_{|a|\le|h|} da\ |a|^{-1}\|\psi_2\|_{L^2}\ \frac{(|a|R)^\alpha}{|\log|a|R|} \\ &\le C|h|^\alpha\ . \end{aligned}$$

6. Finally,

$$\begin{aligned} T_4 &\le C|h| \int_{|h|\le|a|\le 1} da\ |a|^{-3} \left[|a|^\alpha + \frac{(|a|R+|h|)^\alpha}{|\log(|a|R+|h|)|}\right] (|a|R+|h|) \\ &\le C'\ |h|\ \left[1+|h|^{-1+\alpha} + |h|\ (1+|h|^{-2+\alpha})\right] \le C''\ |h|^\alpha\ . \qquad \blacksquare \end{aligned}$$

Similar theorems for higher order local regularity can be proved. These theorems justify the name "mathematical microscope," which is sometimes bestowed on the wavelet transform. In Holschneider and Tchamitchian (1990) these and other results were exploited to study the differentiability properties of the function defined by the Fourier series $\sum_{n=1}^{\infty} n^{-2}\sin(n^2\pi x)$, first studied by Riemann.

Notes.

1. All these Bergman spaces can be further transformed, via a (standard) conformal map, to Hilbert spaces of analytic functions on the unit disk.

2. My main reference for this paragraph are the articles by Grossmann, Morlet, and Paul (1985, 1986). Their results can in fact be generalized to reducible representations as well, as long as they have a cyclic vector (A. Grossmann and T. Paul, private communication). This is useful for the higher dimensional case, where the representations of the $ax+b$-group are reducible but cyclic.

3. The operator H_B obtained by "transporting" H to the Bargmann Hilbert space, via the unitary map T^{win}, is particularly simple:

$$T^{\text{win}} H (T^{\text{win}})^{-1} \left[\exp\left(-\frac{1}{4}(\omega^2+t^2)-\frac{i}{2}\omega t\right)\ \phi(\omega+it)\right]$$

$$= \ \exp\left(-\frac{1}{4}(\omega^2+t^2)-\frac{i}{2}\omega t\right) [2(\omega+it)\phi'(\omega+it)] \ ,$$

or $(H_B\phi)(z) = 2z\ \phi'(z)$. It is obvious that the eigenfunctions of H_B are the monomials $u_n(z) = (2^n\ n!)^{-1/2}\ z^n$. The following argument shows that these are indeed the functions in the Bargmann space corresponding to the Hermite functions. One easily computes

$$T^{\text{win}}\ A^*\ (T^{\text{win}})^{-1} \left[\exp\left(-\frac{1}{4}(\omega^2+t^2)-\frac{i}{2}\omega t\right)\ \phi(\omega+it)\right]$$

$$= \ \exp\left(-\frac{1}{4}(\omega^2+t^2)-\frac{i}{2}\omega t\right) (-i)(\omega+it)\ \phi(\omega+it) \ ,$$

so that $\phi_n = (2^n\ n!)^{-1/2}(A^*)^n g_o$ corresponds to $(2^n\ n!)^{-1/2}(-i)^n z^n = (-i)^n\ u_n(z)$ in the Bargmann space. (We use the normalization $\|\phi\|^2_{\text{Bargm.}} = \frac{1}{2\pi}\int dx\ \int dy\ e^{-\frac{1}{2}(x^2+y^2)}\ |\phi(x+iy)|^2$, so that g_o itself corresponds to the constant function 1 in the Bargmann space.) In particular, this means that

$$\langle \phi_n, g^{\omega,t}\rangle = \exp\left[-\frac{1}{4}(\omega^2+t^2)-\frac{i}{2}\omega t\right] (-i)^n\ (2^n\ n!)^{-1/2}(\omega+it)^n \ .$$

4. Not every unbounded function leads to an unbounded operator; some bounded operators can only be represented in this way if an unbounded weight function is used. In fact, Klauder (1966) proved that even some trace-class operators require a non-tempered distribution as weight function!

5. For real functions w, one requires that W should be essentially self-adjoint on this domain.

6. Yet another application in quantum mechanics can be found in Daubechies and Klauder (1985), where it is shown how to write the (mathematically not well defined) path integral for exp $(-itH)$ as a limit of bona fide Wiener integrals (as the diffusion constant of the underlying diffusion process tends to ∞), provided H is of the form (2.8.2), with a weight function $w(p,q)$ that does not increase too drastically for $p,q\to\infty$. A similar theorem can be proved in the wavelet case (Daubechies, Klauder, and Paul (1987)).

7. Exactly the same arguments hold for all operators W of type (2.8.2) for which $w(\omega,t)$ is rotationally symmetric, even if it is not an indicator function. An example is $w(\omega,t) = \exp[-\alpha(\omega^2+t^2)]$, for which it was first shown in Gori and Guattari (1985) that the Hermite functions are the eigenfunctions (irrespective of α; the eigenvalues depend on α, of course!).

8. It is no coincidence that Fefferman and de la Llave would use a representation of type (2.8.3) for the operator (2.8.4): after all, Calderón's formula (to which (2.4.4) is essentially equivalent) is part of a toolbox developed precisely for the study of singular integral operators (long before wavelets!) so that it is well adapted for treating the singular kernel in (2.8.4). In this particular instance, (2.8.4) makes sense even for nonadmissible ψ (C_ψ cancels out); in Fefferman and de la Llave (1986), ψ was taken to be the indicator function of the unit ball (which is nonadmissible, since its integral does not vanish).

9. If we make an extra transformation, mapping the upper half plane $\{b+ia;\ a\geq 0\}$ to the unit disk (by means of a conformal mapping), then everything becomes more transparent: the flow $z\to z(t)$ then corresponds to a simple rotation around the center of the disk, and H as well as its eigenfunctions are given by simple expressions. See Paul (1985) or Seip (1991).

10. There exist many other choices of ψ for which this analysis works. For each choice the set $\tilde{S}_C$ in time-frequency space corresponding to S_C in (a,b)-space takes on a different shape. Explicit computations and a figure illustrating these different shapes can be found in Daubechies and Paul (1988).

CHAPTER 3

Discrete Wavelet Transforms: Frames

In this, the longest chapter in this book, we discuss various aspects of non-orthonormal, discrete wavelet expansions, together with some parallels with the windowed Fourier transform. The "frames" of the chapter title are sets of non-independent vectors; they can nevertheless be used to write a straightforward and completely explicit expansion for every vector in the space. We will discuss wavelet frames as well as frames for the windowed Fourier transform; in the latter case, the approach can be viewed as "oversampled" with respect to the Nyquist density in time-frequency space.

A lot of the material in this chapter has been taken from Daubechies (1990), updated here and there. A very nicely written review of frames (and of the continuous transforms as well), with some additional original theorems, is Heil and Walnut (1989).

3.1. Discretizing the wavelet transform.

In the continuous wavelet transform, we consider the family

$$\psi^{a,b}(x) \;=\; |a|^{-1/2}\,\psi\left(\frac{x-b}{a}\right) ,$$

where $b \in \mathbb{R}$, $a \in \mathbb{R}_+$ with $a \neq 0$, and ψ is admissible. For convenience, in the discretization we restrict a to positive values only, so that the admissibility condition becomes

$$C_\psi \;=\; \int_0^\infty d\xi\ \xi^{-1}\ |\hat{\psi}(\xi)|^2 \;=\; \int_{-\infty}^0 d\xi\ |\xi|^{-1}\ |\hat{\psi}(\xi)|^2 < \infty\ .$$

(See §2.4.) We would like to restrict a, b to discrete values only. The discretization of the dilation parameter seems natural: we choose $a = a_0^m$, where $m \in \mathbb{Z}$, and the dilation step $a_0 \neq 1$ is fixed. For convenience we will assume $a_0 > 1$ (although it does not matter, since we take negative as well as positive powers m). For $m = 0$, it seems natural as well to discretize b by taking only the integer (positive and negative) multiples of one fixed b_0 (we arbitrarily fix $b_0 > 0$), where b_0 is appropriately chosen so that the $\psi(x - nb_0)$ "cover" the whole line (in a sense to be made precise below). For different values of m, the width of $a_0^{-m/2}\ \psi(a_0^{-m}x)$ is a_0^m times the width of $\psi(x)$ (as measured, e.g., by width $(f) = [\int dx\ x^2|f(x)|^2]^{1/2}$, where we assume that $\int dx\ x|f(x)|^2 = 0$), so that the choice $b = nb_0\ a_0^m$ will ensure that the discretized wavelets at level m "cover" the line in the same way that the $\psi(x - nb_0)$ do. Thus we choose $a = a_0^m$, $b = nb_0 a_0^m$, where m, n range over $\mathbb{Z}$, and $a_0 > 1$, $b_0 > 0$ are fixed; the appropriate choices for a_0, b_0 depend, of course, on the wavelet ψ (see below). This corresponds to

$$\psi_{m,n}(x) = a_0^{-m/2}\psi\left(\frac{x - nb_0 a_0^m}{a_0^m}\right) = a_0^{-m/2}\psi(a_0^{-m}x - nb_0)\ . \tag{3.1.1}$$

We can now ask two questions:

(1) Do the discrete wavelet coefficients $\langle f, \psi_{m,n}\rangle$ completely characterize f? Or, stronger, can we reconstruct f in a numerically stable way from the $\langle f, \psi_{m,n}\rangle$?

(2) Can any function f be written as a superposition of "elementary building blocks" $\psi_{m,n}$?[1] Can we write an easy algorithm to find the coefficients in such a superposition?

In fact, these questions are dual aspects of only one problem. We will see below that, for reasonable ψ and appropriate a_0, b_0, there exist $\widetilde{\psi_{m,n}}$ so that the answer to the reconstruction question is simply

$$f = \sum_{m,n} \langle f, \psi_{m,n}\rangle\ \widetilde{\psi_{m,n}}\ .$$

It then follows that, for any $g \in L^2(\mathbb{R})$

$$\begin{aligned}\langle g, f\rangle = \overline{\langle f, g\rangle} &= \left(\sum_{m,n} \langle f, \psi_{m,n}\rangle\langle \widetilde{\psi_{m,n}}, g\rangle\right)^{\star} \\ &= \sum_{m,n} \langle g, \widetilde{\psi_{m,n}}\rangle\langle \psi_{m,n}, f\rangle\ ,\end{aligned}$$

or $g = \sum_{m,n} \langle g, \widetilde{\psi_{m,n}}\rangle\ \psi_{m,n}$, at least in the weak sense; this is effectively a prescription for the computation of the coefficients in a superposition of $\psi_{m,n}$ leading to g. We will mostly focus on the first set of questions here; for a more detailed discussion of the duality between (1) and (2), see Gröchenig (1991).

In the case of the continuous wavelet transform, both questions were answered immediately by the resolution of the identity, at least if ψ was admissible. In the present discrete case there is no analog of the resolution of the identity,[2] so we have to attack the problem some other way. We can also wonder whether there exists a "discrete admissibility condition," and what it is. Let us first give some mathematical content to the questions in (1). We will restrict ourselves mostly to functions $f \in L^2(\mathbb{R})$, although discrete families of wavelets, like their continuously labelled cousins, can be used in many other function spaces as well.

Functions can then be "characterized" by means of their "wavelet coefficients" $\langle f, \psi_{m,n}\rangle$ if it is true that

$$\langle f_1, \psi_{m,n}\rangle = \langle f_2, \psi_{m,n}\rangle \text{ for all } m,n \in \mathbb{Z}$$
$$\text{implies} \quad f_1 \equiv f_2 ,$$

or, equivalently, if

$$\langle f, \psi_{m,n}\rangle = 0 \text{ for all } m,n \in \mathbb{Z} \Rightarrow f = 0 .$$

But we want more than characterizability: we want to be able to reconstruct f in a numerically stable way from the $\langle f, \psi_{m,n}\rangle$. In order for such an algorithm to exist, we must be sure that if the sequence $(\langle f_1, \psi_{m,n}\rangle)_{m,n\in\mathbb{Z}}$ is "close" to $(\langle f_2, \psi_{m,n}\rangle)_{m,n\in\mathbb{Z}}$, then necessarily f_1 and f_2 were "close" as well. In order to make this precise, we need topologies on the function space and on the sequence space. On the function space $L^2(\mathbb{R})$ we already have its Hilbert space topology; on the sequence space we will choose a similar ℓ^2-topology, in which the distance between sequences $c^1 = (c^1_{m,n})_{m,n\in\mathbb{Z}}$ and $c^2 = (c^2_{m,n})_{m,n\in\mathbb{Z}}$ is measured by

$$\|c^1 - c^2\|^2 = \sum_{m,n\in\mathbb{Z}} |c^1_{m,n} - c^2_{m,n}|^2 .$$

This implicitly assumes that the sequences $(\langle f, \psi_{m,n}\rangle)_{m,n\in\mathbb{Z}}$ are in $\ell^2(\mathbb{Z}^2)$ themselves, i.e., that $\sum_{m,n} |\langle f, \psi_{m,n}\rangle|^2 < \infty$ for all $f \in L^2(\mathbb{R})$. In practice, this is no problem. As we will see below, any reasonable wavelet (which means that ψ has some decay in both time and frequency, and that $\int dx\ \psi(x) = 0$), and any choice for $a_0 > 1$, $b_0 > 0$ leads to

$$\sum_{m,n} |\langle f, \psi_{m,n}\rangle|^2 \le B\ \|f\|^2 . \tag{3.1.2}$$

We will assume (without specifying any restrictions yet on the $\psi_{m,n}$; we will come back to these later) that (3.1.2) holds. With the $\ell^2(\mathbb{Z}^2)$ interpretation of "closeness," the stability requirement means that if $\sum_{m,n} |\langle f, \psi_{m,n}\rangle|^2$ is small, then $\|f\|^2$ should be small. In particular, there should exist $\alpha < \infty$ so that $\sum_{m,n} |\langle f, \psi_{m,n}\rangle|^2 \le 1$ implies $\|f\|^2 \le \alpha$. Now take arbitrary $f \in L^2(\mathbb{R})$, and define $\tilde{f} = [\sum_{m,n} |\langle f, \psi_{m,n}\rangle|^2]^{-1/2} f$. Clearly, $\sum_{m,n} |\langle \tilde{f}, \psi_{m,n}\rangle|^2 \le 1$; hence $\|\tilde{f}\|^2 \le \alpha$. But this means

$$\left[\sum_{m,n} |\langle f, \psi_{m,n}\rangle|^2\right]^{-1} \|f\|^2 \le \alpha$$

or

$$A\ \|f\|^2 \le \sum_{m,n} |\langle f, \psi_{m,n}\rangle|^2 \tag{3.1.3}$$

for some $A = \alpha^{-1} > 0$. On the other hand, if (3.1.3) holds for all f, then the distance $\|f_1 - f_2\|$ cannot be arbitrarily large if $\sum_{m,n} |\langle f_1, \psi_{m,n}\rangle - \langle f_2, \psi_{m,n}\rangle|^2$

is small. It follows that (3.1.3) is equivalent to our stability requirement. Combining (3.1.3) with (3.1.2), we obtain that there should exist $A > 0$, $B < \infty$ so that

$$A \, \|f\|^2 \leq \sum_{m,n} |\langle f, \, \psi_{m,n}\rangle|^2 \leq B \, \|f\|^2 \tag{3.1.4}$$

for all $f \in L^2(\mathbb{R})$. In other words, the $\{\psi_{m,n};\ m, n \in \mathbb{Z}\}$ constitute a *frame*, a concept that we review in the next section. The connection between frames and numerically stable reconstructions from discretized wavelets was first pointed out by A. Grossmann (1985, personal communication).

3.2. Generalities about frames.

Frames were introduced by Duffin and Schaeffer (1952), in the context of nonharmonic Fourier series (i.e., expansions of functions in $L^2([0, 1])$ in complex exponentials $\exp(i\lambda_n x)$, where $\lambda_n \neq 2\pi n$); they are also reviewed in Young (1980). We review here their definition and some of their properties.

DEFINITION. *A family of functions $(\varphi_j)_{j\in J}$ in a Hilbert space $\mathcal{H}$ is called a* frame *if there exist $A > 0$, $B < \infty$ so that, for all f in $\mathcal{H}$,*

$$A \, \|f\|^2 \leq \sum_{j\in J} |\langle f, \, \varphi_j\rangle|^2 \leq B \, \|f\|^2 \, . \tag{3.2.1}$$

We call A and B the frame bounds.

If the two frame bounds are equal, $A = B$, then I will call the frame a *tight frame*. In a tight frame we have, for all $f \in \mathcal{H}$,

$$\sum_{j\in J} |\langle f, \, \varphi_j\rangle|^2 = A \, \|f\|^2 \, ,$$

which implies, by the polarization identity,[3]

$$A \, \langle f, g\rangle = \sum_j \langle f, \, \varphi_j\rangle\langle\varphi_j, \, g\rangle$$

or

$$f = A^{-1} \sum_j \langle f, \, \varphi_j\rangle \, \varphi_j \, , \tag{3.2.2}$$

at least in the weak sense. Formula (3.2.2) is very reminiscent of the expansion of f into an orthonormal basis, but it is important to realize that frames, even tight frames, are *not* orthonormal bases, as illustrated by the following finite-dimensional example.

EXAMPLE. Take $\mathcal{H} = \mathbb{C}^2$, $e_1 = (0, 1)$, $e_2 = (-\frac{\sqrt{3}}{2}, \, -\frac{1}{2})$, $e_3 = (\frac{\sqrt{3}}{2}, \, -\frac{1}{2})$. (See Figure 3.1.) For any $v = (v_1, \, v_2)$ in $\mathcal{H}$, we have

$$\begin{aligned} \sum_{j=1}^{3} |\langle v, \, e_j\rangle|^2 &= |v_2|^2 + \left|-\frac{\sqrt{3}}{2}v_1 - \frac{1}{2}v_2\right|^2 + \left|\frac{\sqrt{3}}{2}v_1 - \frac{1}{2}v_2\right|^2 \\ &= \frac{3}{2} \, \left[|v_1|^2 + |v_2|^2\right] = \frac{3}{2} \, \|v\|^2 \, . \end{aligned}$$

It follows that $\{e_1,\ e_2,\ e_3\}$ is a tight frame, but definitely not an orthonormal basis: the three vectors e_1, e_2, e_3 are clearly not linearly independent. □

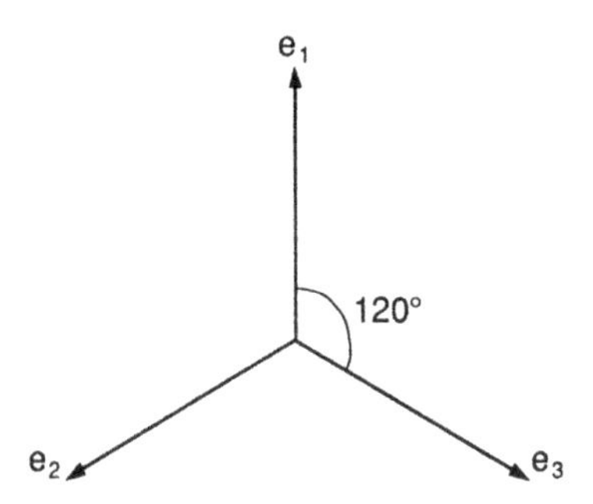

FIG. 3.1. *These three vectors in $\mathbb{C}^2$ constitute a tight frame.*

Note that in this example the frame bound $A = \frac{3}{2}$ gives the "redundancy ratio" (three vectors in a two-dimensional space). If this redundancy ratio, as measured by A, is equal to 1, then the tight frame is an orthonormal basis.

PROPOSITION 3.2.1. *If $(\varphi_j)_{j\in J}$ is a tight frame, with frame bound $A = 1$, and if $\|\varphi_j\| = 1$ for all $j \in J$, then the φ_j constitute an orthonormal basis.*

Proof. Since $\langle f,\ \varphi_j\rangle = 0$ for all $j \in J$ implies $f = 0$, the φ_j span all of $\mathcal{H}$. It remains to check that they are orthonormal. We have, for any $j \in J$,

$$\|\varphi_j\|^2 = \sum_{j'\in J} |\langle\varphi_j,\ \varphi_{j'}\rangle|^2 = \|\varphi_j\|^4 + \sum_{\substack{j'\neq j\\ j'\in J}} |\langle\varphi_j,\ \varphi_{j'}\rangle|^2 \ .$$

Since $\|\varphi_j\| = 1$, this implies $\langle\varphi_j,\ \varphi_{j'}\rangle = 0$ for all $j' \neq j$. ■

Formula (3.2.2) gives a trivial way to recover f from the $\langle f,\ \varphi_j\rangle$, if the frame is tight. Let us return to general frames, and see how things work there. We first introduce the *frame operator.*

DEFINITION. *If $(\varphi_j)_{j\in J}$ is a frame in $\mathcal{H}$, then the frame operator F is the linear operator from $\mathcal{H}$ to $\ell^2(J) = \{c = (c_j)_{j\in J};\ \|c\|^2 = \sum_{j\in J} |c_j|^2 < \infty\}$ defined by*

$$(Ff)_j = \langle f,\ \varphi_j\rangle \ .$$

It follows from (3.2.1) that $\|Ff\|^2 \leq B\ \|f\|^2$, i.e., F is bounded. The adjoint F^* of F is easy to compute:

$$\begin{aligned}\langle F^*c,\ f\rangle &= \langle c,\ Ff\rangle = \sum_{j\in J} c_j\ \overline{\langle f,\ \varphi_j\rangle}\\ &= \sum_{j\in J} c_j\ \langle\varphi_j,\ f\rangle\ ,\end{aligned}$$

so that

$$F^*c = \sum_{j\in J} c_j\varphi_j\ , \tag{3.2.3}$$

at least in the weak sense. (In fact, the series in (3.2.3) converges in norm.[4]) Since $\|F^*\| = \|F\|$, we have

$$\|F^* c\| \leq B^{1/2} \|c\| .$$

The definition of F implies

$$\sum_{j\in J} |\langle f, \varphi_j\rangle|^2 = \|Ff\|^2 = \langle F^* F f, f\rangle ;$$

in terms of F, the frame condition (3.2.1) can therefore be rewritten as

$$A \text{ Id} \leq F^* F \leq B \text{ Id} . \tag{3.2.4}$$

This implies, in particular, that F^*F is invertible, by the following elementary lemma.

LEMMA 3.2.2. *If a positive bounded linear operator S on $\mathcal{H}$ is bounded below by a strictly positive constant α, then S is invertible and its inverse S^{-1} is bounded by α^{-1}.*

Proof.

1. Ran $(S) = \{f \in \mathcal{H};\ f = Sg$ for some $g \in \mathcal{H}\}$ is a closed subspace of $\mathcal{H}$. This means that every Cauchy sequence in Ran (S) has a limit within Ran (S). Let us check this:

$$f_n \in \text{Ran } (S), \quad \text{and} \quad \|f_n - f_m\| \longrightarrow 0 \quad \text{if} \quad n, m \rightarrow \infty .$$

   Then $f_n = Sg_n$, and $\|g_n - g_m\|^2 \leq \alpha^{-1} \langle S(g_n - g_m),\ g_n - g_m\rangle \leq \alpha^{-1} \|S(g_n - g_m)\|\ \|g_n - g_m\|$, where we have used $\alpha\langle h, h\rangle \leq \langle Sh, h\rangle$ in the first inequality. But this implies $\|g_n - g_m\| \leq \alpha^{-1} \|f_n - f_m\|$, so that the g_n necessarily constitute a Cauchy sequence in $\mathcal{H}$. This Cauchy sequence necessarily has a limit g in $\mathcal{H}$. Because S is continuous, we now trivially have $Sg = \lim_{n\to\infty} Sg_n = \lim_{n\to\infty} f_n$, so that $\lim_{n\to\infty} f_n \in$ Ran (S).

2. The orthonormal complement of Ran (S) is $\{0\}$. Indeed, if $\langle f, Sg\rangle = 0$ for all $g \in \mathcal{H}$, then, in particular, $\langle f, Sf\rangle = 0$, which by $\alpha \|f\|^2 \leq \langle Sf, f\rangle$ implies $\|f\| = 0$; hence $f = 0$. Together with point 1 this implies Ran $(S) = \mathcal{H}$. It follows that S is invertible: any $f \in \mathcal{H}$ can be written as $f = Sg$; we define $S^{-1} f = g$. Moreover,

$$\alpha \|S^{-1} f\|^2 \leq \langle SS^{-1} f, S^{-1} f\rangle = \langle f, S^{-1} f\rangle \leq \|f\|\ \|S^{-1} f\| ;$$

   hence $\|S^{-1} f\| \leq \alpha^{-1} \|f\|$, as announced. ∎

Therefore, we have $\|(F^*F)^{-1}\| \leq A^{-1}$. The reader can easily check that we have, in fact,

$$B^{-1} \text{ Id} \leq (F^* F)^{-1} \leq A^{-1} \text{ Id} . \tag{3.2.5}$$

Applying the operator $(F^*F)^{-1}$ to the vectors φ_j leads to an interesting new family of vectors, which we denote by $\tilde{\varphi}_j$,

$$\tilde{\varphi}_j = (F^*F)^{-1}\,\varphi_j\ .$$

The family $(\tilde{\varphi}_j)_{j\in J}$ turns out to be a frame as well.

PROPOSITION 3.2.3. *The $(\tilde{\varphi}_j)_{j\in J}$ constitute a frame with frame constants B^{-1} and A^{-1},*

$$B^{-1}\,\|f\|^2 \le \sum_{j\in J} |\langle f,\,\tilde{\varphi}_j\rangle|^2 \le A^{-1}\,\|f\|^2\ . \tag{3.2.6}$$

The associated frame operator $\tilde{F}:\ \mathcal{H}\longrightarrow \ell^2(J)$, $(\tilde{F}f)_j = \langle f,\,\tilde{\varphi}_j\rangle$ satisfies $\tilde{F} = F(F^*F)^{-1}$, $\tilde{F}^*\tilde{F} = (F^*F)^{-1}$, $\tilde{F}^*F = \mathrm{Id} = F^*\tilde{F}$ and $\tilde{F}F^* = F\tilde{F}^*$ is the orthogonal projection operator, in $\ell^2(J)$, onto Ran (F) = Ran $(\tilde{F})$.

Proof.

1. As an exercise, the reader can check that if a bounded operator S has a bounded inverse S^{-1}, and if $S^* = S$, then $(S^{-1})^* = S^{-1}$. It follows that

$$\langle f,\tilde{\varphi}_j\rangle = \langle f,\,(F^*F)^{-1}\varphi_j\rangle = \langle (F^*F)^{-1}f,\,\varphi_j\rangle\ ;$$

hence

$$\sum_{j\in J}|\langle f,\,\tilde{\varphi}_j\rangle|^2 = \sum_{j\in J}|\langle (F^*F)^{-1}f,\,\varphi_j\rangle|^2 = \|F(F^*F)^{-1}f\|^2$$
$$= \langle (F^*F)^{-1}f,\,F^*F(F^*F)^{-1}f\rangle = \langle (F^*F)^{-1}f,\,f\rangle\ . \tag{3.2.7}$$

By (3.2.5), this implies (3.2.6); the $\tilde{\varphi}_j$ constitute a frame. Moreover, (3.2.7) implies also that the frame operator $\tilde{F}$ satisfies $\tilde{F}^*\tilde{F} = (F^*F)^{-1}$.

2. $(F(F^*F)^{-1}f)_j = \langle (F^*F)^{-1}f,\,\varphi_j\rangle = \langle f,\,\tilde{\varphi}_j\rangle = (\tilde{F}f)_j$,
$\tilde{F}^*F = [F(F^*F)^{-1}]^*F = (F^*F)^{-1}F^*F = \mathrm{Id}$,
$F^*\tilde{F} = F^*F(F^*F)^{-1} = \mathrm{Id}$.

3. Since $\tilde{F} = F(F^*F)^{-1}$, it follows that Ran $(\tilde{F}) \subset$ Ran (F). We have also $F = \tilde{F}(F^*F)$; hence Ran $(F) \subset$ Ran $(\tilde{F})$. Consequently, Ran (F) = Ran $(\tilde{F})$. Let P be the orthogonal projection operator onto Ran (F). We want to prove that $\tilde{F}F^* = P$, which is equivalent to $\tilde{F}F^*(Ff) = Ff$ (i.e., $\tilde{F}F^*$ leaves elements of Ran (F) unchanged) and $\tilde{F}F^*c = 0$ for all c orthogonal to Ran (F). Both assertions are easily checked:

$$\tilde{F}F^*Ff = F(F^*F)^{-1}\,F^*Ff = Ff$$

and

$$c\perp \mathrm{Ran}\,(F)\ \Rightarrow\ \langle c,\,Ff\rangle = 0\ \text{ for all }\ f\in\mathcal{H}$$
$$\Rightarrow F^*c = 0\ \Rightarrow\ \tilde{F}F^*c = 0\ .\ \blacksquare$$

We will call $(\tilde{\varphi}_j)_{j\in J}$ the *dual frame* of $(\varphi_j)_{j\in J}$. It is easy to check that the dual frame of $(\tilde{\varphi}_j)_{j\in J}$ is the original frame $(\varphi_j)_{j\in J}$ back again. We can rewrite some of the conclusions of Proposition 3.2.3 in a slightly less abstract form; $\tilde{F}^*F = \mathrm{Id} = F^*\tilde{F}$ means that

$$\sum_{j\in J} \langle f,\, \varphi_j\rangle\, \tilde{\varphi}_j = f = \sum_{j\in J} \langle f,\, \tilde{\varphi}_j\rangle\, \varphi_j\,. \tag{3.2.8}$$

This means that we have a reconstruction formula for f from the $\langle f,\, \varphi_j\rangle$! At the same time we have also obtained a recipe for writing f as a superposition of φ_j, which demonstrates that the two sets of questions in §3.1 are indeed "dual." When given a frame $(\varphi_j)_{j\in J}$, the only thing we therefore need to do, in order to apply (3.2.8), is to compute the $\tilde{\varphi}_j = (F^*F)^{-1}\, \varphi_j$. We will come back to this soon. First we will address a question that often arises at this point: I have stressed before that frames, even tight frames, are generally *not* (orthonormal) bases because the φ_j are typically not linearly independent. This means that for a given f, there exist many different superpositions of the φ_j which all add up to f. What then singles out the formula in the second half of (3.2.8) as especially interesting? We can get an inkling of the answer with a simple example.

EXAMPLE. We revisit the simple example of Figure 3.1. We had there, for any $v \in \mathbb{C}^2$,

$$v = \frac{2}{3} \sum_{j=1}^{3} \langle v,\, e_j\rangle\, e_j\,. \tag{3.2.9}$$

Since $\sum_{j=1}^{3}\, e_j = 0$ in this example, it follows that the following formulas are also true:

$$v = \frac{2}{3} \sum_{j=1}^{3} [\langle v,\, e_j\rangle + \alpha]\, e_j\,, \tag{3.2.10}$$

where α is arbitrary in $\mathbb{C}$. (In this particular case, one can prove that (3.2.10) gives all the possible superposition formulas valid for arbitrary v.) Somehow, (3.2.9) seems more "economical" than (3.2.10) if $\alpha \neq 0$. This intuitive statement can be made more precise in the following way:

$$\sum_{j=1}^{3} |\langle v,\, e_j\rangle|^2 = \frac{3}{2}\, \|v\|^2,$$

whereas

$$\sum_{j=1}^{3} |\langle v,\, e_j\rangle + \alpha|^2 = \frac{3}{2}\, \|v\|^2 + 3|\alpha|^2 > \frac{3}{2}\, \|v\|^2 \ \text{ if } \alpha \neq 0\,. \quad \Box$$

Likewise, the $\langle f,\, \tilde{\varphi}_j\rangle$ are the most "economical" coefficients for a decomposition of f into φ_j.

PROPOSITION 3.2.4. *If $f = \sum_{j\in J} c_j\, \varphi_j$ for some $c = (c_j)_{j\in J} \in \ell^2(J)$, and if not all c_j equal $\langle f,\, \tilde{\varphi}_j\rangle$, then $\sum_{j\in J} |c_j|^2 > \sum_{j\in J} |\langle f,\, \tilde{\varphi}_j\rangle|^2$.*

Proof.

1. Saying that $f = \sum_{j \in J} c_j \varphi_j$ is equivalent to staying that $f = F^* c$.

2. Write $c = a + b$, where $a \in \text{Ran}(F) = \text{Ran}(\tilde{F})$, and $b \perp \text{Ran}(F)$. In particular, $a \perp b$; hence $\|c\|^2 = \|a\|^2 + \|b\|^2$.

3. Since $a \in \text{Ran}(\tilde{F})$, there exists $g \in \mathcal{H}$ so that $a = \tilde{F}g$, or $c = \tilde{F}g + b$. Hence $f = F^*c = F^*\tilde{F}g + F^*b$. But $b \perp \text{Ran}(F)$, so that $F^*b = 0$, and $F^*\tilde{F} = \text{Id}$. It follows that $f = g$; hence $c = \tilde{F}f + b$, and

$$\sum_{j \in J} |c_j|^2 = \|c\|^2 = \|\tilde{F}f\|^2 + \|b\|^2 = \sum_{j \in J} |\langle f, \tilde{\varphi}_j \rangle|^2 + \|b\|^2 ,$$

which is strictly larger than $\sum_{j \in J} |\langle f, \tilde{\varphi}_j \rangle|^2$, unless $b = 0$ and $c = \tilde{F}f$. ∎

This proposition can also be used to see how the $\tilde{\varphi}_j$ play a special role in the first half of (3.2.8). We typically have nonuniqueness there as well: there may exist many other families $(u_j)_{j \in J}$ so that $f = \sum_{j \in J} \langle f, \varphi_j \rangle u_j$. In our earlier two-dimensional example, such other families are given by $u_j = \frac{2}{3} e_j + a$, where a is an arbitrary vector in $\mathbb{C}^2$. Since $\sum_{j=1}^{3} e_j = 0$, we obviously have

$$\sum_{j=1}^{3} \langle v, e_j \rangle u_j = \frac{2}{3} \sum_{j=1}^{3} \langle v, e_j \rangle e_j + \left[\sum_{j=1}^{3} \langle v, e_j \rangle \right] a = v.$$

Again, however, the u_j are "less economical" than the $\tilde{e}_j$, in the sense that for all v with $\langle v, a \rangle \neq 0$,

$$\begin{aligned} \sum_{j=1}^{3} |\langle v, u_j \rangle|^2 &= \sum_{j=1}^{3} |\langle v, \tilde{e}_j \rangle|^2 + 3 |\langle v, a \rangle|^2 \\ &= \frac{2}{3} \|v\|^2 + 3 |\langle v, a \rangle|^2 > \frac{2}{3} \|v\|^2 = \sum_{j=1}^{3} |\langle v, \tilde{e}_j \rangle|^2 \end{aligned}$$

A similar inequality holds for every frame: if $f = \sum_{j \in J} \langle f, \varphi_j \rangle u_j$, then $\sum_{j \in J} |\langle u_j, g \rangle|^2 \geq \sum_{j \in J} |\langle \tilde{\varphi}_j, g \rangle|^2$ for all $g \in \mathcal{H}$, by Proposition 3.2.4.

Back to the reconstruction issue. If we know $\tilde{\varphi}_j = (F^*F)^{-1} \varphi_j$, then (3.2.8) tells us how to reconstruct f from the $\langle f, \varphi_j \rangle$. So we only need to compute the $\tilde{\varphi}_j$, which involves the inversion of F^*F. If B and A are close to each other, i.e., $r = B/A - 1 \ll 1$, then (3.2.4) tells us that F^*F is "close" to $\frac{A+B}{2}$ Id, so that $(F^*F)^{-1}$ is "close" to $\frac{2}{A+B}$ Id, and $\tilde{\varphi}_j$ "close" to $\frac{2}{A+B} \varphi_j$. More precisely,

$$f = \frac{2}{A+B} \sum_{j \in J} \langle f, \varphi_j \rangle \varphi_j + Rf , \tag{3.2.11}$$

where $R = \mathrm{Id} - \frac{2}{A+B}\, F^*F$; hence $-\frac{B-A}{B+A}\,\mathrm{Id} \le R \le \frac{B-A}{B+A}\,\mathrm{Id}$. This implies[5] $\|R\| \le \frac{B-A}{B+A} = \frac{r}{2+r}$. If r is small, we can drop the rest term Rf in (3.2.11), and we obtain a reconstruction formula for f which is accurate up to an L^2-error of $\frac{r}{2+r}\,\|f\|$. Even if r is not so small, we can write an algorithm for the reconstruction of f with exponential convergence. With the same definition of R, we have

$$F^*F = \frac{A+B}{2}\,(\mathrm{Id} - R)\,;$$

hence $(F^*F)^{-1} = \frac{2}{A+B}\,(\mathrm{Id} - R)^{-1}$. Since $\|R\| \le \frac{B-A}{B+A} < 1$, the series $\sum_{k=0}^{\infty} R^k$ converges in norm, and its limit is $(\mathrm{Id} - R)^{-1}$. It follows that

$$\tilde{\varphi}_j = (F^*F)^{-1}\,\varphi_j = \frac{2}{A+B}\sum_{k=0}^{\infty} R^k\,\varphi_j\,.$$

Using only the zeroth order term in the reconstruction formula leads exactly to (3.2.11) with the rest term dropped. We obtain better approximations by truncating after N terms,

$$\tilde{\varphi}_j^N = \frac{2}{A+B}\sum_{k=0}^{N} R^k\,\varphi_j = \tilde{\varphi}_j - \frac{2}{A+B}\sum_{k=N+1}^{\infty} R^k\varphi_j = [\mathrm{Id} - R^{N+1}]\,\tilde{\varphi}_j\,, \quad (3.2.12)$$

with

$$\begin{aligned}
&\left\| f - \sum_{j\in J} \langle f,\,\varphi_j\rangle\,\tilde{\varphi}_j^N \right\| \\
&\quad = \sup_{\|g\|=1} \left| \left\langle f - \sum_{j\in J}\langle f,\,\varphi_j\rangle\,\tilde{\varphi}_j^N,\, g\right\rangle \right| \\
&\quad = \sup_{\|g\|=1} \left| \sum_{j\in J}\langle f,\,\varphi_j\rangle\langle \tilde{\varphi}_j - \tilde{\varphi}_j^N,\, g\rangle \right| \\
&\quad = \sup_{\|g\|=1} \left| \sum_{j\in J}\langle f,\,\varphi_j\rangle\langle R^{N+1}\tilde{\varphi}_j,\, g\rangle \right| \\
&\quad = \sup_{\|g\|=1} |\langle f,\, R^{N+1}g\rangle| \le \|R\|^{N+1}\,\|f\| \le \left(\frac{r}{2+r}\right)^{N+1}\,\|f\|\,,
\end{aligned}$$

which becomes exponentially small as N increases, since $\frac{r}{2+r} < 1$. In particular, the $\tilde{\varphi}_j^N$ can be computed by an iterative algorithm,

$$\tilde{\varphi}_j^N = \frac{2}{A+B}\,\varphi_j + R\,\tilde{\varphi}_j^{N-1}$$

or

$$\tilde{\varphi}_j^N = \sum_{\ell\in J} \alpha_{j\ell}^N\,\varphi_\ell\,,$$

with

$$\alpha_{j\ell}^N = \frac{2}{A+B}\,\delta_{\ell j} + \alpha_{j\ell}^{N-1} - \frac{2}{A+B}\sum_{m\in J}\alpha_{jm}^{N-1}\,\langle\varphi_m,\,\varphi_\ell\rangle\;.$$

This may look daunting, but it is not so terrible in examples of practical interest, where many $\langle\varphi_m,\,\varphi_\ell\rangle$ are negligibly small. The same iterative technique can be applied directly to f:

$$f \;=\; (F^*F)^{-1}(F^*F)f = \lim_{N\to\infty}\, f_N\;,$$

with

$$\begin{aligned} f_N &= \frac{2}{A+B}\sum_{k=0}^{N} R^k\,(F^*F)f \\ &= \frac{2}{A+B}(F^*F)f + R\,f_{N-1} \\ &= f_{N-1} + \frac{2}{A+B}\sum_{j\in J}\left[\langle f,\varphi_j\rangle - \langle f_{N-1},\varphi_j\rangle\right]\varphi_j\;. \end{aligned}$$

Now that we have thoroughly explored abstract frame questions, we return to discrete wavelets.

3.3. Frames of wavelets.

We saw in §3.1 that in order to have a numerically stable reconstruction algorithm for f from the $\langle f,\,\psi_{m,n}\rangle$, we require that the $\psi_{m,n}$ constitute a frame. In §3.2 we found an algorithm to reconstruct f from the $\langle f,\,\psi_{m,n}\rangle$ if the $\psi_{m,n}$ do constitute a frame; for this algorithm the ratio of the frame bounds is important, and we will come back to ways of computing at least a bound on this ratio, later in this section. First, however, we show that the requirement that the $\psi_{m,n}$ constitute a frame already imposes that ψ is admissible.

3.3.1. A necessary condition: Admissibility of the mother wavelet.

THEOREM 3.3.1. *If the $\psi_{m,n}(x) = a_0^{-m/2}\,\psi(a_0^{-m}x - nb_0)$, $m,n\in\mathbb{Z}$, constitute a frame for $L^2(\mathbb{R})$ with frame bounds A, B, then*

$$\frac{b_0\ln a_0}{2\pi}A \le \int_0^\infty d\xi\;\xi^{-1}\;|\hat\psi(\xi)|^2 \le \frac{b_0\ln a_0}{2\pi}B \tag{3.3.1}$$

and

$$\frac{b_0\ln a_0}{2\pi}A \le \int_{-\infty}^0 d\xi\;|\xi|^{-1}\;|\hat\psi(\xi)|^2 \le \frac{b_0\ln a_0}{2\pi}\;B\;. \tag{3.3.2}$$

Proof.

1. We have, for all $f \in L^2(\mathbb{R})$,

$$A \,\|f\|^2 \leq \sum_{m,n\in\mathbb{Z}} |\langle f,\, \psi_{m,n}\rangle|^2 \leq B\,\|f\|^2 \,. \tag{3.3.3}$$

If we write (3.3.3) for $f = u_\ell$, and add all the resulting inequalities, weighted with coefficients $c_\ell \geq 0$ such that $\sum_\ell c_\ell \|u_\ell\|^2 < \infty$, then we obtain

$$A \sum_\ell c_\ell \|u_\ell\|^2 \leq \sum_\ell c_\ell \sum_{m,n} |\langle u_\ell,\, \psi_{m,n}\rangle|^2 \leq B \sum_\ell c_\ell \|u_\ell\|^2 \,. \tag{3.3.4}$$

In particular, if C is any positive trace-class operator (see Preliminaries), then

$$C = \sum_{\ell\in\mathbb{N}} c_\ell \,\langle \cdot,\, u_\ell\rangle\, u_\ell \,,$$

where the u_ℓ are orthonormal, $c_\ell \geq 0$, and $\sum_{\ell\in\mathbb{N}} c_\ell = \mathrm{Tr}\; C > 0$. For any such operator, we have therefore, by (3.3.4),

$$A \;\mathrm{Tr}\; C \leq \sum_{m,n} \langle C\psi_{m,n},\, \psi_{m,n}\rangle \leq B\;\mathrm{Tr}\; C \,. \tag{3.3.5}$$

2. We now apply (3.3.5) to a very special operator C, constructed via the *continuous* wavelet transform, with a *different* mother wavelet. Take h to be any L^2-function such that support $\hat{h} \subset [0,\infty)$, $\int_0^\infty d\xi\; \xi^{-1}\, |\hat{h}(\xi)|^2 < \infty$, and define, as in Chapter 2, $h^{a,b} = a^{-1/2} h\left(\frac{x-b}{a}\right)$ for $a, b \in \mathbb{R}$, $a > 0$. If $c(a,b)$ is a bounded, positive function, then

$$C = \int_0^\infty \frac{da}{a^2} \int_{-\infty}^\infty db\; \langle \cdot,\, h^{a,b}\rangle\, h^{a,b}\; c(a,b) \tag{3.3.6}$$

is a bounded, positive operator (see §2.8). If, moreover, $c(a,b)$ is integrable with respect to $a^{-2}\, da\, db$, then C is trace-class, and $\mathrm{Tr}\; C = \int_0^\infty \frac{da}{a^2} \int_{-\infty}^\infty db$ $c(a,b)\|h\|^2$.[6] We will in particular choose $c(a,b) = w(|b|/a)$ if $1 \leq a \leq a_0$, 0 otherwise, with w positive and integrable. We then have

$$C = \int_0^\infty \frac{da}{a^2} \int_{-\infty}^\infty db\; \langle \cdot,\, h^{a,b}\rangle\, h^{a,b}\; w\left(\frac{|b|}{a}\right) \,,$$

and

$$\mathrm{Tr}\; C = \int_1^{a_0} \frac{da}{a} \int_{-\infty}^\infty ds\; w(|s|)\; \|h\|^2 = 2 \ln\, a_0 \left[\int_0^\infty ds\; w(s)\right] \|h\|^2 \,.$$

3. The middle term in (3.3.5) becomes, for this C,

$$\sum_{m,n} \langle C\, \psi_{m,n},\, \psi_{m,n}\rangle = \sum_{m,n} \int_1^{a_0} \frac{da}{a^2} \int_{-\infty}^{\infty} db\; w\left(\frac{|b|}{a}\right)\; |\langle \psi_{m,n},\, h^{a,b}\rangle|^2\ .$$

But

$$\begin{aligned}\langle \psi_{m,n},\, h^{a,b}\rangle &= a_0^{-m/2}\, a^{-1/2} \int dx\; \psi(a_0^{-m}x - nb_0)\; \overline{h\left(\frac{x-b}{a}\right)} \\ &= a_0^{m/2}\, a^{-1/2} \int dy\; \psi(y)\; \overline{h\left(\frac{y+nb_0 - ba_0^{-m}}{a\, a_0^{-m}}\right)} \\ &= \langle \psi,\, h^{a_0^{-m}a,\; a_0^{-m}b - nb_0}\rangle\ .\end{aligned}$$

After the change of variables $a' = a_0^{-m}a$, $b' = a_0^{-m}b$ we therefore obtain

$$\begin{aligned}&\sum_{m,n} \langle C\psi_{m,n}, \psi_{m,n}\rangle \\ &\quad = \sum_{m,n} \int_{a_0^{-m}}^{a_0^{-m+1}} \frac{da'}{a'^2} \int_{-\infty}^{\infty} db'\; w\left(\frac{|b'|}{a'}\right) |\langle \psi, h^{a',b'-nb_0}\rangle|^2 \\ &\quad = \int_0^{\infty} \frac{da}{a^2} \int_{-\infty}^{\infty} db\; |\langle \psi,\, h^{a,b}\rangle|^2 \sum_n w\left(\frac{|b+nb_0|}{a}\right)\ .\end{aligned}$$

Take now $w(s) = \lambda\, e^{-\lambda^2\pi^2 s^2}$. This function has only one local maximum, and is monotone decreasing as $|s|$ increases. An elementary approximation argument for integrals (the full details of which can be found in Daubechies (1990), Lemma 2.2) shows that for such functions w and for any $\alpha, \beta \in \mathbb{R}$, $\beta > 0$,

$$\int_{-\infty}^{\infty} dt\; w(t)\; - \beta w_{\max} \le \beta \sum_{n\in\mathbb{Z}} w(\alpha + n\beta) \le \int_{-\infty}^{\infty} dt\; w(t)\; + \beta w_{\max}\ ,$$

or, for our particular w,

$$\sum_n w\left(\frac{|b+nb_0|}{a}\right) = \frac{a}{b_0} + \rho(a,b)\ ,$$

with $|\rho(a,b)| \le w(0) = \lambda$. Consequently,

$$\sum_{m,n} \langle C\psi_{m,n}, \psi_{m,n}\rangle = \frac{1}{b_0} \int_0^{\infty} \frac{da}{a} \int_{-\infty}^{\infty} db\; |\langle \psi,\, h^{a,b}\rangle|^2 + R\ , \tag{3.3.7}$$

where

$$\begin{aligned}|R| &= \int_0^{\infty} \frac{da}{a^2} \int_{-\infty}^{\infty} db\; |\langle \psi,\, h^{a,b}\rangle|^2\; \rho(a,b) \\ &\le \lambda\, C_h\, \|\psi\|^2\ ,\end{aligned}$$

which C_h as defined by (2.4.1). We can rewrite the first term in (3.3.7) as

$$\frac{1}{b_0} \int_0^\infty \frac{da}{a} \int_{-\infty}^\infty db \left| \int_0^\infty d\xi \ \hat{\psi}(\xi) \ a^{1/2} \ \overline{\hat{h}(a\xi)} \ e^{ib\xi} \right|^2$$

$$= \frac{2\pi}{b_0} \int_0^\infty da \int_0^\infty d\xi \ |\hat{\psi}(\xi)|^2 \ |\hat{h}(a\xi)|^2 = \frac{2\pi}{b_0} \ \|h\|^2 \int_0^\infty d\xi \ \xi^{-1} \ |\hat{\psi}(\xi)|^2 \ .$$

4. For the particular weight function w that we have chosen, we have $\int_0^\infty dt \ w(t) = \frac{1}{2}$, hence $\operatorname{Tr} C = \|h\|^2 \ \ln a_0$. Substituting all our results in (3.3.5) we find

$$A \ \|h\|^2 \ \ln a_0 \le \frac{2\pi}{b_0} \ \|h\|^2 \int_0^\infty d\xi \ \xi^{-1} \ |\hat{\psi}(\xi)|^2 + R \le B\|h\|^2 \ \ln a_0,$$

where $|R| \le \lambda \ C_h \ \|\psi\|^2$. If we divide by $\frac{2\pi}{b_0} \ \|h\|^2$ and let λ tend to zero, then this proves (3.3.1). The negative frequency formula (3.3.2) is proved analogously. ∎

REMARKS.

1. Formulas (3.3.1), (3.3.2) impose an a priori restriction on ψ, namely that $\int_0^\infty d\xi \ \xi^{-1} \ |\hat{\psi}(\xi)|^2 < \infty$ and $\int_{-\infty}^0 d\xi \ |\xi|^{-1} \ |\hat{\psi}(\xi)|^2 < \infty$. This is the *same* restriction as in the continuous case (see (2.4.6)).

2. In defining the discretely labelled $\psi_{m,n}$, we only took *positive* dilations a_0^m into consideration (the sign of m affects whether a_0^m is ≥ 1 or ≤ 1, but $a_0^m > 0$ for all m). This is the reason why formulas (3.3.1), (3.3.2) dissociate the positive and negative frequency domains. If we had allowed negative discrete dilations as well, then the condition would have involved only $\int_{-\infty}^\infty d\xi \ |\xi|^{-1} \ |\hat{\psi}(\xi)|^2$ (as is easy to check by mimicking the above proof).

3. If the $\psi_{m,n}$ constitute a tight frame ($A = B$), then (3.3.1), (3.3.2) imply

$$A = \frac{2\pi}{b_0 \ln a_0} \int_0^\infty d\xi \ \xi^{-1} \ |\hat{\psi}(\xi)|^2 = \frac{2\pi}{b_0 \ln a_0} \int_{-\infty}^0 d\xi \ |\xi|^{-1} \ |\hat{\psi}(\xi)|^2 \ .$$

In particular, if the $\psi_{m,n}$ constitute an orthonormal basis of $L^2(\mathbb{R})$ (such as the Haar basis, or other bases we will encounter), then

$$\int_0^\infty d\xi \ \xi^{-1} \ |\hat{\psi}(\xi)|^2 = \int_{-\infty}^0 d\xi \ |\xi|^{-1} \ |\hat{\psi}(\xi)|^2 = \frac{b_0 \ln a_0}{2\pi} \ . \tag{3.3.8}$$

It is an easy exercise that the Haar basis does indeed satisfy (3.3.8). Most of the orthonormal bases we will consider are real, so that the first equality in (3.3.8) is trivially satisfied.

4. A different proof of Proposition 3.3.1 is given in Chui and Shi (1993). □

In all that follows, we will always assume that ψ is admissible.

3.3.2. A sufficient condition and estimates for the frame bounds. Not all choices for ψ, a_0, b_0 lead to frames of wavelets, even if ψ is admissible. In this subsection we derive some fairly general conditions on ψ, a_0, b_0 under which we do indeed obtain a frame, and we estimate the corresponding frame bounds. To do this, we need to estimate $\sum_{m,n} |\langle f,\, \psi_{m,n}\rangle|^2$:

$$\sum_{m,n\in\mathbb{Z}} |\langle f,\, \psi_{m,n}\rangle|^2 = \sum_{m,n} \left| \int_{-\infty}^{\infty} d\xi\, \hat{f}(\xi)\, {a_0}^{m/2}\, \overline{\hat{\psi}(a_0^m \xi)}\, e^{ib_0\, a_0^m\, n\xi} \right|^2$$

$$= \sum_{m,n} a_0^m \left| \int_0^{2\pi b_0^{-1} a_0^{-m}} d\xi\, e^{ib_0 a_0^m n\xi} \sum_{\ell\in\mathbb{Z}} \hat{f}(\xi + 2\pi\ell a_0^{-m} {b_0}^{-1})\, \overline{\hat{\psi}(a_0^m \xi + 2\pi\ell {b_0}^{-1})} \right|^2$$

$$= \frac{2\pi}{b_0} \sum_m \int_0^{2\pi b_0^{-1} a_0^{-m}} d\xi \left| \sum_{\ell\in\mathbb{Z}} \hat{f}(\xi + 2\pi\ell a_0^{-m} {b_0}^{-1})\, \overline{\hat{\psi}(a_0^m \xi + 2\pi\ell {b_0}^{-1})} \right|^2$$

(by Plancherel's theorem for periodic functions)

$$= \frac{2\pi}{b_0} \sum_{m,k\in\mathbb{Z}} \int_{-\infty}^{\infty} d\xi\, \hat{f}(\xi)\, \overline{\hat{f}(\xi + 2\pi k a_0^{-m} {b_0}^{-1})}\, \overline{\hat{\psi}(a_0^m \xi)}\, \hat{\psi}(a_0^m \xi + 2\pi k {b_0}^{-1})$$

$$= \frac{2\pi}{b_0} \int_{-\infty}^{\infty} d\xi\, |\hat{f}(\xi)|^2 \sum_{m\in\mathbb{Z}} |\hat{\psi}(a_0^m \xi)|^2 + \text{Rest}\,(f)\ . \tag{3.3.9}$$

Here Rest (f) is bounded by

$$|\text{Rest}\,(f)| = \left| \frac{2\pi}{b_0} \sum_{\substack{m,k\in\mathbb{Z}\\ k\neq 0}} \int_{-\infty}^{\infty} d\xi\, \hat{f}(\xi) \right.$$
$$\left. \overline{\hat{f}(\xi + 2\pi k a_0^{-m} b_0^{-1})}\ \overline{\hat{\psi}(a_0^m \xi)}\ \hat{\psi}(a_0^m \xi + 2\pi k b_0^{-1}) \right|$$

$$\leq \frac{2\pi}{b_0} \sum_{\substack{m,k\\ k\neq 0}} \left[\int_{-\infty}^{\infty} d\xi\, |\hat{f}(\xi)|^2\, |\hat{\psi}(a_0^m \xi)|\, |\hat{\psi}(a_0^m \xi + 2\pi k b_0^{-1})| \right]^{1/2}$$
$$\cdot \left[\int_{-\infty}^{\infty} d\zeta\, |\hat{f}(\zeta)|^2\, |\hat{\psi}(a_0^m \zeta - 2\pi k b_0^{-1})|\, |\hat{\psi}(a_0^m \zeta)| \right]^{1/2}$$

(use Cauchy–Schwarz, and change variables $\zeta = \xi - 2\pi k b_0^{-1} {a_0}^{-m}$ in the second factor)

$$\leq \frac{2\pi}{b_0} \sum_{k\neq 0} \left[\int_{-\infty}^{\infty} d\xi\, |\hat{f}(\xi)|^2 \sum_m |\hat{\psi}(a_0^m \xi)|\, |\hat{\psi}(a_0^m \xi + 2\pi k b_0^{-1})| \right]^{1/2},$$
$$\cdot \left[\int_{-\infty}^{\infty} d\zeta\, |\hat{f}(\zeta)|^2 \sum_m |\hat{\psi}(a_0^m \zeta)|\, |\hat{\psi}(a_0^m \zeta - 2\pi k b_0^{-1})| \right]^{1/2}$$

(use Cauchy–Schwarz on the sum over m)

$$\leq \frac{2\pi}{b_0} \|f\|^2 \sum_{k\neq 0} \left[\beta\left(\frac{2\pi}{b_0}k\right)\beta\left(-\frac{2\pi}{b_0}k\right)\right]^{1/2}, \tag{3.3.10}$$

where $\beta(s) = \sup_\xi \sum_{m\in\mathbb{Z}} |\hat{\psi}(a_0^m\xi)|\,|\hat{\psi}(a_0^m\xi + s)|$. Putting (3.3.9) and (3.3.10) together, we see that[7]

$$\inf_{\substack{f\in\mathcal{H}\\ f\neq 0}} \|f\|^{-2} \sum_{m,n} |\langle f, \psi_{m,n}\rangle|^2$$
$$\geq \frac{2\pi}{b_0}\left\{\operatorname*{ess\,inf}_\xi \sum_{m\in\mathbb{Z}} |\hat{\psi}(a_0^m\xi)|^2 - \sum_{k\neq 0}\left[\beta\left(\frac{2\pi}{b_0}k\right)\beta\left(-\frac{2\pi}{b_0}k\right)\right]^{1/2}\right\} \tag{3.3.11}$$

$$\sup_{\substack{f\in\mathcal{H}\\ f\neq 0}} \|f\|^{-2} \sum_{m,n} |\langle f, \psi_{m,n}\rangle|^2$$
$$\leq \frac{2\pi}{b_0}\left\{\sup_\xi \sum_{m\in\mathbb{Z}} |\hat{\psi}(a_0^m\xi)|^2 + \sum_{k\neq 0}\left[\beta\left(\frac{2\pi}{b_0}k\right)\beta\left(-\frac{2\pi}{b_0}k\right)\right]^{1/2}\right\}. \tag{3.3.12}$$

If the right-hand sides of (3.3.11), (3.3.12) are strictly positive and bounded, then the $\psi_{m,n}$ constitute a frame, and (3.3.11) gives a lower bound for A, (3.3.12) an upper bound for B. To make this work, we need that, for all $1 \leq |\xi| \leq a_0$ (other values of ξ can be reduced to this range by multiplication with a suitable a_0^m, except for $\xi = 0$, but this constitutes a set of measure zero, and therefore does not matter),

$$0 < \alpha \leq \sum_{m\in\mathbb{Z}} |\hat{\psi}(a_0^m\xi)|^2 \leq \beta < \infty\,;$$

moreover, $\sum_{m\in\mathbb{Z}} |\hat{\psi}(a_0^m\xi)|\,|\hat{\psi}(a_0^m\xi+s)|$ should have sufficient decay at ∞. "Sufficient" in this second condition means that $\sum_{k\neq 0} [\beta(\frac{2\pi}{b_0}k)\beta(-\frac{2\pi}{b_0}k)]^{1/2}$ converges, and that the sum tends to 0 as b_0 tends to 0, ensuring that for small enough b_0 the first terms in (3.3.11), (3.3.12) dominate, so that the $\psi_{m,n}$ do indeed constitute a frame. In order to ensure all this, it is sufficient to require that

- the zeros of $\hat{\psi}$ do not "conspire," so that

$$\sum_{m\in\mathbb{Z}} |\hat{\psi}(a_0^m\xi)|^2 \geq \alpha > 0 \tag{3.3.13}$$

  for all $\xi \neq 0$,

- $|\hat{\psi}(\xi)| \leq C|\xi|^\alpha\,(1 + |\xi|^2)^{-\gamma/2}$, with $\alpha > 0$, $\gamma > \alpha + 1$.[8] (3.3.14)

These decay conditions on $\hat{\psi}$ are very weak, and in practice we will require much more! If $\hat{\psi}$ is continuous, and decays at ∞, then (3.3.13) is a necessary condition: if, for some $\xi_0 \neq 0$, $\sum_{m\in\mathbb{Z}} |\hat{\psi}(a_0^m\xi_0)|^2 \leq \epsilon$, then one can construct $f \in L^2(\mathbb{R})$, with $\|f\| = 1$, so that $(2\pi)^{-1}b_0 \sum_{m,n} |\langle f,\, \psi_{m,n}\rangle|^2 \leq 2\epsilon$, implying $A \leq 4\pi\,\epsilon/b_0$.[9]

If ϵ can be chosen arbitrarily small, then there is no finite lower frame bound. (See also Chui and Shi (1993), where the stronger result $A \le \frac{2\pi}{b_0} \sum_m |\hat{\psi}(a_0^m \xi)|^2 \le B$ is proved.) The following proposition summarizes our findings.

PROPOSITION 3.3.2. *If ψ, a_0 are such that*

$$\inf_{1 \le |\xi| \le a_0} \sum_{m=-\infty}^{\infty} |\hat{\psi}(a_0^m \xi)|^2 > 0 \ ,$$

$$\sup_{1 \le |\xi| \le a_0} \sum_{m=-\infty}^{\infty} |\hat{\psi}(a_0^m \xi)|^2 < \infty \ , \tag{3.3.15}$$

and if $\beta(s) = \sup_\xi \sum_m |\hat{\psi}(a_0^m \xi)| \, |\hat{\psi}(a_0^m \xi + s)|$ decays at least as fast as $(1+|s|)^{-(1+\epsilon)}$, with $\epsilon > 0$, then there exists $(b_0)_{\text{thr}} > 0$ such that the $\psi_{m,n}$ constitute a frame for all choices $b_0 < (b_0)_{\text{thr}}$. For $b_0 < (b_0)_{\text{thr}}$, the following expressions are frame bounds for the $\psi_{m,n}$:

$$A = \frac{2\pi}{b_0} \left\{ \inf_{1 \le |\xi| \le a_0} \sum_{m=-\infty}^{\infty} |\hat{\psi}(a_0^m \xi)|^2 - \sum_{\substack{k=-\infty \\ k \ne 0}}^{\infty} \left[\beta\left(\frac{2\pi}{b_0} k\right) \beta\left(-\frac{2\pi}{b_0} k\right) \right]^{1/2} \right\} ,$$

$$B = \frac{2\pi}{b_0} \left\{ \sup_{1 \le |\xi| \le a_0} \sum_{m=-\infty}^{\infty} |\hat{\psi}(a_0^m \xi)|^2 + \sum_{\substack{k=-\infty \\ k \ne 0}}^{\infty} \left[\beta\left(\frac{2\pi}{b_0} k\right) \beta\left(-\frac{2\pi}{b_0} k\right) \right]^{1/2} \right\} .$$

The conditions on β and (3.3.15) *are satisfied if, e.g., $|\hat{\psi}(\xi)| \le C|\xi|^\alpha \, (1+|\xi|)^{-\gamma}$ with $\alpha > 0$, $\gamma > \alpha + 1$.*

Proof. We have already carried out all the necessary estimates. The decay of β ensures the existence of a $(b_0)_{\text{thr.}}$ so that $\sum_{k \ne 0} [\beta(\frac{2\pi}{b_0}k)\beta(-\frac{2\pi}{b_0}k)]^{1/2} < \inf_{1 \le |\xi| \le a_0} \sum_m |\hat{\psi}(a_0^m \xi)|^2$ if $b_0 < (b_0)_{\text{thr}}$. ∎

The moral of these technical estimates is simple: if ψ is at all "decent" (reasonable decay in time and frequency, $\int dx \ \psi(x) = 0$), then there exists a whole range of a_0, b_0 so that the corresponding $\psi_{m,n}$ constitute a frame. Since our conditions on ψ imply that ψ is admissible in the sense of Chapter 2, this is not so surprising for values of a_0, b_0 close to $1, 0$ respectively: we already know that the resolution of the identity (2.4.4) holds for such ψ, and it is reasonable to expect that a sufficiently fine discretization of the integration variables should not upset the reconstruction too much. Surprisingly enough, for many ψ of practical interest, the range of "good" (a_0, b_0) includes values which are quite far from $(1, 0)$. We will see several examples below. But first we will look at the dual frame for a frame of wavelets, and discuss some variations on the basic scheme.

3.3.3. The dual frame.

As we saw in §3.2, the dual frame is defined by

$$\widetilde{\psi_{m,n}} = (F^* F)^{-1} \, \psi_{m,n} \ , \tag{3.3.16}$$

where $F^*Ff = \sum_{m,n} \langle f, \psi_{m,n}\rangle \psi_{m,n}$. We have an explicit formula for the inverse of F^*F which converges exponentially fast, i.e., like $\sum_{n=0}^{\infty} \alpha^n$, with a convergence ratio α proportional to $\left(\frac{B}{A}-1\right)$. It is therefore useful to have frame bounds A, B which are close to each other. Nevertheless, (3.2.8) necessitates, in principle, an infinite number of $\widetilde{\psi_{m,n}}$ to be computed. The situation is not quite as bad as one might expect: if we introduce the notation

$$(D^m f)(x) = a_0^{-m/2} f(a_0^{-m}x), \qquad (T^n f)(x) = f(x - nb_0) ,$$

then it is easy to check that, for all $f \in L^2(\mathbb{R})$,

$$F^*F\; D^m f = D^m\; F^*F f .$$

It follows that $(F^*F)^{-1}$ and D^m commute as well. In particular, since $\psi_{m,n} = D^m T^n \psi$,

$$\widetilde{\psi_{m,n}} = (F^*F)^{-1}\; D^m T^n \psi = D^m\; (F^*F)^{-1}\; T^n \psi ,$$

or

$$\widetilde{\psi_{m,n}}(x) = a_0^{-m/2}\; \widetilde{\psi_{0,n}}\; (a_0^{-m} x) .$$

Unfortunately, F^*F and T^n do not commute, so that we still have to compute, in principle, infinitely many $\widetilde{\psi_{0,n}}$. In practice, one is interested only in functions "living" on a finite range of scales, on which F^*F can be reasonably approximated by $\sum_{m=m_0}^{m_1} \sum_{n\in\mathbb{Z}} \langle \cdot, \psi_{m,n}\rangle \psi_{m,n}$ (see the time-frequency localization section below, §3.5). If $a_0{}^{m_1-m_0}$ is an integer, $N = a_0{}^{m_1-m_0}$, then one easily checks that this truncated version of F^*F commutes with T^N, so that one only has to compute the N different $\widetilde{\psi_{0,n}}$, $0 \le n \le N-1$ in this case. This number is still very large in many cases of practical interest, however. It is therefore especially advantageous to work with frames which are almost tight ("snug frames"), i.e., which have $\frac{B}{A} - 1 \ll 1$: we can then stop after the zeroth order term of the reconstruction formula (3.2.11), avoid all the complications with the dual frame, and still have a high quality reconstruction of arbitrary f. On the other hand, there exist very special choices of ψ, a_0, b_0 for which the $\psi_{m,n}$ are not close to a tight frame, but it so happens that all the $\widetilde{\psi_{m,n}}$ are generated by a single function,

$$\widetilde{\psi_{m,n}}(x) = \tilde{\psi}_{m,n}(x) = a_0^{-m/2}\; \tilde{\psi}(a_0^{-m}x - n) . \tag{3.3.17}$$

An example is provided by some of the biorthogonal bases that we will encounter in Chapter 8; another example is given by the ϕ-transform of Frazier and Jawerth (1988) (see also Frazier, Jawerth, and Weiss (1991)).

It is important to realize that the $\psi_{m,n}$ and the $\widetilde{\psi_{m,n}}$ may have very different regularity properties. For instance, there exist frames where ψ itself is C^∞ and decays faster than any inverse polynomial, but where some of the $\widetilde{\psi_{0,n}}$ are not in L^p for small p (implying that they have very slow decay). An example, due to P. G. Lemarié, is given in detail in Daubechies (1990), pp. 988–989.[10] Something similar may happen even if all the $\widetilde{\psi_{m,n}}$ are generated by a single function $\tilde{\psi}$: there exist examples where $\psi \in C^k$ (with k arbitrarily large), but where $\tilde{\psi}$ is not continuous. (The biorthogonal bases in Chapter 8 give examples where this happens; the first example was constructed by Tchamitchian (1989).) One can exclude such dissimilarities by imposing extra conditions on ψ, a_0, and b_0 (see Daubechies (1990), § II.D.2, pp. 991–992).

3.3.4. Some variations on the basic scheme. So far, we have not restricted the value of a_0, beyond $a_0 > 1$. In practice, however, it is very convenient to have $a_0 = 2$. Going from one scale to the next then means doubling or halving the translation step, which is much more practical than if another a_0 is used. On the other hand, we have just seen that it is advantageous to use frames with $B/A - 1 \ll 1$. Since our estimates (3.3.11), (3.3.12) for A, B give

$$A \leq \frac{2\pi}{b_0} \sum_{m \in \mathbb{Z}} |\hat{\psi}(a_0^m \xi)|^2 \leq B \tag{3.3.18}$$

for all $\xi \neq 0$, these two requirements together imply that $\sum_{m \in \mathbb{Z}} |\hat{\psi}(2^m \xi)|^2$ should be almost constant for $\xi = 0$, which is a very strong restriction on ψ, not generally satisfied. The Mexican hat function $\psi(x) = (1 - x^2)\ e^{-x^2/2}$, for instance, leads to a frame with B/A close to 1, for $a_0 \leq 2^{1/4}$, but certainly not for $a_0 = 2$ because the amplitude of the oscillations of $\sum_{m \in \mathbb{Z}} |\hat{\psi}(2^m \xi)|^2$ is too large. In order to remedy this situation, without having to give up too much of our freedom in choosing ψ and its width in the frequency domain, we can adopt a method used by A. Grossmann, R. Kronland-Martinet, and J. Morlet, and use different "voices" per octave. This amounts to using several different wavelets, $\psi^1, \cdots, \psi^N$, and to look at the frame $\{\psi^\nu_{m,n};\ m, n \in \mathbb{Z},\ \nu = 1, \cdots, N\}$. One can repeat the analysis of §3.3.2 (see, e.g., Daubechies (1990)), leading to the following estimates for the frame bounds of this multivoice frame:

$$A = \frac{2\pi}{b_0} \left[\inf_{1 \leq |\xi| \leq 2} \sum_{\nu=1}^{N} \sum_{m=-\infty}^{\infty} |\hat{\psi}^\nu(2^m \xi)|^2 - R\left(\frac{2\pi}{b_0}\right) \right] , \tag{3.3.19}$$

$$B = \frac{2\pi}{b_0} \left[\sup_{1 \leq |\xi| \leq 2} \sum_{\nu=1}^{N} \sum_{m=-\infty}^{\infty} |\hat{\psi}^\nu(2^m \xi)|^2 + R\left(\frac{2\pi}{b_0}\right) \right] , \tag{3.3.20}$$

with

$$R(x) = \sum_{k \neq 0} \sum_{\nu=1}^{N} [\beta^\nu(kx)\ \beta^\nu(-kx)]^{1/2} ,$$

and

$$\beta^\nu(s) = \sup_{1 \leq |\xi| \leq 2} \sum_{m=-\infty}^{\infty} |\hat{\psi}^\nu(2^m \xi)|\ |\hat{\psi}^\nu(2^m \xi + s)| .$$

By choosing the $\hat{\psi}^1, \cdots, \hat{\psi}^n$ to have slightly staggered frequency localization centers, coupled with good decay at ∞, one can achieve $B/A - 1 \ll 1$. (See the examples in §3.3.5 below.) The time-frequency lattice corresponding to such a multivoice scheme looks a little different from Figure 1.4a; an example, with

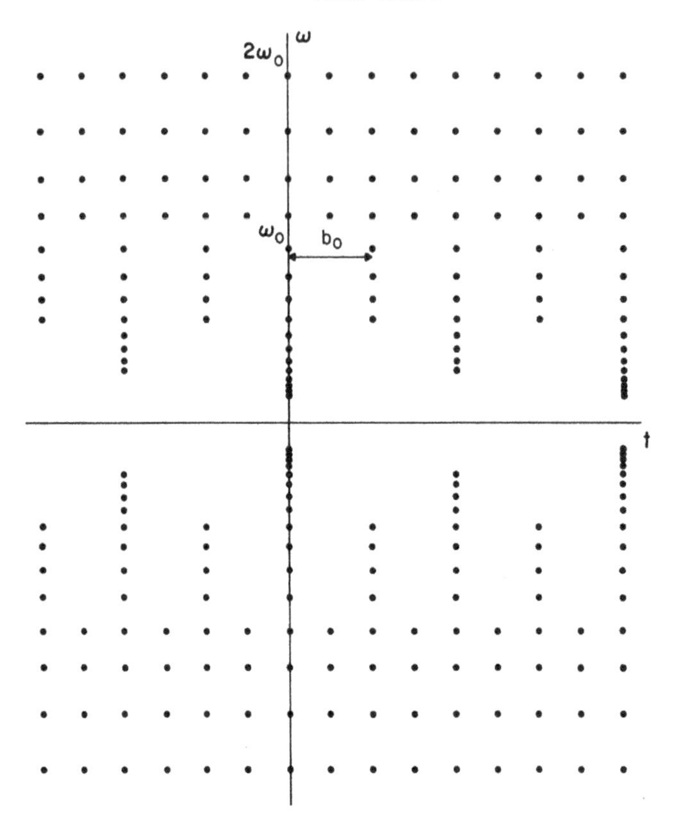

FIG. 3.2. *The time-frequency lattice for a scheme with four voices. In this case the different voice wavelets $\psi^1, \cdots, \psi^4$ are assumed to be dilations of a single function ψ, $\psi^j(x) = 2^{-(j-1)/4}\psi(2^{-(j-1)/4}x)$; if $|\hat{\psi}(\xi)|$ (which we assume to be even) peaks around $\pm\omega_0$, then the $|\hat{\psi}^j|$ will be concentrated around $\pm 2^{-(j-1)/4}\omega_0$.*

four voices per octave, is given in Figure 3.2. For every dilation step, we find four different frequency levels (corresponding to the four different frequency localizations of $\psi^1, \cdots, \psi^4$), all translated by the same translation step. Such a lattice can be viewed as the superposition of four different lattices of the type in Figure 1.4a, stretched by different amounts in the frequency direction. Each of these four sublattices has a different "density," which is reflected by the fact that typically the ψ^ν have different L^2-norms. One choice favored by Grossmann, Kronland-Martinet, and Morlet is to take "fractionally" dilated versions of a single wavelet ψ:

$$\psi^\nu(x) = 2^{-(\nu-1)/N}\ \psi(2^{-(\nu-1)/N}x)\ .$$

(Note that these do indeed have different L^2-norms!) In this case $\sum_{\nu=1}^{N} \sum_{m=-\infty}^{\infty} |\hat{\psi}^\nu(2^m\xi)|^2$ becomes simply $\sum_{m'=-\infty}^{\infty} |\hat{\psi}(2^{m'/N}\xi)|^2$, and this can easily be made to be almost constant, by choosing N large enough.

Fixing $a_0 = 2$ allows also for a modification of the estimation techniques in §3.2, which may be useful in some instances. Let us go back to the estimate for Rest (f). We can rewrite $k \in \mathbb{Z}$, $k \neq 0$, as $k = 2^\ell(2k'+1)$, where $\ell \geq 0$, $k' \in \mathbb{Z}$; the correspondence $k \to (\ell, k')$ is one-to-one. If $a_0 = 2$, then we can regroup different terms, and write

$$\text{Rest }(f) = \frac{2\pi}{b_0} \sum_{m',k' \in \mathbb{Z}} \int d\xi \; \hat{f}(\xi) \overline{\hat{f}(\xi + 2\pi(2k'+1){b_0}^{-1} 2^{-m'})} \cdot \sum_{\ell=0}^{\infty} \overline{\hat{\psi}(2^{m'+\ell}\xi)} \; \hat{\psi}[2^\ell(2^{m'}\xi + 2\pi(2\ell+1){b_0}^{-1})] \, .$$

This leads to

$$A = \frac{2\pi}{b_0} \left\{ \inf_{1 \leq |\xi| \leq 2} \sum_m |\hat{\psi}(2^m\xi)|^2 - \sum_{k'=-\infty}^{\infty} \left[\beta_1\left(\frac{2\pi}{b_0}(2k'+1)\right) \beta_1\left(-\frac{2\pi}{b_0}(2k'+1)\right)\right]^{1/2} \right\} , \tag{3.3.21}$$

$$B = \frac{2\pi}{b_0} \left\{ \sup_{1 \leq |\xi| \leq 2} \sum_m |\hat{\psi}(2^m\xi)|^2 + \sum_{k'=-\infty}^{\infty} \left[\beta_1\left(\frac{2\pi}{b_0}(2k'+1)\right) \beta_1\left(-\frac{2\pi}{b_0}(2k'+1)\right)\right]^{1/2} \right\} , \tag{3.3.22}$$

where

$$\beta_1(s) = \sup_{1 \leq |\xi| \leq 2} \sum_{m \in \mathbb{Z}} \left| \sum_{\ell=0}^{\infty} \hat{\psi}(2^{m+\ell}\xi) \; \overline{\hat{\psi}(2^\ell(2^m\xi + s))} \right| \, . \tag{3.3.23}$$

These estimates are due to Ph. Tchamitchian. (Full details of the derivation can be found in Daubechies (1990).) Note that β_1, unlike β, still takes the phases of $\hat{\psi}$ into account; as a result, the estimates (3.3.21), (3.3.22) are often better than (3.3.11), (3.3.12) when $\hat{\psi}$ is not a positive function. If $\hat{\psi}$ is positive, then (3.3.11), (3.3.12) may be better. The estimates (3.3.21), (3.3.22) hold if we have one single voice per octave; they can of course also be extended to the multivoice case.

3.3.5. Examples.

A. Tight frames. The following construction (first proposed in Daubechies, Grossmann, and Meyer (1986)) leads to a family of tight wavelet frames. Let ν be a C^k (or C^∞) function from $\mathbb{R}$ to $\mathbb{R}$ that satisfies:

$$\nu(x) = \begin{cases} 0 & \text{if} \quad x \leq 0 \, , \\ 1 & \text{if} \quad x \geq 1 \end{cases} \tag{3.3.24}$$

(see Figure 3.3). An example of such a (C^1) function ν is

$$\nu(x) = \begin{cases} 0 , & x \le 0 , \\ \sin^2 \frac{\pi}{2}x , & 0 \le x \le 1 , \\ 1 , & x \ge 1 . \end{cases} \tag{3.3.25}$$

For arbitrary $a_0 > 1$, $b_0 > 0$ we then define $\hat{\psi}^{\pm}(\xi)$ by

$$\hat{\psi}^+(\xi) = [\ln a_0]^{-1/2} \begin{cases} 0 , & \xi \le \ell \text{ or } \xi \ge a_0{}^2\ell , \\ \sin \left[\frac{\pi}{2}\nu\left(\frac{\xi-\ell}{\ell(a_0-1)}\right)\right] , & \ell \le \xi \le a_0\ell , \\ \cos \left[\frac{\pi}{2}\nu\left(\frac{\xi-a_0\ell}{a_0\ell(a_0-1)}\right)\right] , & a_0\ell \le \xi \le a_0{}^2\ell , \end{cases}$$

where $\ell = 2\pi[b_0(a_0{}^2 - 1)]^{-1}$, and $\hat{\psi}^-(\xi) = \hat{\psi}^+(-\xi)$. Figure 3.4 shows $\hat{\psi}^+$ for $a_0 = 2$, $b_0 = 1$, and ν as in (3.3.25). It is easy to check that

$$|\text{support } \hat{\psi}^+| = (a_0{}^2 - 1)\ell = 2\pi/b_0$$

and

$$\sum_{m\in\mathbb{Z}} |\hat{\psi}^+(a_0^m\xi)|^2 = (\ln a_0)^{-1} \, \chi_{(0,\infty)}(\xi) ,$$

where $\chi_{(0,\infty)}$ is the indicator function of the open half line $(0, \infty)$, i.e., $\chi_{(0,\infty)}(\xi) = 1$ if $0 < \xi < \infty$, 0 otherwise.

For any $f \in L^2(\mathbb{R})$, one then has

$$\sum_{m,n\in\mathbb{Z}} |\langle f, \psi^+_{m,n}\rangle|^2$$

$$= \sum_{m,n\in\mathbb{Z}} a_0^m \left| \int_{a_0^{-m}\ell}^{a_0^{-m}\ell a_0{}^2} d\xi \hat{f}(\xi) \, e^{2\pi \, ina_0^m[\ell(a_0{}^2-1)]^{-1}} \, \overline{\hat{\psi}^+(a_0^m\xi)} \right|^2$$

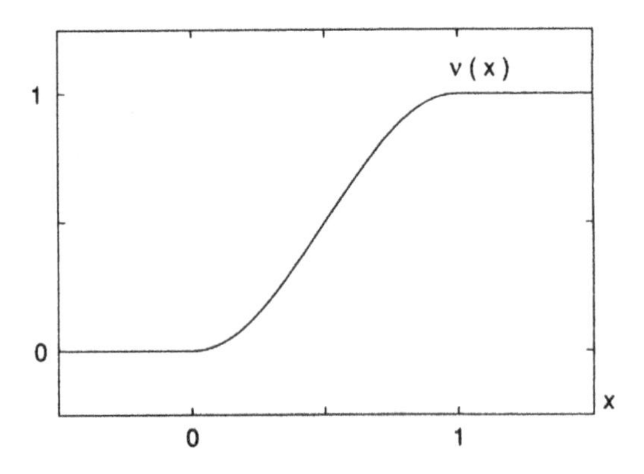

FIG. 3.3. *The function $\nu(x)$ defined by* (3.3.25).

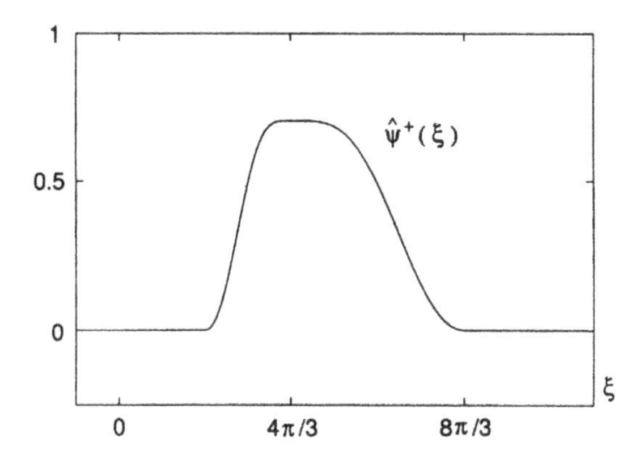

FIG. 3.4. *The function* $\hat{\psi}^+(\xi)$ *with the choices* $a_0 = 2$, $b_0 = 1$.

$$= \frac{2\pi}{b_0} \sum_{m\in\mathbb{Z}} \int d\xi \; |\hat{f}(\xi)|^2 \; |\hat{\psi}^+(a_0^m \xi)|^2$$
$$= \frac{2\pi}{b_0 \ln a_0} \int_0^\infty d\xi \; |\hat{f}(\xi)|^2 \; .$$

Similarly, $\sum_{m,n} |\langle f, \psi^-_{m,n}\rangle|^2 = \frac{2\pi}{b_0 \ln a_0} \int_{-\infty}^0 d\xi \; |\hat{f}(\xi)|^2$. It follows that the collection $\{\psi^\epsilon_{m,n};\; m,n \in \mathbb{Z},\; \epsilon = + \text{ or } -\}$ is a tight frame for $L^2(\mathbb{R})$, with frame bound $\frac{2\pi}{b_0 \ln a_0}$. One can use a variant to obtain a frame consisting of real wavelets: $\psi^1 = \text{Re}\; \psi^+ = \frac{1}{2}[\psi^+ + \psi^-]$ and $\psi^2 = \text{Im}\; \psi^+ = \frac{1}{2i}[\psi^+ - \psi^-]$ generate the tight frame $\{\psi^\lambda_{m,n};\; m,n \in \mathbb{Z},\; \lambda = 1 \text{ or } 2\}$. These frames are not generated by translations and dilations of a *single* function; this is a natural consequence of the decoupling of positive and negative frequencies in the construction. A more serious objection to their practical use is the fact that their Fourier transforms are compactly supported, and that the size of this support is relatively small (for reasonable a_0, b_0). As a result, the decay of the wavelets is numerically rather slow: even though we may choose ν to be C^∞, so that the $\psi^\pm$ decay faster than any inverse polynomial,

$$|\psi^\pm(x)| \le C_N \; (1 + |x|)^{-N} \; ,$$

the value of C_N turns out to be too large to be practical. Note that we did not introduce *any* restriction on a_0, b_0 in this construction.

B. The Mexican hat function. The Mexican hat function is the second derivative of the Gaussian $e^{-x^2/2}$; if we normalize it so that its L^2-norm is 1, and $\psi(0) > 0$ we obtain

$$\psi(x) = \frac{2}{\sqrt{3}} \; \pi^{-1/4} \; (1 - x^2) e^{-x^2/2} \; .$$

This function (and dilated and translated versions of it) was plotted in Figure 1.2b; if you take one such plot, and imagine it rotated around its symmetry axis, then you obtain a shape similar to a Mexican hat. This function is popular in vision analysis (at least in theoretical expositions), where it was also christened. Table 3.1 gives the frame bounds for this function, as computed from (3.3.19), (3.3.20), with $a_0 = 2$, for different values of b_0 and for a number of voices varying from 1 to 4. As soon as we take 2 or more voices, the frame may be considered tight for all $b_0 \leq .75$. Note that $b_0 = .75$ and $(a_0)_{\text{effective}} = 2^{1/2} \simeq 1.41$ (intuitively corresponding to two voices per octave) are not small values for the Mexican hat function: the distance between the maximum of ψ and its zeros is only 1, and the width of the positive frequency bump of $\hat{\psi}$ (as measured by $[\int_0^\infty d\xi\ (\xi - \xi_{\text{av}})^2\ |\hat{\psi}(\xi)||^2]^{1/2}$, with $\xi_{\text{av}} = \int_0^\infty d\xi\ \xi\ |\hat{\psi}(\xi)|^2)$, is $\sqrt{3/2} \simeq 1.23$. For fixed N, and b_0 small enough, so that the frame is almost tight, the table also shows that $A \simeq B$ is inversely proportional to b_0, which fits the intuition that for tight frames of normalized vectors, $A = B$ measures the "redundancy" of the frame (see §3.2), which should indeed double if b_0 is halved. On the other hand, the numbers in the table also show that B/A increases dramatically if b_0 is chosen "too large". For every N, the last value of b_0 shown is the last value (with increments of .25) for which our estimate (3.3.19) for A is positive; from the next b_0 on, the $\psi_{m,n}$ are probably not a frame any more. This very abrupt transition, from a reasonable frame, to a very loose frame and then no more frame, as b_0 increases, was first observed by J. Morlet (1985, personal communication), and was one of the motivations for a more detailed mathematical analysis.

C. A modulated Gaussian. This is the function most often used by R. Kronland-Martinet and J. Morlet. Its Fourier transform is a shifted Gaussian, adjusted slightly so that $\hat{\psi}(0) = 0$,

$$
\begin{aligned}
\hat{\psi}(\xi) &= \pi^{-1/4} \left[e^{-(\xi-\xi_0)^2/2} - e^{-\xi^2/2}\, e^{-\xi_0^2/2} \right], \\
\psi(x) &= \pi^{-1/4} \left(e^{-i\xi_0 x} - e^{-\xi_0^2/2} \right) e^{-x^2/2}.
\end{aligned}
\tag{3.3.26}
$$

Often ξ_0 is chosen so that the ratio of the highest and the second highest maximum of ψ is approximately $\frac{1}{2}$, i.e. $\xi_0 = \pi[2/\ln 2]^{1/2} \simeq 5.3364\ldots$; in practice one often takes $\xi_0 = 5$. For this value of ξ_0, the second term in (3.3.26) is so small that it can be neglected in practice. This Morlet-wavelet is complex, even though most applications in which it is used involve only real signals f. Often (see, e.g., Kronland-Martinet, Morlet, and Grossmann (1989)) the wavelet transform of a real signal with this complex wavelet is plotted in modulus-phase-form, i.e., rather than Re $\langle f,\ \psi_{m,n}\rangle$, Im $\langle f,\ \psi_{m,n}\rangle$, one plots $|\langle f,\ \psi_{m,n}\rangle|$ and $\tan^{-1}$ [Im $\langle f,\ \psi_{m,n}\rangle$/Re $\langle f,\ \psi_{m,n}\rangle$]; the phase plot is particularly suited to the detection of singularities (Grossmann et al. (1987)). For real f, one can exploit $\hat{f}(-\xi) = \overline{\hat{f}(\xi)}$ to derive the following frame bounds (this is analogous to what was done in §2.4 for real f):

$$A\ \|f\|^2 \leq \sum_{m,n\in\mathbb{Z}} |\langle f,\ \psi_{m,n}\rangle|^2 \leq B\ \|f\|^2 \quad \text{for } f \text{ real}$$

TABLE 3.1

Frame bounds for wavelet frames based on the Mexican hat function $\psi(x) = 2/\sqrt{3}\,\pi^{-1/4}(1-x^2)e^{-x^2/2}$. *The dilation parameter* $a_0 = 2$ *in all cases;* N *is the number of voices.*

$N = 1$

b_0	A	B	B/A
.25	13.091	14.183	1.083
.50	6.546	7.092	1.083
.75	4.364	4.728	1.083
1.00	3.223	3.596	1.116
1.25	2.001	3.454	1.726
1.50	.325	4.221	12.986

$N = 2$

b_0	A	B	B/A
.25	27.273	27.278	1.0002
.50	13.673	13.639	1.0002
.75	9.091	9.093	1.0002
1.00	6.768	6.870	1.015
1.25	4.834	6.077	1.257
1.50	2.609	6.483	2.485
1.75	.517	7.276	14.061

$N = 3$

b_0	A	B	B/A
.25	40.914	40.914	1.0000
.50	20.457	20.457	1.0000
.75	13.638	13.638	1.0000
1.00	10.178	10.279	1.010
1.25	7.530	8.835	1.173
1.50	4.629	9.009	1.947
1.75	1.747	9.942	5.691

$N = 4$

b_0	A	B	B/A
.25	54.552	54.552	1.0000
.50	27.276	27.276	1.0000
.75	18.184	18.184	1.0000
1.00	13.586	13.690	1.007
1.25	10.205	11.616	1.138
1.50	6.594	11.590	1.758
1.75	2.928	12.659	4.324

with

$$A = \frac{2\pi}{b_0}\left\{\frac{1}{2}\inf_{\xi}\left[\sum_{m\in\mathbb{Z}} |\hat{\psi}(a_0^m\xi)|^2 + |\hat{\psi}(a_0^{-m}\xi)|^2\right] - R\right\},$$

$$B = \frac{2\pi}{b_0}\left\{\frac{1}{2}\sup_{\xi}\left[\sum_{m\in\mathbb{Z}} |\hat{\psi}(a_0^m\xi)|^2 + |\hat{\psi}(a_0^{-m}\xi)|^2\right] + R\right\},$$

where

$$R = \sum_{\epsilon=+,-}\sum_{k\neq 0}\left[\beta_\epsilon\left(\frac{2\pi}{b_0}k\right)\beta_\epsilon\left(-\frac{2\pi}{b_0}k\right)\right]^{1/2}$$

and

$$\beta_\epsilon(s) = \frac{1}{4}\sup_{\xi}\sum_{m\in\mathbb{Z}} |\hat{\psi}(a_0^m\xi) + \epsilon\hat{\psi}(-a_0^m\xi)|\,|\hat{\psi}(a_0^m\xi + s) + \epsilon\hat{\psi}(-a_0^m\xi - s)|\,.$$

These can, of course, again be generalized to the multivoice case. Table 3.2 gives the frame bounds for $a_0 = 2$, several choices of b_0, and number of voices ranging from 2 to 4. In practice, the number of voices is often even higher.

TABLE 3.2

Frame bounds for wavelet frames based on the modulated Gaussian, $\psi(x) = \pi^{-1/4}\,(e^{-i\xi_0 x} - e^{-\xi_0^2/2})\,e^{-x^2/2}$, *with* $\xi_0 = \pi(2/\ln 2)^{1/2}$. *The dilation constant* $a_0 = 2$ *in all cases;* N *is the number of voices.*

$N = 2$

b_0	A	B	B/A
.5	6.019	7.820	1.299
1.	3.009	3.910	1.230
1.5	1.944	2.669	1.373
2.	1.173	2.287	1.950
2.5	.486	2.282	4.693

$N = 3$

b_0	A	B	B/A
.5	10.295	10.467	1.017
1.	5.147	5.234	1.017
1.5	3.366	3.555	1.056
2.	2.188	3.002	1.372
2.5	1.175	2.977	2.534
3.	.320	3.141	9.824

$N = 4$

b_0	A	B	B/A
.5	13.837	13.846	1.0006
1.	6.918	6.923	1.0008
1.5	4.540	4.688	1.032
2.	3.013	3.910	1.297
2.5	1.708	3.829	2.242
3.	.597	4.017	6.732

D. An example that is easy to implement. So far we have not addressed the question of how the wavelet coefficients $\langle f,\ \psi_{m,n}\rangle$ are computed in practice. In real life, f is not given as a function, but in a sampled version. Computing the integrals $\int dx\ f(x)\ \overline{\psi_{m,n}(x)}$ then requires some quadrature formulas. For the smallest scales (most negative m) of interest, this will not involve many samples of f, and one can do the computation quickly. For larger scales, however, one faces huge integrals, which might considerably slow down the computation of the wavelet transform of any given function. Especially for on-line implementations, one should avoid having to compute these long integrals. A construction achieving this is the so-called "algorithme à trous" (Holschneider et al. (1989)), which uses an interpolation technique to avoid lengthy computations (for details, I refer to their paper). Here I propose an analogous example (although it is not "à trous"), by borrowing a leaf from multiresolution analysis and orthonormal bases (to which we will come back), i.e., by introducing an auxiliary function ϕ. The basic idea is the following: suppose there exists a function ϕ so that

$$\cdot\quad \psi(x) = \sum_k d_k\ \phi(x-k)\ , \tag{3.3.27}$$

$$\cdot\quad \phi(x) = \sum_k c_k\ \phi(2x-k)\ , \tag{3.3.28}$$

where in each case only finitely many coefficients are different from zero.[11] (Such pairs of ϕ, ψ abound; an example is given below. The "algorithme à trous" corresponds to special ϕ for which $c_0 = 1$, all other even-indexed $c_{2n} = 0$.) Here ϕ does not have integral zero (but ψ does!), and we will normalize ϕ so that $\int dx\ \phi(x) = 1$. Define, even though ϕ is not a wavelet, $\phi_{m,n}(x) = 2^{-m/2}\ \phi(2^{-m}x - n)$; we take $a_0 = 2$, $b_0 = 1$. Then it is clear that

$$\langle f,\ \psi_{m,n}\rangle = \sum_k d_k\ \langle f,\ \phi_{m,n+k}\rangle\ ;$$

the problem of finding the wavelet coefficients is reduced to computing the $\langle f,\ \phi_{m,n}\rangle$ (finite combinations of which will give the $\langle f,\ \psi_{m,n}\rangle$). On the other hand,

$$\langle f,\ \phi_{m,n}\rangle = \frac{1}{\sqrt{2}} \sum_k c_k\ \langle f,\ \phi_{m-1,2n+k}\rangle\ ,$$

so that the $\langle f,\ \phi_{m,n}\rangle$ can be computed recursively, working from the smallest scale (where they are easy to compute) to the largest scale. Everything is done by simple, finite convolutions.

An example of a pair of functions satisfying (3.3.27), (3.3.28) is

$$\psi(x) = N\left[-\frac{1}{2}\ \phi(x+1) + \phi(x) - \frac{1}{2}\ \phi(x-1)\right]\ ,$$

$$\hat{\phi}(\xi) = \frac{1}{\sqrt{2\pi}}\ e^{-2i\xi}\ \left(\frac{e^{i\xi}-1}{i\xi}\right)^4 = \frac{1}{\sqrt{2\pi}}\ \left(\frac{\sin\ \xi/2}{\xi/2}\right)^4\ ,$$

which corresponds to

$$\phi(x) = \begin{cases} \frac{1}{6}\ (x+2)^3\ , & -2 \le x \le -1\ , \\ \frac{2}{3} - x^2(1+x/2)\ , & -1 \le x \le 0\ , \\ \frac{2}{3} - x^2(1-x/2)\ , & 0 \le x \le 1\ , \\ -\frac{1}{6}(x-2)^3\ , & 1 \le x \le 2\ , \\ 0 & \text{otherwise}\ . \end{cases}$$

N is a normalization constant chosen so that $\|\psi\| = 1$; one finds $N = 6\sqrt{\frac{70}{1313}}$. Figure 3.5a shows graphs of ϕ, ψ; they are not unlike a Gaussian and its second derivative, plotted in Figure 3.5b for comparison. The function ψ clearly satisfies (3.3.27) with $d_0 = N$, $d_{\pm 1} = -N/2$, all other $d_k = 0$, whereas

$$\begin{aligned} \hat{\phi}(2\xi) &= \frac{1}{\sqrt{2\pi}}\ \left(\frac{\sin\ \xi}{\xi}\right)^4 = \frac{1}{\sqrt{2\pi}}\ \left(\frac{2\ \sin\ \xi/2\ \cos\ \xi/2}{\xi}\right)^4 \\ &= (\cos\ \xi/2)^4\ \hat{\phi}(\xi)\ , \end{aligned}$$

which implies

$$\phi(x) = \tfrac{1}{8}\phi(2x+2) + \tfrac{1}{2}\phi(2x+1) + \tfrac{3}{4}\phi(2x) + \tfrac{1}{2}\phi(2x-1) + \tfrac{1}{8}\phi(2x-2) \ ,$$

or $c_0 = \frac{3}{4}$, $c_{\pm 1} = \frac{1}{2}$, $c_{\pm 2} = \frac{1}{8}$, all other $c_k = 0$. For this ψ, $a_0 = 2$ and $b_0 = 1$, the frame bounds are $A = .73178$, $B = 1.77107$, corresponding to $B/A = 2.42022$; for $a_0 = 2$, $b_0 = .5$ we have $A = 2.33854$, $B = 2.66717$, and $B/A = 1.14053$ (using $b_0 = .5$ means that the recursion formulas, linking the $\psi_{m,n}$ with the $\phi_{m,n}$ and the $\phi_{m,n}$ with the $\phi_{m-1,n}$, have to be adapted, but this is easy). Here we have used only one voice. It is of course possible to choose several different ψ^ν, corresponding to different d_k^ν, that give rise to a multivoice scheme, closer to tight frames.

This concludes our example section; other examples are given in Daubechies (1990) (including one for which the estimates (3.3.21), (3.3.22) outperform (3.3.11), (3.3.12)). Many other examples can of course be constructed. The wavelets used in Mallat and Zhong (1990) are another example of the same type as our last one; in their case ψ is chosen to be the first derivative of a function with non-zero integral (so that $\int dx\ \psi(x) = 0$ but $\int dx\, x\psi(x) \neq 0$).

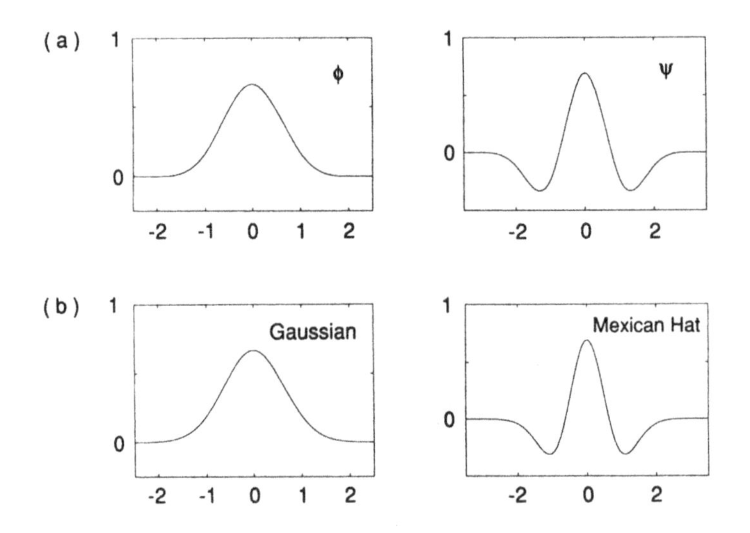

FIG. 3.5. *An example that is easy to implement: graphs of ϕ, ψ (in* a*), and a comparison with a Gaussian and its second derivative (in* b*).*

3.4. Frames for the windowed Fourier transform.

The windowed Fourier transform of Chapter 2 can also be discretized. The natural discretization for ω, t in $g^{\omega,t}(x) = e^{i\omega x}\ g(x-t)$ is $\omega = m\omega_0$, $t = nt_0$, where ω_0, $t_0 > 0$ are fixed, and m, n range over $\mathbb{Z}$; the discretely labelled family is thus

$$g_{m,n}(x) = e^{im\omega_0 x}\ g(x - nt_0) \ .$$

We can again seek answers to the same questions as in the wavelet case: for which choices of g, ω_0, t_0 can a function be characterized by the inner products $\langle f, g_{m,n}\rangle$; when is it possible to reconstruct f in a numerically stable way from these inner products; can an efficient algorithm be given to write f as a linear combination of the $g_{m,n}$? The answers are again provided by the same abstract framework: stable numerical reconstruction of f from its windowed Fourier coefficients

$$\langle f, g_{m,n}\rangle = \int dx \; f(x) \; e^{-im\omega_0 x} \; \overline{g(x - nt_0)}$$

is only possible if the $g_{m,n}$ constitute a frame, i.e., if there exist $A > 0$, $B < \infty$ so that

$$A \int dx \; |f(x)|^2 \le \sum_{m,n\in\mathbb{Z}} |\langle f, g_{m,n}\rangle|^2 \le B \int dx \; |f(x)|^2 \; .$$

If the $g_{m,n}$ constitute a frame, then any function $f \in L^2(\mathbb{R})$ can be written as

$$f = \sum_{m,n} \langle f, g_{m,n}\rangle \; \widetilde{g_{m,n}} = \sum_{m,n} \langle f, \widetilde{g_{m,n}}\rangle \; g_{m,n} \; , \tag{3.4.1}$$

where $\widetilde{g_{m,n}}$ are the vectors in the dual frame; (3.4.1) shows both how to recover f from the $\langle f, g_{m,n}\rangle$ and how to write f as a superposition of the $g_{m,n}$. The detailed analysis of frames of windowed Fourier functions brings out some features different from wavelet frames, due to the differences in their constructions.

3.4.1. A necessary condition: Sufficiently high time-frequency density.

The arguments in the proof of Theorem 3.3.1 can be used for the windowed Fourier case as well (with the obvious modifications), leading to the conclusion

$$A \le \frac{2\pi}{\omega_0 t_0} \; \|g\|^2 \le B \tag{3.4.2}$$

for any frame of windowed Fourier functions, with frame bounds A, B. This does not impose any additional restrictions on g (we always assume $g \in L^2(\mathbb{R})$). A consequence of (3.4.2) is that the frame bound for any tight frame equals $2\pi \; (\omega_0 t_0)^{-1}$ (if we choose g to have norm 1); in particular, if the $g_{m,n}$ constitute an orthonormal basis, then $\omega_0 t_0 = 2\pi$.

The absence of any constraint on g in the inequalities (3.4.2) is similar to the absence of an admissibility condition for the continuous windowed Fourier transform (see Chapter 2), and quite unlike the constraint $\int d\xi \; |\xi|^{-1} \; |\hat{\psi}(\xi)|^2 < \infty$ on the mother wavelet, necessary for wavelet frames as well as for the continuous wavelet transform. Another difference with the wavelet case is that the time and frequency translation steps t_0 and ω_0 are constrained: there exist no windowed Fourier frames for pairs ω_0, t_0 such that $\omega_0 t_0 > 2\pi$. Even more is true: if $\omega_0 t_0 > 2\pi$, then for any choice of $g \in L^2(\mathbb{R})$ there exists a corresponding $f \in L^2(\mathbb{R})$ ($f \ne 0$) so that f is orthogonal to all the $g_{m,n}(x) = e^{im\omega_0 x} \; g(x - nt_0)$. In this case the $g_{m,n}$ not only fail to constitute a frame, but the inner products

$\langle f,\ g_{m,n}\rangle$ are not even sufficient to determine f. We are therefore restricted to $\omega_0 t_0 \le 2\pi$; in order to have good time- and frequency-localization, we even have to choose $\omega_0 t_0 < 2\pi$. Note that no similar restriction on a_0, b_0 existed in the wavelet case! We will come back on these conditions in Chapter 4, where we discuss in much greater detail the role of time-frequency density for windowed Fourier versus wavelet frames; we postpone our proof of the necessity of $\omega_0 t_0 < 2\pi$ until then.

3.4.2. A sufficient condition and estimates for the frame bounds. Even if $\omega_0 t_0 \le 2\pi$, the $g_{m,n}$ do not necessarily constitute a frame. An easy counterexample is $g(x) = 1$ for $0 \le x \le 1$, 0 otherwise; if $t_0 > 1$, then any function f supported in $[1,\ t_0]$ will be orthogonal to all the $g_{m,n}$, however small ω_0 is chosen. In this example $\operatorname{ess\,inf}_x \sum_n\ |g(x - nt_0)|^2 = 0$, and this is what stops the $g_{m,n}$ from being a frame. (Something similar occurs for wavelet frames; see §3.3.)

Computations entirely similar to those in the wavelet case show that

$$\inf_{\substack{f\in\mathcal{H}\\ f\neq 0}} \|f\|^{-2} \sum_{m,n} |\langle f, g_{m,n}\rangle|^2$$
$$\ge \frac{2\pi}{\omega_0}\left\{\inf_x \sum_n |g(x-nt_0)|^2 - \sum_{k\neq 0}\left[\beta\left(\frac{2\pi}{\omega_0}k\right)\beta\left(-\frac{2\pi}{\omega_0}k\right)\right]^{1/2}\right\}, \tag{3.4.3}$$

$$\sup_{\substack{f\in\mathcal{H}\\ f\neq 0}} \|f\|^{-2} \sum_{m,n} |\langle f, g_{m,n}\rangle|^2$$
$$\le \frac{2\pi}{\omega_0}\left\{\sup_x \sum_n |g(x-nt_0)|^2 + \sum_{k\neq 0}\left[\beta\left(\frac{2\pi}{\omega_0}k\right)\beta\left(-\frac{2\pi}{\omega_0}k\right)\right]^{1/2}\right\}, \tag{3.4.4}$$

where β is now defined by

$$\beta(s) = \sup_x \sum_n\ |g(x-nt_0)|\ |g(x-nt_0+s)|\ .$$

As in the wavelet case, sufficiently fast decay on g leads to decay for β, so that by choosing ω_0 small enough, the second terms in the right-hand sides of (3.4.3), (3.4.4) can be made arbitrarily small. If $\sum_n\ |g(x-nt_0)|^2$ is bounded, and bounded below by a strictly positive constant (no "conspiring" of the zeros of g), then this implies that the $g_{m,n}$ constitute a frame for sufficiently small ω_0, with frame bounds given by (3.4.3), (3.4.4). Explicitly, we have the following proposition.

PROPOSITION 3.4.1. *If* g, t_0 *are such that*

$$\inf_{0\le x\le t_0} \sum_{n=-\infty}^{\infty} |g(x-nt_0)|^2 > 0\ ,$$

$$\sup_{0\le x\le t_0} \sum_{n=-\infty}^{\infty} |g(x-nt_0)|^2 < \infty\ , \tag{3.4.5}$$

and if $\beta(s) = \sup_{0\le x\le t_0} \sum_n |g(x-nt_0)|\,|g(x-nt_0+s)|$ *decays at least as fast as* $(1+|s|)^{-(1+\epsilon)}$, *with* $\epsilon > 0$, *then there exists* $(\omega_0)_{\text{thr}} > 0$ *so that the* $g_{m,n}(x) = e^{im\omega_0 x}\, g(x-nt_0)$ *constitute a frame whenever* $\omega_0 < (\omega_0)_{\text{thr}}$. *For* $\omega_0 < (\omega_0)_{\text{thr}}$, *the right-hand sides of* (3.4.3), (3.4.4) *are frame bounds for the* $g_{m,n}$.

The conditions on β and (3.4.5) are met if, e.g., $|g(x)| \le C(1+|x|)^{-\gamma}$ with $\gamma > 1$.

REMARK. The windowed Fourier case exhibits a symmetry under the Fourier transform absent in the wavelet case. We have

$$(g_{m,n})^\wedge(\xi) = e^{-int_0\xi}\, \hat{g}(\xi - m\omega_0)\ ,$$

which implies that (3.4.3), (3.4.4) still hold if we replace g, ω_0, t_0 by $\hat{g}$, t_0, ω_0, respectively, everywhere in the right-hand sides (including in the definition of β). Using this remark, we can therefore compute two estimates each for A and B, and pick the highest one for A, the lowest for B. □

3.4.3. The dual frame.

The dual frame is again defined by

$$\widetilde{g_{m,n}} = (F^*F)^{-1}\, g_{m,n}\ ,$$

where F^*F is now $(F^*F)f = \sum_{mn} \langle f, g_{m,n}\rangle\, g_{m,n}$. In this case one easily checks that F^*F commutes with translations by t_0 as well as with multiplications by $e^{i\omega_0 x}$, i.e., if $(Tf)(x) = f(x-t_0)$, $(Ef)(x) = e^{i\omega_0 x} f(x)$, then

$$F^*F\,T = T\,F^*F, \qquad F^*F\,E = E\,F^*F\ .$$

It follows that $(F^*F)^{-1}$ also commutes with E and T, so that

$$\begin{aligned}\widetilde{g_{m,n}} &= (F^*F)^{-1}\, E^m\, T^n\, g \\ &= E^m\, T^n\, (F^*F)^{-1}\, g\ ,\end{aligned}$$

or

$$\widetilde{g_{m,n}}(x) = e^{im\omega_0 x}\, \tilde{g}(x-nt_0) = \tilde{g}_{m,n}(x)\ ,$$

where $\tilde{g} = (F^*F)^{-1}g$. Unlike the generic wavelet case, the dual frame is *always* generated by a single function $\tilde{g}$. This means that it is not as important in the windowed Fourier case that the frame be close to a tight frame: if $B/A - 1$ is nonnegligible, then one simply computes $\tilde{g}$ to high precision, once and for all, and works with the two dual frames.

3.4.4. Examples.

A. Tight frames with compact support in time or frequency. The following construction, again from Daubechies, Grossmann, and Meyer (1986), and very similar to §3.3.5.A, leads to tight windowed Fourier frames with arbitrarily high regularity if $\omega_0 t_0 < 2\pi$. If support $g \subset \left[-\frac{\pi}{\omega_0}, \frac{\pi}{\omega_0}\right]$, then

$$\begin{aligned}\sum_{m,n} |\langle f, g_{m,n}\rangle|^2 &= \sum_{m,n} \left| \int_0^{2\pi/\omega_0} dx\, e^{im\omega_0 x} \sum_{\ell\in\mathbb{Z}} f\left(x + \ell\frac{2\pi}{\omega_0}\right) \overline{g\left(x + \ell\frac{2\pi}{\omega_0} - nt_0\right)}\right|^2 \\ &= \frac{2\pi}{\omega_0} \sum_n \int_0^{2\pi/\omega_0} dx \left| \sum_{\ell\in\mathbb{Z}} f\left(x + \ell\frac{2\pi}{\omega_0}\right) g\left(x + \ell\frac{2\pi}{\omega_0} - nt_0\right)\right|^2 \\ &= \frac{2\pi}{\omega_0} \sum_{n,\ell} \int_0^{2\pi/\omega_0} dx \left| f\left(x + \ell\frac{2\pi}{\omega_0}\right)\right|^2 \left| g\left(x + \ell\frac{2\pi}{\omega_0} - nt_0\right)\right|^2 ,\end{aligned}$$

where we have used that for any n, at most one value of ℓ can contribute, because of the support property of g.

Consequently,

$$\sum_{m,n} |\langle f,\, g_{m,n}\rangle|^2 = \frac{2\pi}{\omega_0} \int dx\, |f(x)|^2 \sum_n |g(x - nt_0)|^2 ,$$

and the frame is tight if and only if $\sum_n |g(x-nt_0)|^2 =$ constant. For instance, if $\omega_0 t_0 \geq \pi$, then we can start again from a C^k or C^∞-function ν satisfying (3.3.25) and define

$$g(x) = t_0^{-1/2} \begin{cases} \sin\left[\frac{\pi}{2}\,\nu\left(\frac{\pi+\omega_0 x}{2\pi-\omega_0 t_0}\right)\right] , & -\frac{\pi}{\omega_0} \leq x \leq \frac{\pi}{\omega_0} - t_0 , \\ 1 , & \frac{\pi}{\omega_0} - t_0 \leq x \leq -\frac{\pi}{\omega_0} + t_0 , \\ \cos\left[\frac{\pi}{2}\,\nu\left(\frac{\pi-\omega_0 x}{2\pi-\omega_0 t_0}\right)\right] , & -\frac{\pi}{\omega_0} + t_0 \leq x \leq \frac{\pi}{\omega_0} , \\ 0 & \text{otherwise} . \end{cases}$$

Then g is a C^k or C^∞ function (depending on the choice of ν) with compact support, $\|g\| = 1$, and the $g_{m,n}$ constitute a tight frame with frame bound $2\pi(\omega_0 t_0)^{-1}$ (as already followed from (3.4.2)). If $\omega_0 t_0 < \pi$, then this construction can easily be adapted. This construction gives a tight frame with compactly supported g. By taking its Fourier transform, we obtain a frame for which the window function has compactly supported Fourier transform.[12]

B. The Gaussian. In this case $g(x) = \pi^{-1/4}\, e^{-x^2/2}$. Discrete families of windowed Fourier functions starting from a Gaussian window have been discussed extensively in the literature for many reasons. Gabor (1946) proposed their use for communication purposes (he proposed $\omega_0 t_0 = 2\pi$, however, which is inappropriate: see below); because of the importance of the "canonical coherent states" in quantum mechanics (see Klauder and Skagerstam (1985)) they are of interest to physicists; the link between Gaussian coherent states and the Bargmann space of entire function makes it possible to rewrite results concerning the $g_{m,n}$ in terms of sampling properties for the Bargmann space. Exploiting this link with entire functions, it was proved in Bargmann et al. (1971) and independently in Perelomov (1971) that the $g_{m,n}$ span all of $L^2(\mathbb{R})$ if and only if $\omega_0 t_0 \leq 2\pi$; in Bacry, Grossmann, and Zak (1975) a different technique was used to show that if $\omega_0 t_0 = 2\pi$, then

$$\inf_{\substack{f\in\mathcal{H} \\ f\neq 0}} \|f\|^{-2} \sum_{m,n} |\langle f, g_{m,n}\rangle|^2 = 0 ,$$

even though the $g_{m,n}$ are "complete," in the sense that they span $L^2(\mathbb{R})$.[13] (We will see in Chapter 4 that this is a direct consequence of $\omega_0 \cdot t_0 = 2\pi$, and of the regularity of both g and $\hat{g}$.) This is therefore an example of a family of $g_{m,n}$ where the inner products $\langle f, g_{m,n}\rangle$ suffice to characterize the function of f (if $\langle f_1, g_{m,n}\rangle = \langle f_2, g_{m,n}\rangle$ for all m, n, then $f_1 = f_2$), but where there is no numerically stable reconstruction formula for f from the $\langle f, g_{m,n}\rangle$. Bastiaans (1980, 1981) has constructed a dual function $\tilde{g}$ such that

$$f = \sum_{m,n} \langle f, g_{m,n}\rangle \, \tilde{g}_{m,n} , \tag{3.4.6}$$

with $\tilde{g}_{m,n}(x) = e^{im\omega_0 x} \, \tilde{g}(x - nt_0)$, but convergence of (3.4.6) holds only in a very weak sense (in the sense of distributions—see Janssen (1981, 1984)), and *not* even in the weak L^2-sense; in fact, $\tilde{g}$ itself is not in $L^2(\mathbb{R})$.

The case $\omega_0 t_0 = 2\pi$ is thus completely understood; what happens if $\omega_0 t_0 < 2\pi$? Table 3.3 shows the values of the frame bounds A, B and of the ratio B/A, for various values of $\omega_0 \cdot t_0$, computed from (3.4.3), (3.4.4) and the analogous formulas using $\hat{g}$. We find that the $g_{m,n}$ do constitute a frame, even for $\omega_0 \cdot t_0/(2\pi) = .95$, although B/A becomes very large so close to the "critical" density. It turns out that when $\omega_0 \cdot t_0/(2\pi) = 1/N$, $N \in \mathbb{N}$, $N > 1$, the frame bounds can also be computed via another technique, which leads to exact values (within the error of computation) instead of lower, respectively, upper, bounds for A, B.[14] For the choices $\omega_0 \cdot t_0/(2\pi) = \frac{1}{4}$ and $\frac{1}{2}$, Table 3.3 reveals these exact values as well; it is surprising to see how close our bounds on A, B (which are, after all, obtained via a Cauchy–Schwarz inequality, and might therefore be quite coarse) are to the exact values. Substituting these values for A, B into the approximation scheme at the end of §3.2, we can compute $\tilde{g}$ for these different choices of ω_0, t_0. Figure 3.6 shows plots of $\tilde{g}$ for the special case where $\omega_0 = t_0 = (\lambda 2\pi)^{1/2}$, with λ taking the values .25, .375, .5, .75, .95, and 1. Note that Bastiaans' function $\tilde{g}$, which corresponds to $\lambda = 1$ (lower right plot in Figure 3.6), has to be computed differently, since $A = 0$ for $\lambda = 1$. For small λ, the frame is very close to tight, and $\tilde{g}$ is close to g itself, as is illustrated by the near-Gaussian profile of $\tilde{g}$ for $\lambda = .25$. As λ increases, the frame becomes both less redundant (as reflected by the growing maximum amplitude of $\tilde{g}$) and less tight, causing $\tilde{g}$ to deviate more and more from a Gaussian. Because both g and $\hat{g}$ have (faster than) exponential decay, one can easily prove from the converging series representation for $\tilde{g}$ (see §3.2) that $\tilde{g}$ and $\hat{\tilde{g}}$ have exponential decay as well, if $A > 0$. It follows that the $\tilde{g}_{m,n}$ have good time-frequency localization properties, for all the values of $\lambda < 1$ in Figure 3.6, even though it is quite striking how $\tilde{g}$ tends to Bastiaans'

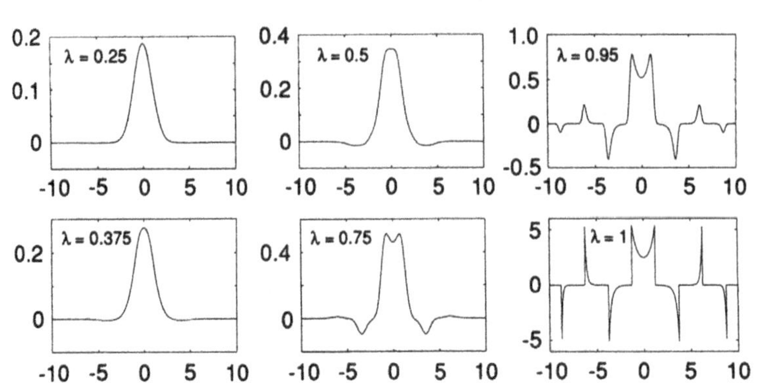

FIG. 3.6. *The dual frame function $\tilde{g}$ for Gaussian g and $\omega_0 = t_0 = (2\pi\lambda)^{1/2}$, with $\lambda = .25, .375, .5, .75, .95$, and 1. As λ increases, $\tilde{g}$ deviates more and more from a Gaussian (reflecting the increase of B/A), and its amplitude increases as well (because $A+B$ decreases). For $\lambda = 1$, $\tilde{g}$ is no longer square integrable.*

pathological $\tilde{g}$ as λ increases. For $\lambda = 1$, all time-frequency localization breaks down.[15] The series of plots in Figure 3.6 suggests the conjecture, first formulated in Daubechies and Grossmann (1988), that, at least for Gaussian g, the $g_{m,n}$ are a frame whenever $\omega_0 t_0 < 2\pi$. In Daubechies (1990) it was shown that this is indeed the case for $\omega_0 t_0/(2\pi) < .996$. Using entire function methods, this conjecture has since been proved, by Lyubarskii (1989) and independently by Seip and Wallstén (1990).

There exist of course many other possible and popular choices for the window function g, but we will stop our list of examples here, and return to wavelets.

3.5. Time-frequency localization.

One of our main motivations for studying wavelet transforms (or windowed Fourier transforms) is that they provide a time-frequency picture, with, hopefully, good localization properties in both variables. We have asserted several times that if ψ itself is well localized in time and in frequency, then the frame generated by ψ will share that property. In this section we want to make this vague statement more precise.

For the sake of convenience, we assume $|\psi|$ and $|\hat{\psi}|$ to be symmetric (true if, e.g., ψ is real and symmetric—a good example is the Mexican hat function)[16]; then ψ is centered around 0 in time and near $\pm\xi_0$ in frequency (with, e.g., $\xi_0 = \int_0^\infty d\xi\, \xi|\hat{\psi}(\xi)|^2/[\int_0^\infty d\xi\, |\hat{\psi}(\xi)|^2]$). If ψ is well localized in time and frequency, then $\psi_{m,n}$ will similarly be well localized around $a_0^m n b_0$ in time and around $\pm a_0^{-m}\xi_0$ in frequency. Intuitively speaking, $\langle f,\ \psi_{m,n}\rangle$ then represents the "information content" in f near time $a_0^m n b_0$ and near the frequencies $\pm a_0^{-m}\xi_0$. If f itself is "essentially localized" on two rectangles in time-frequency space, meaning that, for some $0 < \Omega_0 < \Omega_1 < \infty$, $0 < T < \infty$,

TABLE 3.3

Values for the frame bounds A, B and their ratio B/A for the case $g(x) = \pi^{-1/4} \exp(-x^2/2)$, for different values of ω_0, t_0. For $\omega_0 t_0 = \pi/2$ and π, the exact values can be computed via the Zak transform (see Daubechies and Grossmann (1988)).

$\omega_0 t_0 = \pi/2$

t_0	A_a	A_{exact}	B	B_{exact}	B/A
0.5	1.203	1.221	7.091	7.091	5.896
1.0	3.853	3.854	4.147	4.147	1.076
1.5	3.899	3.899	4.101	4.101	1.052
2.0	3.322	3.322	4.679	4.679	1.408
2.5	2.365	2.365	5.664	5.664	2.395
3.0	1.427	1.427	6.772	6.772	4.745

$\omega_0 t_0 = 3\pi/4$

t_0	A	B	B/A
1.0	1.769	3.573	2.019
1.5	2.500	2.833	1.133
2.0	2.210	3.124	1.414
2.5	1.577	3.776	2.395
3.0	0.951	4.515	4.745

$\omega_0 t_0 = \pi$

t_0	A	A_{exact}	B	B_{exact}	B/A
1.0	0.601	0.601	3.546	3.546	5.901
1.5	1.519	1.540	2.482	2.482	1.635
2.0	1.575	1.600	2.425	2.425	1.539
2.5	1.172	1.178	2.843	2.843	2.426
3.0	0.713	0.713	3.387	3.387	4.752

$\omega_0 t_0 = 3\pi/4$

t_0	A	B	B/A
1.0	0.027	3.545	130.583
1.5	0.342	2.422	7.082
2.0	0.582	2.089	3.592
2.5	0.554	2.123	3.834
3.0	0.393	2.340	5.953
3.5	0.224	2.656	11.880
4.0	0.105	3.014	28.618

$\omega_0 t_0 = 1.9\pi$

t_0	A	B	B/A
1.5	0.031	2.921	92.935
2.0	0.082	2.074	25.412
2.5	0.092	2.021	22.004
3.0	0.081	2.077	25.668
3.5	0.055	2.218	40.455
4.0	0.031	2.432	79.558

$$\int_{\Omega_0 \le |\xi| \le \Omega_1} d\xi \; |\hat{f}(\xi)|^2 \ge (1-\delta) \; \|f\|^2 \; , \tag{3.5.1}$$

$$\int_{|x| \le T} dx \; |f(x)|^2 \ge (1-\delta) \; \|f\|^2 \; , \tag{3.5.2}$$

where δ is some small number, then this intuitive picture suggests that only those $\langle f, \psi_{m,n} \rangle$ corresponding to m, n for which $(a_0^m n b_0, \; \pm a_0^{-m} \xi_0)$ lies within or close to $[-T, T] \times ([-\Omega_1, -\Omega_0] \cup [\Omega_0, \Omega_1])$ are needed to reconstruct f to a very good approximation. The following theorem states that this is indeed true, thereby justifying our intuitive picture.

THEOREM 3.5.1. *Suppose that the* $\psi_{m,n}(x) = a_0^{-m/2} \; \psi(a_0^{-m} x - n b_0)$ *constitute a frame with frame bounds* A, B, *and suppose that*

$$|\psi(x)| \le C(1 + x^2)^{-\alpha/2}, \qquad |\hat{\psi}(\xi)| \le C |\xi|^\beta \; (1 + \xi^2)^{-(\beta+\gamma)/2} \; , \tag{3.5.3}$$

with $\alpha > 1$, $\beta > 0$, $\gamma > 1$. *Then, for any* $\epsilon > 0$, *there exists a finite set* $B_\epsilon(\Omega_0, \; \Omega_1; \; T) \subset \mathbb{Z}^2$ *so that, for all* $f \in L^2(\mathbb{R})$,

$$\left\| f - \sum_{(m,n) \in B_\epsilon(\Omega_0, \Omega_1; T)} \langle f, \; \psi_{m,n} \rangle \; \widetilde{\psi_{m,n}} \right\|$$
$$\le \sqrt{\frac{B}{A}} \left[\left(\int_{\substack{|\xi| < \Omega_0 \\ \text{or } |\xi| > \Omega_1}} d\xi |\hat{f}(\xi)|^2 \right)^{1/2} + \left(\int_{|x| > T} dx |f(x)|^2 \right)^{1/2} + \epsilon \|f\| \right] . \tag{3.5.4}$$

REMARKS.

1. If f satisfies (3.5.1) and (3.5.2), then the first two terms in the right-hand side of (3.5.3) are bounded by $2\delta \sqrt{\frac{B}{A}} \; \|f\|$; choosing $\epsilon = \delta$ then leads to $\|f - \sum_{m,n \in B_\delta} \langle f, \; \psi_{m,n} \rangle \; \widetilde{\psi_{m,n}}\| = O(\delta)$.
2. As $\epsilon \to 0$, $\# \; B_\epsilon(\Omega_0, \; \Omega_1; \; T) \to \infty$ (see proof, below): infinite precision is only possible if infinitely many $\langle f, \; \psi_{m,n} \rangle$ are used. □

Figure 3.7 gives a sketch of the set $B_\epsilon(\Omega_0, \; \Omega_1; \; T)$ for one particular value of ϵ; the proof will show how we obtain this shape.

Proof.

1. We define the set B_ϵ as

$$B_\epsilon(\Omega_0, \; \Omega_1; \; T) = \{(m,n) \in \mathbb{Z}^2; \; m_0 \le m \le m_1, \; |n b_0| \le a_0^{-m} T + t\} \; ,$$

where m_0, m_1 and t, to be defined below, depend on Ω_0, Ω_1, T, and ϵ. The points $(a_0^m n b_0, \; \pm a_0^{-m} \xi_0)$ corresponding to (m, n) in such a set, do indeed fill a shape like Figure 3.7.

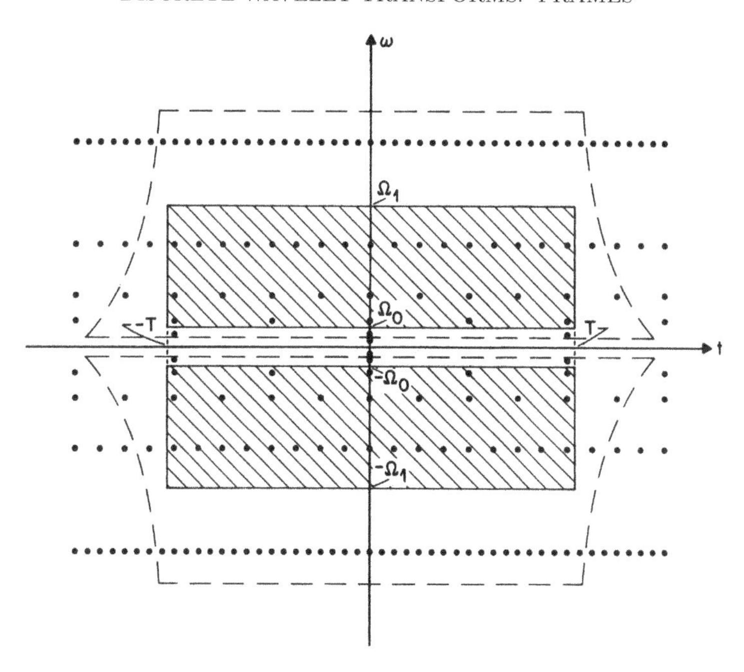

FIG. 3.7. *The set $B_\epsilon(\Omega_0, \Omega_1; T)$ of "wavelet lattice points" needed for an approximate reconstruction of f if f is localized mostly in $[-T, T]$ in time and in $[-\Omega_1, -\Omega_0] \cup [\Omega_0, \Omega_1]$ in frequency.*

$$2. \quad \left\| f - \sum_{m,n \in B_\epsilon} \langle f,\, \psi_{m,n}\rangle\, \widetilde{\psi_{m,n}} \right\|$$

$$= \sup_{\|h\|=1} \left| \langle f, h\rangle - \sum_{(m,n)\in B_\epsilon} \langle f,\, \psi_{m,n}\rangle \langle \widetilde{\psi_{m,n}},\, h\rangle \right|$$

$$= \sup_{\|h\|=1} \left| \sum_{(m,n)\notin B_\epsilon} \langle f,\, \psi_{m,n}\rangle \langle \widetilde{\psi_{m,n}},\, h\rangle \right|$$

$$\le \sup_{\|h\|=1} \sum_{\substack{m<m_0 \\ \text{or } m>m_1}} \sum_{n\in\mathbb{Z}} \left[|\langle P_{\Omega_0,\Omega_1} f,\, \psi_{m,n}\rangle| + |\langle (1-P_{\Omega_0,\Omega_1}) f,\, \psi_{m,n}\rangle| \right] |\langle \widetilde{\psi_{m,n}},\, h\rangle|$$

$$+ \sup_{\|h\|=1} \sum_{m_0\le m\le m_1} \sum_{|nb_0|>a_0^{-m}T+t} \left[|\langle Q_T f,\, \psi_{m,n}\rangle| + |\langle (1-Q_T) f,\, \psi_{m,n}\rangle| \right] |\langle \widetilde{\psi_{m,n}},\, h\rangle| , \tag{3.5.5}$$

where we have introduced $(Q_T f)(x) = f(x)$ for $|x| \le T$, $(Q_T f)(x) = 0$

otherwise, and $(P_{\Omega_0,\Omega_1}f)^\wedge(\xi) = \hat{f}(\xi)$ if $\Omega_0 \leq |\xi| \leq \Omega_1$, $(P_{\Omega_0,\Omega_1}f)^\wedge(\xi) = 0$ otherwise. Since the $\widetilde{\psi_{m,n}}$ constitute a frame with frame bounds B^{-1}, A^{-1}, we have

$$\begin{aligned}
&\sum_{\substack{m<m_0\\ \text{or } m>m_1}} \sum_{n\in\mathbb{Z}} |\langle (1-P_{\Omega_0,\Omega_1})f,\ \psi_{m,n}\rangle|\ |\langle \widetilde{\psi_{m,n}},\ h\rangle| \\
&\quad \leq \left(\sum_{m,n} |\langle (1-P_{\Omega_0,\Omega_1})f,\ \psi_{m,n}\rangle|^2\right)^{1/2} \left(\sum_{m,n} |\langle \tilde{\psi}_{m,n},\ h\rangle|^2\right)^{1/2} \\
&\quad \leq B^{1/2}\ \|(1-P_{\Omega_0,\Omega_1})f\|\ A^{-1/2}\ \|h\| \\
&\quad = \sqrt{\frac{B}{A}} \left[\int_{\substack{|\xi|<\Omega_0\\ \text{or } |\xi|>\Omega_1}} d\xi\ |\hat{f}(\xi)|^2\right]^{1/2} \qquad \text{(because } \|h\| = 1) .
\end{aligned}$$

Similarly,

$$\begin{aligned}
&\sup_{\|h\|=1} \sum_{m_0\leq m\leq m_1} \sum_{|nb_0|>a_0^{-m}T+t} |\langle (1-Q_T)f, \psi_{m,n}\rangle||\langle \widetilde{\psi_{m,n}},\ h\rangle| \\
&\quad \leq \sqrt{\frac{B}{A}} \left[\int_{|x|>T} dx\ |f(x)|^2\right]^{1/2} .
\end{aligned}$$

It remains to check that the other two terms in (3.5.5) can be bounded by $\sqrt{\frac{B}{A}}\epsilon\ \|f\|$.

3. By the same Cauchy–Schwarz trick, we reduce the remaining two terms in (3.5.5) to

$$\begin{aligned}
A^{-1/2} &\left\{ \left[\sum_{\substack{m<m_0\\ \text{or } m>m_1}} \sum_{n\in\mathbb{Z}} |\langle P_{\Omega_0,\Omega_1}f,\ \psi_{m,n}\rangle|^2\right]^{1/2} \right. \\
&\left. + \left[\sum_{m_0\leq m\leq m_1} \sum_{|nb_0|>a_0^{-m}T+t} |\langle Q_Tf,\ \psi_{m,n}\rangle|^2\right]^{1/2} \right\} .
\end{aligned} \tag{3.5.6}$$

It is therefore sufficient to show that for appropriate m_0, m_1, t each of the expressions between square brackets is smaller than $B\epsilon^2\ \|f\|^2/4$.

4. We tackle the first term in (3.5.6) by the same technique as in the proof of

Proposition 3.3.2:

$$\sum_{\substack{m<m_0 \\ \text{or } m>m_1}} \sum_{n\in\mathbb{Z}} |\langle P_{\Omega_0,\Omega_1} f,\ \psi_{m,n}\rangle|^2$$

$$\le \frac{2\pi}{b_0} \sum_{\substack{m<m_0 \\ \text{or } m>m_1}} \sum_{\ell\in\mathbb{Z}} \int\limits_{\substack{\Omega_0\le|\xi|\le\Omega_1 \\ \Omega_0\le|\xi-2\pi\ell a_0^{-m}b_0^{-1}|\le\Omega_1}} d\xi\ |\hat{f}(\xi)|$$

$$\cdot \left|\hat{f}\left(\xi - \frac{2\pi}{b_0}\ell a_0^{-m}\right)\right|\ |\hat{\psi}(a_0^m\xi)| \left|\hat{\psi}\left(a_0^m\xi - \frac{2\pi}{b_0}\ell\right)\right|$$

$$\le \frac{2\pi}{b_0} \sum_{\ell\in\mathbb{Z}} \Bigg[\int\limits_{\substack{\Omega_0\le|\xi|\le\Omega_1 \\ \Omega_0\le|\xi-2\pi\ell a_0^{-m}b_0^{-1}|\le\Omega_1}} d\xi\ |\hat{f}(\xi)|^2$$

$$\cdot \sum_{\substack{m<m_0 \\ \text{or } m>m_1}} |\hat{\psi}(a_0^m\xi)|^{2-\lambda} \left|\hat{\psi}\left(a_0^m\xi - \frac{2\pi}{b_0}\ell\right)\right|^{\lambda}\Bigg]^{1/2}$$

$$\cdot \Bigg[\int\limits_{\substack{\Omega_0\le|\zeta|\le\Omega_1 \\ \Omega_0\le|\zeta+2\pi\ell a_0^{-m}b_0^{-1}|\le\Omega_1}} d\zeta\ |\hat{f}(\zeta)|^2$$

$$\cdot \sum_{\substack{m<m_0 \\ \text{or } m>m_1}} |\hat{\psi}\left(a_0^m\zeta + \frac{2\pi}{b_0}\ell\right)|^{\lambda}\ |\hat{\psi}(a_0^m\zeta)|^{2-\lambda}\Bigg]^{1/2}, \tag{3.5.7}$$

where $0 < \lambda < 1$ will be fixed below. Since $[1 + (u - s)^2]^{-1}$ $(1+s^2)[1+(u+s)^2]^{-1}$ is bounded uniformly in u and s, we have

$$|\hat{\psi}(a_0^m\xi)| \left|\hat{\psi}\left(a_0^m\xi - \frac{2\pi}{b_0}\ell\right)\right|$$

$$\le C[1+(a_0^m\xi)^2]^{-\gamma/2} \left[1+\left(a_0^m\xi - \frac{2\pi}{b_0}\ell\right)^2\right]^{-\gamma/2}$$

$$\le C_1(1+\ell^2)^{-\gamma/2}\ .$$

Substituting this into (3.5.7) we find

$$(3.5.7) \le \frac{2\pi}{b_0} C_2 \|P_{\Omega_0,\Omega_1} f\|^2$$

$$\sum_{\ell\in\mathbb{Z}} (1+\ell^2)^{-\gamma\lambda/2} \sup_{\Omega_0\le|\xi|\le\Omega_1} \sum_{\substack{m<m_0 \\ \text{or } m>m_1}} |\hat{\psi}(a_0^m\xi)|^{2(1-\lambda)}. \tag{3.5.8}$$

The sum over ℓ converges if $\gamma\lambda > 1$, i.e., if $\lambda > 1/\gamma$; we can choose, e.g., $\lambda = \frac{1}{2}(1+\gamma^{-1})$. On the other hand, for $\Omega_0 \le |\xi| \le \Omega_1$,

$$\begin{aligned} \sum_{m>m_1} |\hat{\psi}(a_0^m\xi)|^{2(1-\lambda)} &\le C_3 \sum_{m>m_1} (1+a_0^{2m}\Omega_0^2)^{-\gamma(1-\lambda)} \\ &\le C_4\ \Omega_0^{-2\gamma(1-\lambda)}\ a_0^{-2m_1\gamma(1-\lambda)} \end{aligned} \tag{3.5.9}$$

and

$$\sum_{m<m_0} |\hat{\psi}(a_0^m \xi)|^{2(1-\lambda)} \leq C_5 \sum_{m<m_0} (a_0^m \Omega_1)^{2\beta(1-\lambda)}$$
$$\leq C_6 \, \Omega_1^{2\beta(1-\lambda)} \, a_0^{2m_0\beta(1-\lambda)} . \quad (3.5.10)$$

In all these estimates, the constants C_j may depend on a_0, b_0, λ, β, and γ but are independent of Ω_0, Ω_1, m_0, and m_1. Substituting into (3.5.8), with the choice $\lambda = \frac{1}{2}(1+\gamma^{-1})$, leads to

$$(3.5.8) \leq C_7 \, \|f\|^2 \, \left[(\Omega_0 a_0^{m_1})^{-(\gamma-1)} + (a_0^{m_0}\Omega_1)^{\beta(\gamma-1)/\gamma} \right] .$$

If $m_1 \geq (\ln a_0)^{-1} \, [(\gamma-1)^{-1} \, \ln(4C_7/B\epsilon^2) - \ln \Omega_0]$ and $m_0 \leq (\ln \, a_0)^{-1}$ $[\gamma\beta^{-1}(\gamma-1)^{-1} \, \ln \, (B\epsilon^2/4C_7) - \ln \, \Omega_1]$, then this leads to

$$\sum_{\substack{m<m_0 \\ \text{or } m>m_1}} \sum_{n\in\mathbb{Z}} |\langle P_{\Omega_0,\Omega_1} f, \, \psi_{m,n}\rangle|^2 \leq B\frac{\epsilon^2}{4} \, \|f\|^2 ,$$

as desired.

5. The second term in (3.5.6) is easier to deal with. We have

$$\sum_{|nb_0|>a_0^{-m}T+t} |\langle Q_T f, \, \psi_{m,n}\rangle|^2$$
$$\leq \sum_{|nb_0|>a_0^{-m}T+t} \|f\|^2 \, \|Q_T \, \psi_{m,n}\|^2$$
$$\leq \|f\|^2 \int_{|x|\leq T} dx \; a_0^{-m} \sum_{|nb_0|>a_0^{-m}T+t} |\psi(a_0^m x - nb_0)|^2 .$$

The sum over n splits into two parts, $n > b_0^{-1} \, (a_0^{-m}T + t)$, and $n < -b_0^{-1}(a_0^{-m}T + t)$. Let n_1 be the smallest integer larger than $b_0^{-1}(a_0^{-m}T + t)$. Then

$$a_0^{-m} \int_{|x|\leq T} dx \sum_{nb_0>a_0^{-m}T+t} |\psi(a_0^{-m}x - nb_0)|^2$$
$$\leq a_0^{-m} \int_{|x|\leq T} dx \sum_{n=n_1}^{\infty} C_8 \left\{1 + [t + (n-n_1)b_0 + a_0^{-m}(T-x)]^2\right\}^{-\alpha}$$

(because $|a_0^{-m}x - nb_0| = nb_0 - a_0^{-m}x \geq (n-n_1)b_0 + t + a_0^{-m}(T-x)$)

$$\leq C_9 \sum_{\ell=0}^{\infty} [1 + (t + \ell b_0)^2]^{-\alpha} \leq C_{10} \, t^{-2\alpha} .$$

The sum over $n < -b_0^{-1}(a_0^{-m}T+t)$ is dealt with in the same way. It follows that

$$\sum_{m_0 \le m \le m_1} \sum_{|nb_0| > a_0^{-m}T+t} |\langle Q_T f,\ \psi_{m,n}\rangle|^2 \le 2(m_1 - m_0 + 1)\ C_{10}\ t^{-2\alpha}\ \|f\|^2 ,$$

which can be made smaller than $B\ \epsilon^2 \|f\|^2/4$ by choosing

$$t \ge [8(m_1 - m_0 + 1)C_{10}\ B^{-1}\epsilon^{-2}]^{1/2\alpha} .$$

This concludes the proof. ∎

The estimates for m_0, m_1, t that follow from this proof are very coarse; in practice, one can obtain much less coarse values if ψ and $\hat{\psi}$ have faster decay than stated in the theorem (see, e.g., Daubechies (1990), p. 996).

For later reference, let us estimate $\#\ B_\epsilon(\Omega_0,\ \Omega_1;\ T)$, as a function of $\Omega_0,\ \Omega_1,\ T$, and ϵ. We find

$$\begin{aligned}
\#\ B_\epsilon(\Omega_0,\ \Omega_1;\ T) &\simeq \sum_{m=m_0}^{m_1} 2b_0^{-1}\ (a_0^{-m}T + t) \\
&= 2b_0^{-1}T\ \frac{a_0^{-m_0+1} - a_0^{-m_1}}{a_0 - 1} + 2b_0^{-1}(m_1 - m_0 + 1)t \\
&\simeq 2T\ C_{11}b_0^{-1}(a_0 - 1)^{-1}\ \epsilon^{2/(\gamma-1)}\ (\Omega_1 - \Omega_0) \\
&\quad + 2\epsilon^{-1/\alpha}\ b_0^{-1}(\ln\ a_0)^{-(2\alpha+1)/2\alpha}\ C_{12}\ [C_{13} + \ln\ \Omega_1 - \ln\ \Omega_0]^{(2\alpha+1)/2\alpha} .
\end{aligned}$$

On the other hand, the area of the time-frequency region $[-T,T] \times ([-\Omega_1,\ \Omega_0] \cup [\Omega_0,\ \Omega_1])$ is $4T(\Omega_1 - \Omega_0)$. As $\Omega_0 \to 0$ and $T,\ \Omega_1 \to \infty$, we find

$$\lim \frac{\#\ B_\epsilon(\Omega_0,\ \Omega_1;\ T)}{4T(\Omega_1 - \Omega_0)} = \frac{1}{2}\ C_{11}\ b_0^{-1}(a_0 - 1)^{-1}\ \epsilon^{2/(\gamma-1)} , \tag{3.5.11}$$

which is not independent of ϵ. We will come back to this in Chapter 4.

Theorem 3.5.1 tells us that if ψ has reasonable decay in time and in frequency, then frames generated by ψ do indeed exhibit time-frequency localization features, at least with respect to time-frequency sets of the type $[-T,T] \times ([-\Omega_1,\ -\Omega_0] \cup [\Omega_0,\ \Omega_1])$. In practice, one is interested in localization on many other sets. A chirp signal, for instance, intuitively corresponds to a diagonal region (possibly curved) in the time-frequency plane, and it should be possible to reconstruct it from only those $\psi_{m,n}$ for which $(a_0^m nb_0,\ \pm a_0^{-m}\xi_0)$ is in or close to this region. This turns out to be the case in practice (for chirp signals, and many others). It is harder to formulate this in a precise theorem, mainly because one first has to agree on the meaning of "localization" on a prescribed time-frequency set, when this set is not a union of rectangles as in Theorem 3.5.1.

If we choose the interpretation in terms of the operators L_S defined in §2.8 (i.e., f is mostly localized in S if $\|(1-L_S)f\| \ll \|f\|$), then the theorem is almost trivial if the wavelets in the definition of L_S and in the frame both have good decay properties. For any other time-frequency localization procedure (e.g., using Wigner distributions, or the affine Wigner distributions of Bertrand and Bertrand (1989)), we still expect results analogous to those in Theorem 3.5.1, but the proof will depend on the chosen localization procedure.

An entirely similar localization theorem holds for the windowed Fourier case.

THEOREM 3.5.2. *Suppose that the* $g_{m,n}(x) = e^{imw_0 x} g(x - nt_0)$ *constitute a frame with frame bounds* A, B, *and suppose that*

$$|g(x)| \le C(1+x^2)^{-\alpha/2}, \quad |\hat{g}(\xi)| \le C(1+\xi^2)^{-\alpha/2} ,$$

with $\alpha > 1$. *Then, for any* $\epsilon > 0$, *there exist* $t_\epsilon, \omega_\epsilon > 0$, *such that, for all* $f \in L^2(\mathbb{R})$, *and all* $T, \Omega > 0$,

$$\left\| f - \sum_{\substack{|m\omega_0| \le \Omega + \omega_\epsilon \\ |nt_0| \le T + t_\epsilon}} \langle f, g_{m,n} \rangle \, \tilde{g}_{m,n} \right\|$$

$$\le \sqrt{\frac{B}{A}} \left[\left(\int_{|x|>T} dx \, |f(x)|^2 \right)^{1/2} + \left(\int_{|\xi|>\Omega} d\xi \, |\hat{f}(\xi)|^2 \right)^{1/2} + \epsilon \, \|f\| \right] .$$

Proof.

1. By the same tricks as in points 2, 3 of the proof of Theorem 3.5.1, we have

$$\begin{aligned} \Big\| f - \sum_{\substack{|m\omega_0| \le \Omega + \omega_\epsilon \\ |nt_0| \le T + t_\epsilon}} \langle f, g_{m,n} \rangle \tilde{g}_{m,n} \Big\| &\le \sqrt{\frac{B}{A}} \, [\|(1-Q_T)f\| + \|(1-P_\Omega)f\|] \\ &\quad + A^{-1/2} \left\{ \left[\sum_{n \in \mathbb{Z}} \sum_{|m\omega_0| > \Omega + \omega_\epsilon} |\langle P_\Omega f, g_{m,n} \rangle|^2 \right]^{1/2} \right. \\ &\quad \left. + \left[\sum_{m \in \mathbb{Z}} \sum_{|nt_0| > T + t_\epsilon} |\langle Q_T f, g_{m,n} \rangle|^2 \right]^{1/2} \right\} , \end{aligned} \tag{3.5.12}$$

where $(Q_T f)(x) = f(x)$ for $|x| \le T$, 0 otherwise, and $(P_\Omega f)^\wedge(\xi) = \hat{f}(\xi)$ for $|\xi| \le \Omega$, 0 otherwise. The theorem follows if we can prove that the last two terms in (3.5.12) can be bounded by $B^{1/2} \, \epsilon \|f\|$. Let us first concentrate on the last term.

2. $$\sum_{m\in\mathbb{Z}} \sum_{|nt_0|\geq T+t_\epsilon} |\langle Q_T f,\, g_{m,n}\rangle|^2$$

$$\leq \frac{2\pi}{\omega_0} \sum_{|nt_0|\geq T+t_\epsilon} \sum_{\ell\in\mathbb{Z}} \int_{\substack{|x|\leq T \\ |x-\frac{2\pi}{\omega_0}\ell|\leq T}} dx\, |f(x)| \left|f\left(x-\frac{2\pi}{\omega_0}\ell\right)\right| |g(x-nt_0)| \left|g\left(x-\frac{2\pi}{\omega_0}\ell-nt_0\right)\right|$$

$$\leq \frac{2\pi}{\omega_0} \sum_{\ell\in\mathbb{Z}} \left[\int_{\substack{|x|\leq T \\ |x-\frac{2\pi}{\omega_0}\ell|\leq T}} dx\, |f(x)|^2 \sum_{|nt_0|\geq T+t_\epsilon} |g(x-nt_0)| \left|g\left(x-\frac{2\pi}{\omega_0}\ell-nt_0\right)\right| \right]^{1/2}$$

$$\cdot \left[\int_{\substack{|y|\leq T \\ |y+\frac{2\pi}{\omega_0}\ell|\leq T}} dy\, |f(y)|^2 \sum_{|nt_0|\geq T+t_\epsilon} \left|g\left(y+\frac{2\pi}{\omega_0}\ell-nt_0\right)\right| |g(y-nt_0)| \right]^{1/2}$$

$$\leq \frac{2\pi}{\omega_0} \|Q_T f\|^2 \sum_{\ell\in\mathbb{Z}} \sum_{|nt_0|\geq T+t_\epsilon} \sup_{\substack{|x|\leq T \\ |x-\frac{2\pi}{\omega_0}\ell|\leq T}} |g(x-nt_0)| \left|g\left(x-\frac{2\pi}{\omega_0}\ell-nt_0\right)\right|$$

$$\leq \frac{2\pi}{\omega_0} \|f\|^2\, C^2 \sum_{\ell\in\mathbb{Z}} \sum_{|nt_0|\geq T+t_\epsilon} \sup_{\substack{|x|\leq T \\ |x-\frac{2\pi}{\omega_0}\ell|\leq T}} [1+(x-nt_0)^2]^{-\alpha/2} \left[1+\left(x-\frac{2\pi}{\omega_0}\ell-nt_0\right)^2\right]^{-\alpha/2} . \tag{3.5.13}$$

One easily checks that the contribution for $n > t_0^{-1}(T+t_\epsilon)$ is exactly equal to that for $n < -t_0^{-1}(T+t_\epsilon)$; we may restrict ourselves to negative n only, at the price of a factor 2. By redefining $y = x - \frac{2\pi}{\omega_0}\,\ell$ if ℓ is positive, we see that we may restrict ourselves to negative ℓ as well. Hence

$$(3.5.13) \;\leq\; \frac{4\pi}{\omega_0}\, C^2\, \|f\|^2 \sum_{|nt_0|\geq T+t_\epsilon} \sum_{\ell\geq 0} \sup_{\substack{|x|\leq T \\ |x-\frac{2\pi}{\omega_0}\ell|\leq T}} [1+(x+nt_0)^2]^{-\alpha/2} \left[1+\left(x+\frac{2\pi}{\omega_0}\ell+nt_0\right)^2\right]^{-\alpha/2}$$

$$\leq \frac{4\pi}{\omega_0} C^2 \|f\|^2 \sum_{|nt_0| \geq T+t_\epsilon} \sum_{\ell \geq 0} [1+(nt_0-T)^2]^{-\alpha/2} \left[1+\left(nt_0 - T + \frac{2\pi\ell}{\omega_0}\right)^2\right]^{-\alpha/2} . \quad (3.5.14)$$

However, for any $\mu, \nu > 0$, we have

$$\begin{aligned}
&\sum_{\ell=0}^{\infty} [1+(\mu+\nu\ell)^2]^{-\alpha/2} \\
&\quad \leq (1+\mu^2)^{-\alpha/2} + \int_0^\infty dx\ [1+(\mu+\nu x)^2]^{-\alpha/2} \\
&\quad \leq (1+\mu^2)^{-\alpha/2} + 2^{\alpha/2}\frac{1}{\nu}\int_0^\infty dy\ (\sqrt{1+\mu^2}+y)^{-\alpha} \\
&\qquad \left(\text{use } a^2+b^2 \geq \frac{1}{2}(a+b)^2\right) \\
&\quad \leq (1+\mu^2)^{-\alpha/2} + 2^{\alpha/2}\ \nu^{-1}(\alpha-1)^{-1}(1+\mu^2)^{-\frac{\alpha-1}{2}} .
\end{aligned}$$

It follows that

$$(3.5.14) \leq \frac{4\pi}{\omega_0} C_1 \|f\|^2 \sum_{nt_0 \geq T+t_\epsilon} [1+(nt_0-T)^2]^{-\alpha+1/2} .$$

Let n_1 be the smallest integer larger than $T+t_\epsilon$. Then

$$\begin{aligned}
&\sum_{nt_0 \geq T+t_\epsilon} [1+(nt_0-T)^2]^{-\alpha+1/2} \\
&\quad \leq \sum_{n=n_1}^{\infty} [1+((n-n_1)t_0+t_\epsilon)^2]^{-\alpha+1/2} \\
&\quad \leq C_2\ (1+t_\epsilon^2)^{-\alpha+1}
\end{aligned}$$

by the computation above. Putting it altogether, we have

$$\sum_{m\in\mathbb{Z}} \sum_{|nt_0| \geq T+t_\epsilon} |\langle Q_T f, g_{m,n}\rangle|^2 \leq C_3\ (1+t_\epsilon^2)^{-\alpha+1}\ \|f\|^2 , \quad (3.5.15)$$

where C_3 depends on ω_0, t_0, α, and C, but not on T (or Ω).

3. Similarly, one proves

$$\sum_{n\in\mathbb{Z}} \sum_{|m\omega_0| \geq \Omega+\omega_\epsilon} |\langle P_\Omega f,\ g_{m,n}\rangle|^2 \leq C_4(1+\omega_\epsilon^2)^{-\alpha+1}\ \|f\|^2 . \quad (3.5.16)$$

Since $\alpha > 1$, it is clear that appropriate choices of $t_\epsilon, \omega_\epsilon$ (independent of T or Ω!) make (3.5.15), (3.5.16) smaller than $B\epsilon^2\ \|f\|^2/4$. This concludes the proof. ■

Figure 3.8 gives a sketch of the collection (m, n) such that $|m\omega_0| \leq \Omega + \omega_\epsilon$, $|nt_0| \leq T + t_\epsilon$, as compared to the time-frequency rectangle $[-T, T] \times [-\Omega, \Omega]$. The "$\epsilon$-box" has a different shape from that in Figure 3.7.

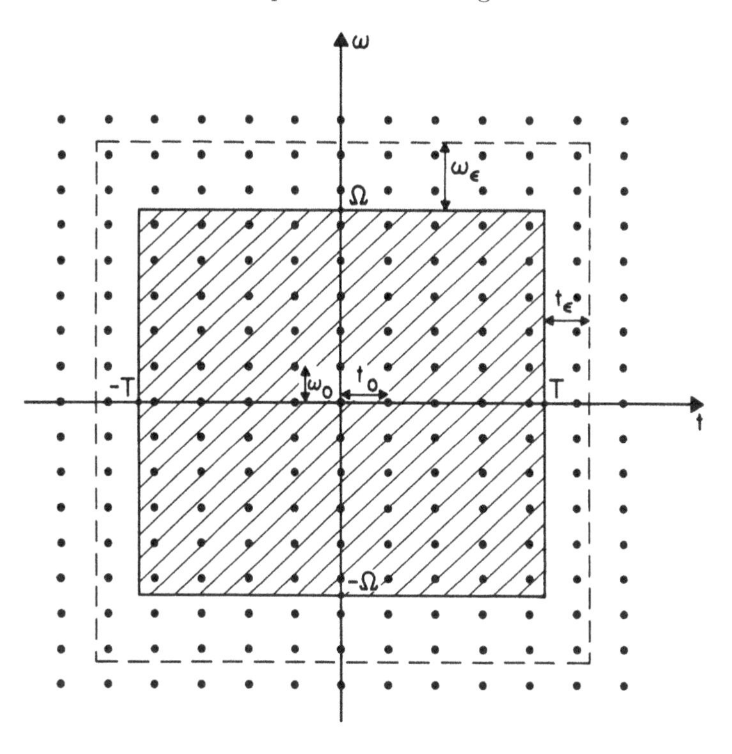

FIG. 3.8. *The set of lattice points B_ϵ needed for approximate reconstruction, via the windowed Fourier transform, of a function localized mostly in $[-T, T]$ in time and in $[-\Omega, \Omega]$ in frequency.*

Let us compare again the number of points in the enlarged box $B_\epsilon = \{(m, n);\ |m\omega_0| \leq \Omega + \omega_\epsilon,\ |nt_0| \leq T + t_\epsilon\}$ with the area of $[-T, T] \times [-\Omega, \Omega]$, in the limit for large T, Ω:

$$\frac{\#\ B_\epsilon(T,\ \Omega)}{4T\Omega} \simeq \frac{2\omega_0^{-1}(\Omega + \omega_\epsilon) \cdot 2t_0^{-1}(T + t_\epsilon)}{4T\Omega} \longrightarrow (\omega_0 t_0)^{-1}\ . \tag{3.5.17}$$

In contrast to the wavelet case, this limit is independent of ϵ. We will come back to the significance of this in Chapter 4.

3.6. Redundancy in frames: What does it buy?

As illustrated by the different tables of frame bounds, frames (wavelets or windowed Fourier functions) can be very redundant (as measured by, e.g., $\frac{A+B}{2}$ if the frame is close to tight, and if all the frame vectors are normalized). In some applications (e.g., as in the work of the Marseille groups—see the papers of

Grossmann, Kronland-Martinet, Torresani) this redundancy is sought, because representations close to the continuous transform are wanted. It was noticed very early on by J. Morlet (private communication, 1986) that this redundancy also leads to robustness, in the sense that he could afford to store the wavelet coefficients $\langle f,\ \psi_{m,n}\rangle$ with low precision (only a couple of bits), and still reconstruct f with comparatively much higher precision. Intuitively, one can understand this phenomenon as follows. Let $(\varphi_j)_{j\in J}$ be a frame (not necessarily of wavelets or windowed Fourier functions). If this frame is an orthonormal basis, then

$$F:\ \mathcal{H}\to\ell^2(J), \qquad (Ff)_j = \langle f,\ \varphi_j\rangle$$

is a unitary map, and the image of $\mathcal{H}$ under F is all of $\ell^2(J)$. If the frame is redundant, i.e., if the φ_j are not independent, then the elements of $F\mathcal{H}$ are sequences with some correlations built into them, and $F\mathcal{H} = \text{Ran}\ (F)$ is a subspace of $\ell^2(J)$, smaller than $\ell^2(J)$ itself. The more redundant the frame is, the "smaller" Ran (F) will be. As shown in §3.2, the reconstruction formula

$$f = \sum_{j\in J} \langle f,\ \varphi_j\rangle\ \tilde{\varphi}_j$$

involves a projection onto Ran (F): it can be rewritten as

$$f = \tilde{F}^*\ Ff\ ,$$

and $\tilde{F}^* c = 0$ if $c \perp \text{Ran}\ (F)$. If the $\langle f,\ \varphi_j\rangle$ are adulterated by adding some α_j to each coefficient (an example would be round-off error), the total effect on the reconstructed function would be

$$f_{\text{approx}} = \tilde{F}^*\ (Ff + \alpha)\ .$$

Since $\tilde{F}^*$ includes a projection onto Ran (F), the component of the sequence α orthogonal to Ran (F) does not contribute, and we expect $\|f - f_{\text{approx}}\|$ to be smaller than $\|\alpha\|$. The effect should become more pronounced the "smaller" Ran (F) is, i.e., the more redundant the frame is.

Let us make this more explicit with the two-dimensional frame used as an example in §3.2, and by comparing it with an orthonormal basis. Define $u_1 = (1,0)$, $u_2 = (0,1)$, $e_1 = u_2$, $e_2 = -\frac{\sqrt{3}}{2}u_1 - \frac{1}{2}u_2$, $e_3 = \frac{\sqrt{3}}{2}u_1 - \frac{1}{2}u_2$; (u_1, u_2) constitutes an orthonormal basis for $\mathbb{C}^2$, (e_1, e_2, e_3) a tight frame with frame bound $\frac{3}{2}$. If we add $\alpha_j\epsilon$ to the coefficients $\langle f,\ u_j\rangle$, where α_j are independent random variables with mean zero and variance 1, then the expected error on the reconstruction will be

$$\mathbb{E}\left(\left\|f - \sum_{j=1}^{2}(\langle f,\ u_j\rangle + \alpha_j\epsilon)u_j\right\|^2\right) = \epsilon^2\mathbb{E}\left(\left\|\sum_{j=1}^{2}\ \alpha_j u_j\right\|^2\right) = \epsilon^2\ \mathbb{E}(\alpha_1^2+\alpha_2^2) = 2\epsilon^2\ .$$

If we add $\alpha_j\epsilon$ to the frame coefficients $\langle f,\ e_j\rangle$, then we find

$$\mathbb{E}\left(\left\|f - \frac{2}{3}\sum_{j=1}^{3}\ (\langle f,\ e_j\rangle + \alpha_j\ \epsilon)e_j\right\|^2\right)$$

$$= \frac{4}{9}\epsilon^2\, \mathbb{E}\left(\left\|\sum_{j=1}^{3} \alpha_j e_j\right\|^2\right) = \frac{4}{9}\epsilon^2\, \mathbb{E}\,(\alpha_1^2 + \alpha_2^2 + \alpha_3^2 - \alpha_1\alpha_2 - \alpha_2\alpha_3 - \alpha_1\alpha_3)$$

$$= \frac{4}{3}\epsilon^2 ,$$

which represents a gain of $\frac{2}{3}$ over the orthonormal case!

A similar argument can be applied to wavelet or windowed Fourier frames. In order to confine ourselves to only finitely many $\psi_{m,n}$ or $g_{m,n}$, we assume f to be "essentially localized" on $[-T,T] \times ([-\Omega_1,-\Omega_0] \cup [\Omega_0,\Omega_1])$ (wavelet case) or $[-T,T] \times [-\Omega,\Omega]$ (windowed Fourier case), so that there exists a finite set B_ϵ (see §3.5) for which

$$\left\| f - \sum_{m,n \in B_\epsilon} \langle f,\, \psi_{m,n}\rangle\, \widetilde{\psi_{m,n}} \right\| \le \epsilon \|f\|$$

(and similarly for the windowed Fourier case). Let us assume the frame is almost tight, $\widetilde{\psi_{m,n}} \simeq A^{-1}\psi_{m,n}$. Adding $\alpha_{m,n}\,\delta$ to every $\langle f,\, \psi_{m,n}\rangle$, under the assumptions $\mathbb{E}(\alpha_{m,n}\, \alpha_{m',n'}) = \delta_{mm'}\delta_{nn'}$, $\mathbb{E}(\alpha_{m,n}) = 0$, leads to

$$\mathbb{E}\left(\left\| f - A^{-1} \sum_{m,n\in B_\epsilon} (\langle f,\, \psi_{m,n}\rangle + \alpha_{m,n}\delta)\psi_{m,n}\right\|^2\right)$$
$$\le \epsilon^2\, \|f\|^2 \,+\, \delta^2(\#\, B_\epsilon)A^{-2} \tag{3.6.1}$$

If we assume $\|\psi_{m,n}\| = 1$). If we "double the redundancy" by halving b_0, then the new frame would again be almost tight (see, e.g., (3.3.11), (3.3.12)), with a frame bound A' twice as large; on the other hand, the new "ϵ-box" B'_ϵ would contain twice as many elements. It follows that

$$(\#\, B'_\epsilon)A'^{-2} = \tfrac{1}{2}\,(\#\, B_\epsilon)A^{-2} ,$$

i.e., doubling the redundancy leads to halving the effect of adding errors to the wavelet coefficients. The same argument can be made for the windowed Fourier case.

As it stands, the argument above is rather heuristic. There are indications also that it can be considerably strengthened: the gain factor observed by Morlet was in fact larger than what would follow from these arguments. Moreover, recently Munch (1992) showed that for tight windowed Fourier frames with $\lambda = (2\pi)^{-1}\omega_0 t_0 = N^{-1}$, $N \in \mathbb{N}$, $N > 1$, the gain factor with respect to the orthonormal case is in fact N^{-2}, and *not* N^{-1}, as would follow from our argument. His proof uses that λ^{-1} is integer in an essential way, but it is hard to believe that the same phenomenon would not exist for noninteger λ^{-1}; maybe it also holds for wavelet frames! I put this as a challenge to the reader....

3.7. Some concluding remarks.

In this chapter we have studied in some depth the reconstruction of f from the sequence $(\langle f,\ \psi_{m,n}\rangle)_{m,n\in\mathbb{Z}}$, where $\psi_{m,n}(x) = a_0^{-m/2}\ \psi(a_0^{-m}x - nb_0)$ (and variations thereof—see §3.3.4). We have seen that numerically stable reconstruction is only possible if the $\psi_{m,n}$ constitute a frame, and we have derived a reconstruction formula if the $\psi_{m,n}$ are a frame. One can, however, use other reconstruction formulas (provided the $\psi_{m,n}$ do constitute a frame: the necessity of that condition remains!). To conclude this chapter, let me sketch the approach of S. Mallat, which addresses moreover the problem of shift-invariance.

The discrete wavelet transform, such as I have described it in this chapter, is highly non-invariant under translations. By this I mean that two functions may be shifted versions of each other, while their wavelet coefficients may be very different. This is already illustrated by the "hyperbolic lattice"[17] in Figure 1.4a, where the axis $t = 0$ plays a unique role. In practice one does not use an infinite number of scales, but cuts off very low and very high frequencies: only those m for which $m_1 \leq m \leq m_0$ are used. The resulting truncated lattice is then invariant under translations by $b_0 2^{m_0}$ (choose $a_0 = 2$ for simplicity), which is, however, very large with respect to, e.g., the sampling time step for f (in most applications, f is given in sampled form). If f_1 is a shifted version of f obtained by a translation $\neq nb_0 2^{m_0}$, then typically the wavelet coefficients of f_1 will be different from those of f_0. Even if the shift is $\bar{n}b_0 2^{\bar{m}}$, with $m_1 \leq \bar{m} \leq m_0$, then $\langle f_1,\ \psi_{m,n}\rangle = \langle f,\ \psi_{m,n-2^{\bar{m}-m}\bar{n}}\rangle$ if $\bar{m} \geq m$, but no such formula can be written for $m > \bar{m}$. For some applications (in particular, all applications that involve "recognizing" f) this can be a real problem. In a first approximation, the solution proposed by S. Mallat is the following:

- Compute all the $\int dx\ f(x)\ \psi(2^{-m}x - n2^{-m}b_0) = \alpha_{m,n}(f)$ (a special ψ such as in §3.3.5.D makes it possible to compute these in $C\ N \log N$ operations, if f consists of N samples). This list of coefficients is invariant for shifts of f by $\bar{n}b_0$.

- At every level m, retain only those $\alpha_{m,n}(f)$ that are local extrema (as a function of n). This effectively corresponds to a subsampling of the highly redundant $\alpha_{m,n}(f)$. In practice, the number of subsamples retained is proportional to 2^{-m} times the original number; this is about the same number as one has in a not too redundant frame of the type described earlier, but the subsampling is now adapted to f, and not imposed a priori by the hyperbolic lattice.

Together with this prescription for decomposition (here described in a simplified form) Mallat then proposes a reconstruction algorithm, which works very well in practice (see Mallat (1991)).[18] In Mallat and Zhong (1992), the procedure was extended to two dimensions, to treat images. One way to view Mallat's approach is to look upon the $2^{-m}\psi(2^{-m}x - n2^{-m}b_0)$ as the underlying frame (note that the change in normalization counters the larger number of frame vectors in every m-level). Again, it is necessary that this family satisfy the frame condition

(3.1.4) for a stable reconstruction algorithm to exist, but once this condition is satisfied, several reconstruction algorithms can be proposed. In this case, Mallat's extrema-algorithm is certainly more sophisticated than the standard frame inversion algorithm.

Notes.

1. If any f can be written as such a superposition, the $\psi_{m,n}$ are also called "atoms," and the corresponding expansions "atomic decompositions." Atomic decompositions (for many spaces besides $L^2(\mathbb{R})$) have been studied and used in harmonic analysis for many years: see, e.g., Coifman and Rochberg (1980) for atomic decompositions in entire function spaces.

2. This is true except for very special ψ. If the $\psi_{m,n}$ constitute an orthonormal basis (see Chapters 4 and later), then the expansion with respect to this orthonormal basis provides a discrete "resolution of the identity."

3. The polarization identity recovers $\langle f, g\rangle$ from $\|f \pm g\|$, $\|f \pm ig\|$:
$$\langle f,g\rangle = \tfrac{1}{4}\ \left[\|f+g\|^2 - \|f-g\|^2 + i\|f+ig\|^2 - i\|f-ig\|^2\right] .$$

4. That is, if $(J_n)_{n\in\mathbb{N}}$ is any increasing sequence of finite subsets of J, i.e., $J_n \subset J_m$ if $n \le m$, tending to J as n tends to ∞, i.e., $\cup_{n\in\mathbb{N}}\ J_n = J$, then $\|F^*c - \sum_{j\in J_n}\ c_j\ \varphi_j\| \longrightarrow 0$ for $n \to \infty$. The proof is in two steps:

   - If $n_2 \ge n_1 \ge n_0$, then
   $$\begin{aligned} &\| \textstyle\sum_{j\in J_{n_2}} c_j\varphi_j - \sum_{j\in J_{n_1}} c_j\varphi_j \| \\ &\quad = \sup_{\|f\|=1} \left|\langle \textstyle\sum_{j\in J_{n_2}\setminus J_{n_1}} c_j\varphi_j, f\rangle\right| \\ &\quad \le \sup_{\|f\|=1} \left(\textstyle\sum_{j\in J_{n_2}\setminus J_{n_1}} |c_j|^2\right)^{1/2} \left(\textstyle\sum_{j\in J} |\langle\varphi_j, f\rangle|^2\right)^{1/2} \\ &\quad \le B^{1/2} \left(\textstyle\sum_{j\in J\setminus J_{n_0}} |c_j|^2\right)^{1/2} , \end{aligned}$$
   which tends to 0 for $n_0 \to \infty$. Hence the $\eta_n = \sum_{j\in J_n} c_j\varphi_j$ constitute a Cauchy sequence, with limit η in $L^2(\mathbb{R})$.
   - For this η, and any $f \in L^2(\mathbb{R})$,
   $$\begin{aligned} \langle \eta, f\rangle &= \lim_{n\to\infty} \langle\eta_n, f\rangle = \lim_{n\to\infty} \textstyle\sum_{j\in J_n} c_j\langle\varphi_j, f\rangle \\ &= \textstyle\sum_{j\in J} c_j\langle\varphi_j, f\rangle = \langle c, Ff\rangle\ . \end{aligned}$$
   Hence $\eta = F^*c$.

5. This is proved as follows:

   - $\langle R(f+g),\ f+g\rangle - \langle R(f-g),\ f-g\rangle$
   $$= 2\langle Rf, g\rangle + 2\langle Rg, f\rangle = 4\,\mathrm{Re}\ \langle Rf, g\rangle$$
   $$\text{(because } R^* = R\text{)} ;$$

- Re $\langle Rf, g\rangle \leq \frac{1}{4}\,\frac{B-A}{B+A}\,[\|f+g\|^2 + \|f-g\|^2] = \frac{1}{2}\,\frac{B-A}{B+A}[\|f\|^2 + \|g\|^2]$;
- $$\begin{aligned} |\langle Rf, g\rangle| &= \langle Rf, g\rangle\ \overline{\langle Rf, g\rangle}/|\langle Rf, g\rangle| \\ &= \langle Rf,\ \langle Rf, g\rangle g/|\langle Rf, g\rangle|\rangle \\ &\leq \frac{1}{2}\,\frac{B-A}{B+A}\,[\|f\|^2 + \|\langle Rf, g\rangle g/|\langle Rf, g\rangle|\ \|^2] \\ &\leq \frac{1}{2}\,\frac{B-A}{B+A}\,[\|f\|^2 + \|g\|^2]\ ; \end{aligned}$$
- $\|R\| = \sup_{\|f\|=1,\ \|g\|=1}\ |\langle Rf, g\rangle| \leq \frac{B-A}{B+A}$.

6. Intuitively, C can be understood as a "superposition" of the rank one trace-class operators $\langle\cdot,\ h^{a,b}\rangle h^{a,b}$, with weights $c(a,b)$. If c is integrable with respect to $a^{-2}\,da\,db$, then the individual traces of $\langle\cdot,\ h^{a,b}\rangle h^{a,b}$ (which are all equal to $\|h\|^2$), weighted by the $c(a,b)$ are "summable," so that the whole superposition has finite trace,
$$\text{Tr } C = \|h\|^2 \int_0^\infty \frac{da}{a^2} \int_{-\infty}^\infty db\ c(a,b)\ .$$
This handwaving argument can be made rigorous by approximation arguments.

7. We use here the "essential infimum" (notation: ess inf) defined by
$$\operatorname*{ess\,inf}_x\ f(x) = \inf\{\alpha;\ |\{y;\ f(y) \geq \alpha\}| > 0\}\ ,$$
where $|A|$ stands for the Lebesgue measure of $A \subset \mathbb{R}$. The difference between $\operatorname{ess\,inf}_x\ f(x)$ and $\inf_x\ f(x)$ lies in the positive measure requirement: if $f(0) = 0$, $f(x) = 1$ for all $x \neq 0$, then $\inf_x\ f(x) = 0$, but $\operatorname{ess\,inf}_x\ f(x) = 1$, because $f \geq 1$ except on a set of measure zero, which "does not count." In fact we could be pedantic, and replace inf or sup by ess inf or ess sup in most of our conditions without invalidating them, but it is usually not worth it: in practice the expressions we are dealing with are continuous functions, for which inf and ess inf coincide. In (3.3.11) the situation is different: even for very smooth $\hat{\psi}$, the sum $\sum_{m\in\mathbb{Z}} |\hat{\psi}(a_0^m\xi)|^2$ is discontinuous at $\xi = 0$, because $\hat{\psi}(0) = 0$. For the Haar function, for instance, $|\hat{\psi}(\xi)| = 4(2\pi)^{-1/2}|\xi|^{-1}\ \sin^2\ \xi/4$, and $\sum_{m\in\mathbb{Z}} |\hat{\psi}(\xi)|^2 = (2\pi)^{-1}$ if $\xi \neq 0$, 0 if $\xi = 0$. We therefore need to take the essential infimum; the infimum is zero.

8. This condition implies both the boundedness of $\sum_{m\in\mathbb{Z}} |\hat{\psi}(a_0^m\xi)|^2$ and the decay of $\beta(s)$:
$$\begin{aligned} \sum_{m\in\mathbb{Z}} |\hat{\psi}(a_0^m\xi)|^2 &\leq \sup_{1\leq|\zeta|\leq a_0}\ \sum_{m\in\mathbb{Z}} |\hat{\psi}(a_0^m\zeta)|^2 \\ &\leq\ C^2 a_0^{2\alpha}\left[\sum_{m=-\infty}^{0}\ a_0^{2m\alpha}\ +\ \sum_{m=1}^{\infty}\ a_0^{2m\alpha}(1+a_0^{2m})^{-\gamma}\right] < \infty \end{aligned}$$

and

$$\begin{aligned}\beta(s) &= \sup_{1\le|\xi|\le a_0} \sum_{m\in\mathbb{Z}} |\hat{\psi}(a_0^m\xi)|\,|\hat{\psi}(a_0^m\xi+s)| \\ &\le C^2 \sup_{1\le|\xi|\le a_0} \left\{ a_0^\alpha \sum_{m=-\infty}^{-1} a_0^{m\alpha}\,(1+|a_0^m 2\xi+s|^2)^{-(\gamma-\alpha)/2} \right. \\ &\qquad \left. + \sum_{m=0}^{\infty} \left[(1+|a_0^m\xi|^2)(1+|a_0^m\xi+s|^2)\right]^{-(\gamma-\alpha)/2} \right\}.\end{aligned}$$

In the first term we can use that, for $|s|\ge 2$, $|a_0^m\xi+s|\ge|s|-1\ge\frac{|s|}{2}$, and hence $(1+|a_0^m\xi+s|^2)^{-1}\le 4(1+|s|^2)^{-1}$; for $|s|\le 2$, $(1+|a_0^m\xi+s|^2)^{-1}\le 1\le 5(1+|s|^2)^{-1}$. It follows that the first term can be bounded by $C'(1+|s|^2)^{-(\gamma-\alpha)/2}$ as soon as $\alpha>0$, $\gamma>\alpha$. For the second term we use that $\sup_{x,y\in\mathbb{R}}\,(1+y^2)[1+(x-y)^2]^{-1}[1+(x+y)^2]^{-1}<\infty$ to bound the sum by $C''(1+|s|^2)^{-(\gamma-\alpha)(1-\epsilon)/2}\sum_{m=0}^{\infty}(1+|a_0^m\xi|^2)^{-\epsilon(\gamma-\alpha)/2}$, where $0<\epsilon<1$ is arbitrary. Since $1\le|\xi|\le a_0$, this can be bounded by $C'''(1+|s|^2)^{-(\gamma-\alpha)(1-\epsilon)/2}$ if $\gamma>\alpha$. We have therefore, for $0<\rho<\gamma-\alpha$,

$$\beta(s)\le C(\rho)(1+|s|^2)^{-\rho/2}\;;$$

hence

$$\sum_{k\ne 0}\left[\beta\left(\frac{2\pi}{b_0}k\right)\,\beta\left(-\frac{2\pi}{b_0}k\right)\right]^{1/2}\le C'(\rho)\;b_0^{-\rho+1}$$

if $\rho>1$.

9. If $\hat{\psi}$ is continuous and has decay at ∞, then $\sum_m|\hat{\psi}(a_0^m\xi)|^2$ is continuous in ξ, except at $\xi=0$. There exists therefore α so that $\sum_{m\in\mathbb{Z}}|\hat{\psi}(a_0^m\xi)|^2\le\frac{3}{2}\epsilon$ if $|\xi-\xi_0|\le\alpha$. Define, for $\alpha'<\alpha$, a function f by $\hat{f}(\xi)=(2\alpha')^{-1/2}$ if $|\xi-\xi_0|\le\alpha'$, $\hat{f}(\xi)=0$ otherwise. Then

$$\begin{aligned}\sum_{m,n\in\mathbb{Z}}|\langle f,\psi_{m,n}\rangle|^2 \;&\le\; \frac{3}{2}\epsilon+\frac{2\pi}{b_0}\sum_{\substack{m,k\in\mathbb{Z}\\ k\ne 0}}(2\alpha')^{-1} \\ &\qquad \int\limits_{\substack{|\xi-\xi_0|\le\alpha'\\ |\xi+2\pi k b_0^{-1}a_0^{-m}-\xi_0|\le\alpha'}} d\xi|\hat{\psi}(a_0^m\xi)|\;|\hat{\psi}(a_0^m\xi+2\pi k b_0^{-1})| \\ &\le\; \frac{3}{2}\epsilon+\frac{2\pi}{b_0}\sum_{m\in\mathbb{Z}}\sum_{\substack{k\ne 0\\ |k|\le\alpha' b_0 a_0^m\pi^{-1}}}(2\alpha')^{-1} \\ &\qquad \int\limits_{|\xi-\xi_0|\le\alpha'} d\xi\;|\hat{\psi}(a_0^m\xi)|^2\end{aligned}$$

(use Cauchy–Schwarz on the integral)

$$\leq \frac{3}{2}\epsilon + \frac{2\pi}{b_0}(2\alpha')^{-1}\, 2\alpha' b_0 \pi^{-1}$$
$$\sum_{m\in\mathbb{Z}} a_0^m \int_{|\xi-\xi_0|\leq\alpha'} d\xi\; |\hat{\psi}(a_0^m\xi)|^2$$
$$\leq \frac{3}{2}\epsilon + 2\alpha' \sup_{\xi} \sum_{m\in\mathbb{Z}} a_0^m |\hat{\psi}(a_0^m\xi)|^2 .$$

If $|\hat{\psi}(\xi)| \leq C(1+|\xi|^2)^{-\gamma/2}$ with $\gamma > 1$, then this infinite sum is uniformly bounded in ξ, and we can choose α' so that the whole right hand-side of the inequality is $\leq 2\epsilon$.

10. Beware of a mistake in the example on pp. 988–989 of Daubechies (1990). The formula for $(h_{00})^\wedge$ should read $(h_{00})^\wedge = \sum_{j=0}^{\infty} \bar{r}^j \psi_{j0}$, and leads to the conclusion that $h_{00} \notin L^p(\mathbb{R})$ for *small* p. I would like to thank Chui and Shi (1993) for pointing this out to me.

11. This is slightly different from multiresolution analysis, where (3.3.27) would also contain a scaling factor 2:

$$\psi(x) = \sum_k d_k\; \phi(2x-k) .$$

12. One can also construct tight frames where neither g nor $\hat{g}$ have compact support. It is, for instance, possible to construct a tight frame in which both g and $\hat{g}$ have exponential decay. One way of doing this is to start from any windowed Fourier frame, with window function g, and to define the function $G = (F^*F)^{-1/2}\, g$, where $F^*F = \sum_{m,n} \langle \cdot, g_{m,n}\rangle g_{m,n}$. The functions $G_{m,n}(x) = e^{im\omega_0 x}\; G(x - nt_0)$ (same ω_0, t_0 as in the $g_{m,n}$) then constitute a tight frame. One has indeed

$$\sum_{m,n} |\langle f, G_{m,n}\rangle|^2 = \sum_{m,n} |\langle f,\; (F^*F)^{-1/2} g_{m,n}\rangle|^2$$
$$= \sum_{m,n} |\langle (F^*F)^{-1/2} f,\; g_{m,n}\rangle|^2 = \langle (F^*F)(F^*F)^{-1/2} f,\; (F^*F)^{-1/2} f\rangle$$
$$= \|f\|^2 .$$

The explicit computation of G can be carried out by a series expansion for $(F^*F)^{-1/2}$ analogous to the series for $(F^*F)^{-1}$ in §3.2. If g and $\hat{g}$ have exponential decay (in particular if g is Gaussian), then the resulting G and its Fourier transform have exponential decay as well. For more details, plots of examples, and an interesting application, see Daubechies, Jaffard, and Journé (1991).

13. The proof in Bacry, Grossmann, and Zak (1975) uses the Zak transform, which we introduce and use in Chapter 4. Full details of their argument are also given in Daubechies (1990). It is interesting that their proof can be extended to show that the $g_{m,n}$ still span all of $L^2(\mathbb{R})$ if one (any one) of the $g_{m,n}$ is deleted, but not if two functions are deleted.

14. These exact formulas use again the Zak transform. Their derivation is given in Daubechies and Grossmann (1988); it is also reviewed in Daubechies (1990).

15. In some applications, Bastiaans' result is interpreted (correctly) to mean that one should "oversample" (i.e., choose $\omega_0 t_0 < 2\pi$) in order to restore stability. Nevertheless, even in such an "oversampled" regime, sometimes Bastiaans' pathological dual function is still used (see, for instance, Porat and Zeevi (1988)). If $\omega_0 t_0 = \pi$, then the $g_{m,n}$ can be split into two families, $g_{m,2n}$ and $g_{m,2n+1}$, each of which can be considered to be a family of Gaussian windowed Fourier functions with $\omega_0 t_0 = 2\pi$, one generated by g itself, the other by $g(x - t_0)$. For both families, the badly convergent (non-convergent in L^2) expansion (3.4.6) can be written, and a function can be viewed as the average of the two expansions. This is of course true in the sense of distributions, and in practice reasonable convergence (probably due to cancellations) seems to be achieved (using a truncated version of Bastiaans' $\tilde{g}$—private communication by Zeevi (1989)), but far better time-frequency localization, and I suspect better convergence, in practice, would be achievable by using the optimal dual function $\tilde{g}$ (corresponding to $\lambda = .5$ in Figure 3.6 in this case).

16. This symmetry is certainly not necessary.

17. It is in fact a true hyperbolic lattice with respect to the hyperbolic geometry on both the positive and negative frequency half plane.

18. Note, however, that Y. Meyer has proved recently that the $\alpha_{m,n}(f)$ which are local extrema in the construction above do not suffice to characterize f completely.

CHAPTER 4

Time-Frequency Density and Orthonormal Bases

This chapter splits naturally into two parts. The first section discusses the role of time-frequency density in wavelet transforms versus windowed Fourier transforms. In particular, for the windowed Fourier transform, orthonormal bases are possible only at the Nyquist density but no such restriction exists for the wavelet case. This leads naturally to the second section, which discusses different possibilities for orthonormal bases in the two cases.

4.1. The role of time-frequency density in wavelet and windowed Fourier frames.

We start with the windowed Fourier case. We mentioned in §3.4.1 that a family of functions $(g_{m,n};\ m,n \in \mathbb{Z})$,

$$g_{m,n}(x) = e^{im\,\omega_0 x}\, g(x - nt_0)\ , \tag{4.1.1}$$

cannot be a frame, whatever the choice of g, if $\omega_0 \cdot t_0 > 2\pi$. In fact, for any choice of $g \in L^2(\mathbb{R})$, one can find $f \in L^2(\mathbb{R})$ so that $f \neq 0$ but $\langle f,\ g_{m,n}\rangle = 0$ for all $m,n \in \mathbb{Z}$. If, for instance, $\omega_0 = 2\pi$, $t_0 = 2$, then such a function f is easy to construct: $\langle f,\ g_{m,n}\rangle = 0$ for all $m,n \in \mathbb{Z}$ leads to

$$\begin{aligned} 0 &= \int dx\ e^{2\pi i m x}\ f(x)\ \overline{g(x-2n)} \\ &= \int_0^1 dx\ e^{2\pi i m x} \sum_{\ell\in\mathbb{Z}} f(x+\ell)\ \overline{g(x+\ell-2n)}\ , \end{aligned}$$

so that it is sufficient to find $f \neq 0$ for which $\sum_{\ell\in\mathbb{Z}} f(x+\ell)\ \overline{g(x+\ell-2n)} = 0$. Define now, for $0 \le x < 1$, $\ell \in \mathbb{Z}$, $f(x+\ell) = (-1)^\ell\ \overline{g(x-\ell-1)}$. Clearly, $\int_{-\infty}^{\infty} dx\ |f(x)|^2 = \int_{-\infty}^{\infty} dx\ |g(x)|^2$, so that $f \in L^2(\mathbb{R})$ and $f \neq 0$. However, $\sum_{\ell\in\mathbb{Z}} f(x+\ell)\ \overline{g(x+\ell-2n)} = \sum_{\ell\in\mathbb{Z}} (-1)^\ell\ \overline{g(x-\ell-1)}\ \ \overline{g(x+\ell-2n)}$, which turns into its negative upon the substitution $\ell = 2n - \ell' - 1$, and therefore equals zero. The same construction can be used for any other pair ω_0, t_0 with product 4π; a generalization of this construction exists if $\omega_0 \cdot t_0 > 2\pi$ and $(2\pi)^{-1}\omega_0 t_0$ is rational (see Daubechies (1990), p. 978). If $\omega_0 t_0 (2\pi)^{-1}$ is larger than 1 but irrational, then I know of no explicit construction for $f \neq 0$,

$f \perp g_{m,n}$. The existence of such an f was proved in Rieffel (1981), using arguments involving von Neumann algebras.[1] If only "reasonably nice" g are considered (i.e., g that have some decay in time as well as in frequency), and if we are only interested in proving that the $g_{m,n}$ cannot constitute a frame (which is weaker than proving that there exists $f \perp g_{m,n}$), then the following very elegant argument by H. Landau does the trick. If $|g(x)| \leq C(1+x^2)^{-\alpha/2}$, $|\hat{g}(\xi)| \leq C(1+\xi^2)^{-\alpha/2}$, and the $g_{m,n}$ constitute a frame, then Theorem 3.5.2 tells us that functions f which are "essentially localized" in $[-T,T] \times [-\Omega,\Omega]$ in the time-frequency plane can be reconstructed, up to a small error, by using only the $\langle f,\, g_{m,n}\rangle$ with $|m\omega_0| \lesssim \Omega$, $|nt_0| \lesssim T$. More precisely, if f is bandlimited to $[-\Omega,\Omega]$ and if $[\int_{|x|\geq T} dx\ |f(x)|^2]^{1/2} \leq \epsilon\ \|f\|$, then

$$\Bigg\| f - \sum_{\substack{|m\omega_0| \leq \Omega + \omega_\epsilon \\ |nt_0| \leq T + t_\epsilon}} \langle f,\, g_{m,n}\rangle\, \tilde{g}_{m,n} \Bigg\| \leq 2\epsilon \sqrt{\frac{B}{A}}\ \|f\|\ .$$

According to this formula, all such functions can therefore be written, up to arbitrarily small error, as a superposition of the $\tilde{g}_{m,n}$ with $|m| \leq \omega_0^{-1}(\Omega+\delta)$, $|n| \leq t_0{}^{-1}(T+\delta)$, where δ depends on the error allowed, but not on Ω or T. However, a corollary of the work of Landau, Pollak, and Slepian (see §2.3) is that the space of functions bandlimited to $[-\Omega,\Omega]$ and satisfying $\int_{|x|\geq T} dx\ |f(x)|^2 \leq \gamma\ \|f\|^2$ $(0 < \gamma < 1,\ \gamma$ fixed) contains at least $\frac{4\Omega T}{2\pi} - O(\log\,(\Omega T))$ different orthogonal functions (the appropriate prolate spheroidal wave functions). All these different orthonormal functions can only be approximated by linear combinations of a finite number of $\tilde{g}_{m,n}$ if the number of $\tilde{g}_{m,n}$ exceeds that of the orthonormal functions, i.e., if $2\pi^{-1}\Omega T - O(\log\,(\Omega T)) \leq 4\omega_0^{-1} t_0{}^{-1}(\Omega+\delta)(T+\delta)$, for any Ω, T. Taking the limit as $\Omega, T \to \infty$ leads to $(2\pi)^{-1} \leq (\omega_0 t_0)^{-1}$ or $\omega_0 t_0 \leq 2\pi$. (This is really only a sketch of the proof. For full technical details, see Landau (1993).)

For all practical purposes we even have to limit ourselves to $\omega_0 \cdot t_0 < 2\pi$ (strict inequality) if we want good time-frequency localization: frames for the limit case $\omega_0 \cdot t_0 = 2\pi$ have necessarily bad localization properties in either time or frequency (or even in both). This is the content of the following theorem.

THEOREM 4.1.1. (Balian–Low) *If the* $g_{m,n}(x) = e^{2\pi i m x} g(x-n)$ *constitute a frame for* $L^2(\mathbb{R})$, *then either* $\int dx\ x^2 |g(x)|^2 = \infty$ *or* $\int d\xi\ \xi^2 |\hat{g}(\xi)|^2 = \infty$.

Before proceeding to the proof of this theorem, let us review its history, and add some remarks. Originally, the theorem was stated for orthonormal bases (instead of frames), independently by Balian (1981) and Low (1985). Their proofs are very similar, but contain a small technical gap which was filled by R. Coifman and S. Semmes; this proof can then be extended to frames as well, as reported in Daubechies (1990), pp. 976–977. Subsequently, a different, very elegant proof for orthonormal bases was found by Battle (1988), which was generalized to frames in Daubechies and Janssen (1993). (This is the proof we give below.)

Two well-known examples of functions g for which the family $e^{2\pi i m x} g(x-n)$ constitutes an orthonormal basis are

$$g(x) = \begin{cases} 1, & 0 \leq x \leq 1, \\ 0, & \text{otherwise} \end{cases}$$

and $g(x) = \frac{\sin \pi x}{\pi x}$. In the first case, $\int d\xi\ \xi^2 |\hat{g}(\xi)|^2 = \infty$, in the second case $\int dx\ x^2 |g(x)|^2 = \infty$. It is shown in Jensen, Hoholdt, and Justesen (1988) that one can choose g with slightly better time-frequency localization: they construct g such that both g and $\hat{g}$ are integrable (i.e., $\int dx\ |g(x)| < \infty$, $\int d\xi\ |\hat{g}(\xi)| < \infty$), but their decay is still rather slow, as dictated by Theorem 4.1.1.

Note that the choice $\omega_0 = 2\pi$, $t_0 = 1$ in our formulation of Theorem 4.1.1 is not a serious restriction: the conclusion holds whenever $\omega_0 \cdot t_0 = 2\pi$. To see this, it suffices to apply the unitary operator $(Uf)(x) = (2\pi\omega_0^{-1})^{1/2}\ g(2\pi\omega_0^{-1}x)$; applying U to $g_{m,n}(x) = e^{im\omega_0 x} g(x - nt_0)$ one finds $(Ug_{m,n})(x) = e^{2\pi i m x}(Ug)(x - n)$.

To prove Theorem 4.1.1, we will use the so-called Zak transform. This transform is defined by

$$(Zf)(s,t) = \sum_{\ell \in \mathbb{Z}} e^{2\pi i t \ell} f(s - \ell)\ . \tag{4.1.2}$$

A priori, this is well defined only for f such that $\sum |f(s - \ell)|$ converges for all s, in particular for $|f(x)| \leq C(1 + |x|)^{-(1+\epsilon)}$. It turns out, however, that this restrictive interpretation of Z can be extended to a unitary map from $L^2(\mathbb{R})$ to $L^2([0,1]^2)$. One way of seeing this is the following:

- $e_{m,n}(x) = e^{2\pi i m x} e(x - n)$, with $e(x) = 1$ for $0 \leq x < 1$, $e(x) = 0$ otherwise, constitutes an orthonormal basis for $L^2(\mathbb{R})$.

- $$\begin{aligned}(Ze_{m,n})(s,t) &= \textstyle\sum_\ell\ e^{2\pi i t \ell}\ e^{2\pi i m (s-\ell)}\ e(s - n - \ell) \\ &= e^{2\pi i m s}\ e^{-2\pi i t n} (Ze)(s,t)\ .\end{aligned}$$

- $(Ze)(s,t) = 1$ almost everywhere on $[0,1]^2$.

It follows that Z maps an orthonormal basis of $L^2(\mathbb{R})$ to an orthonormal basis of $L^2([0,1]^2)$, so that Z is unitary. We can extend the image of $L^2(\mathbb{R})$ under Z to a different space, isomorphic to $L^2([0,1]^2)$. From (4.1.2), we find, if we allow (s,t) outside $[0,1]^2$,

$$(Zf)(s, t+1) = (Zf)(s,t)\ ,$$

$$(Zf)(s+1, t) = e^{2\pi i t}(Zf)(s,t)\ .$$

Let us therefore define the space $\mathcal{Z}$ by

$$\begin{aligned}\mathcal{Z} = \{F :\ \mathbb{R}^2 \to \mathbb{C};\ F(s,t+1) = F(s,t),\ F(s+1,t) = e^{2\pi i t} F(s,t) \\ \text{and}\ \ \|F\|^2_{\mathcal{Z}} = \int_0^1 dt \int_0^1 ds\ |F(s,t)|^2 < \infty\}\ ;\end{aligned}$$

then Z is unitary between $L^2(\mathbb{R})$ and $\mathcal{Z}$. The inverse map is easy as well: for any $F \in \mathcal{Z}$,

$$(Z^{-1}F)(x) = \int_0^1 dt\ F(x,t)\ ,$$

if this integral is well defined (otherwise, we have to work with a limiting argument).

The Zak transform has many beautiful and useful properties. As is often the case with beautiful and useful concepts, it was discovered several times, and it goes by many different names, according to the field in which one first learns of it. It is also known as the Weil–Brezin map, and it is claimed that even Gauss was aware of some of its properties. It was also used by Gel′fand. J. Zak discovered it independently, and studied it systematically, first for applications to solid state physics, later in a wider setting. An interesting review article, geared mainly to applications in signal analysis, is Janssen (1988).

Only two of the many properties of the Zak transform will concern us here. The first is that if $g_{m,n}(x) = e^{2\pi imx}\, g(x-n)$, then

$$(Zg_{m,n})(s,t) = e^{2\pi ims}\, e^{-2\pi itn}(Zg)(s,t)$$

(as we already showed above, in the special case $g = e$). This implies

$$\begin{aligned}\sum_{m,n\in\mathbb{Z}} |\langle f, g_{m,n}\rangle|^2 &= \sum_{m,n\in\mathbb{Z}} |\langle Zf,\ Zg_{m,n}\rangle|^2 \qquad \text{(by unitarity)} \\ &= \sum_{m,n\in\mathbb{Z}} \left| \int_0^1 ds \int_0^1 dt\ e^{-2\pi ims}\, e^{2\pi int}\, (Zf)(s,t)\, \overline{(Zg)(s,t)} \right|^2 \\ &= \int_0^1 ds \int_0^1 dt\ |(Zf)(s,t)|^2\, |(Zg)(s,t)|^2\ .\end{aligned}$$

Equivalently, we have $Z(F^*F)Z^{-1}$ = multiplication by $|(Zg)(s,t)|^2$ on $\mathcal{Z}$, where $F^*Ff = \sum_{m,n} \langle f,\ g_{m,n}\rangle\ g_{m,n}$. The second property we need concerns the relation between Z and the operators Q, P defined by $(Qf)(x) = xf(x)$, $(Pf)(x) = -if'(x)$ (or, more properly, $(Pf)^\wedge(\xi) = \xi\hat{f}(\xi)$). One checks that

$$[Z(Qf)](s,t) = s(Zf)(s,t) - \frac{1}{2\pi i}\, \partial_t(Zf)(s,t)\ ,$$

which means that $\int dx\ x^2|f(x)|^2 < \infty$, i.e., $Qf \in L^2(\mathbb{R})$, if and only if $\partial_t(Zf) \in L^2([0,1]^2)$. Similarly, $\int d\xi\ \xi^2|\hat{f}(\xi)|^2 < \infty$ or $Pf \in L^2(\mathbb{R})$ if and only if $\partial_s(Zf) \in L^2([0,1]^2)$. We are now ready to attack the proof of Theorem 4.1.1.

Proof of Theorem 4.1.1.

1. Assume that the $g_{m,n}$ constitute a frame. Since

$$\sum_{m,n} |\langle f,\ g_{m,n}\rangle|^2 = \int_0^1 ds \int_0^1 dt\ |Zf(s,t)|^2\ |Zg(s,t)|^2$$

and since Z is unitary, this implies

$$0 < A \le |Zg(s,t)|^2 \le B < \infty\ . \tag{4.1.3}$$

2. The dual frame vectors $\tilde{g}_{m,n}$ are given by

$$\tilde{g}_{m,n} = (F^*F)^{-1}\, g_{m,n}$$

(see §3.2, §3.4.3). Since $Z(F^*F)Z^{-1}$ = multiplication by $|Zg|^2$, it follows that

$$Z\tilde{g}_{m,n} = |Zg|^{-2}\, Zg_{m,n}$$

or

$$\begin{aligned}(Z\tilde{g}_{m,n})(s,t) &= |Zg(s,t)|^{-2}\, e^{2\pi ims}\, e^{-2\pi itn}(Zg)(s,t) \\ &= e^{2\pi ims}\, e^{-2\pi itn}\, [\,\overline{Zg(s,t)}\,]^{-1}\,, \qquad (4.1.4)\end{aligned}$$

which is in $\mathcal{Z}$ by (4.1.3). In particular, (4.1.4) implies that

$$\tilde{g}_{m,n}(x) = e^{2\pi imx}\, \tilde{g}(x-n)\,,$$

with $Z\tilde{g} = 1/\overline{Zg}$.

3. Suppose now that $\int dx\; x^2|g(x)|^2 < \infty$, $\int d\xi\; \xi^2|\hat{g}(\xi)|^2 < \infty$, i.e., that Qg, $Pg \in L^2(\mathbb{R})$. This will lead to a contradiction, which will then prove the theorem. Since Qg, $Pg \in L^2(\mathbb{R})$, we have $\partial_s(Zg)$, $\partial_t(Zg) \in L^2([0,1]^2)$. Consequently,

$$\partial_s Z\tilde{g} = (\overline{Zg})^{-2}\, \overline{\partial_s\, Zg} \quad \text{and} \quad \partial_t\, Z\tilde{g} = (\overline{Zg})^{-2}\, \overline{\partial_t\, Zg}$$

are in $L^2([0,1]^2)$; hence $Q\tilde{g}$, $P\tilde{g} \in L^2(\mathbb{R})$.

4. $\langle \tilde{g},\, g_{m,n}\rangle = \langle Z\tilde{g},\, Zg_{m,n}\rangle$

$$= \int_0^1 ds \int_0^1 dt\; Z\tilde{g}(s,t)\, \overline{Zg(s,t)}\, e^{-2\pi ims}\, e^{2\pi itn} = \delta_{m0}\delta_{n0}\;;$$

similarly,

$$\langle g,\, \tilde{g}_{m,n}\rangle = \delta_{m0}\delta_{n0}\,. \qquad (4.1.5)$$

5. Since Qg, $P\tilde{g} \in L^2(\mathbb{R})$, and since the $(g_{m,n})_{m,n\in\mathbb{Z}}$, $(\tilde{g}_{m,n})_{m,n\in\mathbb{Z}}$ constitute dual frames, we have

$$\langle Qg,\, P\tilde{g}\rangle = \sum_{m,n} \langle Qg,\, \tilde{g}_{m,n}\rangle\langle g_{m,n},\, P\tilde{g}\rangle\,.$$

But $\langle Qg,\, \tilde{g}_{m,n}\rangle = \int dx\; xg(x)\, e^{-2\pi imx}\, \overline{\tilde{g}(x-n)}$

$$= \int dx\; g(x)\, e^{-2\pi imx}\, (x-n)\, \overline{\tilde{g}(x-n)}$$

(because $\langle g, \tilde{g}_{m,n}\rangle = \delta_{m0}\delta_{n0}$)

$$= \langle g_{-m,-n},\, Q\tilde{g}\rangle\,.$$

Similarly, $\langle g_{m,n},\ P\tilde{g}\rangle = \langle Pg,\ \tilde{g}_{-m,-n}\rangle$. Consequently,

$$\langle Qg,\ P\tilde{g}\rangle = \sum_{m,n} \langle Pg,\ \tilde{g}_{-m,-n}\rangle\langle g_{-m,-n},\ Q\tilde{g}\rangle = \langle Pg,\ Q\tilde{g}\rangle\ , \qquad (4.1.6)$$

where the last term is again well defined because $Pg,\ Q\tilde{g} \in L^2(\mathbb{R})$.

6. We have now reached our contradiction: $\langle Qg,\ P\tilde{g}\rangle = \langle Pg,\ Q\tilde{g}\rangle$ is impossible. For any two functions $f_1,\ f_2$ satisfying $|f_j(x)| \le C(1+x^2)^{-1}$, $|\hat{f}_j(\xi)| \le C(1+\xi^2)^{-1}$, we have

$$\begin{aligned} \langle Qf_1,\ Pf_2\rangle &= \int dx\ x f_1(x)\ i\ \overline{f_2'(x)} \\ &= -\int dx\ i\ [x f_1'(x) + f_1(x)]\ f_2(x) \\ &= -i\ \langle f_1,\ f_2\rangle + \langle Pf_1,\ Qf_2\rangle\ . \end{aligned}$$

On the other hand, since $Pg,\ Qg \in L^2(\mathbb{R})$, there exist g_n satisfying $|g_n(x)| \le C_n(1+x^2)^{-1}$, $|\hat{g}_n(\xi)| \le C_n(1+\xi^2)^{-1}$ such that $\lim_{n\to\infty} g_n = g$, $\lim_{n\to\infty} Pg_n = Pg$, $\lim_{n\to\infty} Qg_n = Qg$. (Take for instance $g_n = \sum_{k=0}^{n} \langle g,\ H_k\rangle H_k$, where H_k are the Hermite functions.) A similar sequence $\tilde{g}_n$ can be constructed for $\tilde{g}$. Then

$$\begin{aligned} \langle Pg,\ Q\tilde{g}\rangle &= \lim_{n\to\infty} \langle Pg_n,\ Q\tilde{g}_n\rangle \\ &= \lim_{n\to\infty} [\langle Qg_n,\ P\tilde{g}_n\rangle + i\ \langle g_n,\ \tilde{g}_n\rangle] \\ &= \langle Qg,\ P\tilde{g}\rangle + i\ \langle g, \tilde{g}\rangle\ . \end{aligned}$$

Together with (4.1.6) this implies $\langle g, \tilde{g}\rangle = 0$. However, from (4.1.5) we have $\langle g,\ \tilde{g}\rangle = 1$. This contradiction proves the theorem.[2] ∎

We can summarize our findings so far as

- $\omega_0 t_0 > 2\pi \quad \longrightarrow$ no frames.
- $\omega_0 t_0 = 2\pi \quad \longrightarrow$ there exist frames, but they have bad time-frequency localization.
- $\omega_0 t_0 < 2\pi \quad \longrightarrow$ frames (even tight frames) with excellent time-frequency localization are possible (see §3.4.4.A).

This is represented in Figure 4.1, showing the three regions in the ω_0, t_0-plane. As pointed out in §3.4.1, orthonormal bases are only possible in the "border case" $\omega_0 t_0 = 2\pi$. In view of Theorem 4.1.1, this means that all orthonormal bases of

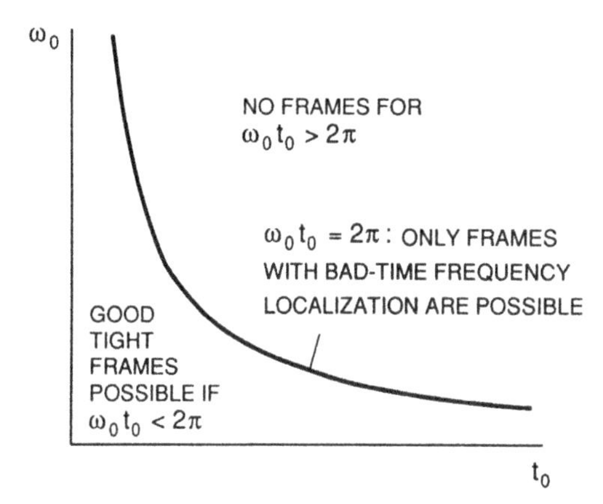

FIG. 4.1. *The regions* $\omega_0 t_0 > 2\pi$ *where no frames are possible, and* $\omega_0 t_0 < 2\pi$*, where tight frames with excellent time-frequency localization exist, are separated by the hyperbola* $\omega_0 t_0 = 2\pi$*, the only region where orthonormal bases are possible.*

the type $\{g_{m,n};\ m, n \in \mathbb{Z}\}$, with $g_{m,n}$ as in (4.1.1), have bad time-frequency localization.

In fact, $\omega_0 \cdot t_0$ is a measure for the "time-frequency density" of the frame constituted by the $g_{m,n}$. We can for instance define this "density" as

$$\lim_{\lambda \to \infty} \frac{\#\ \{(m,n);\ (m\omega_0, nt_0) \in \lambda S\}}{|\lambda S|}\ , \tag{4.1.7}$$

where S is a "reasonable" set in $\mathbb{R}^2$ (with nonzero Lebesgue measure). This limit is independent of S, and equal to $(\omega_0 \cdot t_0)^{-1}$. This "density" also emerged in the time-frequency localization discussion in §3.5; see (3.5.17). The restriction $(\omega_0 \cdot t_0)^{-1} \geq (2\pi)^{-1}$ means that the time-frequency density of the frame has to be at least the Nyquist density (in its "generalized" form; see §2.3). In fact, Theorem 4.1.1 tells us that we have to be strictly *above* the Nyquist density if we want good time-frequency localization with windowed Fourier frames.

Let us now turn to wavelets, where the situation is very different. It turns out that there is no "clean" definition of time-frequency density for wavelet expansions. We already saw a first indication of this in the study of the localization operators in §2.8: for the windowed Fourier case, the number of eigenvalues in the transition region became negligible (as compared to the number of eigenvalues close to 1) when the area of the localization region tended to infinity, whereas these two numbers were of the same order of magnitude in the wavelet case. This made it impossible to make an accurate comparison with the Nyquist density.

Something similar happens with discrete wavelet families. In the discussion of time-frequency localization with frames, in §3.5, we saw that a function that is essentially concentrated in $[-T, T] \times ([-\Omega_1, -\Omega_0] \times [\Omega_0, \Omega_1])$ in the time-frequency plane can be approximated with good precision by a finite number of wavelets. Unlike the windowed Fourier case, the ratio of this number to the time-frequency area $4T(\Omega_1-\Omega_0)$ depends on the desired precision of the approximation (see (3.5.11)), which makes it impossible (as in the continuous case) to make a precise comparison with the Nyquist density. On the other hand, if we try to define the analog of (4.1.7) for the hyperbolic lattices of Figure 1.4a, then we find that

$$R_S(\lambda) = \frac{\#\ \{(m,n);\ (m\omega_0, nt_0) \in \lambda S\}}{|\lambda S|}$$

(where S is chosen so that the numerator is finite) does not tend to a limit as $\lambda \to \infty$. For $S = [-T,T] \times ([-2^{m_1},\ -2^{m_0}] \cup [2^{m_0},\ 2^{m_1}])$ and $a_0 = 2$, for instance, $R_S(\lambda)$ oscillates between $\rho(1-2^{m_0-m_1-1})/(1-2^{m_0-m_1})$ and $2\rho(1-2^{m_0-m_1-1})/(1-2^{m_0-m_1})$, where ρ depends on the chosen wavelet ψ. It might be this phenomenon, rather than the absence of an intrinsic time-frequency density for the frame, that causes the problem in counting the number of wavelets needed for time-frequency localization. So let us probe a little deeper.

As we mentioned before, there is no a priori restriction on the range of dilation and translation parameters in a wavelet frame: *any* choice of a_0, b_0 can be used to define a tight frame with good localization in both time and frequency for ψ (see §3.3.5.A). In fact, from a (tight) frame with discretization parameters a_0, b_0 we can always construct a different (tight) frame with parameters a_0, b_0' (same a_0) with b_0' arbitrary, by simple dilation.[3] It is therefore not surprising that we have no a priori restrictions on a_0, b_0. We can remove this dilational freedom by fixing not only the normalization of ψ, $\|\psi\| = 1$, but by also imposing a fixed value for $\int d\xi\ |\xi|^{-1}\ |\hat{\psi}(\xi)|^2$. For real ψ, we could impose, for instance, $\int_0^\infty d\xi\ \xi^{-1}\ |\hat{\psi}(\xi)|^2 = \int_{-\infty}^0 d\xi\ |\xi|^{-1}\ |\hat{\psi}(\xi)|^2 = 1$. A tight frame generated by a ψ thus restricted would automatically have frame bound $A = \frac{2\pi}{b_0 \ln a_0}$ (see Theorem 3.3.1). A comparison with the formula $A = \frac{2\pi}{\omega_0 t_0}$ for tight windowed Fourier frames suggests that maybe $(b_0 \ln a_0)^{-1}$ could play the role of time-frequency density for the wavelet case. The following example destroys all hope in this direction. In the next section we will encounter the Meyer wavelet ψ; it has a compactly supported Fourier transform $\hat{\psi} \in C^k$ (where k may be ∞, as in §3.3.5.A; the two constructions are related) and the $\psi_{m,n} = 2^{-m/2}\ \psi(2^{-m}x - n)$, $m, n \in \mathbb{Z}$ are an orthonormal basis for $L^2(\mathbb{R})$. Let us, for this chapter only, define

$$\psi^b_{m,n}(x) = 2^{-m/2}\ \psi(2^{-m}x - nb)\ , \tag{4.1.8}$$

where ψ is the Meyer wavelet, and $b > 0$ is arbitrary. Consider the b-dependent families $F(b) = \{\psi^b_{m,n};\ m, n \in \mathbb{Z}\}$. As b changes, the "density" of the associated lattice changes as well. (Note that a_0 and ψ are the same for all the $F(b)$!) If any representation like Figure 4.1 held also for wavelets, then we would expect, since $F(1)$ is an orthonormal basis for $L^2(\mathbb{R})$, that $F(b)$ would not span all of $L^2(\mathbb{R})$ if $b > 1$ ("not enough" vectors), and that $F(b)$ would not be linearly independent ("too many" vectors) if $b < 1$. Yet one can prove (see Theorem 2.10 in Daubechies (1990); we also sketch this proof later in this chapter) that for some $\epsilon > 0$, $F(b)$ is a Riesz basis for $L^2(\mathbb{R})$, for any $b \in]1-\epsilon,\ 1+\epsilon[$. This example shows conclusively that it is not always safe to apply "time-frequency space density intuition" to families of wavelets.

4.2. Orthonormal bases.

4.2.1. Orthonormal wavelet bases. The conclusion of the last paragraph seems a rather negative point for wavelets: no clean time-frequency density concept. In this section we emphasize a much more positive aspect: the existence of orthonormal wavelet bases with good time-frequency localization.

Historically, the first orthonormal wavelet basis is the Haar basis, constructed long before the term "wavelet" was coined. The basic wavelet is then, as we already saw in Chapter 1,

$$\psi(x) = \begin{cases} 1, & 0 \le x < \frac{1}{2}, \\ -1, & \frac{1}{2} \le x < 1, \\ 0 & \text{otherwise.} \end{cases} \tag{4.2.1}$$

We showed in §1.6 that the $\psi_{m,n}(x) = 2^{-m/2}\psi(2^{-m}x - n)$ constitute an orthonormal basis for $L^2(\mathbb{R})$. The Haar function is not continuous, and its Fourier transform decays only like $|\xi|^{-1}$, corresponding to bad frequency localization. It may therefore seem that this basis is no better than the windowed Fourier basis

$$g_{m,n}(x) = e^{2\pi i m x}\, g(x-n) \tag{4.2.2}$$

with

$$g(x) = \begin{cases} 1, & 0 \le x \le 1, \\ 0, & \text{otherwise,} \end{cases}$$

which is also an orthonormal basis for $L^2(\mathbb{R})$. However, the Haar basis already has advantages that this windowed Fourier basis does not have. It turns out, for instance, that the Haar basis is an unconditional basis for $L^p(\mathbb{R})$, $1 < p < \infty$, whereas the windowed Fourier basis (4.2.2) is not if $p \neq 2$.[4] We will come back to this in Chapter 9. For the analysis of smoother functions, the discontinuous Haar basis is ill suited.

An orthonormal wavelet basis with time-frequency properties complementary to the Haar basis is given by the Littlewood–Paley basis,

$$\hat{\psi}(\xi) = \begin{cases} (2\pi)^{-1/2}, & \pi \le |\xi| \le 2\pi, \\ 0, & \text{otherwise} \end{cases}$$

or

$$\psi(x) = (\pi x)^{-1}\,(\sin 2\pi x - \sin \pi x)\ .$$

It is easy to check that the $\psi_{m,n}(x) = 2^{-m/2}\,\psi(2^{-m}x - n)$ constitute indeed an orthonormal basis for $L^2(\mathbb{R})$. We have $\|\psi_{m,n}\| = 1$ for all $m, n \in \mathbb{Z}$, and

$$\sum_{m,n} |\langle f, \psi_{m,n}\rangle|^2 = \sum_{m,n} (2\pi)^{-1}\, 2^m \left| \int_{2^{-m}\pi \le |\xi| \le 2^{-m+1}\pi} d\xi\ \hat{f}(\xi)\, e^{in2^m\xi} \right|^2$$

$$= \sum_{m,n} (2\pi)^{-1} 2^{-m} \left| \int_{\pi \le |\zeta| \le 2\pi} d\zeta \; \hat{f}(2^{-m}\zeta) \; e^{in\zeta} \right|^2$$

$$= \sum_{m,n} (2\pi)^{-1} 2^{-m} \left| \int_0^{2\pi} d\zeta e^{in\zeta} \left[\hat{f}(2^{-m}\zeta)\chi_{[\pi,2\pi]}(\zeta) + \hat{f}(2^{-m}(\zeta - 2\pi))\chi_{[0,\pi]}(\zeta) \right] \right|^2$$

$$= \sum_m 2^{-m} \int_0^{2\pi} d\zeta \; |\hat{f}(2^{-m}\zeta) \; \chi_{[\pi,2\pi]}(\zeta) + \hat{f}(2^{-m}(\zeta - 2\pi)) \; \chi_{[0,\pi]}(\zeta)|^2$$

$$= \sum_m 2^{-m} \int_{\pi \le |\zeta| \le 2\pi} d\zeta \; |\hat{f}(2^{-m}\zeta)|^2 = \int_{-\infty}^{\infty} d\zeta \; |\hat{f}(\zeta)|^2 = \|f\|^2 \; .$$

By Proposition 3.2.1, this implies that $\{\psi_{m,n};\ m, n \in \mathbb{Z}\}$ is an orthonormal basis for $L^2(\mathbb{R})$. The decay of $\psi(x)$ is as bad ($\psi(x) \sim |x|^{-1}$ for $x \to \infty$) as that of the orthonormal windowed Fourier basis used in the Shannon expansion (2.1.1); both have excellent frequency localization, since their Fourier transforms are compactly supported.

In the last ten years, several orthonormal wavelet bases for $L^2(\mathbb{R})$ have been constructed which share the best features of both the Haar basis and the Littlewood–Paley basis: these new constructions have excellent localization properties in both time and frequency. The first construction is due to Stromberg (1982); his wavelets have exponential decay and are in C^k (k arbitrary but finite). Unfortunately, his construction was little noticed at the time. The next example is the Meyer basis mentioned above (Meyer (1985)), in which $\hat{\psi}$ is compactly supported (hence $\psi \in C^\infty$) and C^k (k arbitrary, may be ∞). Unaware at that time of Stromberg's construction, Y. Meyer actually found this basis while trying to prove a wavelet equivalent of Theorem 4.1.1, which would have shown the non-existence of these nice wavelet bases! Soon after, Tchamitchian (1987) constructed the first example of what we shall call biorthogonal wavelet bases (see §8.3). The next year, Battle (1987) and Lemarié (1988) used very different methods to construct identical families of orthonormal wavelet bases with exponentially decaying $\psi \in C^k$ (k arbitrary but finite). (Battle was inspired by techniques in quantum field theory; Lemarié reused some of Tchamitchian's computations.) Despite having similar properties, the Battle–Lemarié wavelets are different from the Stromberg wavelets. In the fall of 1986, S. Mallat and Y. Meyer developed the "multiresolution analysis" framework, which gave a satisfactory explanation for all these constructions, and provided a tool for the construction of yet other bases. But this is for later chapters. Before we get into multiresolution analysis, let us review the construction of Meyer's wavelet basis.

The construction of $|\hat{\psi}|$ is similar to the tight frame in §3.3.5.A. That frame had redundancy 2 (twice "too many" vectors). To get rid of this redundancy, Meyer's construction combines positive and negative frequencies (reducing a pair of functions to a single function). In order to achieve orthonormality, some clever tricks with phase factors are needed. More explicitly, we define ψ by

$$\hat{\psi}(\xi) = \begin{cases} (2\pi)^{-1/2}\, e^{i\xi/2}\, \sin\left[\frac{\pi}{2}\, \nu\left(\frac{3}{2\pi}|\xi| - 1\right)\right], & \frac{2\pi}{3} \le |\xi| \le \frac{4\pi}{3}\,, \\ (2\pi)^{-1/2}\, e^{i\xi/2}\, \cos\left[\frac{\pi}{2}\, \nu\left(\frac{3}{4\pi}|\xi| - 1\right)\right], & \frac{4\pi}{3} \le |\xi| \le \frac{8\pi}{3}\,, \\ 0 & \text{otherwise,} \end{cases} \tag{4.2.3}$$

where ν is a C^k or C^∞ function satisfying (3.3.25), i.e.,

$$\nu(x) = \begin{cases} 0 & \text{if} \quad x \le 0\,, \\ 1 & \text{if} \quad x \ge 1\,, \end{cases} \tag{4.2.4}$$

with the additional property

$$\nu(x) + \nu(1 - x) = 1\,. \tag{4.2.5}$$

The regularity of $\hat{\psi}$ is the same as that of ν. Figure 4.2 shows the shape of a typical ν and $|\hat{\psi}|$. In order to prove that the $\psi_{m,n}(x) = 2^{-m/2}\, \psi(2^{-m}x - n)$ constitute an orthonormal basis, we only have to check that $\|\psi\| = 1$ and that the $\psi_{m,n}$ make up a tight frame with frame constant 1 (see Proposition 3.2.1).

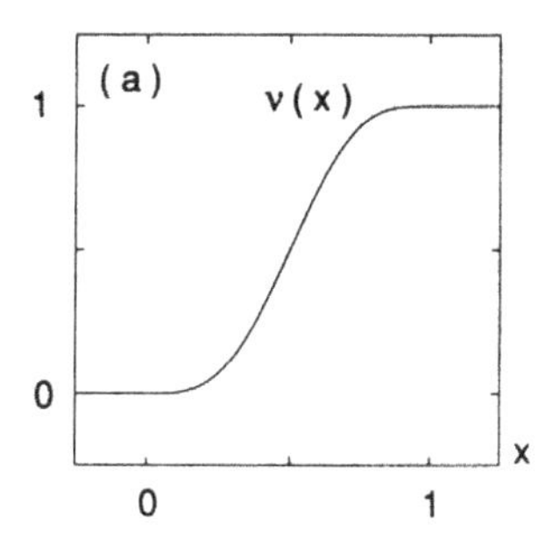

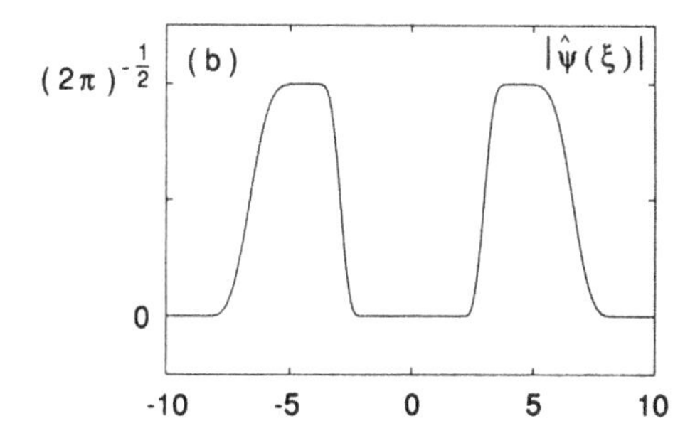

FIG. 4.2. *Functions ν and ψ as given by (4.2.3)–(4.2.5).*

We have

$$\|\psi\|^2 = (2\pi)^{-1}\left\{\int\limits_{\frac{2\pi}{3}\le|\xi|\le\frac{4\pi}{3}} d\xi\ \sin^2\left[\frac{\pi}{2}\,\nu\left(\frac{3}{2\pi}|\xi|-1\right)\right]\right.$$

$$\left.+\int\limits_{\frac{4\pi}{3}\le|\xi|\le\frac{8\pi}{3}} d\xi\ \cos^2\left[\frac{\pi}{2}\,\nu\left(\frac{3}{4\pi}|\xi|-1\right)\right]\right\}$$

$$= (2\pi)^{-1}\left\{2\,\frac{2\pi}{3}\int_0^1 ds\ \sin^2\left[\frac{\pi}{2}\,\nu(s)\right]+2\,\frac{4\pi}{3}\int_0^1 ds\ \cos^2\left[\frac{\pi}{2}\,\nu(s)\right]\right\}$$

$$= \frac{2}{3}\left\{1+\int_0^1 ds\ \cos^2\left[\frac{\pi}{2}\,\nu(s)\right]\right\}.$$

But

$$\int_0^1 ds\ \cos^2\left[\frac{\pi}{2}\,\nu(s)\right] = \int_0^{1/2} ds\ \cos^2\left[\frac{\pi}{2}\,\nu(s)\right]$$

$$+\int_0^{1/2} ds\ \cos^2\left[\frac{\pi}{2}\left(1-\nu\left(\frac{1}{2}-s\right)\right)\right]$$

(because $\nu(s+1/2)=1-\nu(1/2-s)$ by (4.2.5))

$$= \int_0^{1/2} ds\ \cos^2\left[\frac{\pi}{2}\,\nu(s)\right]+\int_0^{1/2} ds'\ \sin^2\left[\frac{\pi}{2}\,\nu(s')\right]$$

$$= \frac{1}{2}\ ;$$

hence $\|\psi\|^2 = 1$.

To evaluate $\sum_{m,n}|\langle f,\,\psi_{m,n}\rangle|^2$, we use Tchamitchian's frame bound estimates (3.3.21), (3.3.22). We first prove that $\beta_1(2\pi(2k+1)) = 0$ for all $k\in\mathbb{Z}$, i.e., for all $\zeta\in\mathbb{R}$,

$$\sum_{\ell=0}^{\infty}\overline{\hat\psi(2^\ell\zeta)}\ \hat\psi[2^\ell(\zeta+2\pi(2k+1))] = 0\ . \tag{4.2.6}$$

Because of the support of $\hat\psi$, nonzero contributions to (4.2.6) are only possible if $|2^\ell\zeta|\le\frac{8\pi}{3}$ and $|2^\ell(\zeta+2\pi(2k+1))|\le\frac{8\pi}{3}$, implying $2^\ell|2k+1|\le 8/3$. The only pairs (ℓ,k) that do not violate this are $(0,0)$, $(0,-1)$, $(1,0)$, and $(1,-1)$. Let us check $k=0$ ($k=-1$ is analogous). Then the left-hand side of (4.2.6) becomes

$$\overline{\hat\psi(\zeta)}\ \hat\psi(\zeta+2\pi)+\overline{\hat\psi(2\zeta)}\ \hat\psi(2\zeta+4\pi)\ . \tag{4.2.7}$$

One easily checks that both terms in (4.2.7) vanish unless $-\frac{4\pi}{3} \le \zeta \le -\frac{2\pi}{3}$. For ζ within this interval, $\zeta = -\frac{4\pi}{3} + \frac{2\pi}{3}\alpha$ with $0 \le \alpha \le 1$, we have

$$\begin{aligned}
(4.2.7) \quad &= \quad e^{-i\zeta/2} \sin\left[\frac{\pi}{2}\, \nu(1-\alpha)\right]\, e^{i(\zeta+2\pi)/2} \sin\left[\frac{\pi}{2}\, \nu(\alpha)\right] \\
&\quad + e^{-i\zeta} \cos\left[\frac{\pi}{2}\, \nu(1-\alpha)\right]\, e^{i(\zeta+2\pi)} \cos\left[\frac{\pi}{2}\, \nu(\alpha)\right] \\
&= \quad -\cos\left[\frac{\pi}{2}\, \nu(\alpha)\right] \sin\left[\frac{\pi}{2}\, \nu(\alpha)\right] + \sin\left[\frac{\pi}{2}\, \nu(\alpha)\right] \cos\left[\frac{\pi}{2}\, \nu(\alpha)\right] \\
&\qquad\qquad\qquad\qquad\qquad\qquad\qquad\qquad \text{(use (4.2.5))} \\
&= \quad 0\,.
\end{aligned}$$

This establishes (4.2.6). On the other hand, one easily checks that $\sum_m |\hat{\psi}(2^m\xi)|^2 = (2\pi)^{-1}$ for all $\xi \neq 0$. It then follows from (3.3.21), (3.3.22) that the $\psi_{m,n}$ constitute a tight frame with frame bound 1. (Similar computations can be used to prove that $F(b)$ (see the end of §4.1) constitutes a Riesz basis for $L^2(\mathbb{R})$ if b is close to 1.[5])

This proof that the Meyer wavelets constitute an orthonormal basis relies on quasi-miraculous cancellations, using the interplay between the phase of $\hat{\psi}$ and the special property (4.2.5) of ν. Using multiresolution analysis, we will be able to explain away most of the miracle (see next chapter). Figure 4.3 shows a graph of $\psi(x)$, with the $C^{4-\epsilon}$ choice $\nu(x) = x^4(35 - 84x + 70x^2 - 20x^3)$ for $0 \le x \le 1$. Note that even if $\nu \in C^\infty$, so that ψ decays faster than any inverse polynomial, i.e., for all $N \in \mathbb{N}$, there exists $C_N < \infty$ so that

$$|\psi(x)| \le C_N\, (1 + |x|^2)^{-N}\,, \tag{4.2.8}$$

the *numerical* decay of ψ may be rather slow (i.e., $\inf\{a;\ |\psi(x)| \le .001\, \|\psi\|_{L^\infty}$ for $|x| > a\}$ may be very large, reflecting a large C_N in (4.2.8)). The exponentially decaying wavelets of Stromberg or Battle–Lemarié have much faster numerical decay, at the price of sacrificing regularity.

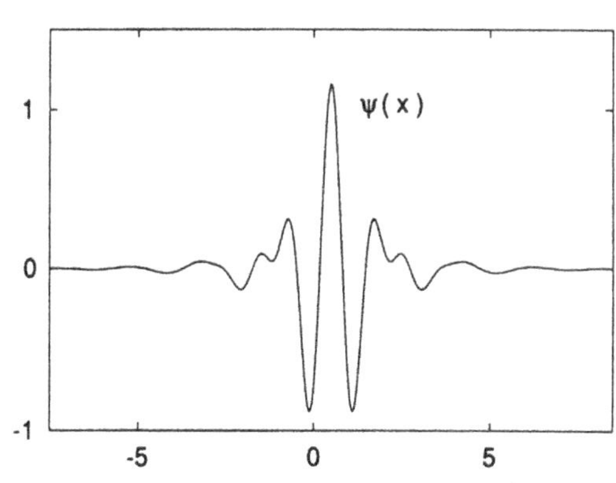

FIG. 4.3. *The Meyer wavelet $\psi(x)$ for the choice $\nu(x) = x^4(35 - 84x + 70x^2 - 20x^3)$.*

In the matter of orthonormal bases then, wavelets seem to do quite a bit better than windowed Fourier functions: there are constructions in which both ψ and $\hat{\psi}$ have fast decay, in stark contrast with Theorem 4.1.1, which forbids simultaneous good decay for g and $\hat{g}$ if g is a window function leading to an orthonormal basis. If I had written this chapter three years ago, this is probably where I would have stopped. But matters are not quite that simple: in the last few years, the windowed Fourier transform has led to a few surprises, which we will discuss briefly in the remainder of this chapter.

4.2.2. The windowed Fourier transform revisited: 'Good" orthonormal bases after all! One way in which one could try to generalize the windowed Fourier construction, so as to get round Theorem 4.1.1, is to consider families $g_{m,n}(x)$ that are not generated by a strict time-frequency lattice. This allows for a little leeway: Bourgain (1988) has constructed an orthonormal basis $(g_j)_{j\in J}$ for $L^2(\mathbb{R})$ such that

$$\int dx\ (x-x_j)^2\ |g_j(x)|^2 \le C\ ,$$
$$\int d\xi\ (\xi-\xi_j)^2\ |\hat{g}_j(\xi)|^2 \le C\ , \tag{4.2.9}$$

uniformly in $j \in J$, where $x_j = \int dx\ x|g_j(x)|^2$, $\xi_j = \int d\xi\ \xi|\hat{g}_j(\xi)|^2$. (Note that wavelet bases do not satisfy such a uniform bound.[6]) Giving up the lattice structure therefore permits better localization than allowed by the Balian–Low theorem. However, Steger (private communication, 1986) proved that even slightly better localization than (4.2.9) is impossible: $L^2(\mathbb{R})$ does not admit an orthonormal basis $(g_j)_{j\in J}$ satisfying

$$\int dx\ (x-x_j)^{2(1+\epsilon)}\ |g_j(x)|^2 \le C\ ,$$
$$\int d\xi\ (\xi-\xi_j)^{2(1+\epsilon)}\ |\hat{g}_j(\xi)|^2 \le C\ , \tag{4.2.10}$$

uniformly in j, if $\epsilon > 0$. This approach can therefore not lead to good time-frequency localization. There is another way in which we can try to break away from the lattice scheme (4.1.1). Note that in (4.2.9), (4.2.10), "time-frequency localization" stands for strong decay properties of the $g_{m,n}$, $(g_{m,n})^\wedge$ away from the average values $x_{m,n}$, $\xi_{m,n}$. This corresponds to a picture in which both $g_{m,n}$ and $(g_{m,n})^\wedge$ have essentially *one* peak. Wilson (1987) proposes instead to construct orthonormal bases $g_{m,n}$ of the type

$$g_{m,n}(x) = f_m(x-n), \qquad m \in \mathbb{N},\ n \in \mathbb{Z}\ , \tag{4.2.11}$$

where $\hat{f}_m$ has *two* peaks, situated near $\frac{m}{2}$ and $-\frac{m}{2}$,

$$\hat{f}_m(\xi) = \phi_m^+\left(\xi - \frac{m}{2}\right) + \phi_m^-\left(\xi + \frac{m}{2}\right)\ , \tag{4.2.12}$$

with ϕ_m^+, ϕ_m^- centered around 0. This ansatz changes the picture completely. Wilson (1987) proposes numerical evidence for the existence of such an orthonormal basis, with uniform exponential decay for f_m and ϕ_m^+, ϕ_m^-. In his numerical construction he further "optimizes" the localization by requiring

$$\int d\xi\ \xi^2 \overline{(\psi_{m,n})^\wedge(\xi)}\ \psi_{m',n'}(\xi) = 0 \qquad \text{if} \quad |m-m'| > 1 \quad \text{or if} \quad \begin{cases} |m-m'| = 1\ , \\ |n-n'| > 1\ . \end{cases} \tag{4.2.13}$$

Sullivan et al. (1987) present arguments explaining both the existence of Wilson's basis and its exponential decay. In both papers there are infinitely many functions $\phi_m^\pm$; as m tends to ∞, the $\phi_m^\pm$ tend to a limit function $\phi_\infty^\pm$.

The moral of Wilson's construction is that orthonormal bases with good phase space localization seem possible after all if bi-modal functions as in (4.2.13) are used.

Note that many of our wavelet constructions, frames as well as the orthonormal bases we saw earlier, have these *two* peaks in frequency (one for $\xi > 0$, one for $\xi < 0$). In the case of frames, or for the continuous wavelet transform, the two frequency regions can be separated (corresponding to one-frequency-peak functions; see §3.3.5.A or (2.4.9)), but this does not seem to be the case for orthonormal bases. We will see later that the two frequency peaks of ψ need not be symmetric: there even exist examples with $\|\psi\|^{-2} \int_{\xi \le 0} d\xi\ |\hat{\psi}(\xi)|^2$ arbitrarily small (but strictly positive!). However, there is no example, so far, of reasonably well-localized functions $\psi^\pm$ with support $(\widehat{\psi^\pm}) \subset \mathbb{R}^\pm$ and such that the $\{\psi_{m,n}^\epsilon;\ m, n \in \mathbb{Z},\ \epsilon = + \text{ or } -\}$ constitute an orthonormal basis for $L^2(\mathbb{R})$, corresponding to wavelet bases with only one "peak" in frequency. (Equivalently, there is no example of a reasonably smooth function $\eta = \widehat{\psi^+}$ such that the functions $2^{m/2} \exp\,(2\pi i\ 2^m\ n\xi)\ \eta(2^m \xi)$, $m, n \in \mathbb{Z}$, are an orthonormal basis of $L^2(\mathbb{R}^+)$.) It is believed, without proof so far, that no such basis exists.[7]

But let us return to Wilson bases. If one gives up the restriction (4.2.13) (if f_m, $\phi_m^\pm$ have exponential decay, then these quantities decay exponentially fast in $|m-m'|$, $|n-n'|$ anyway), then Wilson's ansatz (4.2.11), (4.2.12) can be dramatically simplified.

In Daubechies, Jaffard, and Journé (1991), a construction is proposed that uses only *one* function ϕ. Explicitly, this construction defines

$$g_{m,n}(x) = f_m(x-n), \qquad m \in \mathbb{N} \setminus \{0\}, \quad n \in \mathbb{Z}\ , \tag{4.2.14}$$

with

$$\hat{f}_1(\xi) = \phi(\xi)\ ,$$
$$\hat{f}_2(\xi) = \frac{1}{\sqrt{2}}\ [\phi(\xi - 2\pi) - \phi(\xi + 2\pi)]\ ,$$
$$\hat{f}_3(\xi) = \frac{1}{\sqrt{2}}\ [\phi(\xi - 2\pi) + \phi(\xi + 2\pi)] e^{i\xi/2}\ ,$$

$$\hat{f}_4(\xi) = \frac{1}{\sqrt{2}} \left[\phi(\xi - 4\pi) + \phi(\xi + 4\pi)\right] ,$$

$$\hat{f}_5(\xi) = \frac{1}{\sqrt{2}} \left[\phi(\xi - 4\pi) - \phi(\xi + 4\pi)\right] e^{i\xi/2} ,$$

etc$\cdots$

$$\text{or} \quad \hat{f}_{2\ell+\sigma}(\xi) = \frac{1}{\sqrt{2}} \left[\phi(\xi - 2\pi\ell) + (-1)^{\ell+\sigma} \phi(\xi + 2\pi\ell)\right] e^{i\sigma\xi/2}, \tag{4.2.15}$$

with $\ell \in \mathbb{N}$, $\sigma = 0$ or 1, and $\ell = 0$, $\sigma = 0$ excluded. The result of all these phase factors and alternating signs is that

$$f_1(x) = \check{\phi}(x) ,$$

$$f_{2\ell+\sigma}(x) = \frac{1}{\sqrt{2}} \check{\phi}\left(x + \frac{\sigma}{2}\right) e^{i\pi\sigma\ell} \left[e^{2\pi i\ell x} + (-1)^{\ell+\sigma} e^{-2\pi i\ell x}\right] .$$

If we relabel the $g_{m,n}$ in (4.2.14) by defining $G_{m,n}$, $m \in \mathbb{N}$, $n \in \mathbb{Z}$ as

$$G_{0,n} = g_{1,n} ,$$

$$G_{\ell,2n+\sigma} = g_{2\ell+\sigma,n} ,$$

then

$$G_{0,n}(x) = \check{\phi}(x - n) , \tag{4.2.16}$$

and for $\ell > 0$,

$$G_{\ell,n}(x) = \sqrt{2}\, \check{\phi}\left(x - \frac{n}{2}\right) \begin{cases} \cos\, 2\pi\, \ell x & \text{if} \quad \ell + n \text{ is even} , \\ \sin\, 2\pi\, \ell x & \text{if} \quad \ell + n \text{ is odd} . \end{cases} \tag{4.2.17}$$

This construction (as well as others mentioned below) shows therefore that the key to obtaining good time-frequency localization (ϕ can be chosen so that ϕ, $\check{\phi}$ have exponential decay) *and* orthonormality in the windowed Fourier framework is to use sines and cosines (alternated in an appropriate way) rather than complex exponentials.

But let us get back to (4.2.14), (4.2.15) and show how this construction can lead to an orthonormal basis. As usual, we only need to check $\|g_{m,n}\| = 1$ and $\sum_{m=1}^{\infty} \sum_{n\in\mathbb{Z}} |\langle h,\, g_{m,n}\rangle|^2 = \|h\|^2$. We immediately have $\|g_{1,n}\| = \|f_1\| = \|\phi\|$, and for $m > 1$,

$$\begin{aligned} \|g_{m,n}\|^2 &= \|f_m\|^2 = \|f_{2\ell+\sigma}\|^2 \qquad (m = 2\ell + \sigma, \quad \ell > 0) \\ &= \frac{1}{2} \left[2\|\phi\|^2 + 2(-1)^{\ell+\sigma} \int d\xi\ \phi(\xi)\phi(\xi + 4\pi\ell)\right] \end{aligned}$$

(we assume ϕ is real, for simplicity). Hence $\|g_{m,n}\| = 1$ for all m, n if

$$\int d\xi\ \phi(\xi)\ \phi(\xi + 4\pi\ell) = \delta_{\ell 0} . \tag{4.2.18}$$

On the other hand,

$$\sum_{m=1}^{\infty} \sum_{n\in\mathbb{Z}} |\langle h,\, g_{m,n}\rangle|^2 = 2\pi \sum_{m=1}^{\infty} \sum_{k\in\mathbb{Z}} \int d\xi\ \hat{h}(\xi)\ \overline{\hat{h}(\xi + 2\pi k)\hat{f}_m(\xi)}\ \hat{f}_m(\xi + 2\pi k) .$$

This equals $\|h\|^2$ if

$$\sum_{m=1}^{\infty} \overline{\hat{f}_m(\xi)}\, \hat{f}_m(\xi + 2\pi k) = (2\pi)^{-1}\delta_{k0} \,. \tag{4.2.19}$$

A few simple manipulations lead to

$$\begin{aligned} &\sum_{m=1}^{\infty} \overline{\hat{f}_m(\xi)}\, \hat{f}_m(\xi + 2\pi k) \\ &\quad = \phi(\xi)\, \phi(\xi + 2\pi k) + \frac{1}{2} \sum_{\ell \neq 0} \phi(\xi + 2\pi\ell)\phi(\xi + 2\pi\ell + 2\pi k)\, [1 + (-1)^k] \\ &\qquad + \frac{1}{2} \sum_{\ell \neq 0} (-1)^\ell\, \phi(\xi - 2\pi\ell)\phi(\xi + 2\pi\ell + 2\pi k)[1 - (-1)^k] \,. \end{aligned} \tag{4.2.20}$$

If k is odd, $k = 2k' + 1$, then this reduces to

$$\sum_{\ell \in \mathbb{Z}} (-1)^\ell\, \phi(\xi - 2\pi\ell)\, \phi(\xi + 2\pi(\ell + 2k' + 1)) \,, \tag{4.2.21}$$

which is zero, since the substitution $\ell' = -(\ell + 2k' + 1)$ transforms (4.2.21) in its negative. If k is even, $k = 2k'$, then (4.2.19) reduces to

$$\sum_{\ell \in \mathbb{Z}} \phi(\xi + 2\pi\ell)\, \phi(\xi + 2\pi\ell + 4\pi k') = (2\pi)^{-1}\delta_{k'0} \,. \tag{4.2.22}$$

The $\{g_{m,n};\ m \in \mathbb{N} \setminus \{0\},\ n \in \mathbb{Z}\}$ therefore constitute an orthonormal basis if ϕ is a real function satisfying (4.2.18) and (4.2.22). Note that integrating (4.2.22) over ξ, between 0 and 2π, automatically leads to (4.2.18), so that we really have only the single condition (4.2.22) to satisfy. This turns out to be easy: we can take for instance support $\phi \subset [-2\pi,\ 2\pi]$, so that (4.2.22) is automatically satisfied for $k' \neq 0$, and we only need to check $\sum_{\ell \in \mathbb{Z}} \phi(\xi + 2\pi\ell)^2 = (2\pi)^{-1}$. This is true if, e.g.,

$$\phi(\xi) = \begin{cases} (2\pi)^{-1/2} \sin\left[\frac{\pi}{2}\, \nu\left(\frac{\xi}{2\pi} + 1\right)\right], & -2\pi \le \xi \le 0 \,, \\ (2\pi)^{-1/2} \cos\left[\frac{\pi}{2}\, \nu\left(\frac{\xi}{2\pi}\right)\right], & 0 \le \xi \le 2\pi \,, \\ 0 & \text{otherwise} \,, \end{cases}$$

with ν as in (4.2.4). If ν is C^∞, then the f_m have decay faster than any inverse polynomial, but, as for the Meyer basis, the numerical decay may be slow. Faster decay for the f_m can be obtained with noncompactly supported ϕ. To construct such a ϕ, satisfying (4.2.22), we can again use the Zak transform, now normalized so that

$$(Zh)(s, t) = (4\pi)^{1/2} \sum_{\ell \in \mathbb{Z}} e^{2\pi i t \ell}\, h(4\pi(s - \ell)) \,.$$

With this normalization, Z is again unitary from $L^2(\mathbb{R})$ to $L^2([0,1]^2)$. It is not hard to check that (4.2.22) is equivalent to

$$|(Z\phi)(s,t)|^2 + |(Z\phi)(s+\tfrac{1}{2},\ t)|^2 = 2\ . \tag{4.2.23}$$

(Full details are given in Daubechies, Jaffard, and Journé (1991).) This suggests the following technique for constructing ϕ:

- Take any h such that

$$0 < \alpha \le |Zh(s,t)|^2 + |Zh(s+\tfrac{1}{2},t)|^2 \le \beta < \infty\ ; \tag{4.2.24}$$

- Define ϕ by

$$Z\phi(s,t) = \sqrt{2}\ \frac{Zh(s,t)}{\left[|Zh(s,t)|^2 + \left|Zh\left(s+\frac{1}{2},t\right)\right|^2\right]^{1/2}}\ . \tag{4.2.25}$$

If h and $\hat{h}$ both have exponential decay, then ϕ turns out to have exponential decay as well. Figure 4.4 shows the graph of ϕ and $\check{\phi}$ when h is a Gaussian. (Gaussians do indeed satisfy (4.2.24).) An interesting observation is that (4.2.23) is exactly equivalent to the requirement that the $\check{\phi}_{m,n}(x) = e^{2\pi i m x}\ \check{\phi}\left(x - \frac{n}{2}\right)$, or equivalently, the $\psi_{m,n}(\xi) = e^{\pi i n \xi}\ \phi(\xi - m)$, with $m, n \in \mathbb{Z}$, constitute a tight frame (with necessarily redundancy 2) for $L^2(\mathbb{R})$. The construction (4.2.25) can then be interpreted as the transition from a general frame, generated by h, to a tight frame, by application of $(F^*F)^{-1/2}$ (see note 11 after Chapter 3, or Daubechies, Jaffard, and Journé (1991)). This Wilson basis can therefore be viewed as the result of a clever "weeding" process on a (tight) frame with "twice too many" elements.

Many variations on this Wilson scheme are possible. Laeng (1990) has constructed an extension of the above scheme in which the frequency spacing need not be as regular as here. Auscher (1990) has reformulated the whole construction: starting directly from (4.2.16), (4.2.17) as an ansatz, he derives all the results without use of the Fourier transform, and constructs different examples. In particular, he obtains examples where, in the notations of (4.2.17), the "window" $\check{\phi}$ is compactly supported, which is very useful in applications. (The decay in frequency is less crucial, as long as it is "reasonable.") These examples can also be viewed as the result of a "weeding" procedure on the tight frames with redundancy 2 obtained by taking $\omega_0 t_0 = \pi$ in §3.4.4.A.

Other windowed Fourier bases using cosines and sines rather than complex exponentials, and leading to good time-frequency localization, have been found by Malvar (1990) and Coifman and Meyer (1990). Malvar's paper again uses alternating sines and cosines; he presents applications of his construction to speech coding. Coifman and Meyer's "localized sine basis" starts from a partition of $\mathbb{R}$ in intervals,

$$\mathbb{R} = \bigcup_{j\in\mathbb{Z}} [a_j,\ a_{j+1}]\ ,$$

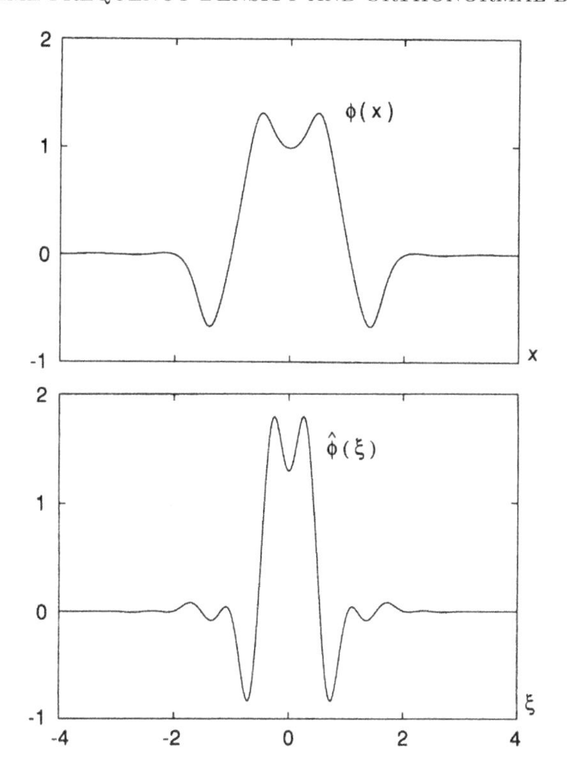

FIG. 4.4. *The functions ϕ and $\check{\phi}$ corresponding to* (4.2.25) *if* $h(x) = \pi^{-1/4} \exp(-x^2/2)$.

with $a_j < a_{j+1}$ and $\lim_{j\to\pm\infty} a_j = \pm\infty$. They then build window functions w_j localized around these $I_j = [a_j,\ a_{j+1}]$, overlapping slightly with the neighboring intervals:

$$0 \le w_j(x) \le 1 \ ,$$

$$w_j(x) = 1 \quad \text{for} \quad a_j + \epsilon_j \le x \le a_{j+1} - \epsilon_{j+1} \ ,$$

$$0 \quad \text{for} \quad x \le a_j - \epsilon_j \quad \text{or} \quad x \ge a_{j+1} + \epsilon_{j+1} \ ,$$

where we assume that the ϵ_k satisfy $a_j + \epsilon_j \ge a_{j+1} - \epsilon_{j+1}$ for all j. Moreover, we require that w_j and w_{j-1} complement each other near a_j: $w_j(x) = w_{j-1}(2a_j - x)$ and $w_j^2(x) + w_{j-1}^2(x) = 1$ if $|x - a_j| \le \epsilon_j$. (All this can be achieved with smooth w_j; one can take, for instance, $w_j(x) = \sin[\frac{\pi}{2}\ \nu(\frac{x-a_j+\epsilon_j}{2\epsilon_j})]$ for $|x - a_j| \le \epsilon_j$, and $w_j(x) = \cos[\frac{\pi}{2}\ \nu(\frac{x-a_{j+1}+\epsilon_{j+1}}{\epsilon_{j+1}})]$ for $|x - a_{j+1}| \le \epsilon_{j+1}$, with ν satisfying (4.2.4) and (4.2.5).) Coifman and Meyer (1990) prove that the family $\{u_{j,k};\ j,k \in \mathbb{Z}\}$, with

$$u_{j,k}(x) = \sqrt{\frac{2}{a_{j+1} - a_j}}\ w_j(x) \sin\left[\pi\left(k + \frac{1}{2}\right)\frac{x - a_j}{a_{j+1} - a_j}\right] \ ,$$

constitutes an orthonormal basis for $L^2(\mathbb{R})$, consisting of compactly supported functions with fast decay in frequency. This basis has moreover a very interesting property: if for any $j \in \mathbb{Z}$ we define P_j to be the orthogonal projection onto the space spanned by the $\{u_{j,k};\ k \in \mathbb{Z}\}$ (P_j is "morally" the projection into $[a_j,\ a_{j+1}]$), then $P_j + P_{j+1}$ is exactly the projection operator $\tilde{P}_j$, associated to $[a_j,\ a_{j+2}]$, that we would have obtained if we had deleted the point a_{j+1} from our "slicing" of $\mathbb{R}$ (i.e. if we had started with the sequence $\tilde{a}_k$, $\tilde{a}_k = a_k$ if $k \leq j$, $\tilde{a}_k = a_{k+1}$ if $k \geq j+1$). This property makes it possible to split and regroup intervals at will, adapted to the application one has in mind. A very nice discussion of this whole construction, with full details, is Auscher, Weiss, and Wickerhauser (1992).

So there is, after all, more to orthonormal windowed Fourier bases than was expected even only a few years ago. None of these bases, however, are unconditional bases for $L^p(\mathbb{R})$ if $p \neq 2$. This is one point where wavelet bases have the advantage: they turn out to be unconditional bases for a much larger family of function spaces than even these "good" windowed Fourier bases. We will come back to this in Chapter 9.

Notes.

1. Rieffel's proof does not produce an explicit f orthogonal to all the $g_{m,n}$. This is a challenge to the reader: find a (simple) construction of $f \perp g_{m,n}$ for all m, n, for arbitrary ω_0, t_0 with $\omega_0 t_0 > 2\pi$.

2. For orthonormal bases the proof is much simpler. In this case we need not bother with the Zak transform, which was only introduced to prove that if $Qg,\ Pg \in L^2$, then $Q\tilde{g},\ P\tilde{g} \in L^2$ as well. For orthonormal bases we can start directly with point 5, establishing $\langle Qg,\ Pg\rangle = \langle Pg,\ Qg\rangle$, which is impossible by point 6. This is the original elegant proof in Battle (1988).

3. If the $\psi_{m,n}(x) = {a_0}^{-m/2}\ \psi({a_0}^{-m}x - nb_0)$ constitute a (tight) frame, then so do the ${\psi_{m,n}}^{\#}(x) = {a_0}^{-m/2}\ \psi^{\#}({a_0}^{-m}x - nb_0')$, with $\psi^{\#}(x) = (b_0/b_0')^{1/2}\ \psi(b_0 x/b_0')$.

4. To illustrate this, the following example shows that the complex exponentials exp $(2\pi i n x)$ do not constitute an unconditional basis for $L^p([0,1])$ if $p \neq 2$. One can show (see Zygmund (1959)) that

$$\left|\sum_{n=2}^{\infty} n^{-1/4}\ e^{2\pi i n x}\right| \underset{|x|\to 0}{\sim} C|x|^{-3/4}$$

$$\left|\sum_{n=2}^{\infty} n^{-1/4}\ e^{i\sqrt{n}}\ e^{2\pi i n x}\right| \begin{array}{l} \underset{\substack{x\to 0\\ >}}{\sim} C|\log x| \\ \underset{\substack{x\to 0\\ <}}{\sim} Cx^{-2}\ . \end{array}$$

In both cases, $x = 0$ is the worst singularity, and the integrability of powers of these functions on $[0,1]$ is determined by their behavior around

0. The first function is in L^p for $p < \frac{4}{3}$, the second is not, even though the absolute values of their Fourier coefficients are the same. This means that the functions $\exp(2\pi i n x)$ do not constitute an unconditional basis for $L^{4/3}([0,1])$.

The Haar basis adapted to the interval $[0,1]$ consists of $\{\phi\} \cup \{\psi_{m,n};\ m,n \in \mathbb{Z},\ m \le 0,\ 0 \le n \le 2^{|m|} - 1\}$, with $\phi(x) \equiv 1$ on $[0,1]$. This basis is orthonormal in $L^2([0,1])$, and is an unconditional basis for $L^p([0,1])$ if $1 < p < \infty$.

5. The following is a sketch of the proof that $F(b) = \{\psi^b_{m,n};\ m,n \in \mathbb{Z}\}$, with $\psi^b_{m,n}$ as in (4.1.8), constitutes a Riesz basis (i.e., a linearly independent frame) for $L^2(\mathbb{R})$ if b is close to 1. First of all, we can still apply (3.3.21), (3.3.22) to find estimates for the frame bounds. For $b \ne 1$, $\beta_1(2\pi(2k+1)/b) \ne 0$, but if $b < 2$, then only $k = 0, \pm 1, \pm 2$ lead to nonzero β_1. In the computation of (4.2.6) (with $(2k+1)$ replaced by $(2k+1)/b$), only a finite number of ℓ contribute, so that this expression is continuous in b. It follows that the "rest terms" in (3.3.21), (3.3.22) are continuous in b as well; since (3.3.21)=(3.3.22)= 1 if $b = 1$, we have $A > 0$, $B < \infty$ for b in a neighborhood of 1. It remains to prove that the $\psi^b_{m,n}$ are independent. To do this, we construct the operator $S(b)$,

$$S(b)f = \sum_{m,n} \langle f,\ \psi^1_{m,n}\rangle\ \psi^b_{m,n}\ .$$

Clearly, $S(b)\psi^1_{m,n} = \psi^b_{m,n}$. To prove independence of the $\psi^b_{m,n}$ it is sufficient to prove that $\|S(b)f\| \ge C\|f\|$, uniformly in $f \in L^2(\mathbb{R})$, for some $C > 0$. But

$$\|S(b)f\|^2 = \|f\|^2 - \sum_{\substack{m,n,m',n' \\ (m,n)\ne(m',n')}} \langle f,\ \psi^1_{m,n}\rangle\langle\psi^b_{m,n},\ \psi^b_{m',n'}\rangle\langle\psi^1_{m',n'},\ f\rangle\ .$$

Using that for $|B_{jk}| = |B_{kj}|$, we have

$$\begin{aligned} \sum_{\substack{j,k \\ j\ne k}} a_j \overline{a_k}\ B_{jk} &\le \left[\sum_{\substack{j,k \\ j\ne k}} |a_j|^2\ |B_{jk}|\right]^{1/2} \left[\sum_{\substack{j,k \\ j\ne k}} |a_k|^2\ |B_{jk}|\right]^{1/2} \\ &\le \|a\|\ \sup_j \sum_{\substack{k \\ k\ne j}} |B_{jk}|\ , \end{aligned}$$

we obtain

$$\begin{aligned} \|S(b)f\|^2 &\ge \|f\|^2 \left[1 - \sup_{m,n} \sum_{\substack{m',n' \\ (m',n')\ne(m,n)}} |\langle\psi^b_{m,n}, \psi^b_{m',n'}\rangle|\right] \\ &= \|f\|^2 \left[1 - \sup_n \sum_{\substack{m',n' \\ (m',n')\ne(0,n)}} |\langle\psi^b_{0,n}, \psi^b_{m',n'}\rangle|\right]\ . \end{aligned} \tag{4.2.26}$$

Because of the support properties of $\hat{\psi}$, only $m' = 0, \pm 1$ contribute in this sum. If $m' = 0$ or -1, then any choice of n gives the same result; if $m' = 1$, then the sum may have one of two possible outcomes, depending on whether n is odd or even. On the other hand, using the decay $|\psi(x)| \le C_N(1 + |x|^2)^{-N}$ of ψ, one easily checks that $\sum_{n' \in \mathbb{Z}} |\langle \psi^b_{0,n}, \psi^b_{m',n'} \rangle|$ converges and is continuous in b for $m' = 0, \pm 1$. It follows that the coefficient of $\|f\|^2$ in the right-hand side of (4.2.26) is continuous in b; since it is 1 for $b = 1$, it is > 0 for b in a neighborhood of 1.

6. They satisfy

$$\int dx \; (x - 2^m n)^2 \; |\psi_{m,n}(x)|^2 \le 2^{2m} \; C \; ,$$
$$\int d\xi \; |\xi|^2 \; |\hat{\psi}_{m,n}(\xi)|^2 \le 2^{-2m} \; C$$

instead.

7. After the first printing of this book, a proof was found by P. Auscher, to be published in the Comptes Rendus de l'Académie Scientifique, Paris, Série 1, 315, pp. 769–772. Explicitly, he proves that it is impossible that $\eta \in C^1$ and $|\eta(\xi)| + |\eta'(\xi)| \le C(1 + |\xi|)^{-\alpha}$ with $\alpha > 1/2$.

CHAPTER 5

Orthonormal Bases of Wavelets and Multiresolution Analysis

The first constructions of smooth orthonormal wavelet bases seemed a bit miraculous, as illustrated by the proof in §4.2.A that the Meyer wavelets constitute an orthonormal basis. This situation changed with the advent of multiresolution analysis, formulated in the fall of 1986 by Mallat and Meyer. Multiresolution analysis provides a natural framework for the understanding of wavelet bases, and for the construction of new examples. The history of the formulation of multiresolution analysis is a beautiful example of applications stimulating theoretical development. When he first learned about the Meyer basis, Mallat was working on image analysis, where the idea of studying images simultaneously at different scales and comparing the results had been popular for many years (see, e.g., Witkin (1983) or Burt and Adelson (1983)). This stimulated him to view orthonormal wavelet bases as a tool to describe mathematically the "increment in information" needed to go from a coarse approximation to a higher resolution approximation. This insight crystallized into multiresolution analysis (Mallat (1989), Meyer (1986)).

5.1. The basic idea.

A multiresolution analysis consists of a sequence of successive approximation spaces V_j. More precisely, the closed subspaces V_j satisfy[1]

$$\cdots V_2 \subset V_1 \subset V_0 \subset V_{-1} \subset V_{-2} \subset \cdots \tag{5.1.1}$$

with

$$\overline{\bigcup_{j\in\mathbb{Z}} V_j} = L^2(\mathbb{R}) , \tag{5.1.2}$$

$$\bigcap_{j\in\mathbb{Z}} V_j = \{0\} . \tag{5.1.3}$$

If we denote by P_j the orthogonal projection operator onto V_j, then (5.1.2) ensures that $\lim_{j\to-\infty} P_j f = f$ for all $f \in L^2(\mathbb{R})$. There exist many ladders of spaces satisfying (5.1.1)–(5.1.3) that have nothing to do with "multiresolution"; the multiresolution aspect is a consequence of the additional requirement

$$f \in V_j \iff f(2^j \cdot) \in V_0 . \tag{5.1.4}$$

That is, all the spaces are scaled versions of the central space V_0. An example of spaces V_j satisfying (5.1.1)–(5.1.4) is

$$V_j = \{f \in L^2(\mathbb{R}); \qquad \forall k \in \mathbb{Z}: \ \ f|_{[2^j k,\, 2^j (k+1)[} = \text{constant}\} \ .$$

We will call this example the Haar multiresolution analysis. (It is associated with the Haar basis; see Chapter 1 or below.) Figure 5.1 shows what the projection of some f on the Haar spaces V_0, V_{-1} might look like. This example also exhibits another feature that we require from a multiresolution analysis: invariance of V_0 under integer translations,

$$f \in V_0 \ \Rightarrow \ f(\cdot - n) \in V_0 \ \text{ for all } n \in \mathbb{Z} \ . \tag{5.1.5}$$

Because of (5.1.4) this implies that if $f \in V_j$, then $f(\cdot - 2^j n) \in V_j$ for all $n \in \mathbb{Z}$. Finally, we require also that there exists $\phi \in V_0$ so that

$$\{\phi_{0,n};\ n \in \mathbb{Z}\} \ \text{ is an orthonormal basis in } V_0 \ , \tag{5.1.6}$$

where, for all $j, n \in \mathbb{Z}$, $\phi_{j,n}(x) = 2^{-j/2}\, \phi(2^{-j}x - n)$. Together, (5.1.6) and (5.1.4) imply that $\{\phi_{j,n};\ n \in \mathbb{Z}\}$ is an orthonormal basis for V_j for all $j \in \mathbb{Z}$. This last requirement (5.1.6) seems a bit more "contrived" than the other ones; we will see below that it can be relaxed considerably. In the example given above, a possible choice for ϕ is the indicator function for $[0, 1]$, $\phi(x) = 1$ if $0 \le x \le 1$, $\phi(x) = 0$ otherwise. We will often call ϕ the "scaling function" of the multiresolution analysis.[2]

The basic tenet of multiresolution analysis is that whenever a collection of closed subspaces satisfies (5.1.1)–(5.1.6), then there exists an orthonormal wavelet basis $\{\psi_{j,k};\ j,k \in \mathbb{Z}\}$ of $L^2(\mathbb{R})$, $\psi_{j,k}(x) = 2^{-j/2}\psi(2^{-j}x - k)$, such that, for all f in $L^2(\mathbb{R})$,

$$P_{j-1}f = P_j f \ + \sum_{k\in\mathbb{Z}} \langle f,\ \psi_{j,k}\rangle\ \psi_{j,k} \ . \tag{5.1.7}$$

(P_j is the orthogonal projection onto V_j.) The wavelet ψ can, moreover, be constructed explicitly. Let us see how.

For every $j \in \mathbb{Z}$, define W_j to be the orthogonal complement of V_j in V_{j-1}. We have

$$V_{j-1} = V_j \oplus W_j \tag{5.1.8}$$

and

$$W_j \perp W_{j'} \ \text{ if } \ j \ne j' \ . \tag{5.1.9}$$

(If $j > j'$, e.g., then $W_j \subset V_{j'} \perp W_{j'}$.) It follows that, for $j < J$,

$$V_j = V_J \oplus \bigoplus_{k=0}^{J-j-1} W_{J-k} \ , \tag{5.1.10}$$

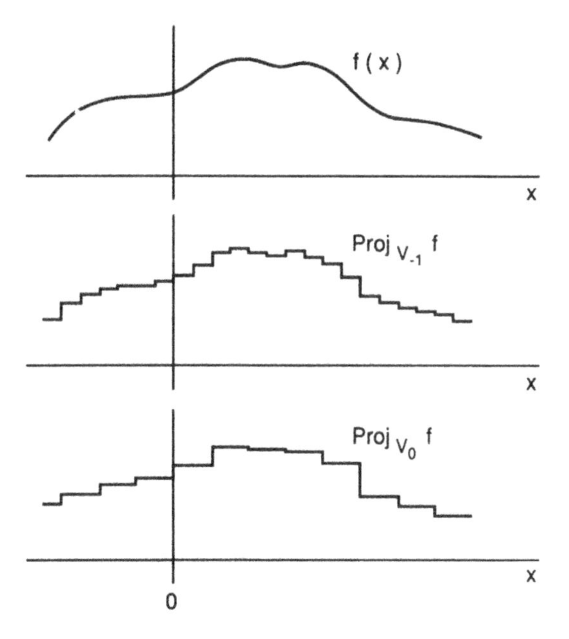

FIG. 5.1. *A function f and its projections onto V_{-1} and V_0.*

where all these subspaces are orthogonal. By virtue of (5.1.2) and (5.1.3), this implies

$$L^2(\mathbb{R}) = \bigoplus_{j \in \mathbb{Z}} W_j \ , \tag{5.1.11}$$

a decomposition of $L^2(\mathbb{R})$ into mutually orthogonal subspaces. Furthermore, the W_j spaces inherit the scaling property (5.1.4) from the V_j:

$$f \in W_j \Longleftrightarrow f(2^j \cdot) \in W_0 \ . \tag{5.1.12}$$

Formula (5.1.7) is equivalent to saying that, for fixed j, $\{\psi_{j,k};\ k \in \mathbb{Z}\}$ constitutes an orthonormal basis for W_j. Because of (5.1.11) and (5.1.2), (5.1.3), this then automatically implies that the whole collection $\{\psi_{j,k};\ j, k \in \mathbb{Z}\}$ is an orthonormal basis for $L^2(\mathbb{R})$. On the other hand, (5.1.12) ensures that if $\{\psi_{0,k};\ k \in \mathbb{Z}\}$ is an orthonormal basis for W_0, then $\{\psi_{j,k};\ k \in \mathbb{Z}\}$ will likewise be an orthonormal basis for W_j, for any $j \in \mathbb{Z}$. Our task thus reduces to finding $\psi \in W_0$ such that the $\psi(\cdot - k)$ constitute an orthonormal basis for W_0.

To construct this ψ, let us write out some interesting properties of ϕ and W_0.

1. Since $\phi \in V_0 \subset V_{-1}$, and the $\phi_{-1,n}$ are an orthonormal basis in V_{-1}, we have

$$\phi = \sum_n h_n \ \phi_{-1,n} \ , \tag{5.1.13}$$

with

$$h_n = \langle \phi,\ \phi_{-1,n} \rangle \ \text{ and } \sum_{n \in \mathbb{Z}} |h_n|^2 = 1 \ . \tag{5.1.14}$$

We can rewrite (5.1.13) as either

$$\phi(x) = \sqrt{2} \sum_n h_n\ \phi(2x - n) \tag{5.1.15}$$

or

$$\hat{\phi}(\xi) = \frac{1}{\sqrt{2}} \sum_n h_n\ e^{-in\xi/2}\ \hat{\phi}(\xi/2)\ , \tag{5.1.16}$$

where convergence in either sum holds in L^2-sense. Formula (5.1.16) can be rewritten as

$$\hat{\phi}(\xi) = m_0(\xi/2)\ \hat{\phi}(\xi/2)\ , \tag{5.1.17}$$

where

$$m_0(\xi) = \frac{1}{\sqrt{2}} \sum_n h_n\ e^{-in\xi}\ . \tag{5.1.18}$$

Equality in (5.1.17) holds pointwise almost everywhere. As (5.1.14) shows, m_0 is a 2π-periodic function in $L^2([0, 2\pi])$.

2. The orthonormality of the $\phi(\cdot - k)$ leads to special properties for m_0. We have

$$\begin{aligned} \delta_{k,0} &= \int dx\ \phi(x)\ \overline{\phi(x-k)} = \int d\xi\ |\hat{\phi}(\xi)|^2\ e^{ik\xi} \\ &= \int_0^{2\pi} d\xi\ e^{ik\xi} \sum_{\ell \in \mathbb{Z}} |\hat{\phi}(\xi + 2\pi\ell)|^2\ , \end{aligned}$$

implying

$$\sum_\ell |\hat{\phi}(\xi + 2\pi\ell)|^2 = (2\pi)^{-1} \quad \text{a.e.} \tag{5.1.19}$$

Substituting (5.1.17) leads to ($\zeta = \xi/2$)

$$\sum_\ell |m_0(\zeta + \pi\ell)|^2\ |\hat{\phi}(\zeta + \pi\ell)|^2 = (2\pi)^{-1}\ ;$$

splitting the sum into even and odd ℓ, using the periodicity of m_0 and applying (5.1.19) once more gives

$$|m_0(\zeta)|^2\ +\ |m_0(\zeta + \pi)|^2 = 1 \quad \text{a.e.} \tag{5.1.20}$$

3. Let us now characterize W_0: $f \in W_0$ is equivalent to $f \in V_{-1}$ and $f \perp V_0$. Since $f \in V_{-1}$, we have

$$f = \sum_n f_n\ \phi_{-1,n}\ ,$$

with $f_n = \langle f, \phi_{-1,n} \rangle$. This implies

$$\hat{f}(\xi) = \frac{1}{\sqrt{2}} \sum_n f_n \, e^{-in\xi/2} \, \hat{\phi}(\xi/2) = m_f(\xi/2) \, \hat{\phi}(\xi/2) \, , \tag{5.1.21}$$

where

$$m_f(\xi) = \frac{1}{\sqrt{2}} \sum_n f_n \, e^{-in\xi} \, ; \tag{5.1.22}$$

m_f is a 2π-periodic function in $L^2([0, 2\pi])$; convergence in (5.1.22) holds pointwise a.e. The constraint $f \perp V_0$ implies $f \perp \phi_{0,k}$ for all k, i.e.,

$$\int d\xi \; \hat{f}(\xi) \; \overline{\hat{\phi}(\xi)} \; e^{ik\xi} \; = \; 0$$

or

$$\int_0^{2\pi} d\xi \; e^{ik\xi} \sum_\ell \hat{f}(\xi + 2\pi\ell) \; \overline{\hat{\phi}(\xi + 2\pi\ell)} \; = \; 0 \; ;$$

hence

$$\sum_\ell \hat{f}(\xi + 2\pi\ell) \; \overline{\hat{\phi}(\xi + 2\pi\ell)} \; = \; 0 \; , \tag{5.1.23}$$

where the series in (5.1.23) converges absolutely in $L^1([-\pi, \pi])$. Substituting (5.1.17) and (5.1.21), regrouping the sums for odd and even ℓ (which we are allowed to do, because of the absolute convergence), and using (5.1.19) leads to

$$m_f(\zeta) \; \overline{m_0(\zeta)} \; + \; m_f(\zeta + \pi) \; \overline{m_0(\zeta + \pi)} \; = \; 0 \quad \text{a.e.} \tag{5.1.24}$$

Since $\overline{m_0(\zeta)}$ and $\overline{m_0(\zeta + \pi)}$ cannot vanish together on a set of nonzero measure (because of (5.1.20)), this implies the existence of a 2π-periodic function $\lambda(\zeta)$ so that

$$m_f(\zeta) = \lambda(\zeta) \; \overline{m_0(\zeta + \pi)} \quad \text{a.e.} \tag{5.1.25}$$

and

$$\lambda(\zeta) + \lambda(\zeta + \pi) \; = \; 0 \quad \text{a.e.} \tag{5.1.26}$$

This last equation can be recast as

$$\lambda(\zeta) \; = \; e^{i\zeta} \; \nu(2\zeta) \; , \tag{5.1.27}$$

where ν is 2π-periodic. Substituting (5.1.27) and (5.1.25) into (5.1.21) gives

$$\hat{f}(\xi) = e^{i\xi/2} \; \overline{m_0(\xi/2 + \pi)} \; \nu(\xi) \; \hat{\phi}(\xi/2) \; , \tag{5.1.28}$$

where ν is 2π-periodic.

4. The general form (5.1.28) for the Fourier transform of $f \in W_0$ suggests that we take

$$\hat{\psi}(\xi) = e^{i\xi/2}\ \overline{m_0(\xi/2+\pi)}\ \hat{\phi}(\xi/2) \tag{5.1.29}$$

as a candidate for our wavelet. Disregarding convergence questions, (5.1.28) can indeed be written as

$$\hat{f}(\xi) = \left(\sum_k \nu_k\ e^{-ik\xi}\right) \hat{\psi}(\xi)$$

or

$$f = \sum_k \nu_k\ \psi(\cdot - k)\ ,$$

so that the $\psi(\cdot - n)$ are a good candidate for a basis of W_0. We need to verify that the $\psi_{0,k}$ are indeed an orthonormal basis for W_0. First of all, the properties of m_0 and $\hat{\phi}$ ensure that (5.1.29) defines indeed an L^2-function $\in V_{-1}$ and $\perp V_0$ (by the analysis above), so that $\psi \in W_0$. Orthonormality of the $\psi_{0,k}$ is easy to check:

$$\begin{aligned}
\int dx\ \psi(x)\ \overline{\psi(x-k)} &= \int d\xi\ e^{ik\xi}\ |\hat{\psi}(\xi)|^2 \\
&= \int_0^{2\pi} d\xi\ e^{ik\xi} \sum_\ell |\hat{\psi}(\xi+2\pi\ell)|^2\ .
\end{aligned}$$

Now

$$\begin{aligned}
\sum_\ell |\hat{\psi}(\xi+2\pi\ell)|^2 &= \sum_\ell |m_0(\xi/2+\pi\ell+\pi)|^2\ |\hat{\phi}(\xi/2+\pi\ell)|^2 \\
&= |m_0(\xi/2+\pi)|^2 \sum_n |\hat{\phi}(\xi/2+2\pi n)|^2 \\
&\quad + |m_0(\xi/2)|^2 \sum_n |\hat{\phi}(\xi/2+\pi+2\pi n)|^2 \\
&= (2\pi)^{-1}\ \left[|m_0(\xi/2)|^2\ +\ |m_0(\xi/2+\pi)|^2\right] \quad \text{a.e.} \qquad \text{(by (5.1.19))} \\
&= (2\pi)^{-1} \quad \text{a.e.} \qquad \text{(by (5.1.20))}\ .
\end{aligned}$$

Hence $\int dx\ \psi(x)\ \overline{\psi(x-k)} = \delta_{k0}$. In order to check that the $\psi_{0,k}$ are indeed a basis for all of W_0, it then suffices to check that any $f \in W_0$ can be written as

$$f = \sum_n \gamma_n\ \psi_{0,n}\ ,$$

with $\sum_n |\gamma_n|^2 < \infty$, or

$$\hat{f}(\xi) = \gamma(\xi)\ \hat{\psi}(\xi)\ , \tag{5.1.30}$$

with γ 2π-periodic and $\in L^2([0, 2\pi])$. Let us go back to (5.1.28). We have $\hat{f}(\xi) = \nu(\xi)\ \hat{\psi}(\xi)$, with $\int_0^{2\pi} d\xi |\nu(\xi)|^2 = 2\int_0^{\pi} d\zeta\ |\lambda(\zeta)|^2$. By (5.1.22),

$$\int_0^{2\pi} d\xi |m_f(\xi)|^2 = \pi \sum_n |f_n|^2 = \pi \|f\|^2 < \infty\ .$$

On the other hand, by (5.1.25),

$$\begin{aligned}\int_0^{2\pi} d\xi\ |m_f(\xi)|^2 &= \int_0^{2\pi} d\xi\ |\lambda(\xi)|^2 |m_0(\xi+\pi)|^2 \\ &= \int_0^{\pi} d\xi\ |\lambda(\xi)|^2\ \left[|m_0(\xi+\pi)|^2 + |m_0(\xi)|^2\right]\ \text{(use (5.1.26))} \\ &= \int_0^{\pi} d\xi\ |\lambda(\xi)|^2\ \ \text{(use (5.1.20))}\ .\end{aligned}$$

Hence $\int_0^{2\pi} d\xi\ |\nu(\xi)|^2 = 2\pi\ \|f\|^2 < \infty$, and f is of the form (5.1.30) with square integrable 2π-periodic γ.

We have thus proved the following theorem.

THEOREM 5.1.1. *If a ladder of closed subspaces $(V_j)_{j\in\mathbb{Z}}$ in $L^2(\mathbb{R})$ satisfies the conditions (5.1.1)–(5.1.6), then there exists an associated orthonormal wavelet basis $\{\psi_{j,k};\ j,k \in \mathbb{Z}\}$ for $L^2(\mathbb{R})$ such that*

$$P_{j-1} = P_j + \sum_k \langle\cdot,\ \psi_{j,k}\rangle\ \psi_{j,k}\ . \tag{5.1.31}$$

One possibility for the construction of the wavelet ψ is

$$\hat{\psi}(\xi) = e^{i\xi/2}\ \overline{m_0(\xi/2+\pi)}\ \hat{\phi}(\xi/2)\ ,$$

(with m_0 as defined by (5.1.18), (5.1.14)), or equivalently

$$\begin{aligned}\psi &= \sum_n (-1)^{n-1}\ \overline{h_{-n-1}}\ \phi_{-1,n}\ , \qquad (5.1.32)\\ \psi(x) &= \sqrt{2}\sum_n (-1)^{n-1}\ \overline{h_{-n-1}}\ \phi(2x-n)\ .\end{aligned}$$

(with convergence of the last series in L^2-sense).

Note that ψ is not determined uniquely by the multiresolution analysis ladder and requirement (5.1.31): if ψ satisfies (5.1.31), then so will any $\psi^{\#}$ of the type

$$\widehat{\psi^{\#}}(\xi) = \rho(\xi)\ \hat{\psi}(\xi)\ , \tag{5.1.33}$$

with ρ 2π-periodic and $|\rho(\xi)| = 1$ a.e.[3] In particular, we can choose $\rho(\xi) = \rho_0\, e^{imp}$ with $m \in \mathbb{Z}$, $|\rho_0| = 1$, which corresponds to a phase change and a shift by m for ψ. We will use this freedom to define, instead of (5.1.32),

$$\psi = \sum_n g_n\ \phi_{-1,n},\quad \text{with}\ \ g_n = (-1)^n\ \overline{h_{-n+1}} \tag{5.1.34}$$

or occasionally

$$g_n = (-1)^n \, h_{-n+1+2N} \, , \tag{5.1.35}$$

with appropriately chosen $N \in \mathbb{Z}$. Of course we can take more general ρ in (5.1.33), but we will generally stick to (5.1.34) or (5.1.35).[4]

Even though every orthonormal wavelet basis of practical interest, known to this date, is associated with a multiresolution analysis, it is possible to construct "pathological" ψ such that the $\psi_{j,k}(x) = 2^{-j/2}\, \psi(2^{-j}x - k)$ constitute an orthonormal basis for $L^2(\mathbb{R})$ but are not derivable from a multiresolution analysis. The following example (due to J. L. Journé) is borrowed from Mallat (1989). Define

$$\hat{\psi}(\xi) = \begin{cases} (2\pi)^{-1/2} & \text{if } \frac{4\pi}{7} \le |\xi| \le \pi \text{ or } 4\pi \le |\xi| \le \frac{32\pi}{7} \, , \\ 0 & \text{otherwise} \, . \end{cases} \tag{5.1.36}$$

We immediately have $\|\psi_{j,k}\| = \|\psi\| = 1$. Furthermore, $2\pi \sum_j |\hat{\psi}(2^j\xi)|^2 = 1$ a.e. By Tchamitchian's criterium (3.3.21)–(3.3.22) the $\psi_{j,k}$ therefore also constitute a tight frame with frame constant 1, provided

$$\sum_{\ell=0}^{\infty} \hat{\psi}(2^\ell \xi)\, \overline{\hat{\psi}(2^\ell(\xi + 2\pi(2k+1)))} \; = \; 0 \;\; \text{a.e.} \tag{5.1.37}$$

One easily checks that support $\hat{\psi} \cap [\text{support } \hat{\psi} + (2k+1)2\pi 2^\ell]$ has zero measure for all $\ell \ge 0$, $k \in \mathbb{Z}$, so that (5.1.37) is indeed satisfied. It then follows from Proposition 3.2.1 that the $\psi_{j,k}$ constitute an orthonormal basis for $L^2(\mathbb{R})$.

If ψ were associated with a multiresolution analysis, then (5.1.29) and (5.1.17) would hold for the corresponding scaling function ϕ (with possibly an extra $\rho(\xi)$, $|\rho(\xi)| = 1$ a.e. in the formula for ψ—see (5.1.33)). It then follows from (5.1.20) that

$$|\hat{\phi}(\xi)|^2 + |\hat{\psi}(\xi)|^2 = |\hat{\phi}(\xi/2)|^2 \, , \tag{5.1.38}$$

which implies, for $\xi \neq 0$,

$$|\hat{\phi}(\xi)|^2 = \sum_{j=1}^{\infty} |\hat{\psi}(2^j\xi)|^2 \, .$$

One easily checks from (5.1.36) that this implies

$$|\hat{\phi}(\xi)| = \begin{cases} (2\pi)^{-1/2} & \text{if } 0 \le |\xi| \le 4\pi/7 \, , \\ & \text{or } \pi \le |\xi| \le 8\pi/7 \, , \\ & \text{or } 2\pi \le |\xi| \le 16\pi/7 \, , \\ 0 & \text{otherwise} \, . \end{cases}$$

If there existed a 2π-periodic m_0 so that (5.1.17) held for this ϕ, then we would have $|m_0(\xi)| = 1$ for $0 \le |\xi| \le 4\pi/7$. By periodicity this would imply $|m_0(\xi)| = 1$

as well for $2\pi \le \xi \le 18\pi/7$; hence $|m_0(\xi)|\,|\hat{\phi}(\xi)| = (2\pi)^{-1/2}$ for $2\pi \le \xi \le 16\pi/7$, even though $|\hat{\phi}(2\xi)| = 0$ on this interval. This contradiction proves that this orthonormal wavelet basis is not derivable from a multiresolution analysis. Note that ψ has very bad decay. This kind of "pathology" cannot persist if some smoothness is imposed upon $\hat{\psi}$ (i.e., decay on ψ).[5] For later convenience, we note that in terms of the h_n, equation (5.1.20) can be rewritten as

$$\sum_n h_n \overline{h_{n+2k}} = \delta_{k,0} . \tag{5.1.39}$$

(This follows easily from writing out the explicit Fourier series for $|m_0(\zeta)|^2 + |m_0(\zeta+\pi)|^2$.)

5.2. Examples.

Let us see what the recipe (5.1.34) gives for the Haar multiresolution analysis. In that case, $\phi(x) = 1$ for $0 \le x < 1$, 0 otherwise; hence

$$h_n = \sqrt{2} \int dx\; \phi(x)\, \overline{\phi(2x-n)} = \begin{cases} 1/\sqrt{2} & \text{if } n = 0, 1, \\ 0 & \text{otherwise.} \end{cases}$$

Consequently, $\psi = \frac{1}{\sqrt{2}}\,\phi_{-1,0} - \frac{1}{\sqrt{2}}\,\phi_{-1,1}$ or

$$\psi(x) = \begin{cases} 1 & \text{if } 0 \le x < \frac{1}{2}, \\ -1 & \text{if } \frac{1}{2} \le x < 1, \\ 0 & \text{otherwise.} \end{cases}$$

This is the Haar basis, which is no surprise: we already saw in §1.6 that this wavelet basis is associated with the Haar multiresolution analysis.

The Meyer basis also fits neatly into this scheme. To see this, define ϕ by

$$\hat{\phi}(\xi) = \begin{cases} (2\pi)^{-1/2}, & |\xi| \le 2\pi/3, \\ (2\pi)^{-1/2}\ \cos\left[\frac{\pi}{2}\nu\left(\frac{3}{2\pi}|\xi| - 1\right)\right], & 2\pi/3 \le |\xi| \le 4\pi/3, \\ 0 & \text{otherwise,} \end{cases}$$

where ν is a smooth function satisfying (4.2.4) and (4.2.5). $\hat{\phi}$ is plotted in Figure 5.2. It is an easy consequence of (4.2.5) that $\sum_{k\in\mathbb{Z}} |\hat{\phi}(\xi+2\pi k)|^2 = (2\pi)^{-1}$, which is equivalent to orthonormality of the $\phi(\cdot - k)$, $k \in \mathbb{Z}$ (see §5.2). We then define V_0 to be the closed subspace spanned by this orthonormal set. Similarly, V_j is the closed space spanned by the $\phi_{j,k}$, $k \in \mathbb{Z}$. The V_j satisfy (5.1.1) if and only if $\phi \in V_{-1}$, i.e., if and only if there exists a 2π-periodic function m_0, square integrable on $[0, 2\pi]$, so that

$$\hat{\phi}(\xi) = m_0(\xi/2)\,\hat{\phi}(\xi/2) .$$

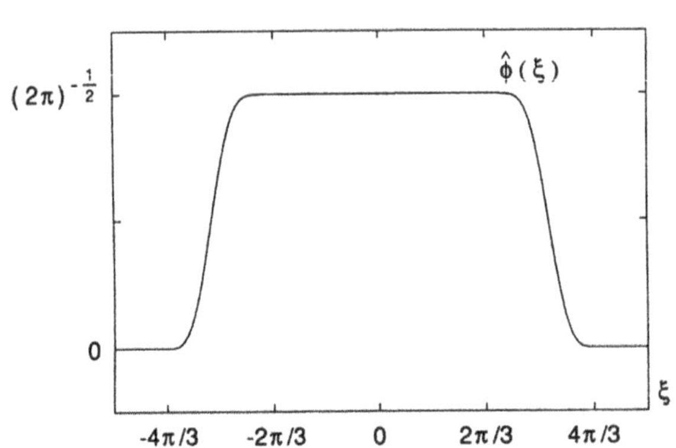

FIG. 5.2. *The scaling function* ϕ *for the Meyer basis, with* $\nu(x) = x^4(35 - 84x + 70x^2 - 20x^3)$.

In this particular case, m_0 can be easily constructed from $\hat{\phi}$ itself: $m_0(\xi) = \sqrt{2\pi}\,\sum_{\ell\in\mathbb{Z}}\,\hat{\phi}(2(\xi + 2\pi\ell))$. This is 2π-periodic and in $L^2([0,\,2\pi])$, and

$$
\begin{aligned}
m_0(\xi/2)\;\hat{\phi}(\xi/2) &= \sqrt{2\pi}\sum_{\ell\in\mathbb{Z}}\hat{\phi}(\xi + 4\pi\ell)\;\hat{\phi}(\xi/2)\\
&= \sqrt{2\pi}\;\hat{\phi}(\xi)\;\hat{\phi}(\xi/2)\\
&\qquad\text{(because [support } \hat{\phi}(\cdot/2)] \text{ and [support } \hat{\phi}(\cdot + 4\pi\ell)]\\
&\qquad\text{do not overlap if } \ell \neq 0)\\
&= \hat{\phi}(\xi)\\
&\qquad\text{(because } \sqrt{2\pi}\;\hat{\phi}(\xi/2) \;=\; 1 \text{ for } \xi \in \text{support } \hat{\phi}).
\end{aligned}
$$

I leave it as an (easy) exercise for the reader to check that the V_j also satisfy properties (5.1.2), (5.1.3) ((5.1.4) and (5.1.5) are trivially satisfied already; see also §5.3.2). Let us now apply the recipe (5.1.29) to find ψ:

$$
\begin{aligned}
\hat{\psi}(\xi) &= e^{i\xi/2}\;\overline{m_0(\xi/2+\pi)}\;\hat{\phi}(\xi/2)\\
&= \sqrt{2\pi}\;e^{i\xi/2}\sum_{\ell\in\mathbb{Z}}\hat{\phi}(\xi + 2\pi(2\ell+1))\;\hat{\phi}(\xi/2)\\
&= \sqrt{2\pi}\;e^{i\xi/2}\;\left[\hat{\phi}(\xi+2\pi)+\hat{\phi}(\xi-2\pi)\right]\;\hat{\phi}(\xi/2)\\
&\qquad\text{(for all other } \ell, \text{ the supports of the two factors do not overlap)}\;.
\end{aligned}
$$

It is easy to check (see also Figure 5.3) that this is equivalent to (4.2.3). The phase factor $e^{i\xi/2}$ which was needed for the "miraculous cancellations" in §4.2 occurs here naturally as a consequence of the general analysis in §5.1.

Before we discuss other examples, we need to relax condition (5.1.6).

5.3. Relaxing some of the conditions.

5.3.1. Riesz bases of scaling functions. The orthonormality of the $\phi(\cdot - k)$ in (5.1.6) can be relaxed: we only need to require that the $\phi(\cdot - k)$ constitute a Riesz basis. The following argument shows how to construct an orthonormal basis $\phi^{\#}(\cdot - k)$ for V_0 starting from a Riesz basis $\{\phi(\cdot - k);\ k \in \mathbb{Z}\}$ of V_0. The $\phi(\cdot - k)$ are a Riesz basis for V_0 if and only if they span V_0 and if, for all $(c_k)_{k\in\mathbb{Z}} \in \ell^2(\mathbb{Z})$,

$$A \sum_k |c_k|^2 \le \left\| \sum_k c_k\ \phi(\cdot - k) \right\|^2 \le B \sum_k |c_k|^2 , \qquad (5.3.1)$$

where $A > 0$, $B < \infty$ are independent of the c_n (see Preliminaries). But

$$\begin{aligned} \left\| \textstyle\sum_k c_k\ \phi(\cdot - k) \right\|^2 &= \int d\xi \left| \textstyle\sum_k c_k\ e^{-ik\xi}\ \hat{\phi}(\xi) \right|^2 \\ &= \int_0^{2\pi} d\xi \left| \sum_k c_k\ e^{-ik\xi} \right|^2 \sum_{\ell\in\mathbb{Z}} |\hat{\phi}(\xi + 2\pi\ell)|^2 \end{aligned}$$

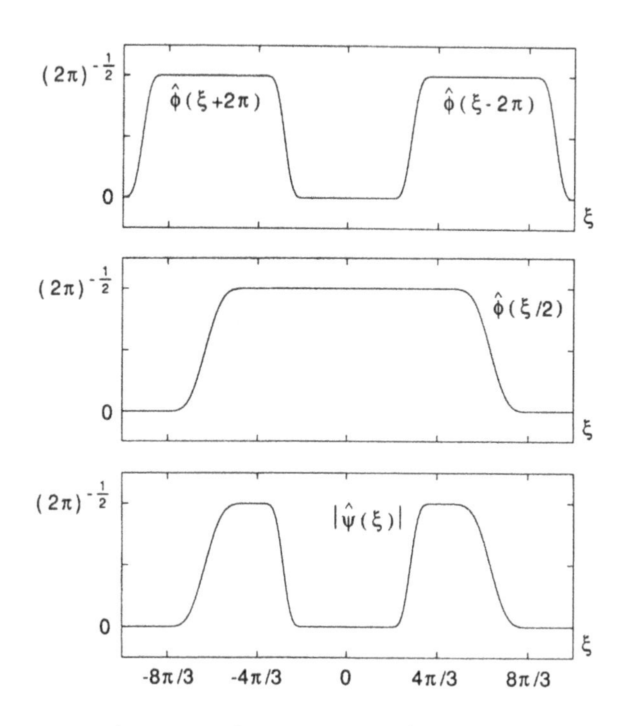

FIG. 5.3. *Graphs of $\hat{\phi}(\xi + 2\pi) + \hat{\phi}(\xi - 2\pi)$ and of $\hat{\phi}(\xi/2)$ for the Meyer multiresolution analysis; their product is $|\hat{\psi}(\xi)|$. (See also Figure 4.2.)*

and

$$\sum_k |c_k|^2 = (2\pi)^{-1} \int_0^{2\pi} d\xi \left| \sum_k c_k\, e^{-ik\xi} \right|^2 ,$$

so that (5.3.1) is equivalent to

$$0 < (2\pi)^{-1} A \le \sum_\ell |\hat{\phi}(\xi + 2\pi\ell)|^2 \le (2\pi)^{-1} B < \infty \text{ a.e.} \tag{5.3.2}$$

We can therefore define $\phi^\# \in L^2(\mathbb{R})$ by

$$\hat{\phi}^\#(\xi) = (2\pi)^{-1/2} \left[\sum_\ell |\hat{\phi}(\xi + 2\pi\ell)|^2 \right]^{-1/2} \hat{\phi}(\xi) . \tag{5.3.3}$$

Clearly, $\sum_\ell |\hat{\phi}^\# (\xi + 2\pi\ell)|^2 = (2\pi)^{-1}$ a.e., which means that the $\phi^\#(\cdot - k)$ are orthonormal. On the other hand, the space $V_0^\#$ spanned by the $\phi^\#(\cdot - k)$ is given by

$$\begin{aligned}
V_0^\# &= \left\{ f;\ f = \sum_n f_n^\# \phi^\#(\cdot - n),\ (f_n^\#)_{n\in\mathbb{Z}} \in \ell^2(\mathbb{Z}) \right\} \\
&= \{ f;\ \hat{f} = \nu\, \hat{\phi}^\# \text{ with } \nu\ 2\pi\text{- periodic, } \nu \in L^2([0,2\pi]) \} \\
&= \{ f;\ \hat{f} = \nu_1\, \hat{\phi} \text{ with } \nu_1\ 2\pi\text{- periodic, } \nu_1 \in L^2([0,2\pi]) \} \\
&\qquad \text{(use (5.3.2) and (5.3.3))} \\
&= \left\{ f;\ f = \sum_n f_n\, \phi(\cdot - n) \text{ with } (f_n)_{n\in\mathbb{Z}} \in \ell^2(\mathbb{Z}) \right\} \\
&= V_0 \quad \text{(since the } \phi(\cdot - n) \text{ are a Riesz basis for } V_0) .
\end{aligned}$$

5.3.2. Using the scaling function as a starting point. As described in §5.1, a multiresolution analysis consists of a ladder of spaces $(V_j)_{j\in\mathbb{Z}}$ and a special function $\phi \in V_0$ such that (5.1.1)–(5.1.6) are satisfied (with (5.1.6) possibly relaxed as in §5.3.1). One can also try to start the construction from an appropriate choice for the scaling function ϕ: after all, V_0 can be constructed from the $\phi(\cdot - k)$, and from there, all the other V_j can be generated. This strategy is followed in many examples. More precisely, we choose ϕ such that

$$\phi(x) = \sum_n c_n\, \phi(2x - n) , \tag{5.3.4}$$

where $\sum_n |c_n|^2 < \infty$, and

$$0 < \alpha \le \sum_{\ell\in\mathbb{Z}} |\hat{\phi}(\xi + 2\pi\ell)|^2 \le \beta < \infty . \tag{5.3.5}$$

We then define V_j to be the closed subspace spanned by the $\phi_{j,k}$, $k \in \mathbb{Z}$, with $\phi_{j,k}(x) = 2^{-j/2}\, \phi(2^{-j}x - k)$. The conditions (5.3.4) and (5.3.5) are necessary and sufficient to ensure that $\{\phi_{j,k};\ k \in \mathbb{Z}\}$ is a Riesz basis in each V_j, and that the V_j satisfy the "ladder property" (5.1.1). It follows that the V_j satisfy (5.1.1), (5.1.4), (5.1.5), and (5.1.6); in order to make sure that we have a multiresolution analysis we need to check whether (5.1.2) and (5.1.3) hold. This is the purpose of the following two propositions.

PROPOSITION 5.3.1. *Suppose* $\phi \in L^2(\mathbb{R})$ *satisfies* (5.3.5), *and define* $V_j = \overline{Span\ \{\phi_{j,k};\ k \in \mathbb{Z}\}}$. *Then* $\cap_{j\in\mathbb{Z}} V_j = \{0\}$.

Proof.

1. By (5.3.5), the $\phi_{0,k}$ constitute a Riesz basis for V_0. In particular, they constitute a frame for V_0, i.e., there exist $A > 0$, $B < \infty$ so that, for all $f \in V_0$,

$$A\ \|f\|^2 \le \sum_{k\in\mathbb{Z}} |\langle f,\ \phi_{0,k}\rangle|^2 \le B\ \|f\|^2 \tag{5.3.6}$$

(see Preliminaries). Since V_j and the $\phi_{j,k}$ are the images of V_0 and the $\phi_{0,k}$ under the unitary map $(D_j f)(x) = 2^{-j/2} f(2^{-j}x)$, it follows that, for all $f \in V_j$,

$$A\ \|f\|^2 \le \sum_{k\in\mathbb{Z}} |\langle f,\ \phi_{j,k}\rangle|^2 \le B\ \|f\|^2\ , \tag{5.3.7}$$

with the same A, B as in (5.3.6).

2. Now take $f \in \cap_{j\in\mathbb{Z}} V_j$. Pick $\epsilon > 0$ arbitrarily small. There exists a compactly supported and continuous $\tilde f$ so that $\|f - \tilde f\|_{L^2} \le \epsilon$. If we denote by P_j the orthogonal projection on V_j, then

$$\|f - P_j\tilde f\| = \|P_j(f - \tilde f)\| \le \|f - \tilde f\| \le \epsilon\ ;$$

hence

$$\|f\| \le \epsilon + \|P_j\tilde f\| \quad \text{for all } j \in \mathbb{Z}\ . \tag{5.3.8}$$

3. $\|P_j\tilde f\| \le A^{-1/2}\left[\sum_{k\in\mathbb{Z}} |\langle \tilde f,\ \phi_{j,k}\rangle|^2\right]^{1/2}$, and

$$\begin{aligned}
\sum_k |\langle \tilde f,\ \phi_{j,k}\rangle|^2 &\le 2^{-j} \sum_k \left[\int_{|x|\le R} dx\ |\tilde f(x)|\ |\phi(2^{-j}x - k)|\right]^2 \\
&\qquad (R \text{ chosen so that } [-R,R] \text{ contains the compact support of } \tilde f) \\
&\le 2^{-j}\ \|\tilde f\|^2_{L^\infty} \sum_k \left(\int_{|x|\le R} dx\ |\phi(2^{-j}x - k)|\right)^2 \\
&\le 2^{-j}\ \|\tilde f\|^2_{L^\infty}\ 2R \sum_k \int_{|x|\le R} dx\ |\phi(2^{-j}x - k)|^2 \\
&= \|\tilde f\|^2_{L^\infty}\ 2R \int_{S_{R,j}} dy\ |\phi(y)|^2\ ,
\end{aligned} \tag{5.3.9}$$

with $S_{R,j} = \cup_{k\in\mathbb{Z}} [k - 2^{-j}R,\ k + 2^{-j}R]$, where we assume that j is large enough so that $2^{-j}R \le \frac{1}{2}$.

4. We can rewrite (5.3.9) as

$$\sum_k |\langle \tilde{f}, \phi_{j,k}\rangle|^2 \le 2R\|\tilde{f}\|^2_{L^\infty} \int_{\mathbb{R}} dy\, \chi_j(y)|\phi(y)|^2 \tag{5.3.10}$$

where χ_j is the indicator function of S_{R_j}, i.e., $\chi_j(y) = 1$ if $y \in S_{R_j}$, $\chi_j(y) = 0$ if $y \notin S_{R_j}$. For $y \notin \mathbb{Z}$, we obviously have $\chi_j(y) \to 0$ for $j \to \infty$. It therefore follows from the dominated convergence theorem that (5.3.10) tends to 0 for $j \to \infty$. In particular, there exists a j such that (5.3.9) $\le \epsilon^2 A$. Putting this together with (5.3.8), we find $\|f\| \le 2\epsilon$. Since ϵ was arbitrarily small to start with, $f = 0$. ■

This proves that (5.1.3) is satisfied. For (5.1.2) we introduce the additional hypotheses that $\hat{\phi}$ is bounded and that $\int dx\, \phi(x) \neq 0$.

PROPOSITION 5.3.2. *Suppose that $\phi \in L^2(\mathbb{R})$ satisfies (5.3.5) and that, moreover, $\hat{\phi}(\xi)$ is bounded for all ξ and continuous near $\xi = 0$, with $\hat{\phi}(0) \neq 0$. Define V_j as above. Then $\overline{\cup_{j\in\mathbb{Z}} V_j} = L^2(\mathbb{R})$.*

Proof.

1. We will use again that (5.3.7) holds, with A, B independent of j.

2. Take $f \in (\cup_{j\in\mathbb{Z}} V_j)^\perp$. Fix $\epsilon > 0$ arbitrarily small. There exists a compactly supported C^∞ function $\tilde{f}$ so that $\|f - \tilde{f}\|_{L^2} \le \epsilon$. Consequently, for all $J = -j \in \mathbb{Z}$

$$\begin{aligned} \|P_{-J}\tilde{f}\| &= \|P_j\tilde{f}\| = \|P_j(\tilde{f} - f)\| \quad (\text{since } P_j f = 0) \\ &\le \epsilon\,. \end{aligned} \tag{5.3.11}$$

On the other hand, by (5.3.7),

$$\|P_{-J}\tilde{f}\|^2 \ge B^{-1} \sum_{k\in\mathbb{Z}} |\langle \tilde{f},\ \phi_{-J,k}\rangle|^2\,. \tag{5.3.12}$$

3. By standard manipulations (see Chapter 3) we have

$$\sum_{k\in\mathbb{Z}} |\langle \tilde{f},\ \phi_{-J,k}\rangle|^2 = 2\pi \int d\xi\ |\hat{\phi}(2^{-J}\xi)|^2\ |\hat{\tilde{f}}(\xi)|^2 + R\,, \tag{5.3.13}$$

with

$$\begin{aligned} |R| &\le 2\pi \sum_{\ell\neq 0} \int d\xi\ |\hat{\tilde{f}}(\xi)|\ |\hat{\tilde{f}}(\xi + 2^J\ 2\pi\ell)|\ |\hat{\phi}(2^{-J}\xi)|\ |\hat{\phi}(2^{-J}\xi + 2\pi\ell)| \\ &\le \|\hat{\phi}\|^2_{L^\infty} \sum_{\ell\neq 0} \int d\xi\ |\hat{\tilde{f}}(\xi)|\ |\hat{\tilde{f}}(\xi + 2^J\ 2\pi\ell)|\,. \end{aligned}$$

Since $\tilde{f}$ is C^∞, we can find C so that

$$|\hat{\tilde{f}}(\xi)| \le C(1 + |\xi|^2)^{-3/2} . \tag{5.3.14}$$

It then follows that

$$\begin{aligned} |R| &\le C^2 \, \|\hat{\phi}\|^2_{L^\infty} \sum_{\ell \neq 0} \int d\xi \, (1 + |\xi + 2^J\pi\ell|^2)^{-3/2} \, (1 + |\xi - 2^J\pi\ell|^2)^{-3/2} \\ &\le C' \, \|\hat{\phi}\|^2_{L^\infty} \sum_{\ell \neq 0} (1 + \pi^2 \, \ell^2 2^{2J})^{-1/2} \int d\zeta \, (1 + |\zeta|^2)^{-1} \\ &\quad \left(\text{use } \sup_{x,y \in \mathbb{R}} \, (1+y^2)[1+(x-y)^2]^{-1}[1+(x+y)^2]^{-1} < \infty \text{ twice} \right) \\ &\le C'' \, 2^{-J} . \end{aligned} \tag{5.3.15}$$

4. Putting (5.3.12), (5.3.13), (5.3.14), and (5.3.15) together, we find

$$2\pi \int d\xi \, |\hat{\phi}(2^{-J}\xi)|^2 \, |\hat{\tilde{f}}(\xi)|^2 \le B\epsilon^2 \, + \, C'' \, 2^{-J} . \tag{5.3.16}$$

   Since $\hat{\phi}(\xi)$ is uniformly bounded as well as continuous in $\xi = 0$, the left-hand side of (5.3.16) converges to $2\pi|\hat{\phi}(0)|^2\|\tilde{f}\|^2_{L^2}$ (by the dominated convergence theorem) for $J \to \infty$. It therefore follows that

$$\|\tilde{f}\|_{L^2} \le |\hat{\phi}(0)|^{-1}C\epsilon, \tag{5.3.17}$$

   with C independent of ϵ. Combining (5.3.17) with $\|f - \tilde{f}\|_{L_2} \le \epsilon$, we obtain

$$\|f\|_{L^2} \le \epsilon + \|\tilde{f}\|_{L^2} \le (1 + C|\hat{\phi}(0)|^{-1})\epsilon.$$

   Since ϵ was arbitrarily small, $f = 0$. ■

REMARKS.

1. If slightly stronger conditions are imposed on ϕ, then Propositions 5.3.1 and 5.3.2 can be proved with easier estimates. In Micchelli (1991), for example, the same conclusions are derived if ϕ is continuous and satisfies $|\phi(x)| \le C(1+|x|)^{-1-\epsilon}$, $\sum_{\ell \in \mathbb{Z}} \phi(x-\ell) = \text{const.} \neq 0$, which implies both $\phi \in L^1$ and $\int dx \, \phi(x) \neq 0$.

2. The extra condition that $\hat{\phi}$ be continuous in 0 in Proposition 5.3.2 is not necessary. The following is an example of a multiresolution analysis in which the scaling function is not absolutely integrable. Let V_j^M, ϕ^M, ψ^M be, respectively, the multiresolution spaces, the scaling function, and the wavelet for the Meyer wavelet basis, with $\nu \in C^\infty$ (see §5.2). Let H be

the Hilbert transform, $(Hf)^\wedge(\xi) = \hat{f}(\xi)$ if $\xi \geq 0$, $-\hat{f}(\xi)$ if $\xi < 0$. Define $V_j = H V_j^M$, $\phi = H\phi^M$. Because the Hilbert transform is unitary and commutes with scaling and with translations (in x), the V_j still constitute a multiresolution analysis, and the $\phi_{0,k}$ are an orthonormal basis in V_0. But $\hat{\phi}$ is not continuous in 0. Because $0 \notin$ support $(\hat{\psi}^M)$, $\hat{\psi} = (H\psi^M)^\wedge$ is a C^∞-function with compact support, so that ψ itself is C^∞ with fast decay. This is therefore an example of a very smooth wavelet with good decay, associated to a multiresolution analysis with bad decay for ϕ.[6] Note also that ϕ^M and ϕ satisfy (5.1.17) with the *same* m_0, illustrating that the c_n in (5.3.4), or equivalently m_0, do not determine ϕ uniquely, and that decay of the c_n as $|n|\to\infty$ does not ensure decay for ϕ.[7]

3. If $\hat{\phi}$ is bounded, and continuous in 0, then the condition $\hat{\phi}(0) \neq 0$ is necessary in Proposition 5.3.2. This can be seen as follows. Take $f \in L^2(\mathbb{R})$, $f \neq 0$, with support $\hat{f} \subset [-R, R]$, $R < \infty$. If $\overline{\cup_{j\in\mathbb{Z}} V_j} = L^2(\mathbb{R})$, then $f = \lim_{J\to\infty} P_{-J}f$. But

$$\begin{aligned} \|P_{-J}f\|^2 &\leq A^{-1} \sum_k |\langle f, \phi_{-J,k}\rangle|^2 \\ &\leq A^{-1} \left[2\pi \int d\xi\, |\hat{\phi}(2^{-J}\xi)|^2\, |\hat{f}(\xi)|^2 + R\right], \end{aligned}$$

as in (5.3.13). Since $\hat{\phi}$ is continuous, the first term tends to $A^{-1}\, 2\pi\, |\hat{\phi}(0)|^2\, \|f\|^2$ for $J\to\infty$, by the dominated convergence theorem. The second term can be bounded exactly as in (5.3.15), so that this term tends to zero for $J\to\infty$. It follows that

$$\|f\|^2 = \lim_{J\to\infty} \|P_{-J}f\|^2 \leq 2\pi\, A^{-1}\, |\hat{\phi}(0)|^2\, \|f\|^2 .$$

Since $\|f\| \neq 0$, this implies $\hat{\phi}(0) \neq 0$.

4. The argument in points 3 and 4 of the proof can also be used to prove $|\hat{\phi}(0)|^2 \leq B/2\pi$. We have indeed

$$\begin{aligned} B\|f\|^2 \geq B\|P_{-J}f\|^2 &\geq \sum_{k\in\mathbb{Z}} |\langle f, \phi_{-J,k}\rangle|^2 \\ &= 2\pi \int d\xi |\hat{\phi}(2^{-J}\xi)|^2\, |\hat{f}(\xi)|^2 + R, \end{aligned}$$

where $|R|$ can be bounded by $C2^{-J}$ for nice f. The other term tends to $2\pi|\hat{\phi}(0)|^2\|f\|^2$ (see 4). Together with remark 3 above, this implies $A/2\pi \leq |\hat{\phi}(0)|^2 \leq B/2\pi$. In particular, if the $\phi_{0,k}$ are orthonormal, then $A = B$ and $|\hat{\phi}(0)| = (2\pi)^{-1/2}$.

5. The conditions $\hat{\phi} \in L^\infty$, $\hat{\phi}(0) \neq 0$ (with $\hat{\phi}$ continuous in 0) imply certain restrictions on the c_n as well. Equation (5.3.4) can be rewritten as

$$\hat{\phi}(\xi) = m_0(\xi/2)\, \hat{\phi}(\xi/2) , \tag{5.3.18}$$

with $m_0(\xi) = \frac{1}{2} \sum_n c_n \, e^{in\xi}$. In particular, $\hat{\phi}(0) = m_0(0) \, \hat{\phi}(0)$, which implies $m_0(0) = 1$ (since $\hat{\phi}(0) \neq 0$) or

$$\sum_n c_n = 2 \, . \tag{5.3.19}$$

Moreover, (5.3.18) implies that m_0 is continuous, except possibly near the zeros of $\hat{\phi}$. In particular, m_0 is continuous in $\xi = 0$. If, furthermore, $|\hat{\phi}(\xi)| \leq C(1 + |\xi|)^{-1/2-\epsilon}$, then the continuity of $\hat{\phi}$ implies that $\sum_\ell |\hat{\phi}(\xi + 2\pi\ell)|^2$ is continuous as well, so that $\hat{\phi}^\#$ (as defined in §5.3.1) is also continuous; consequently, $m_0^\#(\xi) = \hat{\phi}^\#(2\xi)/\hat{\phi}^\#(\xi)$ satisfies $m_0^\#(0) = 1$. Since $|m_0^\#(\xi)|^2 + |m_0^\#(\xi + \pi)|^2 = 1$, it follows that $m_0^\#(\pi) = 0$. This implies $m_0(\pi) = 0$ ($m_0^\#(\xi) = m_0(\xi)[\sum_\ell |\hat{\phi}(\xi + 2\pi\ell)|^2]^{1/2} \cdot [\sum_\ell |\hat{\phi}(2\xi + 2\pi\ell)|^2]^{-1/2}$), or

$$\sum_n c_n (-1)^n = 0 \, . \tag{5.3.20}$$

Together with $\sum_n c_n = 2$, this implies $\sum_n c_{2n} = 1 = \sum_n c_{2n+1}$. This is consistent with the admissibility condition for ψ.[8] Note also that $\sum_n c_{2n} = 1 = \sum_n c_{2n+1}$ is equivalent with Micchelli (1991)'s condition $\sum_\ell \phi(x - \ell) =$ const. $\neq 0$ if $|\phi(x)| \leq C \, (1 + |x|)^{-1-\epsilon}$ and if ϕ is continuous.[9] □

All this suggests the following strategy for the construction of new orthonormal wavelet bases:

- Choose ϕ so that (1) ϕ and $\hat{\phi}$ have reasonable decay,
  (2) (5.3.4) and (5.3.5) are satisfied,
  (3) $\int dx \; \phi(x) \neq 0$

  (by Propositions 5.3.1, 5.3.2 the V_j then constitute a multiresolution analysis);

- If necessary, perform the "orthonormalization trick"

$$\hat{\phi}^\#(\xi) = \hat{\phi}(\xi) \left[2\pi \sum_\ell |\hat{\phi}(\xi + 2\pi\ell)|^2 \right]^{-1/2} ;$$

- Finally, $\hat{\psi}(\xi) = e^{i\xi/2} \overline{m_0^\#(\xi/2 + \pi)} \hat{\phi}^\#(\xi/2)$, with $m_0^\#(\xi) = m_0(\xi) \, [\sum_\ell |\hat{\phi}(\xi + 2\pi\ell)|^2]^{1/2} \, [\sum_\ell |\hat{\phi}(2\xi + 2\pi\ell)|^2]^{-1/2}$, or equivalently

$$\psi(x) = \sum_n (-1)^n \, h^\#_{-n+1} \, \phi^\# \, (2x - n),$$

  with $m_0^\#(\xi) = \frac{1}{\sqrt{2}} \sum_n h_n^\# \, e^{-in\xi}$.

5.4. More examples: The Battle–Lemarié family.

The Battle–Lemarié wavelets are associated with multiresolution analysis ladders consisting of spline function spaces; in each case we take a B-spline with knots at the integers as the original scaling function. If we choose ϕ to be the piecewise constant spline,

$$\phi(x) = \begin{cases} 1, & 0 \le x \le 1, \\ 0 & \text{otherwise}, \end{cases}$$

then we end up with the Haar basis.

The next example is the piecewise linear spline,

$$\phi(x) = \begin{cases} 1 - |x|, & 0 \le |x| \le 1, \\ 0 & \text{otherwise}, \end{cases}$$

plotted in Figure 5.4a. This ϕ satisfies

$$\phi(x) = \tfrac{1}{2}\,\phi(2x+1) + \phi(2x) + \tfrac{1}{2}\,\phi(2x-1)\ ;$$

see Figure 5.4b. Its Fourier transform is

$$\hat{\phi}(\xi) = (2\pi)^{-1/2} \left(\frac{\sin\ \xi/2}{\xi/2}\right)^2 ,$$

and $2\pi \sum_{\ell\in\mathbb{Z}} |\hat{\phi}(\xi + 2\pi\ell)|^2 = \frac{2}{3} + \frac{1}{3}\cos\xi = \frac{1}{3}(1 + 2\cos^2\frac{\xi}{2})$.[10]

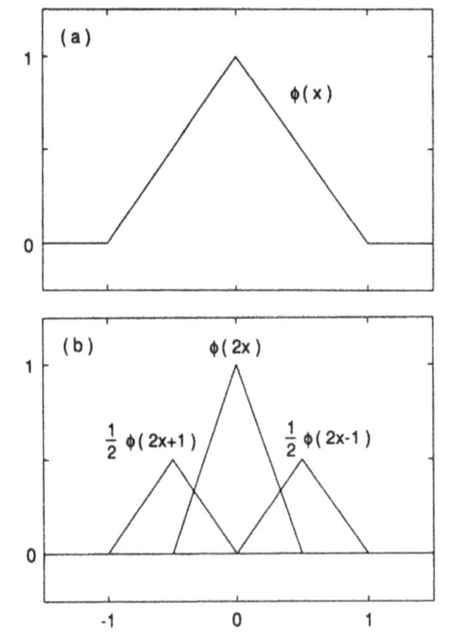

FIG. 5.4. *The piecewise linear B-spline ϕ; it satisfies $\phi(x) = \frac{1}{2}\phi(2x+1) + \phi(2x) + \frac{1}{2}$ $\phi(2x-1)$.*

Both (5.3.4) and (5.3.5) are satisfied, $\phi \in L^1$ and $\int dx\ \phi(x) = 1 \neq 0$. The V_j constitute a multiresolution analysis (consisting of piecewise linear functions with knots at $2^j\mathbb{Z}$). Since ϕ is not orthogonal to its translates, we need to apply the orthogonalization trick (5.3.3)

$$\hat{\phi}^{\#}(\xi) = \sqrt{3}\,(2\pi)^{-1/2}\ \frac{4\sin^2\ \xi/2}{\xi^2\ [1+2\cos^2\ \xi/2]^{1/2}}\ .$$

Unlike ϕ itself, $\phi^{\#}$ is not compactly supported; its graph is plotted in Figure 5.5a. To plot $\phi^{\#}$, the easiest procedure is to compute (numerically) the Fourier coefficients of $[1+2\cos^2\ \xi/2]^{-1/2}$,

$$[1+2\cos^2\ \xi/2]^{-1/2} = \sum_n c_n\ e^{-in\xi}\ ,$$

and to write $\phi^{\#}(x) = \sqrt{3}\ \sum_n\ c_n\ \phi(x-n)$. The corresponding $m_0^{\#}$ is

$$m_0^{\#}(\xi) = \cos^2\ \xi/2\ \left[\frac{1+2\cos^2\ \xi/2}{1+2\cos^2\ \xi}\right]^{1/2}\ ,$$

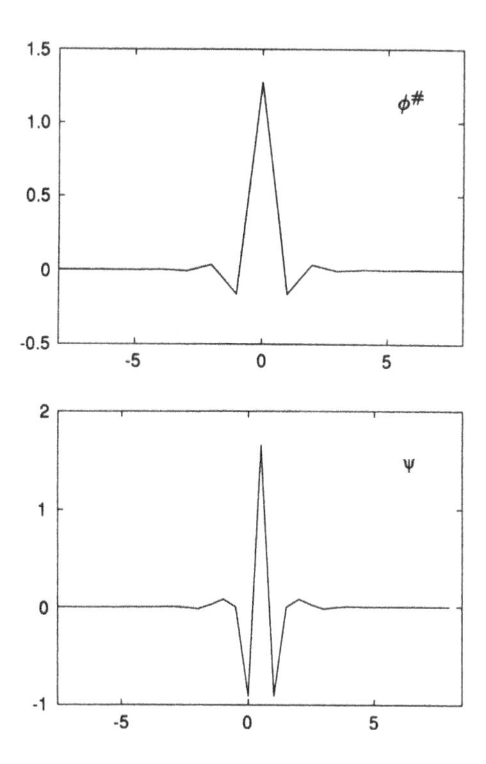

FIG. 5.5. *The scaling function $\phi^{\#}$ and the wavelet ψ for the linear spline Battle–Lemarié construction.*

and $\hat{\psi}$ is given by

$$\begin{aligned}\hat{\psi}(\xi) &= e^{i\xi/2}\sin^2\xi/4\left[\frac{1+2\sin^2\xi/4}{1+2\cos^2\xi/2}\right]^{1/2}\hat{\phi}^{\#}(\xi/2)\\ &= \sqrt{3}e^{i\xi/2}\sin^2\xi/4\left[\frac{1+2\sin^2\xi/4}{(1+2\cos^2\xi/2)(1+2\cos^2\xi/4)}\right]^{1/2}\hat{\phi}(\xi/2)\,.\end{aligned}$$

Again we can compute the Fourier coefficients d_n of $[(1-\sin^2\xi/4)(1+\cos^2\xi/2)^{-1}(1+\cos^2\xi/4)^{-1}]^{1/2}$, and write

$$\psi(x) = \frac{\sqrt{3}}{2}\sum_n (d_{n+1}-2d_n + d_{n-1})\,\phi(2x-n)\,.$$

This function is plotted in Figure 5.5b.

In the next example ϕ is a piecewise quadratic B-spline,

$$\phi(x) = \begin{cases} \frac{1}{2}(x+1)^2\,, & -1\le x\le 0\,,\\ \frac{3}{4}-(x-\frac{1}{2})^2\,, & 0\le x\le 1\,,\\ \frac{1}{2}(x-2)^2\,, & 1\le x\le 2\,,\\ 0 & \text{otherwise}\,,\end{cases}$$

as plotted in Figure 5.6a. Now ϕ satisfies

$$\phi(x) = \tfrac{1}{4}\,\phi(2x+1) + \tfrac{3}{4}\,\phi(2x) + \tfrac{3}{4}\,\phi(2x-1) + \tfrac{1}{4}\,\phi(2x-2)$$

(see Figure 5.6b); we have

$$\hat{\phi}(\xi) = (2\pi)^{-1/2}\,e^{-i\xi/2}\left(\frac{\sin\xi/2}{\xi/2}\right)^3\,,$$

and $2\pi\sum_\ell|\hat{\phi}(\xi+2\pi\ell)|^2 = \frac{11}{20}+\frac{13}{30}\cos\xi+\frac{1}{60}\cos 2\xi = \frac{8}{15}+\frac{13}{30}\cos\xi+\frac{1}{30}\cos^2\xi$.

Again (5.3.4) and (5.3.5) are satisfied, and $\phi\in L^1$, with $\int dx\;\phi(x)\neq 0$. The $\phi(\cdot-k)$ are not orthonormal, and we need to apply the orthogonalization trick (5.3.3) to find $\phi^{\#}$ and $m_0^{\#}$ before we can construct ψ. Graphs of $\phi^{\#}$ and ψ are given in Figure 5.7.

In the general case, ϕ is a B-spline of degree N,

$$\hat{\phi}(\xi) = (2\pi)^{-1/2}\,e^{-i\mathcal{K}\xi/2}\left(\frac{\sin\xi/2}{\xi/2}\right)^{N+1}\,,$$

where $\mathcal{K}=0$ if N is odd, $\mathcal{K}=1$ if N is even. This ϕ satisfies $\int dx\;\phi(x)=1$ and

$$\phi(x) = \begin{cases} 2^{-2M}\displaystyle\sum_{j=0}^{2M+1}\binom{2M+1}{j}\phi(2x-M-1+j) & \text{if } N=2M \text{ is even}\\[2ex] 2^{-2M-1}\displaystyle\sum_{j=0}^{2M+2}\binom{2M+2}{j}\phi(2x-M-1+j) & \text{if } N=2M+1 \text{ is odd.}\end{cases}$$

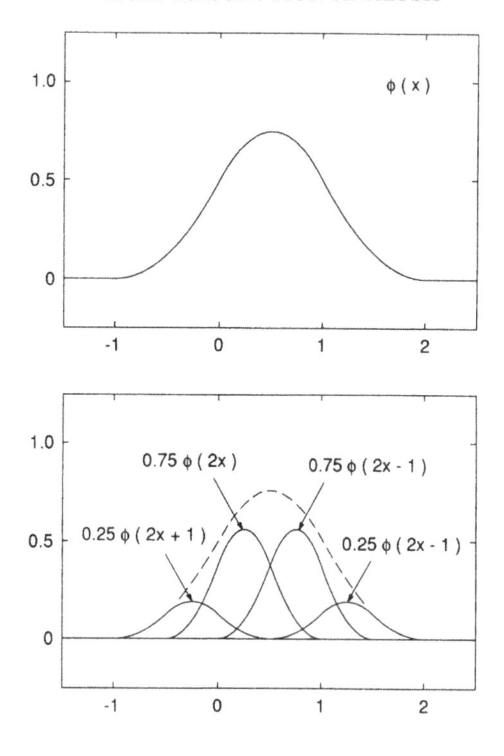

FIG. 5.6. *The quadratic B-spline ϕ, translated so that its knots are at the integers. It satisfies* $\phi(x) = \frac{1}{4}\phi(2x+1) + \frac{3}{4}\phi(2x) + \frac{3}{4}\phi(2x-1) + \frac{1}{4}\phi(2x-2)$.

Explicit formulas for $\sum_\ell |\hat{\phi}(\xi + 2\pi\ell)|^2$, for general N, can be found, e.g., in Chui (1992). In all cases, ϕ satisfies (5.3.4), (5.3.5). For even N, ϕ is symmetric around $x = \frac{1}{2}$, for odd N, around $x = 0$. Except for $N = 0$, the $\phi(\cdot - k)$ are not orthonormal, and the orthogonalization trick (5.3.3) has to be applied. The result is that support $\phi^{\#} = \mathbb{R} =$ support ψ for all the Battle–Lemarié wavelets. The "orthonormalized" $\phi^{\#}$ has the same symmetry axis as ϕ. The symmetry axis of ψ always lies at $x = \frac{1}{2}$. (For N even, ψ is antisymmetric around this axis, for N odd, ψ is symmetric.) Even though the supports of $\phi^{\#}$ and ψ "stretch out" over the whole line, $\phi^{\#}$ and ψ still have very good (exponential) decay. To prove this, we need the following proposition.

PROPOSITION 5.4.1. *Assume that ϕ has exponential decay, $|\phi(x)| \le C\ e^{-\gamma|x|}$, and that, for some $\alpha \le \gamma$ ($\alpha > 0$),*

$$\sup_{|\beta| \le \alpha} |(e^{\beta \cdot}\ \phi)^\wedge(\xi)| \le C\ (1 + |\xi|)^{-1-\epsilon}\ . \tag{5.4.1}$$

Assume also that $0 < a \le \sum_\ell |\hat{\phi}(\xi + 2\pi\ell)|^2$. Define $\phi^{\#}$ by $\hat{\phi}^{\#}(\xi) = \hat{\phi}(\xi)\ [2\pi\ \sum_\ell |\hat{\phi}(\xi + 2\pi\ell)|^2]^{-1/2}$. Then $\phi^{\#}$ has exponential decay as well.

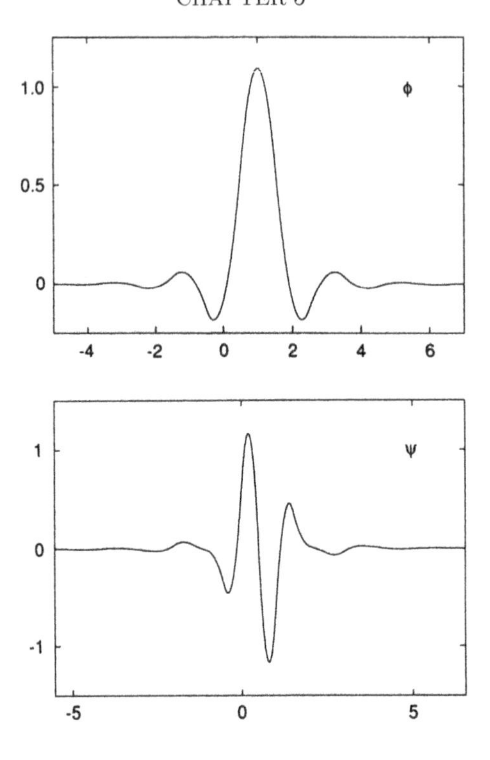

FIG. 5.7. *The scaling function ϕ and the wavelet ψ for the quadratic spline Battle–Lemarié construction.*

Proof.

1. The bound $|\phi(x)| \leq C\ e^{-\alpha|x|}$ implies that $\hat{\phi}(\xi)$ has an analytic extension to the strip $|\mathrm{Im}\ \xi| < \alpha$, and that $\hat{\phi}(\cdot + i\xi_2) \in L^2(\mathbb{R})$ for all $|\xi_2| < \alpha$. The same is true for $\overline{\hat{\phi}(\xi)} = \hat{\bar{\phi}}(-\xi)$.

2. For fixed ξ_2, define $F_{\xi_2}(\xi_1) = \hat{\phi}(\xi_1 + i\xi_2)\ \hat{\bar{\phi}}(-\xi_1 - i\xi_2)$. Then

$$\sum_{\ell \in \mathbb{Z}} |F_{\xi_2}(\xi_1 + 2\pi\ell)|$$

$$\leq \left(\sum_{\ell} |\hat{\phi}(\xi_1 + i\xi_2 + 2\pi\ell|^2\right)^{1/2} \left(\sum_{\ell} |\hat{\bar{\phi}}(-\xi_1 - i\xi_2 - 2\pi\ell)|^2\right)^{1/2},$$

and

$$\sum_{\ell} |\hat{\phi}(\xi_1 + i\xi_2 + 2\pi\ell)|^2$$

$$\leq \frac{1}{2\pi} \int d\xi_1\ |\hat{\phi}(\xi_1 + i\xi_2)|^2 + 2\int d\xi_1\ |\hat{\phi}(\xi_1 + i\xi_2)||\hat{\phi}'(\xi_i + i\xi_2)|$$

$$\leq \frac{1}{2\pi} \int dx e^{2\xi_2 x} |\phi(x)|^2$$
$$+2 \left[\int dx\ e^{2\xi_2 x} |\phi(x)|^2 \right]^{1/2} \left[\int dx\ e^{2\xi_2 x}\ x^2\ |\phi(x)|^2 \right]^{1/2} < \infty.$$

(We have used that $\sum_\ell |f(x+2\pi\ell)| \leq (2\pi)^{-1} \int dx\ |f(x)| + \int dx\ |f'(x)|$.[11]) Consequently, $\sum_\ell F_{\xi_2}(\xi_1 + 2\pi\ell)$ converges absolutely if $|\mathrm{Im}\ \xi_2| < \gamma$. Similar bounds, combined with the dominated convergence theorem, show that $\sum_\ell F_{\xi_2}(\xi_1 + 2\pi\ell)$ is analytic in $\xi = \xi_1 + i\xi_2$ on the strip $|\mathrm{Im}\ \xi| < \gamma$.

3. The function $G(\xi) = \sum_\ell |\hat{\phi}(\xi + 2\pi\ell)|^2$ has thus an analytic extension to $|\mathrm{Im}\ \xi| < \gamma$. Since G is periodic, with period 2π, and $G|_{\mathbb{R}} \geq a > 0$, this implies that there exists $\tilde{\alpha}$, possibly smaller than γ, so that $\mathrm{Re}G(\xi) \geq a/2$ for $|\mathrm{Im}\ \xi| < \tilde{\alpha}$. Consequently, $G^{-1/2}$ can be defined as an analytic function on $|\mathrm{Im}\ \xi| < \tilde{\alpha}$, which means that $\hat{\phi}^\# = G^{-1/2}\ \hat{\phi}$ has an extension to a uniformly bounded analytic function on the strip $|\mathrm{Im}\ \xi| < \tilde{\alpha}$.

4. On the other hand, (5.4.1) implies that

$$|\hat{\phi}(\xi_1 + i\xi_2)| \leq C\ (1 + |\xi_1|)^{-1-\epsilon}$$

   for $|\xi_2| \leq \alpha$. It follows that on $|\mathrm{Im}\ \xi| < \min(\tilde{\alpha},\ \alpha)$, $\hat{\phi}^\#$ is analytic, and is bounded by

$$|\hat{\phi}^\#(\xi)| \leq C\ (1 + |\mathrm{Re}\ \xi|)^{-1-\epsilon}\ .$$

   Consequently,

$$\begin{aligned}
|\phi^\#(x)| &= \lim_{R\to\infty} (2\pi)^{-1/2} \left| \int_{-R}^{R} d\xi\ e^{i\xi x}\ \hat{\phi}^\#(\xi) \right| \\
&= \lim_{R\to\infty} (2\pi)^{-1/2} \left| \int_{-R}^{R} d\xi_1\ e^{i\xi_1 x}\ e^{-\xi_2 x}\ \hat{\phi}^\#\ (\xi_1 + i\xi_2) \right. \\
&\qquad + \int_0^{\xi_2} ds\, e^{-i\xi_1 R}\, e^{-sx} \phi^\#(-R + is) \\
&\qquad \left. - \int_0^{\xi_2} ds\, e^{i\xi_1 R}\, e^{-sx} \phi^\#(R + is) \right| \\
&\leq C\ e^{-\xi_2 x}\ , \quad \text{for} \quad |\xi_2| < \min(\tilde{\alpha}, \alpha)\ . \qquad \blacksquare
\end{aligned}$$

COROLLARY 5.4.2. *All the Battle–Lemarié wavelets ψ and the corresponding orthonormal scaling functions $\phi^\#$ have exponential decay.*

Proof.

1. If the degree N of the B-spline ϕ is zero, then we are in the Haar case and we have nothing to prove. Take $N > 1$. Then $|\hat{\phi}(\xi)| = |2(\sin\ \xi/2)/\xi|^{N+1}$; hence
$$|\hat{\phi}(\xi)| \le C_N\ (1+|\xi|)^{-N-1}\ .$$

2. The condition $|\phi(x)| \le C\ e^{-\gamma|x|}$ is trivially satisfied, with γ arbitrarily large. Moreover, for any $\alpha > 0$, we can construct $f_\beta(x)$, $0 \le |\beta| \le \alpha$ so that $f_\beta \in \mathcal{S}(\mathbb{R})$, $\sup_{\xi\in\mathbb{R}} \sup_{0\le|\beta|\le\alpha} |(1+|\xi|)^M\ \hat{f}_\beta(\xi)| = C'_M < \infty$ for all $M \in \mathbb{N}$, and $f_\beta(x) = e^{\beta x}$ on support (ϕ). Then
$$\begin{aligned} |(e^{\beta\cdot}\phi)^\wedge(\xi)| &= |(f_\beta\phi)^\wedge(\xi)| = |(\hat{f}_\beta * \hat{\phi})(\xi)| \\ &\le C \int d\zeta\ (1+|\zeta-\xi|)^{-M}\ (1+|\zeta|)^{-N-1} \\ &\le C'(1+|\xi|)^{-N-1} \quad \text{for } M \text{ large enough}\ , \end{aligned}$$
so that (5.4.1) is satisfied as well, for α arbitrarily large.

3. It follows that $\phi^\#$ has exponential decay, with its decay rate determined completely by the complex zero of Re $[\sum_\ell \hat{\phi}(\xi+2\pi\ell)\ \overline{\hat{\phi}}(-\xi-2\pi\ell)]$ closest to the real axis.

4. Since $\phi^\#$ has exponential decay, $|\phi^\#(x)| \le C_\#\ e^{-\gamma_\#|x|}$, we have $|h_n^\#| \le \sqrt{2}\ \int dx\ |\phi^\#(x)|\ |\phi^\#(2x-n)| \le C\ e^{-\gamma_\#|n|/2}$ (use $|x+a|+|x-a| \ge 2\max(|x|,|a|)$). Consequently,
$$\begin{aligned} |\psi(x)| \le \sqrt{2} \sum_n |h_{-n+1}||\phi^\#(2x-n)| &\le C \sum_n e^{-\gamma_\# n/2}\ e^{-\gamma_\#|2x-n|} \\ &\le C_\epsilon\ e^{-\gamma_\#|x|(1-\epsilon)}\ . \quad \blacksquare \end{aligned}$$

REMARK. Battle's construction of the Battle–Lemarié wavelets is completely different from the construction given here. His analysis is inspired by techniques from constructive quantum field theory; see, e.g., Battle (1992) for a very readable review. □

Among the "smoother" examples so far, we have

- The Meyer wavelet, which is C^∞ and decays faster than any inverse polynomial (but not exponentially fast);
- The Battle–Lemarié wavelets, which can be chosen to be C^k (i.e., $N \ge k+1$), with k finite, and which have exponential decay (the decay rate decreases as k increases).

In the next section we will see that orthonormal wavelets cannot have the best of both worlds: they cannot be C^∞ *and* have exponential decay. (Note that frames of wavelets do not suffer from this restriction: an example is the Mexican hat function.)

5.5. Regularity of orthonormal wavelet bases.

For wavelet bases (orthonormal or not—see Chapter 8) there is a link between the regularity of ψ and the multiplicity of the zero at $\xi = 0$ of $\hat{\psi}$. This is a consequence of the following theorem (stated and proved in greater generality than needed here, for later convenience).

THEOREM 5.5.1. *Suppose $f, \tilde{f}$ are two functions, not identically constant, such that*

$$\langle f_{j,k}, \tilde{f}_{j',k'} \rangle = \delta_{jj'} \delta_{kk'} ,$$

with $f_{j,k}(x) = 2^{-j/2}\, f(2^{-j}x - k)$, $\tilde{f}_{j,k}(x) = 2^{-j/2} \tilde{f}(2^{-j}x - k)$. Suppose that $|\tilde{f}(x)| \leq C(1 + |x|)^{-\alpha}$, with $\alpha > m + 1$, and suppose that $f \in C^m$, with $f^{(\ell)}$ bounded for $\ell \leq m$. Then

$$\int dx\ x^\ell\ \tilde{f}(x) = 0 \quad \textit{for}\ \ell = 0, 1, \cdots, m\ . \tag{5.5.1}$$

Proof.

1. The idea of the proof is very simple. Choose j, k, j', k' so that $f_{j,k}$ is rather spread out, and $\tilde{f}_{j',k'}$ very much concentrated. (For this expository point only, we assume that $\tilde{f}$ has compact support.) On the tiny support of $\tilde{f}_{j',k'}$ the slice of $f_{j,k}$ "seen" by $\tilde{f}_{j',k'}$ can be replaced by its Taylor series, with as many terms as are well defined. Since, however, $\int dx\ \overline{f_{j,k}(x)}\ \tilde{f}_{j',k'}(x) = 0$, this implies that the integral of the product of $\tilde{f}$ and a polynomial of order m is zero. We can then vary the locations of $\tilde{f}_{j',k'}$, as given by k'. For each location the argument can be repeated, leading to a whole family of different polynomials of order m which all give zero integral when multiplied with $\tilde{f}$. This leads to the desired moment condition. But let us be more precise as follows.

2. We prove (5.5.1) by induction on ℓ. The following argument works for both the initial step and the inductive step. Assume $\int dx\ x^n \tilde{f}(x) = 0$ for $n \in \mathbb{N}$, $n < \ell$. (If $\ell = 0$, then this amounts to no assumption at all.) Since $f^{(\ell)}$ is continuous ($\ell \leq m$), and since the dyadic rationals $2^{-j}k$, $(j, k \in \mathbb{Z})$ are dense in $\mathbb{R}$, there exist J, K so that $f^{(\ell)}(2^{-J}K) \neq 0$. (Otherwise $f^{(\ell)} \equiv 0$ would follow, implying $f \equiv$ constant if $\ell = 0$ or 1, which we know not to be the case, or, if $\ell \geq 2$, $f =$ polynomial of order $\ell - 1 \geq 1$, which would imply that f is not bounded and is therefore also excluded.) Moreover, for any $\epsilon > 0$ there exists $\delta > 0$ so that

$$\left| f(x) - \sum_{n=0}^{\ell} (n!)^{-1}\ f^{(n)}(2^{-J}K)\ (x - 2^{-J}K)^n \right| \leq \epsilon |x - 2^{-J}K|^\ell$$

if $|x - 2^{-J}K| \leq \delta$. Take now $j > J$, $j > 0$. Then

$$0 = \int dx\ f(x)\ \overline{\tilde{f}(2^j x - 2^{j-J}K)}$$

$$= \sum_{n=0}^{\ell} (n!)^{-1} f^{(n)}(2^{-J}K) \int dx\, (x - 2^{-J}K)^n \overline{\tilde{f}(2^j x - 2^{j-J}K)}$$

$$+ \int dx \left[f(x) - \sum_{n=0}^{\ell} (n!)^{-1} f^{(n)}(2^{-J}K)(x - 2^{-J}K)^n \right]$$

$$\cdot \overline{\tilde{f}(2^j x - 2^{j-J}K)} \,. \tag{5.5.2}$$

Since $\int dx\, x^n \tilde{f}(x) = 0$ for $n < \ell$, the first term is equal to

$$(\ell!)^{-1} f^{(\ell)}(2^{-J}K) 2^{-(\ell+1)j} \int dx\, x^\ell\, \overline{\tilde{f}(x)} \,. \tag{5.5.3}$$

Using the boundedness of the $f^{(n)}$, the second term can be bounded by

$$\epsilon \int_{|y|<\delta} dy\, |y|^\ell\, |\tilde{f}(2^j y)| + C' \int_{|y|>\delta} dy\, (1 + |y|^\ell)\, |\tilde{f}(2^j y)|$$

$$\le 2\epsilon C\, 2^{-j(\ell+1)} \int_0^{2^j\delta} dt\, t^\ell (1+t)^{-\alpha}$$

$$+ 2C'\, C \int_\delta^\infty dt\, (1+t)^\ell (1 + 2^j t)^{-\alpha}$$

$$\le C_1\, \epsilon\, 2^{-j(\ell+1)} + C_2\, 2^{-j\alpha} \delta^{-\alpha} (1+\delta)^{\ell+1} \,, \tag{5.5.4}$$

where we replaced the upper integration bound by ∞ in the first term, and where we used in the second term that $(1+2^j t)^{-1} \le \frac{1+\delta}{1+2^j\delta}\, (1+t)^{-1} \le 2^{-j} \frac{1+\delta}{\delta}\, (1+t)^{-1}$ for $t \ge \delta$. Note that C_1, C_2 only depend on C, α, and ℓ; they are independent of ϵ, δ, and j. Combining (5.5.2), (5.5.3), and (5.5.4) leads to

$$\left| \int dx\, x^\ell \tilde{f}(x) \right| \le (\ell!)\, [f^{(\ell)}(2^{-J}K)]^{-1}\, [\epsilon C_1 + \delta^{-\alpha}(1+\delta)^{\ell+1}\, 2^{-j(\alpha-\ell-1)}\, C_2] \,.$$

Here ϵ can be made arbitrarily small, and for the corresponding δ we can choose j sufficiently large to make the second term arbitrarily small as well. It follows that $\int dx\, x^\ell \tilde{f}(x) = 0$. ■

When applied to orthonormal wavelet bases, this theorem has the following corollaries:

COROLLARY 5.5.2. *If the $\psi_{j,k}(x) = 2^{-j/2}\, \psi(2^{-j}x - k)$ constitute an orthonormal set in $L^2(\mathbb{R})$, with $|\psi(x)| \le C\, (1+|x|)^{-m-1-\epsilon}$, $\psi \in C^m(\mathbb{R})$ and $\psi^{(\ell)}$ bounded for $\ell \le m$, then $\int dx\, x^\ell\, \psi(x) = 0$ for $\ell = 0, 1, \cdots, m$.*

Proof. Follows immediately from Theorem 5.5.1, with $f = \tilde{f} = \psi$. ■

REMARKS.

1. Other proofs can be found in Meyer (1990), Battle (1989). Both proofs work with the Fourier transform, unlike this one. Similar links between zero moments and regularity also constituted part of the "folk wisdom" among Calderón–Zygmund theorists, prior to wavelets.

2. Note that we have not used multiresolution analysis to prove Corollary 5.5.2 or Theorem 5.5.1, nor even that the $\psi_{j,k}$ form a basis: orthonormality is the only thing that matters. Battle's proof (which inspired this one) also uses only orthonormality; Meyer's proof uses the full framework of multiresolution analysis. □

COROLLARY 5.5.3. *Suppose the $\psi_{j,k}$ are orthonormal. Then it is impossible that ψ has exponential decay and that $\psi \in C^\infty$, with all derivatives bounded, unless $\psi \equiv 0$.*

Proof.

1. If $\psi \in C^\infty$ with bounded derivatives, then by Theorem 5.5.1, $\int dx\, x^\ell \psi(x) = 0$ for all $\ell \in \mathbb{N}$; hence $\frac{d^\ell}{d\xi^\ell} \hat{\psi}\Big|_{\xi=0} = 0$ for all $\ell \in \mathbb{N}$.

2. If ψ has exponential decay, then $\hat{\psi}$ is analytic on some strip $|\mathrm{Im}\ \xi| < \lambda$. Together with $\frac{d^\ell}{d\xi^\ell} \hat{\psi}\Big|_{\xi=0} = 0$ for all $\ell \in \mathbb{N}$, this implies $\psi \equiv 0$. ■

This is the trade-off announced at the end of the last section: we have to choose for exponential (or faster) decay in either time or frequency; we cannot have both. In practice, decay in x is often preferred over decay in ξ.

A last consequence of Theorem 5.5.1 is the following factorization.

COROLLARY 5.5.4. *Assume that the $\psi_{j,k}$ constitute an orthonormal basis of wavelets, associated with a multiresolution analysis as described in §5.1. If $|\phi(x)|,\ |\psi(x)| \le C\ (1+|x|)^{-m-1-\epsilon}$ and $\psi \in C^m$ with $\psi^{(\ell)}$ bounded for $\ell \le m$, then m_0, as defined by (5.1.18), (5.1.14), factorizes as*

$$m_0(\xi) = \left(\frac{1+e^{-i\xi}}{2}\right)^{m+1} \mathcal{L}(\xi)\ , \tag{5.5.5}$$

where $\mathcal{L}$ is 2π-periodic and $\in C^m$.

Proof.

1. By Corollary 5.5.2, $\frac{d^\ell}{d\xi^\ell} \hat{\psi}\Big|_{\xi=0} = 0$ for $\ell \le m$.

2. On the other hand, $\hat{\psi}(\xi) = e^{-i\xi/2}\ \overline{m_0(\xi/2+\pi)}\ \hat{\phi}(\xi/2)$. Since both $\hat{\psi}$ and $\hat{\phi}$ are in C^m, and $\hat{\phi}(0) \neq 0$ (see Remark 3 at the end of §5.3.2), this means

that m_0 is m times differentiable in $\xi = \pi$, and

$$\frac{d^\ell}{d\xi^\ell} m_0 \bigg|_{\xi=\pi} = 0 \quad \text{for } \ell \leq m \ .$$

3. This implies that m_0 has a zero of order $m+1$ at $\xi = \pi$, or

$$m_0(\xi) = \left(\frac{1+e^{-i\xi}}{2}\right)^{m+1} \mathcal{L}(\xi) \ .$$

Since $m_0 \in C^m$, $\mathcal{L} \in C^m$ as well. ∎

We will come back to the regularity of wavelet bases in Chapter 7.

5.6. Connection with subband filtering schemes.

Multiresolution analysis leads naturally to a hierarchical and fast scheme for the computation of the wavelet coefficients of a given function. Suppose that we have computed, or are given, the inner products of f with the $\phi_{j,k}$ at some given, fine scale.[12] By rescaling our "units" (or rescaling f) we can assume that the label of this fine scale is $j = 0$. It is then easy to compute the $\langle f, \psi_{j,k}\rangle$ for $j \geq 1$. First of all, we have (see (5.1.34))

$$\psi = \sum_n g_n \ \phi_{-1,n} \ ,$$

where $g_n = \langle \psi, \phi_{-1,n}\rangle = (-1)^n \ h_{-n+1}$. Consequently,

$$\begin{aligned} \psi_{j,k}(x) &= 2^{-j/2} \ \psi(2^{-j}x - k) \\ &= 2^{-j/2} \sum_n g_n \ 2^{1/2} \ \phi(2^{-j+1}x - 2k - n) \\ &= \sum_n g_n \ \phi_{j-1,2k+n}(x) \\ &= \sum_n g_{n-2k} \ \phi_{j-1,n}(x) \ . \end{aligned} \tag{5.6.1}$$

It follows that

$$\langle f, \psi_{1,k}\rangle = \sum_n \overline{g_{n-2k}} \ \langle f, \phi_{0,n}\rangle \ ,$$

i.e., the $\langle f, \psi_{1,k}\rangle$ are obtained by convolving the sequence $(\langle f, \phi_{0,n}\rangle)_{h\in\mathbb{Z}}$ with $(\overline{g}_{-n})_{n\in\mathbb{Z}}$, and then retaining only the even samples. Similarly, we have

$$\langle f, \psi_{j,k}\rangle = \sum_n \overline{g_{n-2k}} \ \langle f, \phi_{j-1,n}\rangle \ , \tag{5.6.2}$$

which can be used to compute the $\langle f, \psi_{j,k}\rangle$ by means of the *same operation* (convolution with $\overline{g}$, decimation by factor 2) from the $\langle f, \phi_{j-1,k}\rangle$, if these are known. But, by (5.1.15)

$$\begin{aligned} \phi_{j,k}(x) &= 2^{-j/2}\, \phi(2^{-j}x - k) \\ &= \sum_n h_{n-2k}\, \phi_{j-1,n}(x) \,, \end{aligned} \tag{5.6.3}$$

whence

$$\langle f,\, \phi_{j,k}\rangle = \sum_n \overline{h_{n-2k}}\, \langle f,\, \phi_{j-1,n}\rangle \,. \tag{5.6.4}$$

The procedure to follow is now clear: starting from the $\langle f,\, \phi_{0,n}\rangle$, we compute the $\langle f,\, \psi_{1,k}\rangle$ by (5.6.2), and the $\langle f,\, \phi_{1,k}\rangle$ by (5.6.4). We can then apply (5.6.2), (5.6.4) again to compute the $\langle f,\, \psi_{2,k}\rangle$, $\langle f,\, \phi_{2,k}\rangle$ from the $\langle f,\, \phi_{1,n}\rangle$, etc.. . . : at every step we compute not only the wavelet coefficients $\langle f,\, \psi_{j,k}\rangle$ of the corresponding j-level, but also the $\langle f,\, \phi_{j,k}\rangle$ for the same j-level, which are useful for the computation of the next level wavelet coefficients.

The whole process can also be viewed as the computation of successively coarser approximations of f, together with the difference in "information" between every two successive levels. In this view we start out with a fine-scale approximation to f, $f^0 = P_0 f$ (recall that P_j is the orthogonal projection onto V_j; we will denote the orthogonal projection onto W_j by Q_j), and we decompose $f^0 \in V_0 = V_1 \oplus W_1$ into $f^0 = f^1 + \delta^1$, where $f^1 = P_1 f^0 = P_1 f$ is the next coarser approximation of f in the multiresolution analysis, and $\delta^1 = f^0 - f^1 = Q_1 f^0 = Q_1 f$ is what is "lost" in the transition $f^0 \to f^1$. In each of these V_j, W_j spaces we have the orthonormal bases $(\phi_{j,k})_{k\in\mathbb{Z}}$, $(\psi_{j,k})_{k\in\mathbb{Z}}$, respectively, so that

$$f^0 = \sum_n c_n^0\, \phi_{0,n},\ f^1 = \sum_n c_n^1\, \phi_{1,n},\ \delta^1 = \sum_n d_n^1\, \psi_{1,n} \,.$$

Formulas (5.6.2), (5.6.4) give the effect on the coefficients of the orthogonal basis transformation $(\phi_{0,n})_{n\in\mathbb{Z}} \to (\phi_{1,n}, \psi_{1,n})_{n\in\mathbb{Z}}$ in V_0:

$$c_k^1 = \sum_n \overline{h_{n-2k}}\, c_n^0, \qquad d_k^1 = \sum_n \overline{g_{n-2k}}\, c_n^0 \,. \tag{5.6.5}$$

With the notation $a = (a_n)_{n\in\mathbb{Z}}$, $\overline{a} = (\overline{a_{-n}})_{n\in\mathbb{Z}}$ and $(Ab)_k = \sum_n a_{2k-n}\, b_n$, we can rewrite this as

$$c^1 = \overline{H}\, c^0 \,, \qquad d^1 = \overline{G}\, c^0 \,.$$

The coarser approximation $f^1 \in V_1 = V_2 \oplus W_2$ can again be decomposed into $f^1 = f^2 + \delta^2$, $f^2 \in V_2$, $\delta^2 \in W_2$, with

$$f^2 = \sum_n c_n^2\, \phi_{2,n} \quad \delta^2 = \sum_n d_n^2\, \psi_{2,n} \,.$$

We again have

$$c^2 = \overline{H}\, c^1, \qquad d^2 = \overline{G}\, c^1 \,.$$

Schematically, all this can be represented as in Figure 5.8.

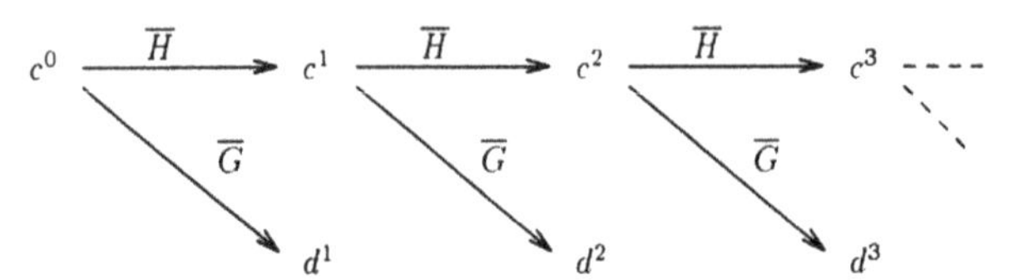

FIG. 5.8. *Schematic representation of* (5.6.5).

In practice, we will stop after a finite number of levels, which means we have rewritten the information in $(\langle f,\ \phi_{0,n}\rangle)_{n\in\mathbb{Z}} = c^0$ as $d^1, d^2, d^3, \cdots, d^J$ and a final coarse approximation c^J, i.e., $(\langle f,\ \psi_{j,k}\rangle)_{k\in\mathbb{Z},\ j=1,\cdots,J}$ and $(\langle f,\ \phi_{J,k}\rangle)_{k\in\mathbb{Z}}$. Since all we have done is a succession of orthogonal basis transformations, the inverse operation is given by the adjoint matrices. Explicitly,

$$\begin{aligned} f^{j-1} &= f^j + \delta^j \\ &= \sum_k c_k^j\ \phi_{j,k}\ +\ \sum_k d_k^j\ \psi_{j,k}\ ; \end{aligned}$$

hence

$$\begin{aligned} c_n^{j-1} &= \langle f^{j-1},\ \phi_{j-1,n}\rangle \\ &= \sum_k c_k^j\ \langle \phi_{j,k},\ \phi_{j-1,n}\rangle\ +\ \sum_k d_k^j\ \langle \psi_{j,k},\ \phi_{j-1,n}\rangle \\ &= \sum_k \left[h_{n-2k}\ c_k^j\ +\ g_{n-2k}\ d_k^j \right] \qquad (5.6.6) \\ &\qquad \text{(use (5.6.1), (5.6.3))}\ . \end{aligned}$$

In electrical engineering terms (5.6.5) and (5.6.6) are the analysis and synthesis steps of a *subband filtering scheme* with exact reconstruction. In a two-channel subband filtering scheme, an incoming sequence $(c_n^0)_{n\in\mathbb{Z}}$ is convolved with two different filters, one low-pass and one high-pass. The two resulting sequences are then subsampled, i.e., only the even (or only the odd) entries are retained. This is exactly what happens in (5.6.5). For readers unfamiliar with this "filtering" terminology, let me explain briefly what it means. Any square summable sequence $(c_n)_{n\in\mathbb{Z}}$ can be interpreted as the sequence of sampled values $\gamma(n)$ of a bandlimited function γ with support $\hat{\gamma} \subset [-\pi, \pi]$ (see Chapter 2),

$$\gamma(x) = \sum_n c_n\ \frac{\sin\ \pi(x-n)}{\pi(x-n)}$$

or

$$\hat{\gamma}(\xi) = \frac{1}{\sqrt{2\pi}} \sum_{n\in\mathbb{Z}} c_n\ e^{-in\xi}\ .$$

A filtering operation corresponds to the multiplication of $\hat{\gamma}$ with a 2π-periodic function, e.g.,

$$\hat{\alpha}(\xi) = \sum_{n\in\mathbb{Z}} a_n\ e^{-in\xi}\ . \qquad (5.6.7)$$

The result is another bandlimited function, $\alpha * \gamma$,

$$(\alpha * \gamma)^\wedge(\xi) = \frac{1}{\sqrt{2\pi}} \sum_{n\in\mathbb{Z}} e^{-in\xi} \sum_{m\in\mathbb{Z}} a_{n-m}\, c_m ,$$

or

$$(\alpha * \gamma)(x) = \sum_n \left(\sum_m a_{n-m}\, c_m \right) \frac{\sin\, \pi(x-n)}{\pi(x-n)} .$$

The filter is *low-pass* if $\hat{\alpha}|_{[-\pi,\pi]}$ is mostly concentrated on $[-\pi/2,\ \pi/2]$, *high-pass* if $\hat{\alpha}|_{[-\pi,\pi]}$ is mostly concentrated on $\{\xi;\ \pi/2 \le |\xi| \le \pi\}$; see Figure 5.9. The "ideal" low-pass and high-pass filters are $\hat{\alpha}_L(\xi) = 1$ if $|\xi| < \pi/2$, 0 if $\pi/2 < |\xi| < \pi$, and $\hat{\alpha}_H(\xi) = 0$ if $|\xi| < \pi/2$, 1 if $\pi/2 < |\xi| < \pi$, respectively. The corresponding a_n (as in (5.6.7)) are given by

$$a_n^L = \begin{cases} \frac{1}{2} & \text{for } n = 0 , \\ 0 & \text{for } n = 2k,\ k \neq 0 , \\ \dfrac{(-1)^k}{(2k+1)\pi} & \text{for } n = 2k+1 ; \end{cases}$$

$$a_n^H = \begin{cases} \frac{1}{2} & \text{for } n = 0 , \\ 0 & \text{for } n = 2k,\ k \neq 0 , \\ \dfrac{(-1)^{k+1}}{(2k+1)\pi} & \text{for } n = 2k+1 . \end{cases}$$

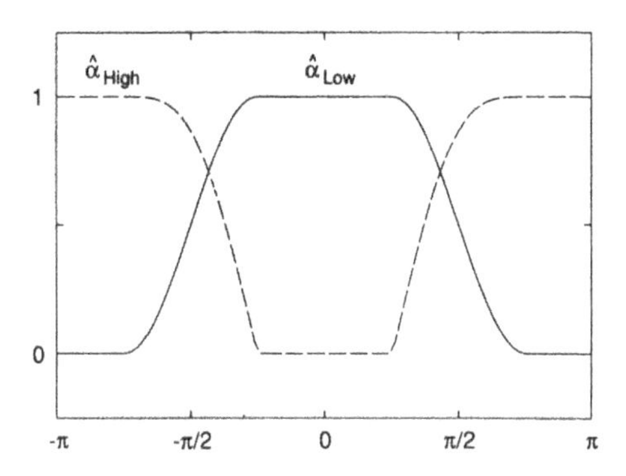

FIG. 5.9. *A low-pass filter (solid line) and a high-pass filter (dashed line).*

When the ideal low-pass filter is applied to γ, the result is a bandlimited function with support $\subset [-\pi/2,\ \pi/2]$. Such a function is completely determined by its sampled values in $2\mathbb{Z}$, and we have (see (2.1.2))

$$(\alpha_L * \gamma)(x) = \sum_n \left(\sum_m a^L_{2n-m}\, c_m \right) \frac{\sin\, [\pi(x-2n)/2]}{\pi(x-2n)/2} .$$

Similarly, the result of applying the ideal high-pass filter to γ is a frequency-shifted version of a bandlimited function with support $\subset [-\pi/2,\ \pi/2]$. Such a function is again completely determined by its sampled values on $2\mathbb{Z}$,

$$\begin{aligned}(\alpha_H * \gamma)(x) &= \frac{1}{\pi} \int_{\frac{\pi}{2}\leq|\xi|\leq\pi} d\xi\ e^{ix\xi} \sum_n \left(\sum_m a^L_{2n-m}\ c_m \right) e^{-2in\xi} \\ &= \sum_n \left(\sum_m a^H_{2n-m}\ c_m \right) \frac{\sin\left[\pi(x-2n)/2\right]}{\pi(x-2n)/2} \left\{2\cos[\pi(x-2n)/2]-1\right\} .\end{aligned}$$

Since the even-indexed entries of the convolutions of the a^L_n, a^H_n with the c_k suffice to characterize $\alpha_L * \gamma$ and $\alpha_H * \gamma$ completely, it makes sense to retain only them after the convolution. This is the rationale behind the decimation by a factor 2 in subband filtering, also called "downsampling." Reconstructing the original c_m from the two filtered and decimated sequences,

$$c^L_n = \sum_m a^L_{2n-m}\ c_m, \qquad c^H_n = \sum_m a^H_{2n-m}\ c_m\ , \tag{5.6.8}$$

is easy:

$$\begin{aligned} c_m &= \gamma(m) = (\alpha_L * \gamma)(m) + (\alpha_H * \gamma)(m) \\ &\qquad\qquad (\text{since } \hat{\alpha}_L + \hat{\alpha}_H = 1) \\ &= \sum_k \frac{\sin[\pi(m-2k)/2]}{\pi(m-2k)/2} \left\{c^L_k + c^H_k (2\cos(\pi(m-2k)/2] - 1)\right\} .\end{aligned}$$

Distinguishing between even and odd m, we find

$$\begin{aligned} c_{2m} &= c^L_m + c^H_m\ , \\ c_{2m+1} &= \sum_\ell \frac{2(-1)^\ell}{\pi(2\ell+1)} \left(c^L_{m-\ell} - c^H_{m-\ell}\right) .\end{aligned}$$

This can also be rewritten as

$$c_m = 2\sum_n \left(a^L_{m-2n} c^L_n + a^H_{m-2n} c^H_n\right) . \tag{5.6.9}$$

This last operation can be seen as the result of

- interleaving both the c^L_n and c^H_n with zeros (i.e., constructing new sequences with zero odd entries, and with even entries given by the consecutive c^L_n, c^H_n);
- convolving these interleaved ("upsampled") sequences with the filters a^L, a^H, respectively;
- adding the two results.

Schematically, (5.6.8) and (5.6.9) can be represented as in Figure 5.10.

The filter coefficients a_n^L, a_n^H for the ideal low-pass and high-pass filters a^L, a^H decay much too slowly to be useful. In practice, one prefers to use the scheme in Figure 5.10 with filters $a^0, a^1, \tilde{a}^0, \tilde{a}^1$ with much faster decaying coefficients. This can only be achieved if the corresponding 2π-periodic functions $\alpha^0, \alpha^1, \tilde{\alpha}^0, \tilde{\alpha}^1$ are smoother than α^L, α^H. This means aliasing can occur: $|\alpha^0|$, $|\alpha^1|$ look like "rounded" versions of α^L, α^H (as in Figure 5.9), which means their support is larger than $[-\pi/2,\ \pi/2]$ and $\{\xi;\ \pi/2 \le |\xi| \le \pi\}$, respectively. Consequently, $\alpha^0 * \gamma$, $\alpha^1 * \gamma$ are not truly bandlimited with maximal frequency $\pi/2$, and sampling them as if they were leads to aliasing, as explained in §2.1. This has to be remedied in the reconstruction stage: $\tilde{a}^0$, $\tilde{a}^1$ need to match a^0 and a^1 to get rid again of the aliasing present after the decomposition. And even this "matching" is only possible if a^0 and a^1 are already matched in some way. To find the appropriate conditions on these filters, it is convenient to use the "z-notation," in which a sequence $(a_n)_{n\in\mathbb{Z}}$ is represented by the formal series $a(z) = \sum_{n\in\mathbb{Z}} a_n\, z^n$. If $z = e^{-i\xi}$ is on the unit circle, then this is nothing but a Fourier series; sometimes it is convenient to consider general $z \in \mathbb{C}$ rather than $|z| = 1$. The decomposition stage of the subband filtering scheme in Figure 5.10 can then be written as

$$\begin{aligned} c^0(z^2) &= \tfrac{1}{2}\left[a^0(z)c(z) + a^0(-z)c(-z)\right] , \\ c^1(z^2) &= \tfrac{1}{2}\left[a^1(z)c(z) + a^1(-z)c(-z)\right] . \end{aligned}$$

Here $a^0(z)c(z)$ is the z-notation for the convolution of a^0 and c; $\frac{1}{2}[b(z)+b(-z)]$ is equal to the formal sequence $\sum_n\ b_{2n}\, z^{2n}$, i.e., $b(z)$ with all the odd entries removed.

The reconstruction stage is

$$\tilde{c}(z) = \tilde{a}^0(z)c^1(z^2) + \tilde{a}^1(z)c^2(z^2) ,$$

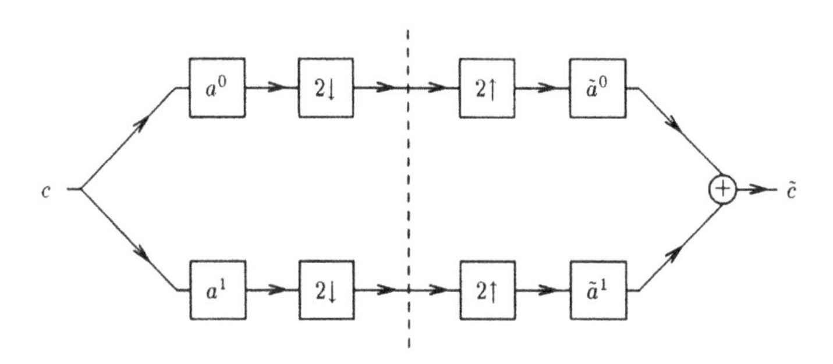

FIG. 5.10. *Schematic representation of the decomposition and reconstruction stages (separated by the vertical dashed line) in a subband filtering scheme. Every letter* $(a^0, a^1, \cdots)$ *in a box represents convolution with the corresponding sequence;* $2\downarrow$ *stands for the downsampling by* 2 *(retaining only the even entries),* $2\uparrow$ *for upsampling by* 2 *(interleaving with zeros). In the "ideal" case,* $a^0 = a^L$, $a^1 = a^H$, $\tilde{a}^0 = 2a^L$, *and* $\tilde{a}^1 = 2a^H$; *the final result is identical to the input,* $\tilde{c} = c$.

where $c^j(z^2)$ is the z-notation for the upsampled version of c^j (zeros have been interleaved: $c^j(z^2) = \sum_n c^j_n z^{2n}$). The total effect is

$$\begin{aligned}\tilde{c}(z) &= \tfrac{1}{2}\left[\tilde{a}^0(z)a^0(z) + \tilde{a}^1(z)a^1(z)\right] c(z) \\ &\quad +\tfrac{1}{2}\left[\tilde{a}^0(z)a^0(-z) + \tilde{a}^1(z)a^1(-z)\right] c(-z) . \end{aligned} \tag{5.6.10}$$

In this expression, the second term contains the aliasing effects: $c(-z)$ corresponds to a shifting of the Fourier series $\sum_n c_n e^{-in\xi}$ by π, exactly what you would expect from aliasing due to sampling at half the Nyquist rate. In order to eliminate aliasing, we therefore need

$$\tilde{a}^0(z)a^0(-z) + \tilde{a}^1(z)a^1(-z) = 0 . \tag{5.6.11}$$

The first subband coding schemes without aliasing date back to Esteban and Galand (1977). In their work, as in most schemes that will be considered in these notes, the sequences are real; they choose

$$\begin{aligned} a^1(z) &= a^0(-z) , \\ \tilde{a}^0(z) &= a^0(z) , \\ \tilde{a}^1(z) &= -a^0(-z) , \end{aligned} \tag{5.6.12}$$

so that (5.6.11) is indeed satisfied, and (5.6.10) simplifies to

$$\tilde{c}(z) = \tfrac{1}{2}\left[a^0(z)^2 - a^0(-z)^2\right] c(z) .$$

If a^0 is symmetric, $a^0_{-n} = a^0_n$, then $\alpha^1(\xi) = \sum_n a^1_n e^{-in\xi}$ is the "mirror" of α^0 with respect to the "half-band" value $\xi = \pi/2$, since $\alpha^1(\xi) = \sum_n a^0_n (-1)^n e^{-in\xi} = \alpha^0(\pi - \xi)$. Filters chosen as in (5.6.12) are therefore called "quadrature mirror filters" (QMF). In practice, one likes to work with FIR filters (FIR = finite impulse response; this means that only finitely many a_n are nonzero). Unfortunately, there exist no FIR a^0 so that $a^0(z)^2 - a^0(-z)^2 = 2$, so that $\tilde{c}$ cannot be identical to c in this scheme. It is nevertheless possible to find a^0 so that $a^0(z)^2 - a^0(-z)^2$ is close to 2, so that the output of the scheme is indeed close to the input. There is by now an immense literature on the design of various QMF; see the issues of IEEE *Trans. Acoust. Speech Signal Process.* for the last 15 years. There also exist many generalizations to splitting into more than 2 bands (GQMF—generalized QMF).

In Mintzer (1985), Smith and Barnwell (1986), and Vetterli (1986) a scheme different from (5.6.12) was proposed:

$$\begin{aligned} a^1(z) &= z^{-1}a^0(-z^{-1}) , \\ \tilde{a}^0(z) &= a^0(z^{-1}) , \\ \tilde{a}^1(z) &= a^1(z^{-1}) = z\,a^0(-z) . \end{aligned} \tag{5.6.13}$$

It is easy to check that this satisfies again (5.6.11), and that (5.6.10) becomes

$$\tilde{c}(z) = \tfrac{1}{2}\left[a^0(z)a^0(z^{-1}) + a^0(-z)a^0(-z^{-1})\right] c(z) .$$

For $z = e^{i\xi}$ and real a_n^0, the expression between square brackets becomes $\frac{1}{2}[|a^0(e^{-i\xi})|^2 + |a^0(-e^{-i\xi})|^2] = \frac{1}{2}[|\alpha^0(\xi)|^2 + |\alpha^0(\xi+\pi)|^2]$. There now exist FIR choices of a^0 for which this is exactly 1, so that we have exact reconstruction in the subband filtering scheme. Smith and Barnwell (1986) named filters chosen as in (5.6.13)[13] "conjugate quadrature filters" (CQF), but the term has not become as generally popular as QMF.

One last remark before we return to wavelets. The whole purpose of subband filtering is of course not to just decompose and reconstruct: a simple wire, instead of the scheme in Figure 5.10, would be far cheaper and more efficient. The goal of the game is to do some compression or processing between the decomposition and reconstruction stages. For many applications (image analysis, for example), compression after subband filtering is more feasible than without filtering. Reconstruction after such compression schemes (quantization) is then not perfect any more,[14] but it is hoped that with specially designed filters, the distortion due to quantization can be kept small, although significant compression ratios are attained. We will come back to this (albeit briefly) in the next chapter.

Back to orthonormal wavelet bases. Formulas (5.6.5), (5.6.6) have *exactly* the same structure as (5.6.8), (5.6.9), respectively. Going from one level in a multiresolution analysis to the next coarser level and the corresponding level of wavelets, and then doing the reverse operation, can therefore be represented by a diagram similar to Figure 5.11. Here $(\overline{h})_n = \overline{h_{-n}}$, $(\overline{g})_n = \overline{g_{-n}}$ (see above). If we assume that the h_n are real, and if we also take into account that $g_n = (-1)^n h_{-n+1}$, then we can identify Figure 5.11 with Figure 5.10 with the choices

$$a^0(z) = h(z^{-1}) \,, \qquad \tilde{a}^0(z) = h(z) \,,$$

$$a^1(z) = g(z^{-1}) = -z^{-1}h(-z) \,, \qquad \tilde{a}^1(z) = g(z) = -z\, h(-z^{-1}) \,.$$

Up to a trivial sign change in a^1 and $\tilde{a}^1$, this corresponds exactly with (5.6.13). This means that every orthonormal wavelet bases associated with a multiresolution analysis gives rise to a pair of CQF filters, i.e., to a subband filtering scheme with exact reconstruction. The reverse is not true: in an orthonormal basis construction we necessarily have $a^0(1) = \sum_n h_n = 2^{1/2}$ (see Remark 5 at the end of §5.3.2), but there exist CQF for which $a^0(1)$ is close to, but not equal to, $2^{1/2}$. Moreover, all the examples of orthonormal bases we have seen so far correspond to infinitely supported $\phi^{\#}$, and hence to non-finite sequences h_n; for applications, FIR filters are preferred. Is it possible to construct orthonormal wavelet bases corresponding to finite filters? What does it mean for these filters to correspond to, e.g., regular wavelets? How can wavelets be useful in filtering contexts? All these are questions that will be addressed in the next chapter.

Notes.

1. I choose here the same nesting order (the more negative the index, the larger the space) as in the ladder of Sobolev spaces. This is also the order

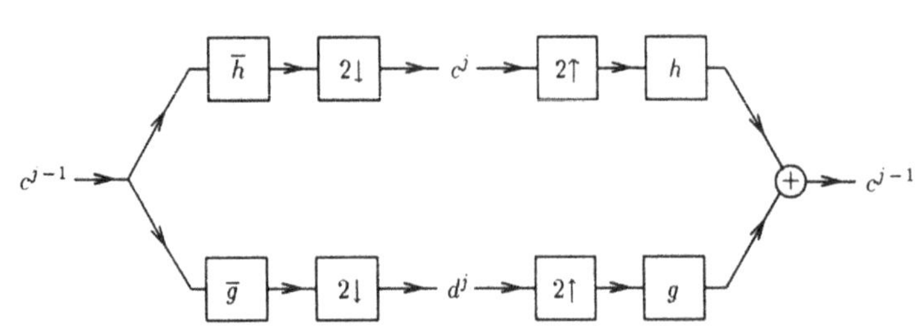

FIG. 5.11. *Subband filtering scheme for one decomposition + reconstruction step in multiresolution analysis.*

that follows naturally from the notation of non-orthogonal wavelets as initiated by A. Grossmann and J. Morlet. It is non-standard, however: Meyer (1990) uses the reverse ordering, more in accordance with established practices in harmonic analysis. For applications in numerical analysis, Beylkin, Coifman, and Rokhlin (1989) find the ordering presented here the most practical.

2. We impose here no a priori regularity or decay on ϕ, unlike, e.g., Meyer (1990).

3. Equation (5.1.33) characterizes *all* the possible $\psi^{\#}$. This follows from Lemma 8.1.1 in Chapter 8.

4. In case ϕ has compact support, and we would like ψ to have the same compact support, (5.1.35) is the only possible choice.

5. It is generally believed that there is no such "pathological" example with continuous ψ. Another challenge for the reader!

   When this book was in its last stages, I heard that Lemarié (1991) has proved that if ψ is compactly supported (continuous or not), then it is automatically associated with a multiresolution analysis. This solves the open problem for one very important special case.

6. Note that there exists $\phi^{\#} \in V_0$ so that the $\phi^{\#}_{0,k}$ are an orthonormal basis for V_0 and $\phi^{\#} \in L^1(\mathbb{R})$, unlike ϕ. It suffices to take $\hat{\phi}^{\#}(\xi) = \lambda(\xi)\,\hat{\phi}(\xi)$, where λ is 2π-periodic and $\lambda(\xi) = \text{sign}(\xi) \cdot e^{i\xi/2}$ for $|\xi| \leq \pi$. This $\hat{\phi}^{\#}$ is again a Schwartz function. The same Hilbert transform trick can be applied to other multiresolution analyses, such as the Battle–Lemarié case or the constructions with compactly supported ψ in the next chapter.

7. If we impose that $\hat{\phi}$ is continuous at $\xi = 0$, then m_0 does determine $\hat{\phi}$ uniquely.

8. The continuity of $\hat{\phi}$, together with $\hat{\phi}(0) \neq 0$, implies that $m_0(0) = 1$ and that m_0 is continuous in $\xi = 0$. It follows that $m_0^{\#}$ is continuous in $\xi = 0$.

Since $|m_0^{\#}(\xi)|^2 + |m_0^{\#}(\xi+\pi)|^2 = 1$, it follows that $|m_0^{\#}|$ is continuous in $\xi = \pi$. Consequently, $|\hat{\psi}(\xi)| = |m_0^{\#}(\xi/2)+\pi)|\,|\hat{\phi}(\xi/2)|$ is continuous in $\xi = 0$ since ψ has to be admissible, this implies $m_0^{\#}(\pi) = 0$; hence $m_0(\pi) = 0$. This provides another derivation for (5.3.20).

9. *Proof.* We prove that $\sum c_{2n} = 1 = \sum c_{2n+1} \Longleftrightarrow \sum_\ell \phi(x-\ell) = \text{const.} \neq 0$ if $|\phi(x)| \le C(1+|x|)^{-1-\epsilon}$ and ϕ is continuous.

   $\Rightarrow$ Define $f(x) = \sum_\ell \phi(x-\ell)$. The conditions on ϕ ensure that f is well defined and continuous. We have

$$\begin{aligned} f(x) &= \sum_\ell \sum_n c_n\, \phi(2x-2\ell-n) = \sum_\ell \sum_m c_{m-2\ell}\, \phi(2x-m) \\ &= \sum_m \left(\sum_j c_{m-2j}\right) \phi(2x-m) = \sum_m \phi(2x-m) = f(2x)\,. \end{aligned}$$

   Hence f is continuous, periodic with period 1, and

$$f(x) = f(2x) = \cdots = f(2^n x) = \cdots \quad .$$

   It follows that f is constant.

   $\Leftarrow$ $\sum_\ell \phi(x-\ell) = c$ implies $\hat{\phi}(2\pi n) = \delta_{n0}(2\pi)^{-1/2}\, c$. But $\hat{\phi}(\xi) = m_0(\xi/2)\, \hat{\phi}(\xi/2)$; hence

$$0 = \hat{\phi}(2\pi(2n{+}1)) = m_0(\pi(2n{+}1))\, \hat{\phi}(\pi(2n{+}1)) = m_0(\pi)\, \hat{\phi}(\pi(2n{+}1))\,.$$

   If $m_0(\pi) \neq 0$, then $\hat{\phi}(\pi(2n+1)) = 0$ would follow for all $n \in \mathbb{Z}$, in contradiction with $\sum |\hat{\phi}(\pi+2\pi n)|^2 > 0$. Hence $m_0(\pi) = 0$, or $\sum c_{2n} = 1 = \sum c_{2n+1}$. ■

10. An easy way to compute the Fourier coefficients of $\sum_\ell |\hat{\phi}(\xi+2\pi\ell)|^2$ is the following:

$$\begin{aligned} \frac{1}{2\pi}\int_0^{2\pi} d\xi\; e^{in\xi} \sum_\ell |\hat{\phi}(\xi+2\pi\ell)|^2 &= \frac{1}{2\pi}\int_{-\infty}^{\infty} d\xi\; e^{in\xi}\, |\hat{\phi}(\xi)|^2 \\ &= \frac{1}{2\pi}\int_{-\infty}^{\infty} dx\; \phi(x)\, \overline{\phi(x-n)}\,. \end{aligned}$$

   For B-spline ϕ, these are easy to compute; see also Chui (1992) for an explicit formula.

11. *Proof.* $f(y) = f(x) + \int_x^y dz\; f'(z)$

   $\Rightarrow$ for $0 \le y \le 2\pi$,

$$\begin{aligned} 2\pi\; f(y+2\pi\ell) &= \int_0^{2\pi} dx\; f(x+2\pi\ell) \;+\; \int_0^y dx \int_x^y dz\; f'(z+2\pi\ell) \\ &\quad - \int_y^{2\pi} dx \int_y^x dz\; f'(z+2\pi\ell) \end{aligned}$$

$\Rightarrow |f(y+2\pi\ell)| \leq \frac{1}{2\pi} \int_{2\pi\ell}^{2\pi(\ell+1)} dx\ |f(x)| + \int_{2\pi\ell}^{2\pi(\ell+1)} dz\ |f'(z)|$

$\Rightarrow \sum_\ell |f(y+2\pi\ell)| \leq (2\pi)^{-1} \int dx\ |f(x)| + \int dx\ |f'(x)|$. ■

12. If f is given in "sampled" form, i.e., if we only know the $f(n)$, then the inner products $\langle f,\ \phi_{0,n}\rangle$ can be computed by a convolution (or filtering) operation, under the assumption that $f \in V_0$ to start with (components of f orthogonal to V_0 cannot be recovered). We have $f = \sum_k \langle f,\ \phi_{0,k}\rangle\, \phi_{0,k}$; hence $f(n) = \sum_k \langle f,\ \phi_{0,k}\rangle\, \phi(n-k)$. Consequently,

$$\sum_n f(n)\, e^{-in\xi} = \left(\sum_k \langle f,\ \phi_{0,k}\rangle\, e^{-ik\xi}\right) \cdot \left(\sum_m \phi(m)\, e^{-im\xi}\right) ,$$

i.e., the $\langle f, \phi_{0,k}\rangle$ are the Fourier coefficients of $\left(\sum_n f(n)\, e^{-in\xi}\right)$ $\left(\sum_m \phi(m)\, e^{-im\xi}\right)^{-1}$. It follows that $\langle f,\ \phi_{0,k}\rangle = \sum_n a_{k-n}\, f(n)$, where

$$a_m = (2\pi)^{-1} \int_0^{2\pi} d\xi\ e^{im\xi} \left(\sum_\ell \phi(\ell)\, e^{-i\ell\xi}\right)^{-1} .$$

13. For convenience, they chose $a'(z) = z^{2N-1} a^0(-z^{-1})$, $\tilde{a}^0(z) = z^{2N}\, a^0(z^{-1})$, $\tilde{a}^1(z) = z\, a^0(-z)$ rather than (5.6.13), with $N \in \mathbb{Z}$ picked so that all the a^j, $\tilde{a}^j$ are polynomials in z (no negative powers). One finds then $\tilde{c}(z) = z^{2N}\, c(z)$, which corresponds to a pure delay in the reconstruction.

14. This is an argument used by fans of the Esteban–Galand-type QMF filters: these fail to give exact reconstruction from the start, but their deviation from exact reconstruction can be made small in comparison with distortions introduced by quantization.

CHAPTER 6

Orthonormal Bases of Compactly Supported Wavelets

Except for the Haar basis, all the examples of orthonormal wavelet bases in the previous chapter consisted of infinitely supported functions, as a result of the orthogonalization trick (5.3.3). To construct orthonormal examples in which ψ is compactly supported, it pays to start from m_0 (or, equivalently, from the subband filtering scheme—see §5.6) rather than from ϕ or the V_j. In §6.1 we show how to construct m_0 so that (5.1.20) is satisfied as well as (5.5.5) for some $N > 0$ (a necessary condition to have some regularity for ψ). Not every such m_0 is associated to an orthonormal wavelet basis, however, an issue addressed in §§6.2 and 6.3. The main results of these two sections are summarized in Theorem 6.3.6, at the end of §6.3. Section 6.4 contains examples of compactly supported wavelets generating orthonormal bases. The orthonormal wavelet bases thus obtained cannot, in general, be written in a closed analytic form. Their graph can be computed with arbitrarily high precision, via an algorithm that I call the "cascade algorithm," which is in fact a "refinement scheme" as used in computer aided design. All this is discussed in §6.5.

A lot of this material goes back to Daubechies (1988b); for many of the results, better, simpler, or more general proofs have been found since, and I have given preference to these new ways of looking at things. These different approaches are borrowed mainly from Mallat (1989), Cohen (1990), Lawton (1990, 1991), Meyer (1990), and Cohen, Daubechies, and Feauveau (1992); for the link with refinement equations the references are Cavaretta, Dahmen, and Micchelli (1991) and Dyn and Levin (1990), as well as earlier papers by these authors (see §6.5).

6.1. Construction of m_0.

In this chapter we are mainly interested in constructing compactly supported wavelets ψ. The easiest way to ensure compact support for the wavelet ψ is to choose the scaling function ϕ with compact support (in its orthogonalized version). It then follows from the definition of the h_n,

$$h_n = \sqrt{2} \int dx\ \phi(x)\ \overline{\phi(2x-n)}\ ,$$

that only finitely many h_n are nonzero, so that ψ reduces to a finite linear combination of compactly supported functions (see (5.1.34)), and therefore automatically has compact support itself. Choosing both ϕ and ψ with compact support also has the advantage that the corresponding subband filtering scheme (see §5.6) uses only FIR filters.

For compactly supported ϕ the 2π-periodic function m_0,

$$m_0(\xi) = \frac{1}{\sqrt{2}} \sum_n h_n \, e^{-in\xi} ,$$

becomes a trigonometric polynomial. As shown in Chapter 5 (see (5.1.20)), orthonormality of the $\phi_{0,n}$ implies

$$|m_0(\xi)|^2 + |m_0(\xi + \pi)|^2 = 1 , \tag{6.1.1}$$

where we have dropped the "almost everywhere" because m_0 is necessarily continuous, so that (6.1.1) has to hold for all ξ if it holds a.e.

We are also interested in making ψ and ϕ reasonably regular. By Corollary 5.5.4, this means that m_0 should be of the form

$$m_0(\xi) = \left(\frac{1 + e^{-i\xi}}{2} \right)^N \mathcal{L}(\xi) , \tag{6.1.2}$$

with $N \geq 1$, and $\mathcal{L}$ a trigonometric polynomial. Note that even without regularity constraint, we need (6.1.2) with N at least 1.[1] Putting (6.1.1), (6.1.2) together, it follows that we are looking for

$$M_0(\xi) = |m_0(\xi)|^2 , \tag{6.1.3}$$

a polynomial in $\cos\xi$, satisfying

$$M_0(\xi) + M_0(\xi + \pi) = 1 \tag{6.1.4}$$

and

$$M_0(\xi) = \left(\cos^2 \frac{\xi}{2} \right)^N L(\xi) , \tag{6.1.5}$$

where $L(\xi) = |\mathcal{L}(\xi)|^2$ is also a polynomial in $\cos\xi$. For our purpose it is convenient to rewrite $L(\xi)$ as a polynomial in $\sin^2 \xi/2 = (1 - \cos\xi)/2$,

$$M_0(\xi) = \left(\cos^2 \frac{\xi}{2} \right)^N P\left(\sin^2 \frac{\xi}{2} \right) . \tag{6.1.6}$$

In terms of P, the constraint (6.1.4) becomes

$$(1-y)^N \, P(y) + y^N \, P(1-y) = 1 , \tag{6.1.7}$$

which should hold for all $y \in [0,1]$, hence for all $y \in \mathbb{R}$. To solve (6.1.7) for P we use Bezout's theorem.[2]

THEOREM 6.1.1. *If p_1, p_2 are two polynomials, of degree n_1, n_2, respectively, with no common zeros, then there exist unique polynomials q_1, q_2, of degree $n_2 - 1$, $n_1 - 1$, respectively, so that*

$$p_1(x)\, q_1(x) + p_2(x)\, q_2(x) = 1 \; . \tag{6.1.8}$$

Proof.

1. We first prove existence; uniqueness follows later. We can assume that $n_1 \geq n_2$ (by renumbering, if necessary). Since degree$(p_2) \leq$ degree(p_1), we can find polynomials $a_2(x)$, $b_2(x)$, with degree(a_2) = degree(p_1) − degree(p_2), degree$(b_2) <$ degree(p_2), so that

$$p_1(x) = a_2(x)\, p_2(x) + b_2(x) \; .$$

2. Similarly, we can find $a_3(x)$, $b_3(x)$, with degree(a_3) = degree(p_2) −degree(b_2), degree$(b_3) <$ degree(b_2), so that

$$p_2(x) = a_3(x)\, b_2(x) + b_3(x) \; .$$

We keep going with this procedure, with b_{n-1} taking the role of p_2 in this last equation, and b_n the role of b_2,

$$b_{n-1}(x) = a_{n+1}(x)\, b_n(x) + b_{n+1}(x) \; .$$

Since degree(b_n) is strictly decreasing, this has to stop at some point, which is only possible if $b_{N+1} = 0$ for some N, with $b_N \neq 0$,

$$b_{N-1}(x) = a_{N+1}(x)\, b_N(x) \; .$$

3. Since

$$b_{N-2} = a_N\, b_{N-1} + b_N \; ,$$

it follows that b_N divides b_{N-2} as well. By induction b_N divides all the previous b_n, and p_2, so that b_N divides both p_1 and p_2. Since p_1 and p_2 have no zeros in common, it follows that b_N is a constant different from zero.

4. We have now

$$\begin{aligned} b_N &= b_{N-2} - a_N\, b_{N-1} = b_{N-2} - a_N(b_{N-3} - a_{N-1}\, b_{N-2}) \\ &= (1 + a_N\, a_{N-1}) b_{N-2} - a_N\, b_{N-3} \\ &\quad \text{etc} \ldots \; . \end{aligned}$$

By induction

$$b_N = \tilde{a}_{N,k}\, b_{N-k} + \tilde{\tilde{a}}_{N,k}\, b_{N-k-1} \; ,$$

with $\tilde{a}_{N,1} = -a_N$, $\tilde{\tilde{a}}_{N,1} = 1$, $\tilde{a}_{N,k+1} = \tilde{\tilde{a}}_{N,k} - \tilde{a}_{N,k}\, a_{N-k}$, $\tilde{\tilde{a}}_{N,k+1} = \tilde{a}_{N,k}$. It follows, again by induction, that degree$(\tilde{a}_{N,k})$ = degree(b_{N-k-1}) −

degree (b_{N-1}), degree $(\tilde{\tilde{a}}_{N,k})$ = degree (b_{N-k}) − degree (b_{N-1}). For $k = N-1$, we find

$$b_N = \tilde{a}_{N,N-1}\, p_2 + \tilde{\tilde{a}}_{N,N-1}\, p_1 \ ,$$

with degree $(\tilde{a}_{N,N-1})$ = degree (p_1) − degree (b_{N-1}) < degree (p_1), degree $(\tilde{\tilde{a}}_{N,N-1})$ = degree (p_2) − degree (b_{N-1}) < degree (p_2). (We have used that degree $(b_{N-1}) \geq 1$; if degree (b_{N-1}) were 0, then b_N would be zero.) It follows that $q_1 = \tilde{\tilde{a}}_{N,N-1}/b_N$, $q_2 = \tilde{a}_{N,N-1}/b_N$ solve (6.1.8) and satisfy the desired degree constraints.

5. It remains to establish uniqueness. Suppose q_1, q_2 and $\tilde{q}_1$, $\tilde{q}_2$ are two solution pairs to (6.1.8), both satisfying the degree restrictions. Then

$$p_1(q_1 - \tilde{q}_1) + p_2(q_2 - \tilde{q}_2) = 0 \ .$$

Since p_1, p_2 have no zeros in common, this implies that every zero of p_2 is a zero of $q_1 - \tilde{q}_1$, with at least the same multiplicity. If $q_1 \neq \tilde{q}_1$, then this means degree $(q_1 - \tilde{q}_1) \geq$ degree (p_2), which is impossible since degree (q_1), degree $(\tilde{q}_1)$ < degree (p_2). Hence $q_1 = \tilde{q}_1$. It then follows immediately that $q_2 = \tilde{q}_2$. ∎

REMARKS.

1. For later convenience (Chapter 8), we have stated Bezout's theorem in greater generality than needed in the present chapter. In fact, it holds under even more general conditions: if p_1 and p_2 have zeros in common, then (6.1.8) can still be solved if its right-hand side is divisible by the greatest common denominator (g.c.d.) of p_1, p_2. The proof is still the same, but b_N is now the g.c.d. of p_1, p_2 instead of a constant. The argument in the proof is nothing but the construction of the g.c.d. by Euclid's algorithm; it works in many other frameworks than the polynomials (in any graded ring, in algebraic terminology).

2. It is clear from the construction of p_1, p_2, that if p_1 and p_2 have only rational coefficients, then so will q_1 and q_2. This will be useful in Chapter 8. □

Let us now apply this to the problem at hand, i.e., (6.1.7). By Theorem 6.1.1 there exist unique polynomials q_1, q_2, of degree $\leq N-1$, so that

$$(1-y)^N q_1(y) + y^N q_2(y) = 1 \ . \tag{6.1.9}$$

Substituting $1-y$ for y in (6.1.9) leads to

$$(1-y)^N q_2(1-y) + y^N q_1(1-y) = 1 \ ;$$

the uniqueness of q_1, q_2 thus implies $q_2(y) = q_1(1-y)$. It follows that $P(y) = q_1(y)$ is a solution of (6.1.7). In this case, we can find the explicit form of q_1

without even using Euclid's algorithm:

$$\begin{aligned} q_1(y) &= (1-y)^{-N}\,[1-y^N q_1(1-y)] \\ &= \sum_{k=0}^{N-1} \binom{N+k-1}{k}\, y^k + O(y^N)\,, \end{aligned}$$

where we have written out explicitly the first N terms of the Taylor expansion for $(1-y)^{-N}$. Since degree $(q_1) \leq N-1$, q_1 is equal to its Taylor expansion truncated after N terms, or

$$q_1(y) = \sum_{k=0}^{N-1} \binom{N+k-1}{k}\, y^k \,.$$

This gives an explicit solution to (6.1.7). (Fortunately, it is positive for $y \in [0,1]$, so it is a good candidate for $|\mathcal{L}(\xi)|^2$.) It is the unique lowest degree solution, which we will denote by P_N.[3] There exist however many solutions of higher degree. For any such higher degree solution, we have

$$(1-y)^N\,[P(y) - P_N(y)] + y^N[P(1-y) \;-\; P_N(1-y)] = 0\,.$$

This already implies that $P - P_N$ is divisible by y^N,

$$P(y) \;-\; P_N(y) = y^N \tilde{P}(y)\,.$$

Moreover,

$$\tilde{P}(y) + \tilde{P}(1-y) = 0\,,$$

i.e., $\tilde{P}$ is antisymmetric with respect to $\frac{1}{2}$. We can summarize all our findings as follows.

PROPOSITION 6.1.2. *A trigonometric polynomial m_0 of the form*

$$m_0(\xi) = \left(\frac{1+e^{-i\xi}}{2}\right)^N \mathcal{L}(\xi) \tag{6.1.10}$$

satisfies (6.1.1) *if and only if $L(\xi) = |\mathcal{L}(\xi)|^2$ can be written as*

$$L(\xi) = P(\sin^2\,\xi/2)\,,$$

with

$$P(y) = P_N(y) + y^N\, R(\tfrac{1}{2} - y)\,, \tag{6.1.11}$$

where

$$P_N(y) = \sum_{k=0}^{N-1} \binom{N-1+k}{k}\, y^k \tag{6.1.12}$$

and R is an odd polynomial, chosen such that $P(y) \geq 0$ for $y \in [0,1]$.

This proposition completely characterizes $|m_0(\xi)|^2$. For our purposes we need however m_0 itself, not $|m_0|^2$. So how do we "extract the square root" from L? Here a lemma by Riesz (see Polya and Szegö (1971)) comes to our help.

LEMMA 6.1.3. *Let A be a positive trigonometric polynomial invariant under the substitution $\xi \to -\xi$; A is necessarily of the form*

$$A(\xi) = \sum_{m=0}^{M} a_m \cos m\xi, \quad \textit{with} \quad a_m \in \mathbb{R} .$$

Then there exists a trigonometric polynomial B of order M, i.e.,

$$B(\xi) = \sum_{m=0}^{M} b_m \, e^{im\xi}, \quad \textit{with} \quad b_m \in \mathbb{R} ,$$

such that $|B(\xi)|^2 = A(\xi)$.

Proof.

1. We can write $A(\xi) = p_A(\cos\xi)$, where p_A is a polynomial of degree M with real coefficients. This polynomial can be factored,

$$p_A(c) = \alpha \prod_{j=1}^{M} (c - c_j) ,$$

where the zeros c_j of p_A appear either in complex duplets c_j, $\bar{c}_j$, or in real singlets. We can also write

$$A(\xi) = e^{+iM\xi} \, P_A(e^{-i\xi}) ,$$

where P_A is a polynomial of degree $2M$. For $|z| = 1$, we have

$$\begin{aligned} P_A(z) &= z^M \, \alpha \prod_{j=1}^{M} \left(\frac{z + z^{-1}}{2} - c_j \right) \\ &= \alpha \prod_{j=1}^{M} \left(\frac{1}{2} - c_j z + \frac{1}{2} z^2 \right) ; \end{aligned} \tag{6.1.13}$$

the two polynomials in the right- and left-hand sides of (6.1.13) therefore agree on all of $\mathbb{C}$.

2. If c_j is real, then the zeros of $\frac{1}{2} - c_j z + \frac{1}{2} z^2$ are $c_j \pm \sqrt{c_j^2 - 1}$. For $|c_j| \geq 1$, these are two real zeros (degenerate if $c_j = \pm 1$) of the form r_j, r_j^{-1}. For $|c_j| < 1$, the two zeros are complex conjugate and of absolute value 1, i.e., they are of the form $e^{i\alpha_j}$, $e^{-i\alpha_j}$. Since $|c_j| < 1$, such zeros correspond to "physical" zeros of A (i.e., to values of ξ for which $A(\xi) = 0$). In order not to cause any contradiction with $A \geq 0$, these zeros must have even multiplicity.

3. If c_j is not real, then we consider it together with $c_k = \overline{c_j}$. The polynomial $\left(\frac{1}{2} - c_j z + \frac{1}{2}z^2\right)\left(\frac{1}{2} - \overline{c}_j z + \frac{1}{2}z^2\right)$ has four zeros, $c_j \pm \sqrt{c_j^2 - 1}$ and $\overline{c}_j \pm \sqrt{\overline{c}_j^2 - 1}$. One easily checks that the four zeros are all different, and form a quadruplet z_j, z_j^{-1}, $\overline{z}_j$, $\overline{z}_j^{-1}$.

4. We therefore have

$$\begin{aligned} P_A(z) &= \frac{1}{2}\, a_M \left[\prod_{j=1}^{J} (z - z_j)(z - \overline{z}_j)(z - z_j^{-1})(z - \overline{z}_j^{-1})\right] \\ &\quad \cdot \left[\prod_{k=1}^{K} (z - e^{i\alpha_k})^2 (z - e^{-i\alpha_k})^2\right] \cdot \left[\prod_{\ell=1}^{L} (z - r_\ell)(z - r_\ell^{-1})\right], \end{aligned}$$

where we have regrouped the three different kinds of zeros.

5. For $z = e^{-i\xi}$ on the unit circle, we have

$$|(e^{-i\xi} - z_0)(e^{-i\xi} - \overline{z}_0^{-1})| = |z_0|^{-1}\, |e^{-i\xi} - z_0|^2 .$$

Consequently,

$$\begin{aligned} A(\xi) &= |A(\xi)| = |P_A(e^{-i\xi})| \\ &= \left[\frac{1}{2}|a_M| \prod_{j=1}^{J} |z_j|^{-2} \prod_{k=1}^{K} |r_k|^{-1}\right] \left|\prod_{j=1}^{J} (e^{-i\xi} - z_j)(e^{-i\xi} - \overline{z}_j)\right|^2 \\ &\quad \cdot \left|\prod_{k=1}^{K} (e^{-i\xi} - e^{i\alpha_k})(e^{-i\xi} - e^{-i\alpha_k})\right|^2 \cdot \left|\prod_{\ell=1}^{L} (e^{-i\xi} - r_\ell)\right|^2 \\ &= |B(\xi)|^2 , \end{aligned}$$

where

$$\begin{aligned} B(\xi) &= \left[\frac{1}{2}|a_M| \prod_{j=1}^{J} |z_j|^{-2} \prod_{k=1}^{K} |r_k|^{-1}\right]^{1/2} \cdot \prod_{j=1}^{J} (e^{-2i\xi} - 2e^{-i\xi}\mathrm{Re} z_j + |z_j|^2) \\ &\quad \cdot \prod_{k=1}^{K} (e^{-2i\xi} - 2e^{-i\xi}\cos\alpha_j + 1) \cdot \prod_{\ell=1}^{L} (e^{-i\xi} - r_\ell) \end{aligned}$$

is clearly a trigonometric polynomial of order M with real coefficients. ∎

REMARKS.

1. This proof is constructive. It uses factorization of a polynomial of degree M, however, which has to be done numerically and may lead to problems if M is large and some zeros are close together. Note that in this proof we need to factor a polynomial of degree only M, unlike some other procedures, which factor directly P_A, a polynomial of degree $2M$.

2. This procedure of "extracting the square root" is also called *spectral factorization* in the engineering literature.

3. The polynomial B is not unique! For M odd, for instance, P_A may have $\frac{M-1}{2}$ quadruplets of complex zeros, and 1 pair of real zeros. In each quadruplet we can choose to retain either z_j, $\overline{z}_j$ to make up B, or z_j^{-1}, $\overline{z}_j^{-1}$; in each duplet we can choose either r_ℓ or r_ℓ^{-1}. This makes already for $2^{(M+1)/2}$ different choices for B. Moreover, we can always multiply B with $e^{in\xi}$, n arbitrary in $\mathbb{Z}$. □

Together, Proposition 6.1.2 and Lemma 6.1.3 tell us how to construct all the possible trigonometric polynomials m_0 satisfying (6.1.1) and (6.1.2). It is not yet clear, however, whether any such m_0 leads to an orthonormal wavelet basis. In fact, some do not. This will be discussed in the next two sections. Readers who would like to skip most of the technicalities can find the main results summed up in Theorem 6.3.6 at the end of §6.3.

6.2. Correspondence with orthonormal wavelet bases.

We start by deriving a formula for a candidate scaling function ϕ. Once this is done, we will check when this candidate defines indeed a bona fide multiresolution analysis.

If a trigonometric polynomial m_0 is associated with a multiresolution analysis as in §5.1, and if the corresponding scaling function ϕ is in $L^1(\mathbb{R})$, then we know that for all ξ,

$$\hat{\phi}(\xi) = m_0(\xi/2)\, \hat{\phi}(\xi/2)\ . \tag{6.2.1}$$

(See (5.1.17). Continuity of $\hat{\phi}$ and m_0 allows us to drop the "a.e.") Moreover, we know from Remark 3 following Proposition 5.3.2 that necessarily $\hat{\phi}(0) \neq 0$, hence $m_0(0) = 1$. Because of (6.1.1) this in turn implies $m_0(\pi) = 0$. It follows that, for all $k \in \mathbb{Z}$, $k \neq 0$,

$$\begin{aligned}
\hat{\phi}(2k\pi) &= \hat{\phi}(2\, 2^\ell(2m+1)\pi) \qquad\qquad (\text{for some } \ell \geq 0,\quad m \in \mathbb{Z}) \\
&= \left[\prod_{j=1}^{\ell} m_0(2^{\ell+1-j}(2m+1)\pi)\right] m_0((2m+1)\pi)\, \hat{\phi}((2m+1)\pi) \\
&= m_0(\pi)\, \hat{\phi}((2m+1)\pi) = 0\ .
\end{aligned}$$

Since $\sum_\ell |\hat{\phi}(\xi+2\pi\ell)|^2 = (2\pi)^{-1}$ (see (5.1.19)), this fixes the normalization of ϕ: $|\hat{\phi}(0)| = (2\pi)^{-1/2}$, or $|\int dx\ \phi(x)| = 1$. It is convenient to choose the phase of ϕ so that $\int dx\ \phi(x) = 1$. Taking all this into account, it follows from (6.2.1) that

$$\hat{\phi}(\xi) = (2\pi)^{-1/2} \prod_{j=1}^{\infty} m_0(2^{-j}\xi) . \tag{6.2.2}$$

This infinite product makes sense: since $\sum_n |h_n|\,|n| < \infty$, and $m_0(0) = 1$, $m_0(\xi) = 2^{-1/2} \sum_n h_n\ e^{-in\xi}$ satisfies

$$|m_0(\xi)| \le 1 + |m_0(\xi) - 1| \le 1 + \sqrt{2} \sum_n |h_n|\,|\sin\ n\xi/2| \le 1 + C|\xi| \le e^{C|\xi|}$$

hence

$$\prod_{j=1}^{\infty} |m_0(2^{-j}\xi)| \le \exp\left(\sum_{j=1}^{\infty} C|2^{-j}\xi|\right) \le e^{C|\xi|} .$$

The infinite product in the right hand side of (6.2.2) therefore converges absolutely and uniformly on compact sets.[4]

All this applies generally whenever $\phi \in L^1$, and the h_n have sufficient decay. In our present case, m_0 is a trigonometric polynomial (only finitely many of the h_n are different from zero), and we are looking for ϕ with compact support. Together with the obvious constraint $\phi \in L^2$, compact support for ϕ means $\phi \in L^1$, so that the above discussion applies. It follows that (6.2.2) is the only possible candidate (up to a constant phase factor) for the scaling function corresponding to a trigonometric polynomial m_0 constructed as in §6.1. We now need to check that ϕ satisfies some basic requirements for a scaling function. First of all, ϕ is square integrable:

LEMMA 6.2.1. (Mallat (1989)) *If m_0 is a 2π-periodic function satisfying (6.1.1), and if $(2\pi)^{-1/2} \prod_{j=1}^{\infty} m_0(2^{-j}\xi)$ converges pointwise a.e., then its limit $\hat{\phi}(\xi)$ is in $L^2(\mathbb{R})$, and $\|\phi\|_{L^2} \le 1$.*

Proof.

1. Define $f_k(\xi) = (2\pi)^{-1/2} \left[\prod_{j=1}^{k} m_0(2^{-j}\xi)\right] \chi_{[-\pi,\pi]}(2^{-k}\xi)$, where $\chi_{[-\pi,\pi]}(\zeta) = 1$ if $|\zeta| \le \pi$, 0 otherwise. Then $f_k \longrightarrow \hat{\phi}$ pointwise a.e.

2. Moreover,

$$\int d\xi\ |f_k(\xi)|^2 = (2\pi)^{-1} \int_{-2^k\pi}^{2^k\pi} d\xi \prod_{j=1}^{k} |m_0(2^{-j}\xi)|^2$$

$$= (2\pi)^{-1} \int_0^{2^{k+1}\pi} d\xi \prod_{j=1}^{k} |m_0(2^{-j}\xi)|^2 \quad \text{(by the } 2\pi\text{-periodicity of } m_0)$$

$$= (2\pi)^{-1} \int_0^{2^k \pi} d\xi \left[\prod_{j=1}^{k-1} |m_0(2^{-j}\xi)|^2 \right] \left[|m_0(2^{-k}\xi)|^2 + |m_0(2^{-k}\xi + \pi)|^2 \right]$$

$$= (2\pi)^{-1} \int_0^{2^k \pi} d\xi \prod_{j=1}^{k-1} |m_0(2^{-j}\xi)|^2 \qquad \text{(by (6.1.1))}$$

$$= \|f_{k-1}\|^2 .$$

3. It follows that, for all k,

$$\|f_k\|^2 = \|f_{k-1}\|^2 = \cdots = \|f_0\|^2 = 1 .$$

Consequently, by Fatou's lemma,

$$\int d\xi \, |\hat{\phi}(\xi)|^2 \leq \limsup_{k \to \infty} \int d\xi \, |f_k(\xi)|^2 \leq 1 . \quad \blacksquare$$

Second, since m_0 is a trigonometric polynomial, the following lemma borrowed from Deslauriers and Dubuc (1987) proves that ϕ has compact support.

LEMMA 6.2.2. *If* $\Gamma(\xi) = \sum_{n=N_1}^{N_2} \gamma_n \, e^{-in\xi}$, *with* $\sum_{n=N_1}^{N_2} \gamma_n = 1$, *then* $\prod_{j=1}^{\infty} \Gamma(2^{-j}\xi)$ *is an entire function of exponential type. In particular, it is the Fourier transform of a distribution with support in* $[N_1, N_2]$.

Proof. By the Paley–Wiener theorem for distributions, it is sufficient to prove that $\prod_{j=1}^{\infty} \Gamma(2^{-j}\xi)$ is an entire function of exponential type with bounds

$$\left| \prod_{j=1}^{\infty} \Gamma(2^{-j}\xi) \right| \leq C_1 (1 + |\xi|)^{M_1} \exp\left(N_1 \, |\text{Im}\, \xi|\right) \quad \text{for Im}\, \xi \geq 0 ,$$

$$\left| \prod_{j=1}^{\infty} \Gamma(2^{-j}\xi) \right| \leq C_2 (1 + |\xi|)^{M_2} \exp\left(N_2 \, |\text{Im}\, \xi|\right) \quad \text{for Im}\, \xi \leq 0 ,$$

for some C_1, C_2, M_1, M_2. We will only prove the first bound; the second is entirely analogous. Define

$$\Gamma_1(\xi) = e^{iN_1\xi} \, \Gamma(\xi) = \sum_{n=0}^{N_2 - N_1} \gamma_{n+N_1} \, e^{-in\xi} .$$

Then

$$\prod_{j=1}^{\infty} \Gamma(2^{-j}\xi) = e^{-iN_1\xi} \prod_{j=1}^{\infty} \Gamma_1(2^{-j}\xi) ,$$

so we only need to prove a polynomial bound for $\prod_{j=1}^{\infty} \Gamma_1(2^{-j}\xi)$ for Im $\xi \geq 0$. For Im $\zeta \geq 0$ we have

$$|\Gamma_1(\zeta) - 1| \leq \sum_{n=0}^{N_2 - N_1} |\gamma_{n+N_1}| \, |e^{-in\zeta} - 1|$$

$$\leq 2 \sum_{n=0}^{N_2-N_1} |\gamma_{n+N_1}| \min(1, n|\zeta|)$$
$$\leq C \min(1, |\zeta|) .$$

Take ξ arbitrary, with Im $\xi \geq 0$. If $|\xi| \leq 1$, then

$$\left|\prod_{j=1}^{\infty} \Gamma_1(2^{-j}\xi)\right| \leq \prod_{j=1}^{\infty}[1 + C\, 2^{-j}]$$
$$\leq \prod_{j=1}^{\infty} \exp(2^{-j}C) \leq e^C . \tag{6.2.3}$$

If $|\xi| \geq 1$, then there exists $j_0 \geq 0$ so that $2^{j_0} \leq |\xi| < 2^{j_0+1}$, and

$$\left|\prod_{j=1}^{\infty} \Gamma_1(2^{-j}\xi)\right| \leq \prod_{j=1}^{j_0+1} (1+C) \left|\prod_{j=1}^{\infty} \Gamma_1(2^{-j}\, 2^{-j_0-1}\xi)\right|$$
$$\leq (1+C)^{j_0+1}\, e^C$$
$$\leq e^C\, (1+C) \exp\left[\ln(1+C)\, \ln|\xi|/\ln 2\right]$$
$$\leq (1+C)\, e^C\, |\xi|^{\ln(1+C)/\ln 2} . \tag{6.2.4}$$

Combining (6.2.3) for $|\xi| \leq 1$ and (6.2.4) for $|\xi| \geq 1$ establishes the desired polynomial bound. ∎

So far, so good. All this is not sufficient, however, to define a bona fide scaling function. A counterexample is

$$m_0(\xi) = \left(\frac{1+e^{-i\xi}}{2}\right)(1 - e^{-i\xi} + e^{-2i\xi})$$
$$= \frac{1+e^{-3i\xi}}{2} = e^{-3i\xi/2} \cos\frac{3\xi}{2} .$$

This satisfies (6.1.1), as well as $m_0(0) = 1$. Substituting it into (6.2.2) leads to[5]

$$\hat{\phi}(\xi) = (2\pi)^{-1/2}\, e^{-3i\xi/2}\, \frac{\sin 3\xi/2}{3\xi/2}$$

or

$$\phi(x) = \begin{cases} \frac{1}{3}, & 0 \leq x \leq 3 , \\ 0 & \text{otherwise.} \end{cases}$$

This is not a "good" scaling function: the $\phi_{0,n}(x) = \phi(x-n)$ are not orthonormal, even though m_0 satisfies (6.1.1). Another way of looking at this is to see that

(5.1.19) is not satisfied:

$$\sum_{\ell} |\hat{\phi}(\xi + 2\pi\ell)|^2 = (2\pi)^{-1} \left[\frac{1}{3} + \frac{4}{9}\cos\, \xi + \frac{2}{9}\cos 2\xi\right] .$$

Note that this means that $\sum_{\ell} |\hat{\phi}(\xi + 2\pi\ell)|^2 = 0$ for $\xi = \frac{2\pi}{3}$, so that even (5.3.2) is not satisfied: the $\phi_{0,n}$ are not even a Riesz basis for the space they span.[6]

In order to avoid this kind of mishap, we have to impose extra conditions on m_0 to make sure that ϕ generates a true multiresolution analysis. These conditions ensure that

$$\sum_{\ell} |\hat{\phi}(\xi + 2\pi\ell)|^2 = (2\pi)^{-1} \tag{6.2.5}$$

for all ξ. Once (6.2.5) is satisfied, everything else works: the spaces $V_j = \overline{\mathrm{Span}\{\phi_{j,n};\ n \in \mathbb{Z}\}}$ constitute a multiresolution analysis (by §5.3.2); in each V_j, the $(\phi_{j,n})_{n\in\mathbb{Z}}$ constitute an orthonormal basis. We define ψ by

$$\psi(x) = \sqrt{2} \sum_{n} (-1)^n\, \overline{h_{-n+1}}\, \phi(2x - n); \tag{6.2.6}$$

this is automatically compactly supported because ϕ is and because only finitely many h_n differ from zero. The $(\psi_{j,k})_{j,k\in\mathbb{Z}}$ constitute then an orthonormal basis of compactly supported wavelets for $L^2(\mathbb{R})$.

Before we go into the conditions on m_0 that ensure (6.2.5), it is interesting to remark that even if (6.2.5) is not satisfied, the function ψ defined by (6.2.6) still generates a tight frame, as proved in Lawton (1990).

PROPOSITION 6.2.3. *Let m_0 be a trigonometric polynomial satisfying* (6.1.1) *and $m_0(0) = 1$, and let ϕ, ψ be the compactly supported L^2-functions defined by* (6.2.2), (6.2.6). *Define, as usual, $\psi_{j,k}(x) = 2^{-j/2}\, \psi(2^{-j}x - k)$. Then, for all $f \in L^2(\mathbb{R})$,*

$$\sum_{j,k\in\mathbb{Z}} |\langle f,\, \psi_{j,k}\rangle|^2 = \|f\|^2 \ ,$$

i.e., the $(\psi_{j,k};\ j,k \in \mathbb{Z})$ constitute a tight frame for $L^2(\mathbb{R})$.

Proof.

1. First remember that (6.1.1) can also be written as

$$\sum_{m} h_m\, \overline{h_{m+2k}} = \delta_{k,0} \ . \tag{6.2.7}$$

(see (5.1.39)).

2. Take f compactly supported and C^∞. Then $\sum_k |\langle f,\, \phi_{j,k}\rangle|^2$ converges for all j:

$$\sum_{k} |\langle f,\, \phi_{j,k}\rangle|^2 \le 2^{-j} \sum_{k} \left[\int dx\ |f(x)|\, |\phi(2^{-j}x - k)|\right]^2$$

$$\le \|f\|_\infty^2 \, |\text{support}(f)| \, 2^{-j} \sum_k \int_{x \in \text{support}(f)} dx \, |\phi(2^{-j}x - k)|^2$$

$$\le \|f\|_\infty^2 |\text{support}(f)| \sum_k \int_{y \in 2^{-j} \text{ support}(f)} dy \, |\phi(y-k)|^2 . \qquad (6.2.8)$$

Choose K so that 2^{-j} support$(f) \cap [2^{-j}$ support$(f) + k]$ is empty if $k \ge K$. Then

$$\begin{aligned}
&\sum_{k \in \mathbb{Z}} \int_{y \in 2^{-j} \text{ support}(f)} dy \; |\phi(y-k)|^2 \\
&\quad = \sum_{m \in \mathbb{Z}} \sum_{\ell=0}^{K-1} \int_{y \in 2^{-j} \text{ support}(f)} dy \; |\phi(y - mK - \ell)|^2 \\
&\quad \le \sum_{\ell=0}^{K-1} \int dy \; |\phi(y-\ell)|^2 \\
&\qquad\quad \text{(because, for every } \ell, \text{ the sets} \\
&\qquad\qquad (2^{-j}\text{support}(f) + \ell + mK)_{m \in \mathbb{Z}} \text{ do not overlap)} \\
&\quad \le K \; \|\phi\|^2 .
\end{aligned}$$

Similarly, $\sum_k |\langle f, \psi_{j,k} \rangle|^2$ converges for all j.

3. Because $\phi = \sum_n h_n \phi_{-1,n}$, $\psi = \sum_n (-1)^n \; \overline{h_{-n+1}} \; \phi_{-1,n}$, we have

$$\begin{aligned}
&\sum_k \left[|\langle f, \phi_{0,k} \rangle|^2 + |\langle f, \psi_{0,k} \rangle|^2 \right] \\
&\quad = \sum_k \sum_{m,n} \left[h_{n-2k} \; \overline{h_{m-2k}} + (-1)^{n+m} \; \overline{h_{-n+1+2k}} \; h_{-m+1+2k} \right] \\
&\qquad\qquad \cdot \langle f, \phi_{-1,n} \rangle \langle \phi_{-1,m}, f \rangle \; .
\end{aligned} \qquad (6.2.9)$$

It is easy to check that the right-hand side of (6.2.9) is absolutely summable (use that only finitely many h_n are nonzero), so that we may invert the order of the summations.

4. If n, m are even, $n = 2r$, $m = 2s$, we have

$$\begin{aligned}
&\sum_k \left[h_{2r-2k} \; \overline{h_{2s-2k}} + \overline{h_{-2r+2k+1}} \; h_{-2s+2k+1} \right] \\
&\quad = \sum_k h_{2r-2k} \; \overline{h_{2s-2k}} + \sum_\ell \overline{h_{2s-2\ell+1}} \; h_{2r-2\ell+1} \\
&\qquad\qquad \text{(substitute } k = s + r - \ell \text{)} \\
&\quad = \sum_p h_{2r-p} \; \overline{h_{2s-p}} = \delta_{r,s} = \delta_{n,m} \quad \text{(by (6.2.7))} \; .
\end{aligned}$$

Similarly, for $n = 2r+1$, $m = 2s+1$ both odd,

$$\sum_k \left[h_{2r+1-2k} \; \overline{h_{2s+1-2k}} + \overline{h_{-2r+2k}} \; h_{-2s+2k} \right] = \delta_{r,s} = \delta_{n,m} \; .$$

5. If $n = 2r$ is even and $m = 2s + 1$ is odd, then

$$\sum_k \left[h_{2r-2k} \, \overline{h_{2s+1-2k}} - \overline{h_{-2r+2k+1}} \, h_{-2s+2k} \right]$$
$$= \sum_k h_{2r-2k} \, \overline{h_{2s+1-2k}} - \sum_\ell \overline{h_{2s+1-2\ell}} \, h_{2r-2\ell}$$
$$\text{(substitute } k = s + r - \ell)$$
$$= 0 = \delta_{n,m} \ .$$

6. This establishes

$$\sum_k \left[h_{n-2k} \, \overline{h_{m-2k}} + (-1)^{n+m} \, \overline{h_{-n+1+2k}} \, h_{-m+1+2k} \right] = \delta_{m,n}$$

for all m, n. Consequently,

$$\sum_k \left[|\langle f, \phi_{0,k} \rangle|^2 + |\langle f, \psi_{0,k} \rangle|^2 \right] = \sum_m |\langle f, \phi_{-1,m} \rangle|^2 \ .$$

By "telescoping," we have

$$\sum_{j=-J+1}^{J} \sum_{k \in \mathbb{Z}} |\langle f, \psi_{j,k} \rangle|^2 = \sum_k |\langle f, \phi_{-J,k} \rangle|^2 - \sum_k |\langle f, \phi_{J,k} \rangle|^2 \ . \tag{6.2.10}$$

7. The same estimates as in points 3 and 4 of the proof of Proposition 5.3.1 show that, for fixed continuous and compactly supported f, $\sum_k |\langle f, \phi_{J,k} \rangle|^2 \le \epsilon$ if J is large enough, with ϵ arbitrarily small (J depending on f and ϵ). Similarly, the estimate in point 3 of the proof of Proposition 5.3.2 leads to

$$\sum_k |\langle f, \phi_{-J,k} \rangle|^2 = 2\pi \int d\xi \, |\hat{\phi}(2^{-J}\xi)|^2 \, |\hat{f}(\xi)|^2 + R \ , \tag{6.2.11}$$

with $|R| \le \epsilon$ if J is sufficiently large. Since $\hat{\phi}$ is continuous at $\xi = 0$, and $\hat{\phi}(0) = (2\pi)^{-1/2}$, the first term in the right-hand side of (6.2.11) converges to $\int d\xi \, |\hat{f}(\xi)|^2$ for $J \to \infty$ (by dominated convergence: $|\hat{\phi}(\xi)| \le (2\pi)^{-1/2}$ for all ξ, because $|m_0| \le 1$ by (6.1.1)). Combining all this with (6.2.10), we have

$$\sum_{j,k \in \mathbb{Z}} |\langle f, \psi_{j,k} \rangle|^2 = \|f\|^2$$

for all compactly supported C^∞ functions f. Since these form a dense set in L^2, the result extends to all of $L^2(\mathbb{R})$ by the standard density argument. ∎

Without any extra conditions on m_0, we therefore already have a tight frame with frame constant 1. By Proposition 3.2.1, this frame is an orthonormal basis if and only if $\|\psi\| = 1$ (use that $\|\psi_{j,k}\| = \|\psi\|$ for all $j, k \in \mathbb{Z}$), or equivalently, if $\int dx\ \psi(x)\ \overline{\psi(x-k)} = \delta_{k,0}$ for all $k \in \mathbb{Z}$.[7] This in turn is equivalent to $\sum_\ell |\hat{\psi}(\xi + 2\pi\ell)|^2 = (2\pi)^{-1}$. Using $|\hat{\psi}(\xi)| = |m_0(\xi/2 + \pi)|\ |\hat{\phi}(\xi/2)|$ (a consequence of (6.2.6)), this can be rewritten as

$$|m_0(\xi/2 + \pi)|^2\ \alpha\ (\xi/2) + |m_0(\xi/2)|^2\ \alpha(\xi/2 + \pi) = 1\ , \tag{6.2.12}$$

with $\alpha(\zeta) = 2\pi \sum_\ell |\hat{\phi}(\zeta + 2\pi\ell)|^2$. This is equivalent to

$$|m_0(\zeta)|^2\ [\alpha(\zeta + \pi) - 1] + |m_0(\zeta + \pi)|^2\ [\alpha(\zeta) - 1] = 0\ . \tag{6.2.13}$$

We have $m_0(\zeta) = \frac{1}{\sqrt{2}} \sum_{n=N_1}^{N_2} h_n\ e^{-in\zeta}$, with $h_{N_1} \neq 0 \neq h_{N_2}$, so that $|m_0(\zeta)|^2$ is a polynomial in $\cos\zeta$ of degree $N_2 - N_1$. On the other hand, $\alpha(\zeta) = \sum_\ell\ \alpha_\ell\ e^{-i\ell\zeta}$, with $\alpha_\ell = (2\pi)^{-1} \int d\xi\ e^{i\ell\xi}\ |\hat{\phi}(\xi)|^2 = (2\pi)^{-1} \int dx\ \phi(x)\ \overline{\phi(x-\ell)} = 0$ if $\ell \geq N_2 - N_1$, since support $\phi \subset [N_1, N_2]$. Consequently, $\alpha(\zeta) - 1$ is a polynomial in $\cos\zeta$ of degree $N_2 - N_1 - 1$. However, by (6.2.13), $\alpha(\zeta) - 1$ is zero whenever $|m_0(\zeta)|^2$ is ($|m_0(\zeta)|^2$ and $|m_0(\zeta + \pi)|^2$ have no zeros in common), so that this polynomial has at least $N_2 - N_1$ zeros (counting multiplicity). Since it is of degree $N_2 - N_1 - 1$, it therefore has to vanish identically, i.e., $\alpha(\zeta) \equiv 1$, or $\sum_\ell |\hat{\phi}(\zeta + 2\pi\ell)|^2 = (2\pi)^{-1}$. This is another way to derive that (6.2.5) is necessary and sufficient for the $\psi_{j,k}$ to constitute an orthonormal basis.

In the non-orthonormal example we saw above, with $m_0(\xi) = \frac{1}{2}(1 + e^{-3i\xi})$, the recipe (6.2.6) for ψ leads to

$$\psi(x) = \begin{cases} \frac{1}{3}\ , & 0 \leq x < \frac{3}{2}\ , \\ -\frac{1}{3}\ , & \frac{3}{2} \leq x < 3\ , \\ 0 & \text{otherwise}\ . \end{cases}$$

In this case ψ is indeed not normalized, $\|\psi\| = 3^{-1/2}$. If we define $\tilde{\psi} = \|\psi\|^{-1}\psi$, then the $\tilde{\psi}_{j,k}$ are normalized, and constitute a tight frame with frame constant 3: the "redundancy factor" of the frame is 3. This is not so surprising once one realizes that the family $(\tilde{\psi}_{j,k})_{j,k\in\mathbb{Z}}$ can be considered as the union of three shifted copies of a "stretched" Haar basis:

$$\begin{aligned} \tilde{\psi}_{j,3k} &= D_3\ \psi_{j,k}^{\text{Haar}}\ , \\ (\tilde{\psi}_{j,3k+1})(x) &= \left(D_3\ \psi_{j,k}^{\text{Haar}}\right)(x - 1/3)\ , \\ (\tilde{\psi}_{j,3k+2})(x) &= \left(D_3\ \psi_{j,k}^{\text{Haar}}\right)(x - 2/3)\ , \end{aligned}$$

with $(D_3 f)(x) = 3^{1/2}\ f(3x)$.

But let us return to the condition (6.2.5),

$$\sum_\ell |\hat{\phi}(\xi + 2\pi\ell)|^2 = (2\pi)^{-1}\ ,$$

or its equivalent

$$\int dx\ \phi(x)\ \overline{\phi(x-n)} = \delta_{n,0}\ . \tag{6.2.14}$$

Several strategies have been developed, corresponding to conditions on m_0, to ensure that (6.2.5) or (6.2.14) hold. Most of these strategies involve proving that the truncated functions f_k introduced in the proof of Lemma 6.2.1 (or some other truncated family) converge to $\hat{\phi}$ not only pointwise, but also in $L^2(\mathbb{R})$. Since it is not hard to show that, for every fixed k, the $\{f_k(\cdot - n);\ n \in \mathbb{Z}\}$ are orthonormal, this L^2-convergence then automatically implies (6.2.14). Conditions on m_0 sufficient to ensure this L^2-convergence are, e.g.,

- $$\inf_{|\xi| \le \pi/2} |m_0(\xi)| > 0 \quad \text{(Mallat (1989))} \tag{6.2.15}$$

or

- $$m_0(\xi) = \left(\frac{1+e^{i\xi}}{2}\right)^N \mathcal{L}(\xi)\ ,$$

with

$$\sup_{\xi} |\mathcal{L}(\xi)| \le 2^{N-1/2} \quad \text{(Daubechies (1988b))}\ . \tag{6.2.16}$$

Neither of these conditions is necessary, but both cover many interesting examples. Better bounds in $|\mathcal{L}|$ than (6.2.16) lead to regularity for ϕ and ψ; we will come back to this in Chapter 7. Subsequently, necessary and sufficient conditions on m_0 were found. We discuss these at length in the next section.

6.3. Necessary and sufficient conditions for orthonormality.

Cohen (1990) identified a first *necessary and sufficient* condition on m_0 ensuring L^2-convergence of the f_k. Cohen's condition involves the structure of the zero-set of m_0. Before starting his result, it is convenient to introduce a new concept.

DEFINITION. *A compact set K is called congruent to $[-\pi, \pi]$ modulo 2π if*

1. $|K| = 2\pi$;
2. *For all ξ in $[-\pi, \pi]$, there exists $\ell \in \mathbb{Z}$ so that $\xi + 2\ell\pi \in K$.*

Typically, such a compact set K congruent to $[-\pi, \pi]$ can be viewed as the result of some "cut-and-paste-work" on $[-\pi, \pi]$. An example is given in Figure 6.1. We are now ready to state and prove Cohen's theorem.

THEOREM 6.3.1. (Cohen (1990)) *Assume that m_0 is a trigonometric polynomial satisfying* (6.1.1), *with $m_0(0) = 1$, and define ϕ as in* (6.2.2). *Then the following are equivalent:*

1. $$\int dx\ \phi(x)\ \overline{\phi(x-n)} = \delta_{n,0}\ . \tag{6.3.1}$$

2. *There exists a compact set K congruent to $[-\pi, \pi]$ modulo 2π and containing a neighborhood of 0 so that*

$$\inf_{k>0}\ \inf_{\xi \in K}\ |m_0(2^{-k}\xi)| > 0\ . \tag{6.3.2}$$

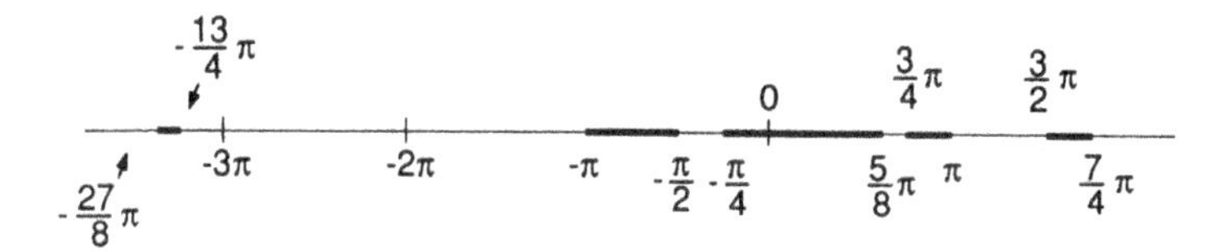

FIG. 6.1. $K = \left[-\frac{27}{8}\pi, -\frac{13}{4}\pi\right] \cup \left[-\pi, -\frac{\pi}{2}\right] \cup \left[-\frac{\pi}{4}, \frac{5\pi}{8}\right] \cup \left[\frac{3\pi}{4}, \pi\right] \cup \left[\frac{3\pi}{2}, \frac{7\pi}{4}\right]$ *is a compact set congruent to* $[-\pi, \pi]$ *modulo* 2π*; it can be viewed as the result of cutting* $[-\pi/2, -\pi/4]$ *and* $[5\pi/8, 3\pi/4]$ *out of* $[-\pi, \pi]$ *and moving the first piece to the right by* 2π*, the second to the left by* 4π*.*

REMARK. The condition (6.3.2) may seem a bit technical, and hard to verify in practice. Remember however that K is compact, and is therefore bounded: $K \subset [-R, R]$. By the continuity of m_0 and $m_0(0) = 1$, it follows that $|m_0(2^{-k}\xi)| > \frac{1}{2}$, uniformly for all $|\xi| \le R$, if k is larger than some k_0. This means that (6.3.2) reduces to requiring that the k_0 functions $m_0(\xi/2)$, $m_0(\xi/4), \cdots, m_0(2^{-k_0}\xi)$ have no zero on K, or equivalently, that m_0 has no zero in $K/2$, $K/4$, $\cdots$, $2^{-k_0}K$. This is already much more accessible! □

Proof of Theorem 6.3.1.

1. We start by proving (1) $\Rightarrow$ (2).
   Assume that (6.3.1) holds, or equivalently, $\sum_\ell |\hat{\phi}(\xi + 2\pi\ell)|^2 = (2\pi)^{-1}$. Then, for all $\xi \in [-\pi, \pi]$, there exists $\ell_\xi \in \mathbb{N}$ so that

$$\sum_{|\ell| \le \ell_\xi} |\hat{\phi}(\xi + 2\pi\ell)|^2 \ge (4\pi)^{-1} .$$

   Since $\hat{\phi}$ is continuous, the finite sum $\sum_{|\ell| \le \ell_\xi} |\hat{\phi}(\cdot + 2\pi\ell)|^2$ is continuous as well. Therefore there exists, for every ξ in $[-\pi, \pi]$, a neighborhood $\{\zeta; |\zeta - \xi| \le R_\xi\}$ so that, for all ζ in this neighborhood,

$$\sum_{|\ell| \le \ell_\xi} |\hat{\phi}(\zeta + 2\pi\ell)|^2 \ \ge\ (8\pi)^{-1} .$$

   Since $[-\pi, \pi]$ is compact, there exists a finite subset of the collection of intervals $\{\zeta; |\zeta - \xi| \le R_\xi\}$ which still covers $[-\pi, \pi]$. Take ℓ_0 to be the maximum of the ℓ_{ξ_j} associated to this finite covering. Then, for all $\zeta \in [-\pi, \pi]$,

$$\sum_{|\ell| \le \ell_0} |\hat{\phi}(\zeta + 2\pi\ell)|^2 \ge (8\pi)^{-1} . \tag{6.3.3}$$

2. It follows that for every $\xi \in [-\pi, \pi]$, there exists ℓ between $-\ell_0$ and ℓ_0 so that $|\hat{\phi}(\xi + 2\pi\ell)| \ge [8\pi(2\ell_0 + 1)]^{-1/2} = C$. Define now sets S_ℓ, $-\ell_0 \le \ell \le \ell_0$

by

$$S_0 = \{\xi \in [-\pi, \pi];\ |\hat{\phi}(\xi)| \geq C\}$$

and, for $\ell \neq 0$,

$$S_\ell = \left\{\xi \in [-\pi, \pi] \setminus \left(\bigcup_{k=-\ell_0}^{\ell-1} S_k \cup S_0\right);\ |\hat{\phi}(\xi + 2\pi\ell)| \geq C\right\} .$$

The S_ℓ, $-\ell_0 \leq \ell \leq \ell_0$ form a partition of $[-\pi, \pi]$. Since $|\hat{\phi}(0)| = (2\pi)^{-1/2} > C$, and since $\hat{\phi}$ is continuous, S_0 contains a neighborhood of 0. Now define

$$K = \bigcup_{\ell=-\ell_0}^{\ell_0} \overline{(S_\ell + 2\pi\ell)} .$$

Then K is clearly compact and congruent to $[-\pi, \pi]$ modulo 2π. By construction, $|\hat{\phi}(\xi)| \geq C$ on K, and K contains a neighborhood of 0.

3. Next we show that K satisfies (6.3.2). As pointed out in the remark before the proof, we only need to check that $\inf_{\xi \in K} |m_0(2^{-k}\xi)| > 0$ for a finite number of k, $1 \leq k \leq k_0$. For $\xi \in K$, we have that

$$|\hat{\phi}(\xi)| = \left(\prod_{k=1}^{k_0} |m_0(2^{-k}\xi)|\right) |\hat{\phi}(2^{-k_0}\xi)| \tag{6.3.4}$$

is bounded below away from zero. Since $|\hat{\phi}|$ is also bounded, the first factor in the right-hand side of (6.3.4) has therefore no zeros on the compact set K. As a finite product of continuous functions it is itself continuous, so that

$$\prod_{k=1}^{k_0} |m_0(2^{-k}\xi)| \geq C_1 > 0 \quad \text{for } \xi \in K .$$

Since $|m_0| \leq 1$, we therefore have, for any k, $1 \leq k \leq k_0$,

$$|m_0(2^{-k}\xi)| \geq \prod_{k'=1}^{k_0} m_0(2^{-k'}\xi)| \geq C_1 > 0 .$$

This proves that (6.3.2) is satisfied, and finishes the proof (1) $\Rightarrow$ (2).

4. We now prove the converse, (2) $\Rightarrow$ (1).
Define $\mu_k(\xi) = (2\pi)^{-1/2} \left[\prod_{j=1}^{k} m_0(2^{-j}\xi)\right] \cdot \chi_K(2^{-k}\xi)$, where χ_K is the indicator function of K, $\chi_K(\xi) = 1$ if $\xi \in K$, 0 otherwise. Since K contains a neighborhood of 0, $\mu_k \to \hat{\phi}$ pointwise for $k \to \infty$.

5. By assumption, $|m_0(2^{-k}\xi)| \geq C > 0$ for $k \geq 1$ and $\xi \in K$. On the other hand, we also have, for any ξ, $|m_0(\xi) - m_0(0)| \leq C'|\xi|$; hence $|m_0(\xi)| \geq 1 - C'|\xi|$. Since K is bounded we can find k_0 so that $2^{-k}C'|\xi| < \frac{1}{2}$ if $\xi \in K$ and $k \geq k_0$. Using $1 - x \geq e^{-2x}$ for $0 \leq x \leq \frac{1}{2}$, we find therefore, for $\xi \in K$,

$$\begin{aligned} |\hat{\phi}(\xi)| &= (2\pi)^{-1/2} \prod_{k=1}^{k_0} |m_0(2^{-k}\xi)| \prod_{k=k_0+1}^{\infty} |m_0(2^{-k}\xi)| \\ &\geq (2\pi)^{-1/2} C^{k_0} \prod_{k=k_0+1}^{\infty} \exp\left[-2C'\, 2^{-k}|\xi|\right] \\ &\geq (2\pi)^{-1/2}\, C^{k_0} \exp\left[-C' 2^{-k_0+1} \max_{\xi \in K} |\xi|\right] = C'' > 0\,. \end{aligned}$$

We can rephrase this as

$$\chi_K(\xi) \leq |\hat{\phi}(\xi)|/C''\,.$$

This implies

$$\begin{aligned} |\mu_k(\xi)| &= (2\pi)^{-1/2} \prod_{j=1}^{k} |m_0(2^{-j}\xi)|\, \chi_K(2^{-k}\xi) \\ &\leq (C'')^{-1}\, (2\pi)^{-1/2} \prod_{j=1}^{k} |m_0(2^{-j}\xi)|\, |\hat{\phi}(2^{-k}\xi)| \\ &= (C'')^{-1}\, (2\pi)^{-1/2}\, |\hat{\phi}(\xi)|\,. \end{aligned} \tag{6.3.5}$$

We can therefore apply the dominated convergence theorem and conclude that $\mu_k \to \hat{\phi}$ in L^2.

6. The congruence of K with $[-\pi, \pi]$ modulo 2π means that for any 2π-periodic function f, $\int_{\xi \in K} d\xi\ f(\xi) = \int_{-\pi}^{\pi} d\xi\ f(\xi) = \int_0^{2\pi} d\xi\ f(\xi)$. In particular,

$$\begin{aligned} &\int d\xi\ |\mu_k(\xi)|^2\ e^{-in\xi} = (2\pi)^{-1}\ 2^k \int_{\zeta \in K} d\zeta \prod_{\ell=0}^{k-1} |m_0(2^\ell \zeta)|^2\ e^{-in2^k\zeta} \\ &= (2\pi)^{-1} 2^k \int_0^{2\pi} d\zeta\ e^{-in2^k\zeta} \left[\prod_{\ell=1}^{k-1} |m_0(2^\ell\zeta)|^2\right] |m_0(\zeta)|^2 \\ &= (2\pi)^{-1} 2^k \int_0^{\pi} d\zeta\ e^{-in2^k\zeta} \left[\prod_{\ell=1}^{k-1} |m_0(2^\ell\zeta)|^2\right] [|m_0(\zeta)|^2 + |m_0(\zeta+\pi)|^2] \\ &= (2\pi)^{-1} 2^k \int_0^{\pi} d\zeta\ e^{-in2^k\zeta} \prod_{\ell=1}^{k-1} |m_0(2^\ell\zeta)|^2 \end{aligned}$$

$$= (2\pi)^{-1} 2^{k-1} \int_0^{2\pi} d\xi \, e^{-in2^{k-1}\xi} \prod_{\ell=0}^{k-2} |m_0(2^\ell \xi)|^2$$

$$= \int d\xi \, |\mu_{k-1}(\xi)|^2 e^{-in\xi} .$$

Since

$$\int d\xi \, |\mu_1(\xi)|^2 \, e^{-in\xi} = (2\pi)^{-1} \, 2 \int_0^{\pi} d\zeta \, e^{-2in\zeta} = \delta_{n,0} \, ,$$

this implies $\int d\xi \, |\mu_k(\xi)|^2 \, e^{-in\xi} = \delta_{n,0}$ for all k. Hence

$$\int d\xi \, |\hat{\phi}(\xi)|^2 \, e^{-in\xi} = \lim_{k\to\infty} \int |\mu_k(\xi)|^2 \, e^{-in\xi}$$

(because $\mu_k \to \hat{\phi}$ pointwise, together with the dominance (6.3.5))

$$= \delta_{n,0} \, ,$$

which is equivalent with (6.2.5) and therefore with (6.3.1). ∎

REMARK. The "truncated" functions μ_k are not the same as the f_k introduced in the proof of Lemma 6.2.1, but the following argument shows that L^2-convergence of the μ_k implies L^2-convergence of the f_k. First of all, K contains a neighborhood of 0, $K \supset [-\alpha, \alpha]$ for some α, $0 < \alpha < \pi$. Define $\nu_k = (2\pi)^{-1/2} \prod_{j=1}^{k} m_0(2^{-j}\xi) \, \chi_{[-\alpha,\alpha]}(2^{-k}\xi)$. Since $\chi_{[-\alpha,\alpha]} \le \chi_K$, the same dominated convergence argument as for the μ_k applies, and $\nu_k \to \hat{\phi}$ in L^2. Consequently, $\|\mu_k - \nu_k\|_{L^2} \to 0$ for $k\to\infty$. Using the congruence of K to $[-\pi, \pi]$ modulo 2π, one shows that $\|\mu_k - \nu_k\|_{L^2} = \|f_k - \nu_k\|_{L^2}$. Hence $\|f_k - \hat{\phi}\|_{L^2} \le \|f_k - \nu_k\|_{L^2} + \|\nu_k - \hat{\phi}\|_{L^2} \to 0$ for $k\to\infty$. □

Note that if Mallat's condition (6.2.15) is satisfied, then we can simply take $K = [-\pi, \pi]$; Cohen's condition is then trivially satisfied, and the $\hat{\phi}_{0,n}$ are indeed orthonormal. The following corollary gives another example of how to apply Cohen's condition.

COROLLARY 6.3.2. (Cohen (1990)) *Assume that m_0 is a trigonometric polynomial satisfying* (6.1.1), *with $m_0(0) = 1$, and define ϕ as is* (6.2.2). *If m_0 has no zeros in $[-\pi/3, \pi/3]$, then the $\phi_{0,n}$ are orthonormal.*

Proof. We need only to construct a satisfactory compact set K. Since m_0 may have zeros in $\pi/3 < |\xi| \le \pi/2$, $K = [-\pi, \pi]$ is no longer a good choice. But we can start with this choice as an ansatz, and "cut out" the zeros. More precisely, assume that the zeros of m_0 in $\pi/3 < \xi \le \pi/2$ are $\xi_1^+ < \cdots < \xi_{L_+}^+$. (They are necessarily finite in number, since m_0 is a trigonometric polynomial.) Similarly, we have $\xi_{L_-}^- < \cdots < \xi_1^-$ for the zeros of m_0 in $-\pi/2 \le \xi < -\pi/3$. For every ℓ choose $I_\ell^\pm$ to be the intersection with $[-\pi, \pi]$ of a small open interval around $\xi_\ell^\pm$, small enough so that the $I_\ell^\pm$ do not overlap with each other or with $[-\pi/3, \pi/3]$ and so that $|m_0|_{I_\ell^\pm} < \frac{1}{2}$. (If $\xi_{L_+}^+ = \pi/2$, then $I_{L_+}^+$ will be of the form $]\pi/2 - \epsilon,\ \pi/2]$.) We define K by excising the intervals $2I_\ell^\pm$ from $[-\pi,\ \pi]$ and moving them to the right or left by 2π before including them again:

$$K = [-\pi,\pi] \backslash \left\{ \left(\bigcup_{\ell=1}^{L_+} 2I_\ell^+ \right) \cup \left(\bigcup_{\ell=1}^{L_-} 2I_\ell^- \right) \right\}$$
$$\cdot \cup \left\{ \bigcup_{\ell=1}^{L_+} \overline{(2I_\ell^+ - 2\pi)} \right\} \cup \left\{ \bigcup_{\ell=1}^{L_-} \overline{(2I_\ell^- + 2\pi)} \right\} . \tag{6.3.6}$$

(See Figure 6.2.)

Let us now check whether m_0 has any zeros on $K/2,\ K/4, \cdots$. Write K as $K_0 \cup K_1$, where K_0 is $[-\pi,\ \pi]$ with the $2I_\ell^\pm$ excised, and K_1 is the rest. By construction m_0 has no zeros on $K_0/2$. On the other hand,

$$K_1/2 = \left[\bigcup_{\ell=1}^{L_+} \overline{(I_\ell^+ - \pi)} \right] \cup \left[\bigcup_{\ell=1}^{L_-} \overline{(I_\ell^- + \pi)} \right] .$$

Since $|m_0(\xi)| \le 1/2$ for $\xi \in \overline{I_\ell^\pm}$, and m_0 satisfies (6.1.1), $|m_0(\xi \pm \pi)| \ge \sqrt{3}/2$ for $\xi \in \overline{I_\ell^\pm}$, so that m_0 has no zeros on $K_1/2$ either. For all $n \ge 2$, the following argument shows that $2^{-n}K \subset [-\pi/3,\ \pi/3]$, so that m_0 has no zeros on $2^{-n}K$, which proves that K satisfies (6.3.2). By construction, the "left-most" piece of K is $\overline{2I_1^+ - 2\pi}$, the "right-most" piece $\overline{2I_1^- + 2\pi}$. But $I_1^+ \subset [\pi/3,\ \pi/2]$, and hence $\overline{2I_1^+ - 2\pi} \subset [-\frac{4\pi}{3},\ -\pi]$; similarly, $\overline{2I_1^- + 2\pi} \subset [\pi,\ \frac{4\pi}{3}]$. Hence $K \subset [-\frac{4\pi}{3},\ \frac{4\pi}{3}]$ and $2^{-n}K \subset [-2^{-n+2}\pi/3,\ 2^{-n+2}\pi/3]$. ∎

Corollary 6.3.2 is optimal in the following sense: it is not possible to find $\alpha < \frac{1}{3}$ so that the absence of zeros of m_0 on $[-\alpha\pi,\ \alpha\pi]$ guarantees orthonormality of the $\phi_{0,n}$. (This is illustrated by the counterexample $m_0(\xi) = \frac{1}{2}(1 + e^{-3i\xi})$ discussed above.) The points $\xi = \pm\frac{\pi}{3}$ play a special role for the following reason: $m_0(\pm\frac{\pi}{3}) = 0$ implies $\hat{\phi}(\frac{2\pi}{3} + 2k\pi) = 0$ for all $k \in \mathbb{Z}$, contradicting (6.2.5). This implication can be checked as follows. Take any $k \in \mathbb{N}$ (negative k can be treated similarly). Then k has a binary representation $k = \sum_{j=0}^{n} \epsilon_j 2^j$, with $\epsilon_j = 0$ or 1; for good measure we can add a couple of zeros at the front end of $k = \epsilon_n \epsilon_{n-1} \cdots \epsilon_1 \epsilon_0$, so that we can assume $\epsilon_n = \epsilon_{n-1} = 0$. If k is even, $k = 2\ell$, then

$$\hat{\phi}\left(\frac{2\pi}{3} + 2k\pi\right) = m_0\left(\frac{\pi}{3} + 2\ell\pi\right)\, \hat{\phi}\left(\frac{\pi}{3} + 2\ell\pi\right) = 0$$
$$\left(\text{because } m_0\left(\frac{\pi}{3}\right) = 0\right) .$$

We therefore need to check only what happens if k is odd, $k = 2\ell + 1$, or $\epsilon_0 = 1$. Then $\frac{2\pi}{3} + 2k\pi = \frac{8\pi}{3} + 4\ell\pi$; hence

$$\hat{\phi}\left(\frac{2\pi}{3} + 2k\pi\right) = m_0\left(\frac{4\pi}{3}\right)\, m_0\left(\frac{2\pi}{3} + \ell\pi\right)\, \hat{\phi}\left(\frac{2\pi}{3} + \ell\pi\right) .$$

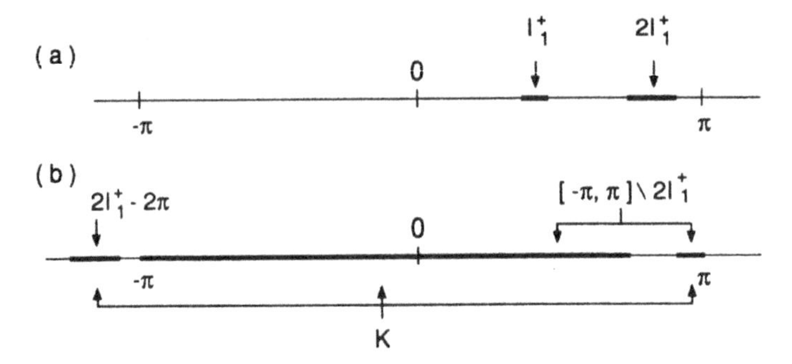

FIG. 6.2. *This figure assumes that m_0 has only one zero in $\pi/3 < |\xi| \leq \pi/2$, namely in $\xi_1^+ = \frac{5\pi}{12}$. We choose $I_1^+ =]\frac{9\pi}{24}, \frac{11\pi}{24}[$; hence $2I_1^+ =]\frac{9\pi}{12}, \frac{11\pi}{12}[$. According to* (6.3.6), *the compact set K is then $[-\frac{15\pi}{12}, -\frac{13\pi}{12}] \cup [-\pi, \frac{9\pi}{12}] \cup [\frac{11\pi}{12}, \pi]$.*

If ℓ is odd, i.e., $\epsilon_1 = 1$, then $m_0(\frac{2\pi}{3} + \ell\pi) = m_0(\frac{5\pi}{3}) = m_0(-\frac{\pi}{3}) = 0$. It follows that we only need to investigate further what happens for $\epsilon_1 = 0$, or ℓ even. We can continue this further, showing that only those k with binary representation ending in $010101\cdots01$ do not automatically lead to $\hat{\phi}(\frac{2\pi}{3} + 2k\pi) = 0$. But if we work back far enough, then we will hit $\epsilon_n\epsilon_{n-1} = 00$, so that we indeed have $\hat{\phi}(\frac{2\pi}{3} + 2k\pi) = 0$.

This whole argument uses that the zero set of m_0 contains $\{\frac{\pi}{3}, -\frac{\pi}{3}\} = [\{\frac{2\pi}{3}, \frac{-2\pi}{3}\} + \pi] \bmod (2\pi)$ and that $\{\frac{2\pi}{3}, \frac{-2\pi}{3}\}$ is an invariant cycle under the operation $\xi \mapsto 2\xi \bmod (2\pi)$, mapping $[-\pi, \pi]$ into itself. In his Ph.D. Thesis, Cohen (1990b) proves that such invariant cycles are the root of the problem.

THEOREM 6.3.3. *Assume that m_0 is a trigonometric polynomial satisfying* (6.1.1) *and $m_0(0) = 1$, and define ϕ as in* (6.2.2). *Then conditions* (1) *and* (2) *in Theorem* 6.3.1 *are also equivalent to*

3. there is no non-trivial cycle $\{\xi_1, \cdots, \xi_n\}$ in $[-\pi, \pi]$ for the operation $\xi \mapsto 2\xi \bmod (2\pi)$ such that $|m_0(\xi_j)| = 1$ for all $j = 1, \cdots, n$.

REMARKS.

1. Because of (6.1.1), $|m_0(\xi_j)| = 1$ is of course equivalent to $|m_0(\xi_j + \pi)| = 0$.

2. Non-trivial means different from $\{0\}$, which is always an invariant cycle.

3. In our example above, $\xi_1 = \frac{2\pi}{3}$, $\xi_2 = -\frac{2\pi}{3}$. □

For a proof of this theorem and related results, we refer to Cohen (1990b); one of the two implications is in fact proved in step 6 of the proof of Theorem 6.3.5 below.

A very different approach to the derivation of conditions on m_0 that ensure (6.2.5) was initiated by Lawton (1990). Let us assume that m_0 is of the form

$$m_0(\xi) = \frac{1}{\sqrt{2}} \sum_{n=0}^{N} h_n \, e^{-in\xi} , \tag{6.3.7}$$

i.e., $h_n = 0$ for $n < 0$ or $n > N$; m_0 can always be brought into this form by multiplication by $e^{iN_1\xi}$, corresponding to a shift of ϕ by N_1. Define $\alpha_\ell = \int dx \; \phi(x) \; \overline{\phi(x-\ell)}$. Since support$(\phi) \subset [0, N]$, $\alpha_\ell = 0$ if $|\ell| \geq N$, and we can regroup the non-trivial α_ℓ, $|\ell| < N$ into a $(2N-1)$-dimensional vector $(\alpha_{-N+1}, \cdots, \alpha_0, \cdots, \alpha_{N-1})$. Because $\phi(x) = \sqrt{2} \, \sum_n \; h_n \; \phi(2x-n)$, the α_ℓ satisfy

$$\begin{aligned}
\alpha_\ell &= 2 \sum_{n,m=0}^{N} h_n \, \overline{h_m} \int dx \; \phi(2x-n) \; \overline{\phi(2x-2\ell-m)} \\
&= \sum_{n,m=0}^{N} h_n \, \overline{h_m} \; \alpha_{2\ell+m-n} \\
&= \sum_{k=-N+1}^{N-1} \left(\sum_{n=0}^{N} h_n \; \overline{h_{k-2\ell+n}} \right) \alpha_k \; .
\end{aligned} \tag{6.3.8}$$

If follows that if we define the $(2N-1) \times (2N-1)$ matrix A by

$$A_{\ell k} = \sum_{n=0}^{N} h_n \; \overline{h_{k-2\ell+n}} \; , \qquad -N+1 \leq \ell, k \leq N-1, \tag{6.3.9}$$

where implicitly $h_m = 0$ if $m < 0$ or $m > N$, then

$$A\alpha = \alpha \; , \tag{6.3.10}$$

i.e., α is an eigenvector of A with eigenvalue 1. Note that 1 is always an eigenvalue of A: if we define β by $\beta = (0, \cdots, 0, 1, 0, \cdots, 0)$ (1 in the central position), or $\beta_\ell = \delta_{\ell,0}$, then

$$(A\beta)_\ell = \sum_k A_{\ell k} \; \delta_{k,0} = \sum_n h_n \; \overline{h_{n-2\ell}} = \delta_{\ell,0} = \beta_\ell$$

by (6.2.7), i.e., $A\beta = \beta$. If the eigenvalue 1 of A is nondegenerate, then α necessarily has to be a multiple of β, i.e., $\int dx \; \phi(x) \; \overline{\phi(x-\ell)} = \gamma \delta_{\ell,0}$ for some $\gamma \in \mathbb{C}$. This implies $\sum_k |\hat{\phi}(\xi + 2\pi k)|^2 = (2\pi)^{-1}\gamma$; since $|\hat{\phi}(2\pi k)| = 0$ for $k \neq 0$ (see the start of §6.2) and $\hat{\phi}(0) = (2\pi)^{-1/2}$ by definition, it follows that $\gamma = 1$, so that $\int dx \; \phi(x) \; \overline{\phi(x-\ell)} = \delta_{\ell,0}$. We have thus a very simple sufficient criterion for orthonormality of the $\phi_{0,n}$.

THEOREM 6.3.4. (Lawton (1990)) *Assume that m_0 is a trigonometric polynomial of the form* (6.3.7), *satisfying* (6.1.1) *and* $m_0(0) = 1$; *define ϕ as in* (6.2.2). *If the eigenvalue* 1 *of the* $(2N-1) \times (2N-1)$ *matrix A defined by* (6.3.9) *is nondegenerate, then the $\phi_{0,n}$ are orthonormal.*

Orthonormality of the $\phi_{0,n}$ can only fail if the characteristic equation for A has a multiple zero at 1. This indicates that among all the possible choices for the h_n, $n = 0, \cdots, N$ (keep N fixed), the "bad" choices (leading to non-orthonormal $\phi_{0,n}$) constitute a very "thin" set. (This statement is made more precise in Lawton (1990).) For $N = 3$, for instance, the only non-orthonormal choice (up to an overall phase factor) is $h_0 = h_3 = 1/2$, $h_1 = h_2 = 0$.

Lawton's condition can be recast in terms of trigonometric polynomials. Define, as before, $M_0(\xi) = |m_0(\xi)|^2$, and define the following operator P_0, acting on 2π-periodic functions f.

$$(P_0 f)(\xi) = M_0(\xi/2)\ f(\xi/2) + M_0(\xi/2 + \pi)\ f(\xi/2 + \pi)\ .$$

Clearly the constant polynomial 1 is invariant under P_0 by (6.1.1). Writing everything out in terms of Fourier coefficients, we have

$$\begin{aligned} M_0(\xi) &= \frac{1}{2} \sum_k \left(\sum_n h_n\, \overline{h_{n-k}} \right) e^{-ik\xi}\ , \\ M_0(\xi)\ f(\xi) &= \frac{1}{2} \sum_\ell \left(\sum_{k,n} h_n\, \overline{h_{n-k}}\, f_{\ell-k} \right) e^{-i\ell\xi}\ ; \end{aligned}$$

hence

$$(P_0 f)(\xi) = \sum_\ell \left(\sum_{k,n} h_n\, \overline{h_{n-k}}\, f_{2\ell-k} \right) e^{-i\ell\xi}$$

or

$$(P_0 f)_\ell = \sum_{k,n} h_n\, \overline{h_{n-k}}\, f_{2\ell-k} = \sum_m \left(\sum_n h_n\, \overline{h_{n-2\ell+m}} \right) f_m\ .$$

This is essentially the same expression as (6.3.8)! (We have not assumed that $f_m = 0$ for $|m| > N$, however, so it is not quite the same.) It follows that Lawton's condition is satisfied if we know that the only trigonometric polynomials invariant under P_0 are the constants.

A priori it is not clear whether Lawton's condition is sufficient or not: it is conceivable that A has an eigenvector different from β with eigenvalue 1, but that α nevertheless happens to be equal to β. However, in the spring of 1990 both Cohen and Lawton proved, independently, that their two conditions are equivalent (a generalization appears in Cohen, Daubechies, and Feauveau (1992) as Theorem 4.3; see also Lawton (1991)), implying the sufficiency of Lawton's condition.

THEOREM 6.3.5. *Assume that m_0 is a trigonometric polynomial such that (6.1.1) is satisfied and $m_0(0) = 1$. If there exists a compact set K congruent to $[-\pi, \pi]$ modulo 2π, containing a neighborhood of 0, such that $\inf_{k\geq 1} \inf_{\xi\in K} |m_0(2^{-k}\xi)| > 0$, then the only trigonometric polynomials invariant under P_0 are the constants.*

REMARK. This is sufficient to prove equivalence. If we denote Lawton's original condition by (L), Cohen's condition by (C), Lawton's condition rephrased in terms of P_0 by (P) and the orthonormality of the $\phi_{0,n}$ by (O), then we already know

$$(P) \Rightarrow (L) \Rightarrow (O) \Rightarrow (C) .$$

The implication $(C) \Rightarrow (P)$ suffices to prove equivalence of all four conditions. □

Proof of Theorem 6.3.5.

1. We will prove that the existence of a nonconstant trigonometric polynomial f invariant for P_0 contradicts the existence of a compact set K with all the desired properties. Suppose f is such a nonconstant trigonometric polynomial, invariant for P_0. Define $f_1(\xi) = f(\xi) - \min_\zeta f(\zeta)$, $f_2(\xi) = -f(\xi) + \max_\zeta f(\zeta)$. Since f is not constant, at least one of f_1, f_2 has $f_j(0) \neq 0$. Pick j so that $f_j(0) \neq 0$, and define $f_0 = f_j$. Then f_0 is nonnegative, $f_0(0) \neq 0$, f_0 has at least one zero, and f_0 is invariant under P_0.

2. Next we explore the zero set of f_0, which turns out to have a very particular structure. If $f_0(\xi) = 0$ for $0 \neq \xi \in [0, 2\pi[$, then

$$0 = f_0(\xi) = (P_0 f_0)(\xi) = M_0(\xi/2)\; f_0(\xi/2) + M_0(\xi/2 + \pi)\; f_0(\xi/2 + \pi) .$$

   Here M_0, f_0 are both nonnegative, and $M_0(\xi/2)$, $M_0(\xi/2+\pi)$ cannot vanish simultaneously, by (6.1.1). Therefore, either $f_0(\xi/2)$ or $f_0(\xi/2 + \pi) = 0$. It follows that if we pick one zero $0 \neq \xi_1 \in [0, 2\pi[$ of f_0, then we can associate to it a chain of zeros in $[0, 2\pi[$, $\xi_2, \cdots, \xi_k, \cdots$, with the property that ξ_{j+1} equals either $\frac{\xi_j}{2}$ or $\frac{\xi_j}{2} + \pi$, or, equivalently, $\xi_j = \tau\xi_{j+1}$, where τ is the transformation $\xi \mapsto 2\xi \bmod (2\pi)$, which maps $[0, 2\pi[$ into itself. Being a trigonometric polynomial, f_0 has only finitely many zeros, so that this chain cannot go on ad infinitum. Note that the chain has at least two elements, since $\xi_2 = \xi_1$ would imply $\xi_1 = 0$. Let r be the first index for which recurrence occurs, i.e., $\xi_r = \xi_k$ for some $k < r$. Then necessarily $k = 1$, because $k > 1$ would lead to $\xi_1 = \tau^{k-1}\xi_k = \tau^{k-1}\xi_r = \xi_{r-k+1}$ with $1 < r - k + 1 < r$, so that r would not be the first index for recurrence. It follows that we have a cycle of zeros, $\xi_1, \cdots, \xi_{r-1}$, with $\tau\xi_{j+1} = \xi_j$ for $j = 1, \cdots, r-2$, and $\tau\xi_1 = \xi_{r-1}$. Note that $\tau^{r-1}\xi_j = \xi_j$ for every zero in this cycle.

3. If this cycle of zeros does not exhaust the set of zeros different from 0, then we can find $0 \neq \zeta_1 \neq \xi_j$, $j = 1, \cdots, r-1$, for which $f_0(\zeta_1) = 0$. This can again be taken as a seed for a chain of zeros, $\zeta_1, \zeta_2, \cdots, \zeta_\ell, \cdots$. Every element of this new chain is necessarily different from all the ξ_j, since $\zeta_\ell = \xi_j$ would imply $\zeta_1 = \tau^{\ell-1}\zeta_\ell = \tau^{\ell-1}\xi_j$, i.e., ζ_1 would equal some ξ_k. By the same argument as above, ζ_1 generates therefore a cycle of zeros

for f, invariant under τ, and disjoint from the first cycle. We can keep on constructing such cycles until exhaustion of the finite set of zeros of f_0. The zero set of f_0 therefore consists of a union of finite invariant cycles for τ.

4. Now note that if $f_0(\xi) = 0$, then necessarily $f_0(\xi + \pi) \neq 0$. Indeed, since $\tau\xi = \tau(\xi + \pi)$, both ξ and $\xi + \pi$ would belong to the same cycle of zeros if $f_0(\xi) = 0 = f_0(\xi + \pi)$. If this cycle has length n, then it would follow that $\xi = \tau^n \xi = \tau^{n-1}\tau\xi = \tau^{n-1}\tau(\xi + \pi) = \xi + \pi$, which is impossible.

5. Finally, we remark that if $f_0(\xi) = 0$, then $M_0(\xi + \pi) = 0$. Indeed, for any ξ so that $f_0(\xi) = 0$, $\tau\xi$ is also a zero for f_0, and it follows that

$$0 = f_0(\tau\xi) = (P_0 f)(\tau\xi) = M_0(\xi)\ f(\xi) + M_0(\xi + \pi)\ f(\xi + \pi)\ .$$

Since $f_0(\xi) = 0$ and $f_0(\xi + \pi) \neq 0$, this implies $M_0(\xi + \pi) = 0$; hence $m_0(\xi + \pi) = 0$. Therefore, the existence of f_0 implies the existence of a cyclic set $\xi_1, \cdots, \xi_n$ for τ, with $\xi_{j+1} = \tau\xi_j$, $j = 1, \cdots, n-1$, $\xi_1 = \tau\xi_n$, so that $m_0(\xi_j + \pi) = 0$ for all j. Since $f_0(0) \neq 0$, we have $\xi_j \neq 0$.

6. We now show how these zeros $\xi_j + \pi$ for m_0 are incompatible with the existence of K. Since $\tau\xi_j = \xi_{j+1}$, $\tau\xi_n = \xi_1$, and, in particular, $\xi_j = \tau^n \xi_j$, we have $\xi_j = 2\pi x_j$, where the $x_j \in [0, 1[$ have the following binary representations:

$$\begin{aligned} x_1 &= .d_1 d_2 \cdots d_n d_1 \cdots d_n d_1 \cdots d_n \cdots \qquad (d_j = 0 \text{ or } 1) \\ x_2 &= .d_2 \cdots d_n d_1 \cdots d_n d_1 \cdots d_n \cdots \\ &\vdots \\ x_n &= .d_n d_1 \cdots d_n d_1 \cdots d_n \cdots \quad . \end{aligned}$$

Since $\xi_1 \neq 0$, not all the d_j are zero. Let us, for this point only, define $\overline{d} = 1 - d$ for $d = 0$ or 1. Then $\xi_j + \pi = 2\pi y_j$ modulo 2π, with y_j given by

$$\begin{aligned} y_1 &= .\overline{d}_1 d_2 d_3 \cdots d_n d_1 \cdots d_n d_1 \cdots d_n \cdots \\ y_2 &= .\overline{d}_2 d_3 \cdots d_n d_1 \cdots d_n d_1 \cdots d_n \cdots \\ &\vdots \\ y_n &= .\overline{d}_n d_1 \cdots d_n d_1 \cdots d_n \cdots \quad . \end{aligned}$$

We have $m_0(2\pi y_j) = 0$, $j = 1, \cdots, n$. Suppose a compact set K existed with all the desired properties. Then there would be an integer ℓ, with a binary expansion with at most a certain preassigned number L of digits (L depends only on the size of K), so that $2\pi y = 2\pi(2y_1 + \ell)$ has the property that $m_0(2\pi 2^{-k} y) \neq 0$ for all $k \geq 0$. We have

$$y = e_L \cdots e_2 e_1\ .d_2 d_3 \cdots d_n d_1 \cdots d_n d_1 \cdots d_n \cdots\ ,$$

with $e_j = 1$ or 0 for $j = 1, \cdots, L$. We can also rewrite this as

$$y = e_{L+n} \cdots e_{L+1} e_L \cdots e_2 e_1\ .d_2 d_3 \cdots d_n d_1 \cdots d_n d_1 \cdots d_n \cdots\ ,$$

where $e_j = 1$ or 0 for $j = 1, \cdots, L$ and $e_j = 0$ if $j > L$. The $2^{-k}y$ are obtained by shifting the decimal point to the left. Since m_0 is 2π-periodic, only the "tail," i.e., the part of the expansion of $2^{-k}y$ to the right of the decimal point, decides whether $m_0(2\pi 2^{-k}y)$ vanishes or not. If $e_1 = \overline{d}_1$, then $y/2$ would have the same decimal part as y_1, hence $m_0(2\pi y/2) = 0$ would follow. Since $m_0(2\pi y/2) \neq 0$, we therefore have $e_1 = d_1$. Similarly, we conclude $e_2 = d_n$, $e_3 = d_{n-1}$, etc. It follows that $e_{L+1}, \cdots, e_{L+n}$ are also successively equal to $d_k, d_{k-1}, \cdots, d_1, d_n, \cdots, d_{k+1}$ for some $k \in \{1, 2, \cdots, n\}$. Since the d_j are not all equal to 0, whereas $e_{L+1} = \cdots = e_{L+n} = 0$, this is a contradiction. This finishes the proof. ∎

With Theorem 6.3.5 we end our discussion of necessary and sufficient conditions on m_0. The following theorem summarizes the main results of §§6.2 and 6.3.

THEOREM 6.3.6. *Suppose m_0 is a trigonometric polynomial such that $|m_0(\xi)|^2 + |m_0(\xi + \pi)|^2 = 1$ and $m_0(0) = 1$. Define ϕ, ψ by*

$$\hat{\phi}(\xi) = (2\pi)^{-1/2} \prod_{j=1}^{\infty} m_0(2^{-j}\xi) ,$$

$$\hat{\psi}(\xi) = -e^{-i\xi/2} \, \overline{m_0(\xi/2 + \pi)} \, \hat{\phi}(\xi/2) .$$

Then ϕ, ψ are compactly supported L^2-functions, satisfying

$$\phi(x) = \sqrt{2} \sum_n h_n \, \phi(2x - n) ,$$

$$\psi(x) = \sqrt{2} \sum_n (-1)^n \, h_{-n+1} \, \phi(2x - n) ,$$

where h_n is determined by m_0 via $m_0(\xi) = \frac{1}{\sqrt{2}} \sum_n h_n \, e^{-in\xi}$. Moreover, the $\psi_{j,k}(x) = 2^{-j/2} \, \psi(2^{-j}x - k)$, $j, k \in \mathbb{Z}$ constitute a tight frame for $L^2(\mathbb{R})$ with frame constant 1. This tight frame is an orthonormal basis if and only if m_0 satisfies one of the following equivalent conditions:

- *There exists a compact set K, congruent to $[-\pi, \pi]$ modulo 2π, containing a neighborhood of 0, so that*

$$\inf_{k>0} \inf_{\xi \in K} |m_0(2^{-k}\xi)| > 0 .$$

- *There exists no nontrivial cycle $\{\xi_1, \cdots \xi_n\}$ in $[0, 2\pi[$, invariant under τ: $\xi \mapsto 2\xi$ modulo 2π, such that $m_0(\xi_j + \pi) = 0$ for all $j = 1, \cdots n$.*

- *The eigenvalue 1 of the $[2(N_2 - N_1) - 1] \times [2(N_2 - N_1) - 1]$-dimensional matrix A defined by*

$$A_{\ell k} = \sum_{n=N_1}^{N_2} h_n \, \overline{h_{k-2\ell+n}}, \quad -(N_2 - N_1) + 1 \leq \ell, k \leq (N_2 - N_1) + 1$$

(where we assume $h_n = 0$ for $n < N_1$, $n > N_2$) is nondegenerate.

From the point of view of subband filtering, this theorem tells us that, provided the high-pass filter has a null at DC ($m_0(\pi) = 0$, hence $m_0(0) = 1$ with the appropriate phase choice), we "almost always" have a corresponding orthonormal wavelet basis. The correspondence only fails "accidentally," as is illustrated by the last two equivalent necessary and sufficient conditions. In practice, one likes to work with filter pairs in which the low-pass filter has no zeros in the band $|\xi| \leq \pi/2$, which is sufficient to ensure that the $\psi_{j,k}$ are an orthonormal basis. But it is time to look at some examples!

6.4. Examples of compactly supported wavelets generating an orthonormal basis.

All the examples we give in this section are obtained by spectral factorization of (6.1.11), with different choices of N and R. Except for the Haar basis, we have no closed-form formula for $\phi(x)$, $\psi(x)$; we will explain in the next section how the plots for ϕ, ψ are obtained.

A first family of examples, constructed in Daubechies (1988b), corresponds to $R \equiv 0$ in (6.1.11). In the spectral factorization needed to extract $\mathcal{L}(\xi)$ from $L(\xi) = P_N(\sin^2\ \xi/2)$, we retain systematically the zeros within the unit circle. For each N, the corresponding ${}_N m_0$ has $2N$ non-vanishing coefficients; we can choose the phase of ${}_N m_0$ so that

$$ {}_N m_0(\xi) = \frac{1}{\sqrt{2}} \sum_{n=0}^{2N-1} {}_N h_n\ e^{-in\xi}\ . $$

Table 6.1 lists the ${}_N h_n$ for $N = 2$ through 10. For faster implementation, it makes sense to keep the factorization (6.1.10) explicit: the filter $\mathcal{L}$ is much shorter than m_0 (N taps instead of $2N$), and the filters $\frac{1+e^{-i\xi}}{2}$ are very easy to implement. Table 6.2 lists the coefficients of $\mathcal{L}(\xi)$, for $N = 2$ to 10. Figure 6.3 shows the plots of the corresponding ${}_N\phi$, ${}_N\psi$ for $N = 2, 3, 5, 7$, and 9. Both ${}_N\phi$ and ${}_N\psi$ have support width $2N - 1$; their regularity clearly increases with N. In fact, one can prove (see Chapter 7) that for large N ${}_N\phi$, ${}_N\psi \in C^{\mu N}$, with $\mu \simeq .2$.

Retaining systematically the zeros within the unit circle in the spectral factorization procedure amounts to choosing the "minimum phase filter" m_0 among all the possibilities once $|m_0|^2$ is fixed. This corresponds to a very marked asymmetry in ϕ and ψ, as illustrated by Figure 6.3. Other choices may lead to less asymmetric ϕ, ψ, although, as we will see in detail in Chapter 8, complete symmetry for ϕ, ψ can not be achieved (except by the Haar basis) within the framework of compactly supported orthonormal wavelet bases. Table 6.3 lists the h_n for the "least asymmetric" ϕ, ψ, for $N = 4$ through 10, corresponding to the same $|m_0|^2$ as in Table 6.1, with a different "square root" m_0. We will come back in Chapter 8 on how this "least asymmetric square root" is determined. Figure 6.4 shows the corresponding ϕ and ψ functions.

Figure 6.5 shows plots of $|m_0|$ as functions of ξ, for the above examples, for $N = 2$, 6, and 10. These plots show that the subband filters for these

TABLE 6.1

The filter coefficients $_N h_n$ *(low-pass filter) for the compactly supported wavelets with extremal phase and highest number of vanishing moments compatible with their support width. The* $_N h_n$ *are normalized so that* $\sum_n \, _N h_n = \sqrt{2}$.

	n	$_N h_n$
$N = 2$	0	.4829629131445341
	1	.8365163037378077
	2	.2241438680420134
	3	−.1294095225512603
$N = 3$	0	.3326705529500825
	1	.8068915093110924
	2	.4598775021184914
	3	−.1350110200102546
	4	−.0854412738820267
	5	.0352262918857095
$N = 4$	0	.2303778133088964
	1	.7148465705529154
	2	.6308807679298587
	3	−.0279837694168599
	4	−.1870348117190931
	5	.0308413818355607
	6	.0328830116668852
	7	−.0105974017850690
$N = 5$	0	.1601023979741929
	1	.6038292697971895
	2	.7243085284377726
	3	.1384281459013203
	4	−.2422948870663823
	5	−.0322448695846381
	6	.0775714938400459
	7	−.0062414902127983
	8	−.0125807519990820
	9	.0033357252854738
$N = 6$	0	.1115407433501095
	1	.4946238903984533
	2	.7511339080210959
	3	.3152503517091982
	4	−.2262646939654400
	5	−.1297668675672625
	6	.0975016055873225
	7	.0275228655303053
	8	−.0315820393174862
	9	.0005538422011614
	10	.0047772575109455
	11	−.0010773010853085
$N = 7$	0	.0778520540850037
	1	.3965393194818912
	2	.7291320908461957
	3	.4697822874051889
	4	−.1439060039285212
	5	−.2240361849938412
	6	.0713092192668272
	7	.0806126091510774
	8	−.0380299369350104
	9	−.0165745416306655
	10	.0125509985560986
	11	.0004295779729214
	12	−.0018016407040473
	13	.0003537137999745
$N = 8$	0	.0544158422431072
	1	.3128715909143166
	2	.6756307362973195
	3	.5853546836542159
	4	−.0158291052563823
	5	−.2840155429615824
	6	.0004724845739124
	7	.1287474266204893
	8	−.0173693010018090
	9	−.0440882539307971
	10	.0139810279174001
	11	.0087460940474065
	12	−.0048703529934520
	13	−.0003917403733770
	14	.0006754494064506
	15	−.0001174767841248
$N = 9$	0	.0380779473638778
	1	.2438346746125858
	2	.6048231236900955
	3	.6572880780512736
	4	.1331973858249883
	5	−.2932737832791663
	6	−.0968407832229492
	7	.1485407493381256
	8	.0307256814793385
	9	−.0676328290613279
	10	.0002509471148340
	11	.0223616621236798
	12	−.0047232047577518
	13	−.0042815036824635
	14	.0018476468830563
	15	.0002303857635232
	16	−.0002519631889427
	17	.0000393473203163
$N = 10$	0	.0266700579005473
	1	.1881768000776347
	2	.5272011889315757
	3	.6884590394534363
	4	.2811723436605715
	5	−.2498464243271598
	6	−.1959462743772862
	7	.1273693403357541
	8	.0930573646035547
	9	−.0713941471663501
	10	−.0294575368218399
	11	.0332126740593612
	12	.0036065535669870
	13	−.0107331754833007
	14	.0013953517470688
	15	.0019924052951925
	16	−.0006858566949564
	17	−.0001164668551285
	18	.0000935886703202
	19	−.0000132642028945

TABLE 6.2

The coefficients ℓ_n *of* $\sqrt{2}\ \mathcal{L}(\xi) = \sum_n \ell_n e^{-in\xi}$, *for* $N = 2$ *to* 10. *Normalization:* $\sum_n \ell_n = \sqrt{2}$.

N	ℓ_n
$N = 2$	1.93185165258
	− 0.517638090205
$N = 3$	2.6613644236
	− 1.52896119631
	0.281810335086
$N = 4$	3.68604501294
	− 3.30663492292
	1.20436190091
	− 0.169558428561
$N = 5$	5.12327673517
	− 6.29384704236
	3.41434077007
	− 0.936300109646
	0.106743209135
$N = 6$	7.13860757441
	− 11.1757164609
	8.04775526289
	− 3.24691364198
	0.719428097459
	− 0.0689472694597
$N = 7$	9.96506292288
	− 18.9984075665
	17.0514392132
	− 9.03858510919
	2.93696631047
	− 0.547537574895
	0.0452753663967
$N = 8$	13.9304556142
	− 31.3485176398
	33.6968524121
	− 22.07104076339
	.38930245651
	− 2.56627196249
	0.413507501939
	− 0.0300740567359
$N = 9$	19.4959090503
	− 50.6198280511
	63.3951659783
	− 49.3675482281
	25.8600363319
	− 9.24491588775
	2.18556614566
	− 0.310317604756
	0.0201458280019
$N = 10$	27.3101392901
	− 80.408349622
	114.98124563
	− 103.671381722
	64.3509475067
	− 28.2911921431
	8.74937688138
	− 1.82464995075
	0.231660236047
	− 0.013582543764

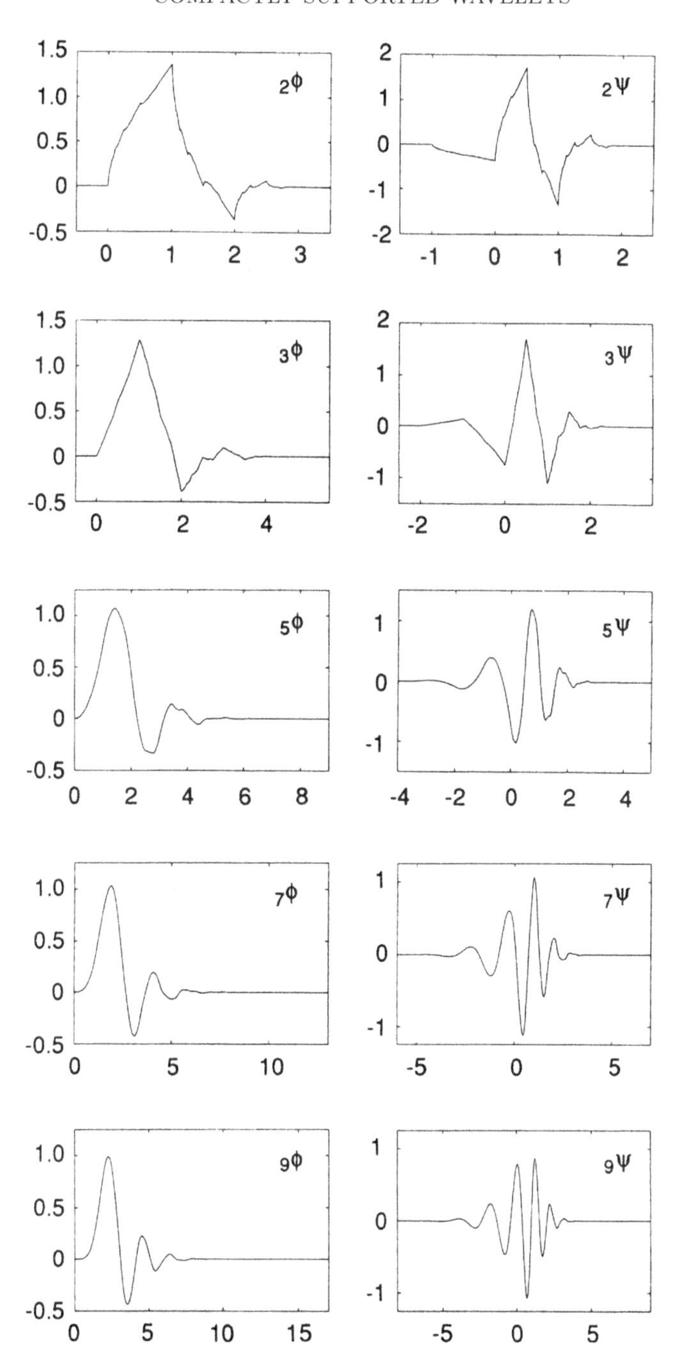

FIG. 6.3. *Plots of the scaling functions* $_N\phi$ *and wavelets* $_N\psi$ *for the compactly supported wavelets with maximum number of vanishing moments for their support width, and with the extremal phase choice, for* $N = 2, 3, 5, 7,$ *and* 9.

TABLE 6.3

The low-pass filter coefficients for the "least asymmetric" compactly supported wavelets with maximum number of vanishing moments, for $N = 4$ to 10. Listed here are the $c_{N,n} = \sqrt{2}\, h_{N,n}$; one has $\sum_n c_{N,n} = 2$.

	n	$c_{N,n}$
$N=4$	0	− 0.107148901418
	1	− 0.041910965125
	2	0.703739068656
	3	1.136658243408
	4	0.421234534204
	5	− 0.140317624179
	6	− 0.017824701442
	7	0.045570345896
$N=5$	0	0.038654795955
	1	0.041746864422
	2	− 0.055344186117
	3	0.281990696854
	4	1.023052966894
	5	0.896581648380
	6	0.023478923136
	7	− 0.247951362613
	8	− 0.029842499869
	9	0.027632152958
$N=6$	0	0.021784700327
	1	0.004936612372
	2	− 0.166863215412
	3	− 0.068323121587
	4	0.694457972958
	5	1.113892783926
	6	0.477904371333
	7	− 0.102724969862
	8	− 0.029783751299
	9	0.063250562660
	10	0.002499922093
	11	− 0.011031867509
$N=7$	0	0.003792658534
	1	− 0.001481225915
	2	− 0.017870431651
	3	0.043155452582
	4	0.096014767936
	5	− 0.070078291222
	6	0.024665659489
	7	0.758162601964
	8	1.085782709814
	9	0.408183939725
	10	− 0.198056706807
	11	− 0.152463871896
	12	0.005671342686
	13	0.014521394762
$N=8$	0	0.002672793393
	1	− 0.000428394300
	2	− 0.021145686528
	3	0.005386388754
$N=8$	4	0.069490465911
	5	− 0.038493521263
	6	− 0.073462508761
	7	0.515398670374
	8	1.099106630537
	9	0.680745347190
	10	− 0.086653615406
	11	− 0.202648655286
	12	0.010758611751
	13	0.044823623042
	14	− 0.000766690896
	15	− 0.004783458512
$N=9$	0	0.001512487309
	1	− 0.000669141509
	2	− 0.014515578553
	3	0.012528896242
	4	0.087791251554
	5	− 0.025786445930
	6	− 0.270893783503
	7	0.049882830959
	8	0.873048407349
	9	1.015259790832
	10	0.337658923602
	11	− 0.077172161097
	12	0.000825140929
	13	0.042744433602
	14	− 0.016303351226
	15	− 0.018769396836
	16	0.000876502539
	17	0.001981193736
$N=10$	0	0.001089170447
	1	0.000135245020
	2	− 0.012220642630
	3	− 0.002072363923
	4	+ 0.064950924579
	5	0.016418869426
	6	− 0.225558972234
	7	− 0.100240215031
	8	0.667071338154
	9	1.088251530500
	10	0.542813011213
	11	− 0.050256540092
	12	− 0.045240772218
	13	0.070703567550
	14	0.008152816799
	15	− 0.028786231926
	16	− 0.001137535314
	17	0.006495728375
	18	0.000080661204
	19	− 0.000649589896

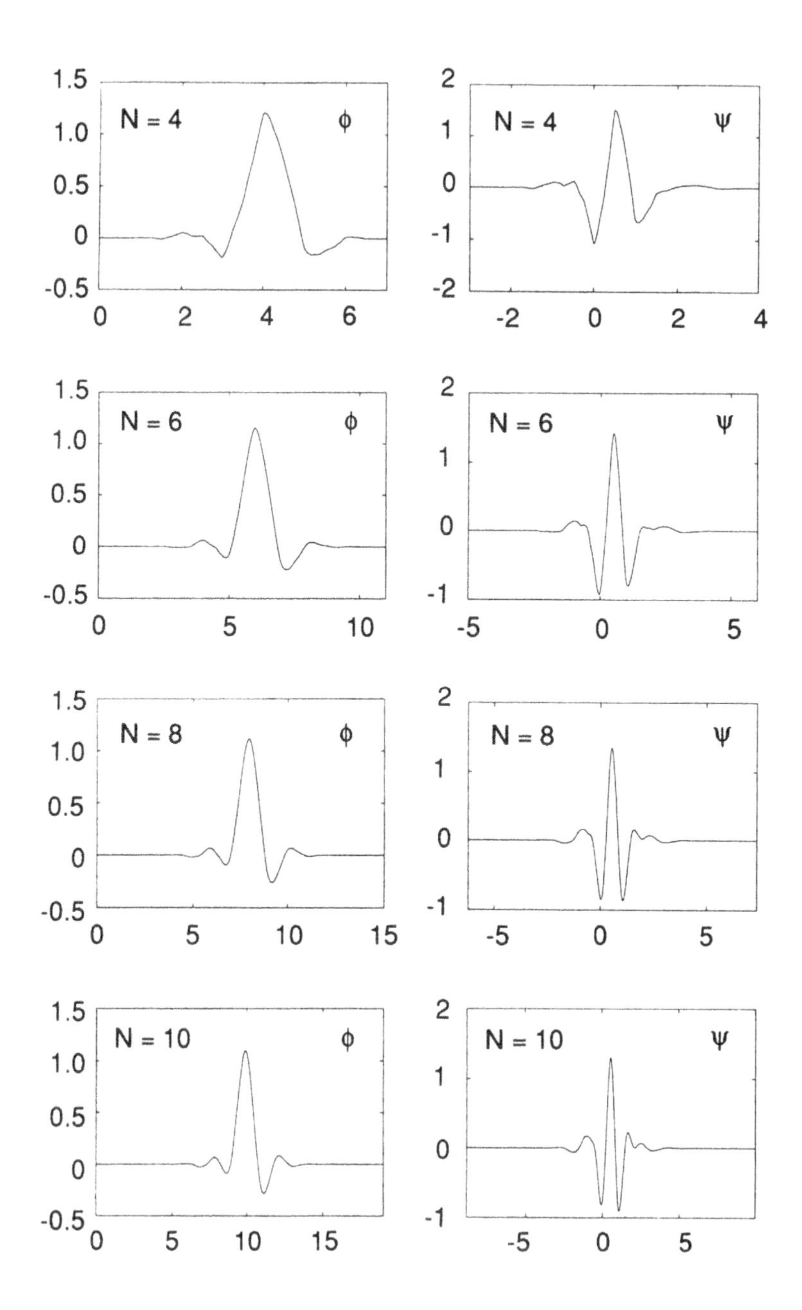

FIG. 6.4. *Plots of the scaling function* ϕ *and the wavelet* ψ *for the "least asymmetric" compactly supported wavelets with maximum number of vanishing moments, for* $N = 4, 6, 8,$ *and* 10.

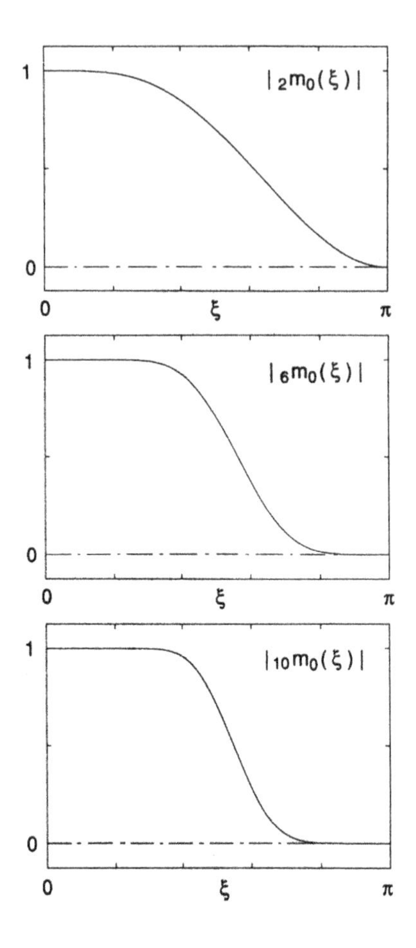

FIG. 6.5. *$|m_0(\xi)|$ for $N = 2, 6$ and 10, corresponding to the filters in Table 6.1 or 6.3.*

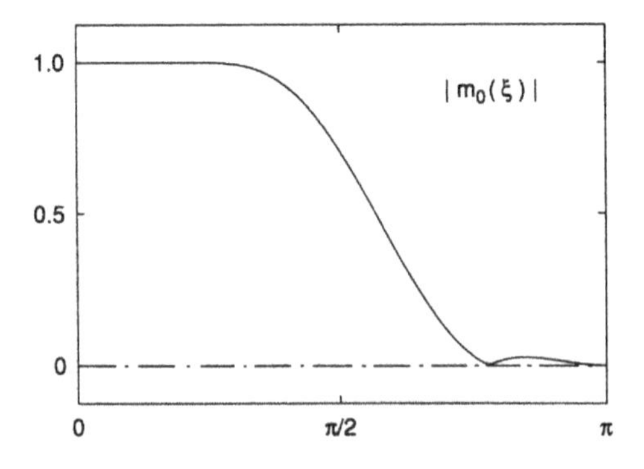

FIG. 6.6. *Plot of $|m_0(\xi)|$ for the 8-tap filter corresponding to $N = 2$ and $m_0(7\pi/9) = 0$.*

orthonormal bases are indeed very flat at 0 and π, but very "round" in the transition region, near $\pi/2$. The filters can be made "steeper" in this transition region by a judicious choice of R in (6.1.11). Figure 6.6 shows the plot of $|m_0|$ corresponding to $N = 2$ and R of degree 3 chosen such that $|m_0(\xi)|^2$ has a zero at $\xi = 7\pi/9$ ($= 140°$). This is much closer to a "realistic" subband coding filter. The corresponding "least asymmetric" function ϕ is shown in Figure 6.7; it is less smooth than ${}_4\phi$ (which has the same support width, but corresponds to $N = 4$ and $R \equiv 0$), but turns out to be smoother than ${}_2\phi$ (for which m_0 has a zero of the same multiplicity, i.e., 2, at $\xi = \pi$). In Chapter 7 we will come back in greater detail to these regularity and flatness issues. The h_n corresponding to Figure 6.7 are listed in Table 6.4.

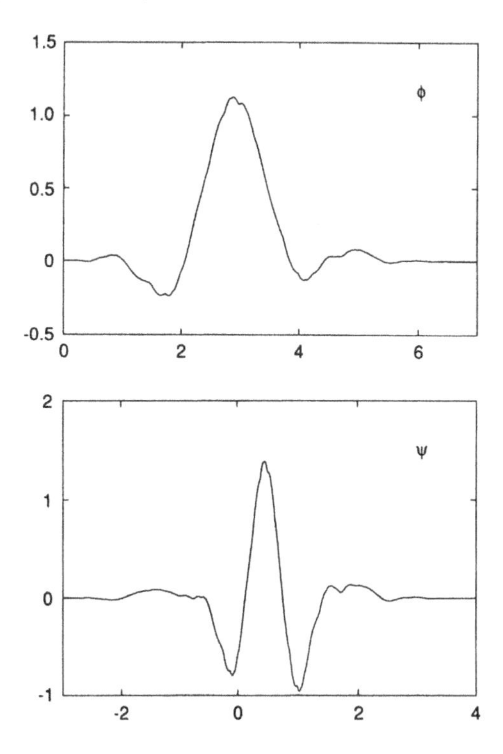

FIG. 6.7. *The "least asymmetric" scaling function ϕ and wavelet ψ corresponding to $|m_0|$ as plotted in Figure* 6.6.

All these examples correspond to real h_n, ϕ and ψ, i.e., to $|\hat{\phi}|$ and $|\hat{\psi}|$ symmetric around $\xi = 0$. It is also possible to construct (complex) examples with $|\hat{\phi}|$, $|\hat{\psi}|$ concentrated much more on $\xi > 0$ than on $\xi < 0$. Take for instance the m_0 of the previous example, which satisfies $m_0(\pm\frac{2\pi}{9}) = 1$, and define $m_0^\#(\xi) = m_0(\xi - \frac{2\pi}{9})$. This $m_0^\#$ obviously satisfies (6.1.1), since m_0 does, and $m_0^\#(0) = 1$. We can therefore construct $\hat{\phi}^\#(\xi) = \prod_{j=1}^\infty m_0^\#(2^{-j}\xi)$, $\hat{\psi}^\#(\xi) = e^{-i\xi/2}\, \overline{m_0^\#(\xi/2 + \pi)}\, \hat{\phi}^\#(\xi/2)$;

TABLE 6.4

The coefficients for the low-pass filter corresponding to the scaling function in Figure 6.7.

n	h_n
0	−0.0802861503271
1	−0.0243085969067
2	0.362806341592
3	0.550576616156
4	0.229036357075
5	−0.0644368523121
6	−0.0115565483406
7	0.0381688330633

these are compactly supported L^2-functions, and the $\psi^{\#}_{j,k}$, $j,k \in \mathbb{Z}$ constitute a tight frame for $L^2(\mathbb{R})$, by Proposition 6.2.3. Moreover, since the only zeros of m_0 on $[-\pi,\pi]$ are in $\xi = \pm\frac{7\pi}{9}$, $\pm\pi$, it follows that $m_0^{\#}(\xi) = 0$ only for $\xi = \pm\pi$, $-\frac{5\pi}{9}$ or $-\frac{7\pi}{9}$. Consequently, $|m_0^{\#}(\xi)| \geq C > 0$ for $|\xi| \leq \frac{\pi}{3}$, and the $\psi^{\#}_{j,k}$ constitute an orthonormal wavelet basis, by Corollary 6.3.2. Figure 6.8 plots $|m_0^{\#}(\xi)|$, $|\hat{\phi}^{\#}(\xi)|$ and $|\hat{\psi}^{\#}(\xi)|$; it is clear that $\int_0^\infty d\xi\ |\hat{\psi}^{\#}(\xi)|^2$ is much larger than $\int_{-\infty}^0 d\xi\ |\hat{\psi}^{\#}(\xi)|^2$. Note that the negative frequency part of $\hat{\psi}^{\#}$ is much closer to the origin than the positive frequency part, as required by the necessary condition $\int_0^\infty d\xi\ |\xi|^{-1}|\hat{\psi}^{\#}(\xi)|^2 = \int_{-\infty}^0 d\xi\ |\xi|^{-1}|\hat{\psi}^{\#}(\xi)|^2$ (see §3.4). The existence of such "asymmetric" $\hat{\psi}$ was first pointed out in Cohen (1990); in fact, for any $\epsilon > 0$ one can find an orthonormal wavelet basis such that $\int_{-\infty}^0 d\xi\ |\hat{\psi}(\xi)|^2 < \epsilon$.

6.5. The cascade algorithm: The link with subdivision or refinement schemes.

It can already be suspected from the figures in §6.4 that there is no closed-form analytic formula for the compactly supported $\phi(x)$, $\psi(x)$ constructed here (except for the Haar case). Nevertheless, we can, if ϕ is continuous, compute $\phi(x)$ with arbitrarily high precision for any given x; we also have a fast algorithm to compute the plot of ϕ.[8] Let us see how this works.

First of all, since ϕ has compact support, and $\phi \in L^1(\mathbb{R})$ with $\int dx\ \phi(x) = 1$, we have

PROPOSITION 6.5.1. *If f is a continuous function on $\mathbb{R}$, then, for all $x \in \mathbb{R}$,*

$$\lim_{j\to\infty} 2^j \int dy\ f(x+y)\ \overline{\phi(2^j y)} = f(x)\ . \tag{6.5.1}$$

If f is uniformly continuous, then this pointwise convergence is uniform as well. If f is Hölder continuous with exponent α,

$$|f(x) - f(y)| \leq C|x-y|^\alpha\ ,$$

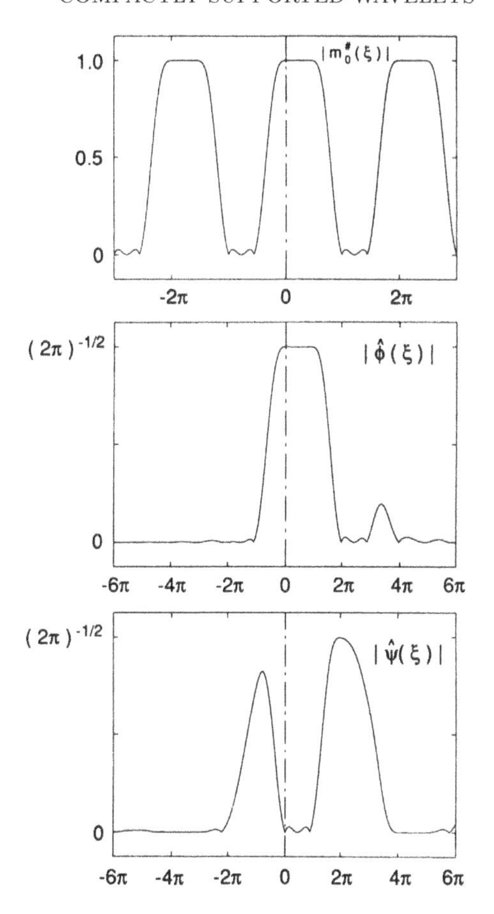

FIG. 6.8. *Plots of* $|m_0|$, $|\hat{\phi}|$, *and* $|\hat{\psi}|$ *for an orthonormal wavelet basis where* $\hat{\psi}$ *is concentrated more on positive than on negative frequencies.*

then the convergence is exponentially fast in j*:*

$$|f(x) - 2^j \int dy\ f(x+y)\ \overline{\phi(2^j y)}| \le C2^{-j\alpha}\ . \tag{6.5.2}$$

Proof. All the assertions follow from the fact that $2^j\phi(2^j\cdot)$ is an "approximate δ-function" as j tends to ∞. More precisely,

$$\begin{aligned}
&\left| f(x) - 2^j \int dy\ f(x+y)\ \overline{\phi(2^j y)} \right| \\
&\quad = \left| 2^j \int dy\ \ [f(x) - f(x+y)]\ \overline{\phi(2^j y)} \right| \\
&\quad = \left| \int dz\ [f(x) - f(x+2^{-j}z)]\ \overline{\phi(z)} \right|
\end{aligned}$$

$$\leq \|\phi\|_{L^1} \cdot \sup_{|u|\leq 2^{-j}R} |f(x) - f(x+u)|$$
$$\text{(where we suppose support } \phi \subset [-R,R]) .$$

If f is continuous, then this can be made arbitrarily small by choosing j sufficiently large. If f is uniformly continuous, then the choice of j can be made independently of x, and the convergence is uniform. If f is Hölder continuous, then (6.5.2) follows immediately as well. ∎

Assume now that ϕ itself is continuous, or even Hölder continuous with exponent α. (We will see many techniques to compute the Hölder exponent of ϕ in the next chapter.) Take x to be any dyadic rational, $x = 2^{-J}K$. Then Proposition 6.5.1 tells us that

$$\begin{aligned}
\phi(x) &= \lim_{j\to\infty} 2^j \int dy\ \phi(2^{-J}K + y)\ \overline{\phi(2^j y)} \\
&= \lim_{j\to\infty} 2^{j/2} \int dz\ \phi(z)\ \overline{\phi_{-j,2^{j-J}K}(z)} \\
&= \lim_{j\to\infty} 2^{j/2} \langle \phi,\ \phi_{-j,2^{j-J}K}\rangle \ .
\end{aligned}$$

Moreover, for j larger than some j_0,

$$|\phi(2^{-J}K) \ - \ 2^{j/2}\ \langle \phi, \phi_{-j,2^{j-J}K}\rangle| \leq C\, 2^{-j\alpha} \ , \tag{6.5.3}$$

where C, j_0 are dependent on J or K. If $2^{j-J}K$ is integer, which is automatically true if $j \geq J$, then the inner products $\langle \phi, \phi_{-j,2^{j-J}K}\rangle$ are easy to compute. Under the assumption that the $\phi_{0,n}$ are orthonormal (which can be checked with any of the necessary and sufficient conditions on m_0 listed in Theorem 6.3.5), ϕ is the unique function f characterized by

$$\langle f,\, \phi_{0,n}\rangle \ = \ \delta_{0,n} \ , \tag{6.5.4}$$
$$\langle f,\, \psi_{-j,k}\rangle \ = \ 0 \quad \text{for } \ j > 0,\ k \in \mathbb{Z} \ . \tag{6.5.5}$$

We can use this as input for the reconstruction algorithm of the subband filtering associated with m_0 (see §5.6). More specifically, we start with a low pass sequence $c_n^0 = \delta_{0,n}$ and a highpass sequence $d_n^0 = 0$, and we "crank the machine" to obtain

$$c_n^{-1} = \sum_k \ h_{n-2k}\ c_k^0 \ . \tag{6.5.6}$$

We then use $d_n^{-1} = 0$, to obtain, after another cranking,

$$c_m^{-2} = \sum_n \ h_{m-2n}\ c_n^{-1} \ , \tag{6.5.7}$$

etc. At every stage, the c_n^{-j} are equal to $\langle \phi, \phi_{-j,n}\rangle$. Together with (6.5.3), this means that we have an algorithm with exponentially fast convergence to compute the values of ϕ at dyadic rationals. We can interpolate these values and thus obtain a sequence of functions η_j approximating ϕ.[9] We can, for instance, define $\eta_j^0(x)$ to be the function, piecewise constant on the intervals $[2^{-j}(n-1/2),\ 2^{-j}(n+1/2)[$, $n \in \mathbb{Z}$, such that $\eta_j^0(2^{-j}k) = 2^{j/2}\,\langle\phi,\,\phi_{-j,k}\rangle$. Another possible choice is $\eta_j^1(x)$, piecewise linear on the $[2^{-j}n, 2^{-j}(n+1)], n \in \mathbb{Z}$, so that $\eta_j^1(2^{-j}k) = 2^{j/2}\langle\phi, \phi_{-j,k}\rangle$.

For both choices we have the following proposition.

PROPOSITION 6.5.2. *If ϕ is Hölder continuous with exponent α, then there exists $C > 0$ and $j_0 \in \mathbb{N}$ so that, for $j \geq j_0$,*

$$\|\phi - \eta_j^0\|_{L^\infty} \leq C\,2^{-\alpha j}\ , \qquad \|\phi - \eta_j^1\|_{L^\infty} \leq C\,2^{-\alpha j}\ . \tag{6.5.8}$$

Proof. Take any $x \in \mathbb{R}$. For any j, choose n so that $2^{-j}n \leq x < 2^{-j}(n+1)$. By the definition of η_j^ϵ, $\eta_j^\epsilon(x)$ is necessarily a convex linear combination of $2^{j/2}\langle\phi, \phi_{-j,n}\rangle$ and $2^{j/2}\langle\phi, \phi_{-j,n+1}\rangle$, whether $\epsilon = 0$ or 1. On the other hand, if j is larger than some j_0,

$$\begin{aligned}
&|\phi(x) - 2^{j/2}\langle\phi, \phi_{-j,n}\rangle| \\
&\quad \leq |\phi(x) - \phi(2^{-j}n)| + |\phi(2^{-j}n) - 2^{j/2}\langle\phi, \phi_{-j,n}\rangle| \\
&\quad \leq C\,|x - 2^{-j}n|^\alpha + C\,2^{-j\alpha} \leq C\,2^{-j\alpha}\ ;
\end{aligned}$$

the same is true if we replace n by $n+1$. It follows that a similar estimate holds for any convex combination, or $|\phi(x) - \eta_j^\epsilon(x)| \leq C\ 2^{-j\alpha}$. Here C can be chosen independently of x, so that (6.5.8) follows. ∎

This then is our fast algorithm to compute approximate values of $\phi(x)$ with arbitrarily high precision:

1. Start with the sequence $\cdots 0 \cdots 010 \cdots 0 \cdots$, representing the $\eta_0^\epsilon(n)$, $n \in \mathbb{Z}$.

2. Compute the $\eta_j^\epsilon\ (2^{-j}n)$, $n \in \mathbb{Z}$, by "cranking the machine" as in (6.5.7). At every step of this cascade, twice as many values are computed: values at "even points" $2^{-j}(2k)$ are refined from the precious step,

$$\eta_j^\epsilon(2^{-j}2k) = \sqrt{2}\sum_\ell h_{2(k-\ell)}\ \eta_{j-1}^\epsilon(2^{-j+1}\ell)\ , \tag{6.5.9}$$

   and values at the "odd points" $2^{-j}(2k+1)$ are computed for the first time,

$$\eta_j^\epsilon(2^{-j}(2k+1)) = \sqrt{2}\sum_\ell h_{2(k-\ell)+1}\ \eta_{j-1}^\epsilon(2^{-j+1}\ell)\ . \tag{6.5.10}$$

   Both (6.5.9) and (6.5.10) can be viewed as convolutions.

3. Interpolate the $\eta_j^\epsilon(2^{-j}n)$ (piecewise constant if $\epsilon = 0$, piecewise linear if $\epsilon = 1$) to obtain $\eta_j^\epsilon(x)$ for non-dyadic x.

The whole algorithm was called the *cascade algorithm* in Daubechies and Lagarias (1991), where $\epsilon = 1$ was chosen; in Daubechies (1988b), the choice $\epsilon = 0$ was made.[10] All the plots of ϕ, ψ in §6.4 and in later chapters are, in fact, plots of η_j^1, with $j = 7$ or 8; at the resolution of these figures, the difference between ϕ and these η_j^1 is imperceptible. A particularly attractive feature of the cascade algorithm is that it allows one to "zoom in" on particular features of ϕ. Suppose we already have computed all the $\eta_5^\epsilon(2^{-5}n)$, but we would like to look at a blowup, with much better resolution, of ϕ in the interval $[\frac{15}{16}, \frac{17}{16}]$ centered around 1. We could do this by computing all the $\eta_J^\epsilon(2^{-J}n)$ for very large J, and then plotting $\eta_J^\epsilon(x)$ only on the small interval of interest, corresponding to $2^{J-4} \cdot 15 \leq n \leq 2^{J-4} \cdot 17$. But we do not need to: by the "local" nature of (6.5.9), (6.5.10) much fewer computations suffice. Suppose $h_n = 0$ for $n < 0$, $n > 3$. The computation of $\eta_J^\epsilon(2^{-J}n)$ only involves those $\eta_{J-1}^\epsilon(2^{-J+1}k)$ for which $(n-3)/2 \leq k \leq n/2$. Computation of these, in turn, involves only the $\eta_{J-2}^\epsilon(2^{-J+2}\ell)$ with $(k-3)/2 \leq \ell \leq k/2$, or $n/4 - 3/2 - 3/4 \leq \ell \leq n/4$. Working back to $j = J - 4$, we see that to compute η_9^ϵ on $[\frac{15}{16}, \frac{17}{16}]$ we only need the $\eta_5^\epsilon(2^{-5}m)$ for $28 \leq m \leq 34$. We can therefore start the cascade from $\cdots 0 \cdots 010 \cdots 0 \cdots$, go five steps, select the seven values $\eta_5^\epsilon(2^{-5}m)$, $28 \leq m \leq 34$, use only these as the input for a new cascade, with four steps, and end up with a graph of η_9^ϵ on $[\frac{15}{16}, \frac{17}{16}]$. For larger blowups on even smaller intervals, we simply repeat the process; the blowup graphs in Chapter 7 have all been computed in this way.[11]

The arguments leading to the cascade algorithm have implicitly used the orthonormality of the $\psi_{j,k}$, or equivalently (see §6.2, 6.3), of the $\phi_{0,n}$: we have characterized ϕ as the unique function f satisfying (6.5.4), (6.5.5). The cascade algorithm can also be viewed differently, without emphasizing orthonormality at all, as a special case of a stationary subdivision or refinement scheme.

Refinement schemes are used in computer graphics to design smooth curves or surfaces going through or passing near a discrete, often rather sparse, set of points. An excellent review is Cavaretta, Dahmen, and Micchelli (1991). We will restrict ourselves, in this short discussion, to one-dimensional subdivision schemes.[12] Suppose that we want a curve $y = f(x)$ taking on the preassigned values $f(n) = f_n$. One possibility is simply to construct the piecewise linear graph through the points (n, f_n); this graph has the peculiarity that, for all n,

$$f\left(\frac{2n+1}{2}\right) = \frac{1}{2}f(n) + \frac{1}{2}f(n+1) , \tag{6.5.11}$$

which gives a quick way to compute f at half-integer points. The values of f at quarter-integer points can be computed similarly,

$$f\left(\frac{n}{2}+\frac{1}{4}\right) = \frac{1}{2}f\left(\frac{n}{2}\right) + \frac{1}{2}f\left(\frac{n}{2}+\frac{1}{2}\right) , \tag{6.5.12}$$

and so on for $\mathbb{Z}/4 + \mathbb{Z}/8$, etc. This provides a fast recursive algorithm for the computation of f at all dyadic rationals. If we choose to have a smoother spline interpolation than by piecewise linear splines (quadratic, cubic or even higher order splines), then the formulas analogous to (6.5.9), (6.5.10), computing the $f(2^{-j}n + 2^{-j-1})$ from the $f(2^{-j}k)$, would contain an infinite number of terms. It is possible to opt for smoother than linear spline approximation, with interpolation formulas of the type

$$f(2^{-j}n + 2^{-j-1}) = \sum_k a_k \, f(2^{-j}(n-k)) \,, \tag{6.5.13}$$

with only finitely many a_k are nonzero; the resulting curves are no longer splines. An example is

$$\begin{aligned} f(2^{-j}n + 2^{-j-1}) &= -\frac{1}{16}\,[f(2^{-j}(n-1)) + f(2^{-j}(n+2)] \\ &\quad + \frac{9}{16}\,[f(2^{-j}n) + f(2^{-j}(n+1))] \,. \end{aligned} \tag{6.5.14}$$

This example was studied in detail in Dubuc (1986), Dyn, Gregory, and Levin (1987), and generalized in, for example, Deslauriers and Dubuc (1989) and Dyn and Levin (1989); it leads to an almost C^2-function f. (For details on methods to determine the regularity of f, see Chapter 7.) Formula (6.5.14) describes an interpolation refinement scheme, in which, at every stage of the computation, the values computed earlier remain untouched, and only values at intermediate points need to be computed. One can also consider schemes where at every stage the values computed at the previous stage are further "refined," corresponding to a more general refinement scheme of the type

$$f_{j+1}(2^{-j-1}n) = \sum_k w_{n-2k} \, f_j(2^{-j}k) \,. \tag{6.5.15}$$

Formula (6.5.15) corresponds in fact to two convolution schemes (with two *masks*, in the terminology of the refinement literature),

$$f_{j+1}(2^{-j}n) = \sum_k w_{2(n-k)} \, f_j(2^{-j}k) \tag{6.5.16}$$

(the refinement of already computed values), and

$$f_{j+1}(2^{-j}n + 2^{-j-1}) = \sum_k w_{2(n-k)+1} \, f_j(2^{-j}k) \tag{6.5.17}$$

(computation of values at new intermediate points). In a sensible refinement scheme, the f_j converge, as j tends to ∞, to a continuous (or smoother; see Chapter 7) function f_∞. Note that (6.5.15) defines the f_j only on the discrete set $2^{-j}\mathbb{Z}$. A precise statement of the "convergence" of f_j to the continuous function f_∞ is that

$$\lim_{m\to\infty} \left\{ \sup_{j\geq 0, k\in\mathbb{Z}} \left| f^\lambda_\infty(2^{-m}2^{-j}k) - f^\lambda_{m+j}(2^{-m-j}k) \right| \right\} = 0 \,, \tag{6.5.18}$$

where the superscript λ indicates the initial data, $f_0^\lambda(n) = \lambda_n$. The refinement scheme is said to converge if (6.5.18) holds for all $\lambda \in \ell^\infty(\mathbb{Z})$; see Cavaretta, Dahmen, and Micchelli (1991). (It is also possible to rephrase (6.5.18) by first introducing continuous functions f_j, interpolating the $f_j(2^{-j}k)$; see below.) A general refinement scheme is an interpolation scheme if $w_{2k} = \delta_{k,0}$, leading to $f_{j+\ell}(2^{-j}n) = f_j(2^{-j}n)$.

In both cases, general refinement scheme or more restrictive interpolation scheme, it is easy to see that the linearity of the procedure implies that the limit function f_∞ (which we suppose continuous[13]) is given by

$$f_\infty(x) = \sum_n f_0(n)\ F(x-n)\ , \tag{6.5.19}$$

where $F = F_\infty$ is the "fundamental solution," obtained by the same refinement scheme from the initial data $F_0(n) = \delta_{n,0}$. This fundamental solution obeys a particular functional equation. To derive this equation, we first introduce functions $f_j(x)$ interpolating the discrete $f_j(2^{-j}k)$:

$$f_j(x) = \sum_k f_j(2^{-j}k)\ \omega(2^j x - k)\ , \tag{6.5.20}$$

where ω is a "reasonable"[14] function so that $\omega(n) = \delta_{n,0}$. Two obvious choices are $\omega(x) = 1$ for $-\frac{1}{2} \le x < \frac{1}{2}$, 0 otherwise, or $\omega(x) = 1 - |x|$ for $|x| \le 1$, 0 otherwise. (These correspond to the two choices in the exposition of the cascade algorithm above.) The convergence requirement (6.5.18) can then be rewritten as $\|f_j^\lambda - f_\infty^\lambda\|_{L^\infty} \to 0$ for $j \to \infty$. For the fundamental solution F_∞, we start from $F_0(x) = \omega(x)$. The next two approximating functions F_1, F_2 satisfy

$$\begin{aligned}
F_1(x) &= \sum_n F_1(n/2)\ \omega(2x-n) && \text{(by (6.5.20))}\\
&= \sum_n w_n\ \omega(2x-n) && \text{(use (6.5.15) and } F_0(n) = \delta_{n,0})\\
&= \sum_n w_n\ F_0(2x-n)\ , && (6.5.21)\\
F_2(x) &= \sum_n F_2(n/4)\ \omega(4x-n) &&\\
&= \sum_{n,k} w_{n-2k}\ F_1(k/2)\ \omega(4x-n) && \text{(use (6.5.15))}\\
&= \sum_k w_k \sum_\ell w_\ell\ \omega(4x-2k-\ell) && \text{(because} F_1(k/2) = W_k)\\
&= \sum_k w_k\ F_1(2x-k)\ . &&
\end{aligned}$$

This suggests that a similar formula should hold for all F_j, i.e.,

$$F_j(x) = \sum_k w_k\ F_{j-1}(2x-k)\ . \tag{6.5.22}$$

Induction shows that this is indeed the case:

$$\begin{aligned}
F_{j+1}(x) &= \sum_n F_{j+1}(2^{-j-1}n)\ \omega(2^{j+1}x-n) \\
&= \sum_{n,k} w_{n-2k}\ F_j(2^{-j}k)\ \omega(2^{j+1}x-n) \\
&= \sum_{n,k,\ell} w_{n-2k}\ w_\ell\ F_{j-1}(2^{-j+1}k-\ell)\ \omega(2^{j+1}x-n) \\
&\qquad\qquad \text{(by the induction hypothesis)} \\
&= \sum_\ell w_\ell \sum_{m,n} F_{j-1}(2^{-j+1}m)\ w_{n-2m-2^j\ell}\ \omega(2^{j+1}x-n) \\
&= \sum_\ell w_\ell \sum_{m,r} F_{j-1}(2^{-j+1}m)\ w_{r-2m}\ \omega(2^{j+1}x-2^j\ell-r) \\
&= \sum_\ell w_\ell \sum_m F_j(2^{-j}r)\omega(2^j(2x-\ell)-r) \qquad \text{(by (6.5.15))} \\
&= \sum_\ell w_\ell\ F_j(2x-\ell) \qquad \text{(by (6.5.20))}\ .
\end{aligned}$$

Since $F = F_\infty = \lim_{j\to\infty} F_j$, (6.5.22) implies that the fundamental solution F satisfies

$$F(x) = \sum_k w_k\ F(2x-k)\ . \tag{6.5.23}$$

It is now clear how our compactly supported scaling functions ϕ and the cascade algorithm fit into refinement schemes: on the one hand, ϕ satisfies an equation of the type (6.5.23) (basically as a consequence of the multiresolution requirement $V_0 \subset V_{-1}$), and on the other hand the cascade algorithm corresponds exactly to (6.5.15), (6.5.20). Orthonormality in the underlying multiresolution framework made our life a little easier in the proof of Proposition 6.5.2, but similar results can be proved for refinement schemes, without orthonormality of the $F(x-n)$. Some basic results for refinement schemes are:

- If the refinement scheme (6.5.15) converges, then $\sum_n w_{2n} = \sum_n w_{2n+1} = 1$, and the associated functional equation (6.5.23) admits a unique continuous solution of compact support (up to normalization).

- If (6.5.23) admits a continuous compactly supported solution F, and if the $F(x-n)$ are independent (i.e., the mapping $\ell^\infty(\mathbb{Z}) \ni \lambda \mapsto \sum_n \lambda_n F(x-n)$ is one-to-one[15]), then the subdivision algorithm converges.

For proofs of this and many other results, we refer to Cavaretta, Dahmen, and Micchelli (1991) and papers cited there. Note that the condition $\sum_n w_{2n} = \sum_n w_{2n+1} = 1$ corresponds exactly to the requirements $m_0(0) = 1$, $m_0(\pi) = 0$.

In a sense, constructions of compactly supported scaling functions and wavelets can therefore be viewed as special cases of refinement schemes. I feel that there is a difference in emphasis, however. A general refinement scheme is associated with a scale of multiresolution spaces V_j, generated by the $F(2^{-j}x-n)$, but typically no attention is paid at all to the complementary subspaces of the V_j in V_{j-1}. Refining a sequence of data points in j steps corresponds to finding a function in V_{-j} of which the projection onto V_0, as given by the adjoint of the refinement scheme (usually not an orthonormal projection), corresponds to the data sequence. There are many such functions in V_{-j}, corresponding to the same data sequence, but the refinement scheme picks out the "minimal" one. There is no interest in the study of the other non-minimal solutions in V_{-j}, and how they differ from the unique refinement solution. This is natural: refinement schemes are meant to build more "complicated" structures from simple ones (they go from V_0 to V_{-j}). In contrast, wavelet analysis wants to decompose *arbitrary* elements of V_{-j} into building blocks in V_0 and its complement. Here it is absolutely necessary to stress the importance of all the complement spaces $W_\ell = V_{\ell-1} \ominus V_\ell$, and to have fast algorithms to compute the coefficients in those spaces as well. This is where the wavelets enter, for which there is typically no analog in general refinement schemes.

There is another, amusing link between orthonormal wavelet bases with compact support and refinement schemes: the mask associated with an orthonormal wavelet basis is always the "square root" of the mask of some interpolation scheme. More explicitly, define $M_0(\xi) = |m_0(\xi)|^2 = \frac{1}{2}\sum_n w_n\, e^{-in\xi}$, i.e., $w_n = \sum_k \overline{h_k}\, h_{k+n}$. Then the w_n are the mask coefficients for an interpolation refinement scheme, since $w_{2n} = \sum_k \overline{h_k}\, h_{k+2n} = \delta_{n,0}$ (see (5.1.39)). In particular, as noticed by Shensa (1991), the interpolation refinement schemes obtained from the choice $R \equiv 0$ in (6.1.11) are the so-called Lagrange interpolation schemes studied in detail by Deslauriers and Dubuc (1989),[16] of which (6.5.14) is an example.

Note that it is impossible (except for the Haar case) for a finite orthonormal wavelet filter m_0 to be itself an interpolation filter as well: orthonormality implies $|m_0(\xi)|^2 + |m_0(\xi+\pi)|^2 = 1$, while the interpolation requirement is equivalent to $h_{2n} = \frac{1}{\sqrt{2}}\delta_{n,0}$, or $m_0(\xi) + m_0(\xi+\pi) = 1$. If both requirements are met, then

$$1 = |m_0(\xi)|^2 + |1 - m_0(\xi)|^2$$

or

$$\sum_n h_n \overline{h_{k+n}} = \frac{1}{\sqrt{2}}\left[\overline{h_k} + h_{-k}\right]. \tag{6.5.24}$$

Assume that $h_n \equiv 0$ for $n < N_1$, $n > N_2$, and $h_{N_1} \neq 0 \neq h_{N_2}$. Then (6.5.24) already implies that either $N_1 = 0$ or $N_2 = 0$. Suppose $N_1 = 0$ ($N_2 = 0$ is analogous); N_2 is necessarily odd, $N_2 = 2L+1$. Take $k = 2L$ in (6.5.24). Then

$$h_0\, \overline{h_{2L}} + h_1\, \overline{h_{2L+1}} = \frac{1}{\sqrt{2}}\, \overline{h_{2L}}\,.$$

Since $h_0 = 2^{-1/2}$, and $h_{2L+1} \neq 0$, this implies $h_1 = 0$. Similarly $k = 2L - 2$

leads to

$$h_0 \, \overline{h_{2L-2}} + h_1 \, \overline{h_{2L-1}} + h_2 \, \overline{h_{2L}} + h_3 \, \overline{h_{2L+1}} = \frac{1}{\sqrt{2}} \, \overline{h_{2L-2}} \, ,$$

which, together with $h_1 = 0$, $h_{2n} = 2^{-1/2}\delta_{n,0}$ implies $h_3 = 0$. It follows eventually that only h_0 and h_{2L+1} are nonzero; they are both equal to $1/\sqrt{2}$, so that the mask is a "stretched" Haar mask. Orthonormality of the $\phi_{0,n}$ then forces $L = 0$, or $m_0(\xi) = \frac{1}{2}(1 + e^{-i\xi})$, i.e., the Haar basis. If we lift the restriction that m_0 is a trigonometric polynomial, i.e., if ϕ, ψ can be supported on the whole real line, then $m_0(\xi) + m_0(\xi + \pi) = 1$ and $|m_0(\xi)|^2 + |m_0(\xi + \pi)|^2 = 1$ can be satisfied simultaneously by non-trivial m_0; examples can be found in Evangelista (1992) or Lemarié–Malgouyres (1992).

Notes.

1. A compactly supported $\phi \in L^2(\mathbb{R})$ is automatically in $L^1(\mathbb{R})$. It then follows from Remark 5 at the end of §5.3 that $m_0(0) = 1$, $m_0(\pi) = 0$, i.e., that m_0 has a zero of multiplicity at least 1 in π.

2. In Daubechies (1988b), the solutions P of (6.1.7) were found via two combinatorial lemmas. The present more natural approach, using Bezout's theorem, was pointed out to me by Y. Meyer.

3. This formula for P_N was already obtained in Hermann (1971), where maximally flat FIR filters were designed (without any perfect reconstruction schemes, however).

4. This convergence also holds if infinitely many h_n are nonzero, but if they decay sufficiently fast so that $\sum |h_n|(1 + |n|)^\epsilon < \infty$ for some $\epsilon > 0$. In that case $|\sin\, n\zeta| \le |n\zeta|^{\min(1,\epsilon)}$ leads to a similar bound.

5. We use here the classical formula

$$\frac{\sin x}{x} = \prod_{j=1}^{\infty} \cos(2^{-j}x) \, .$$

An easy proof uses $\sin 2\alpha = 2 \cos\alpha \, \sin\alpha$ to write

$$\prod_{j=1}^{J} \cos(2^{-j}x) = \prod_{j=1}^{J} \frac{\sin(2^{-j+1}x)}{2\sin(2^{-j}x)} = \frac{\sin x}{2^J \sin(2^{-J}x)} \, ,$$

which tends to $\frac{\sin x}{x}$ for $J \to \infty$. In Kac (1959) this formula is credited to Vieta, and used as a starting point for a delightful treatise on statistical independence.

6. This is true in general: if m_0 satisfies (6.1.1) and ϕ, as defined by (6.2.2), generates a non-orthonormal family of translates $\phi_{0,n}$, then necessarily $\sum_\ell |\hat{\phi}(\xi + 2\pi\ell)|^2 = 0$ for some ξ. (See Cohen (1990b).)

7. The condition $\int dx\, \psi(x)\, \overline{\psi(x-k)} = \delta_{k,0}$ may seem stronger than $\|\psi\| = 1$, but since the $\psi_{j,k}$ constitute a tight frame with frame constant 1, the two are equivalent, by Proposition 3.2.1.

8. Since $\psi(x)$ is a finite linear combination of translates of $\phi(2x)$, fast algorithms to plot ϕ also lead to fast plots for ψ. Throughout this section, we restrict our attention to ϕ only.

9. If ϕ is not continuous, then the η_j still converge to ϕ in L^2 (see §6.3). Moreover, they converge to ϕ pointwise in every point where ϕ is continuous.

10. The choice $\epsilon = 1$ was used in the proof of Proposition 3.3 in Daubechies (1988b), because the $\hat{\eta}_j^1$ are absolutely integrable, whereas the $\hat{\eta}_j^0$ are not. In Daubechies (1988b) the convergence of the η_j^ϵ to ϕ was actually proved first (using some extra technical conditions), and orthonormality of the $\phi_{0,n}$ was then deduced from this convergence.

11. Note that there exist many other procedures for plotting graphs of wavelets. Instead of a refinement cascade one can also start from appropriate $\phi(n)$ and then compute the $\phi(2^{-j}k)$ directly from $\phi(x) = \sqrt{2}\sum_n h_n \phi(2x-n)$. (In fact, when ϕ is not continuous, the cascade algorithm may diverge, while this direct use of the 2-scale equation with appropriate $\phi(n)$ still converges. I would like to thank Wim Sweldens for pointing this out to me.) This more direct computation can be done in a tree-like procedure; a different way of looking at this, avoiding the tree construction and leading to faster plots, uses a dynamical systems framework, as developed by Berger and Wang (see Berger (1992) for a review). The "zoom in" feature is lost, however.

12. Many experts on refinement or subdivision schemes find the multidimensional case much more interesting!

13. This is not a presentation with fullest generality! We merely suppose that the w_k are such that there exists a continuous limit. This already implies $\sum w_{2n} = \sum w_{2n+1} = 1$.

14. For example, any compactly supported ω with bounded variation would be "reasonable" here.

15. The following stretched Haar function shows how the $F(x-n)$ can fail to be independent. Take $w_0 = w_2 = 1$, all other $w_n = 0$. The solution to (6.5.23) is then (up to normalization) $F(x) = 1$ for $0 \leq x < 2$, 0 otherwise. In this case the ℓ^∞-sequence λ defined by $\lambda_n = (-1)^n$ leads to $\sum_n \lambda_n F(x-n) = 0$ a.e.

16. This is no coincidence. If we fix the length of the symmetric filter $M_0 = |m_0|^2$, then the choice $R \equiv 0$ means that M_0 is divisible by $(1+\cos\xi)$ with the highest possible multiplicity compatible with its length and the constraint $M_0(\xi)+M_0(\xi+\pi) = 1$. On the other hand, Lagrange refinement

schemes of order $2N-1$ are the interpolation schemes with the shortest length that reproduce all polynomials of order $2N-1$ (or less) exactly from their integer samples. In terms of the filter $W(\xi) = \frac{1}{2}\sum_n w_n\, e^{in\xi}$, this means

$$W(\xi) + W(\xi+\pi) = 1 \qquad \text{(interpolation filter: } w_{2n} = \delta_{n,0}\text{)}$$

and

$$W(\xi) = 1 + O(\xi^{2N}) \quad = \quad 1 + O((1-\cos\xi)^N)$$

(see Cavaretta, Dahmen, and Micchelli (1991), or Chapter 8).

The two requirements together mean that $W(\xi+\pi)$ has a zero of order $2N$ in $\xi = 0$, i.e., that $W(\xi+\pi)$ is divisible by $(1-\cos\xi)^N$; hence $W(\xi)$ by $(1+\cos\xi)^N$. It follows that $W = M_0$.

CHAPTER 7

More About the Regularity of Compactly Supported Wavelets

The regularity of the Meyer or the Battle–Lemarié wavelets is easy to assess: the Meyer wavelet has compact Fourier transform, so that it is C^∞, and the Battle–Lemarié wavelets are spline functions, more precisely, piecewise polynomial of degree k, with $(k-1)$ continuous derivatives at the knots. The regularity of compactly supported orthonormal wavelets is harder to determine. Typically, they have a non-integer Hölder exponent; moreover, they are more regular in some points than in others, as is already illustrated by Figure 6.3. This chapter presents a collection of tools that have been developed over the past few years to study the regularity of these wavelets. All of these techniques rely on the fact that

$$\phi(x) = \sum c_n \, \phi(2x-n) , \tag{7.0.1}$$

where only finitely many c_n are nonzero; the wavelet ψ, as a finite linear combination of translates of $\phi(2x)$, then inherits the same regularity properties. It follows that the techniques exposed in this chapter are not restricted to wavelets alone; they apply as well to the basic functions in subdivision schemes (see §6.5). Some of the tools discussed here were in fact first developed for subdivision schemes, and not for wavelets.

The different techniques fall into two groups: those that prove decay for the Fourier transform $\hat{\phi}$, and those that work directly with ϕ itself. We will illustrate each method by applying it to the family of examples ${}_N\phi$ constructed in §6.4. It turns out that Fourier-based methods are better suited for asymptotic estimates (rate of regularity increase as N is increased in the examples, for instance); the second method gives more accurate local estimates, but is often harder to use.

References for the results in this chapter are Daubechies (1988b) and Cohen (1990b) for §7.1.1; Cohen (1990b) and Cohen and Conze (1992) for §7.1.2; Cohen and Daubechies (1991) for §7.1.3; Daubechies and Lagarias (1991, 1992), Micchelli and Prautzsch (1989), Dyn and Levin (1990), and Rioul (1992) for §7.2; Daubechies (1990b) for §7.3.

7.1. Fourier-based methods.

The Fourier transform of equation (7.0.1) is

$$\hat{\phi}(\xi) = m_0(\xi/2) \, \hat{\phi}(\xi/2) , \tag{7.1.1}$$

where $m_0(\xi) = \frac{1}{2}\sum_n c_n e^{-in\xi}$ is a trigonometric polynomial. As we have seen many times before, (7.1.1) leads to

$$\hat{\phi}(\xi) = (2\pi)^{-1/2} \prod_{j=1}^{\infty} m_0(2^{-j}\xi) , \tag{7.1.2}$$

where we have assumed $m_0(0) = 1$ and $\int dx\ \phi(x) = 1$, as usual. Moreover, m_0 can be factorized as

$$m_0(\xi) = \left(\frac{1+e^{-i\xi}}{2}\right)^N \mathcal{L}(\xi) , \tag{7.1.3}$$

where $\mathcal{L}$ is a trigonometric polynomial as well; this leads to

$$\hat{\phi}(\xi) = (2\pi)^{-1/2} \left(\frac{1-e^{-i\xi}}{i\xi}\right)^N \prod_{j=1}^{\infty} \mathcal{L}(2^{-j}\xi) . \tag{7.1.4}$$

A first method is based on a straightforward estimate of the growth of the infinite product of the $\mathcal{L}(2^{-j}\xi)$ as $|\xi|\to\infty$.

7.1.1. Brute force methods. For $\alpha = n + \beta$, $n \in \mathbb{N}$, $0 \le \beta < 1$, we define C^α to be the set of functions f which are n times continuously differentiable and such that the nth derivative $f^{(n)}$ is Hölder continuous with exponent β, i.e.,

$$|f^{(n)}\ (x) - f^{(n)}\ (x+t)| \le C|t|^\beta \ \text{ for all } x, t .$$

It is well known and easy to check that if

$$\int d\xi\ |\hat{f}(\xi)|\ (1+|\xi|)^\alpha < \infty ,$$

then $f \in C^\alpha$. In particular, if $|\hat{f}(\xi)| \le C(1+|\xi|)^{-1-\alpha-\epsilon}$, then it follows that $f \in C^\alpha$. It follows that, if the growth for $|\xi|\to\infty$ of $\prod_{j=1}^\infty \mathcal{L}(2^{-j}\xi)$ in (7.1.4) can be kept in check, then the factor $\left((1-e^{-i\xi})/i\xi\right)^N$ ensures smoothness for ϕ.

LEMMA 7.1.1. *If* $q = \sup_\xi |\mathcal{L}(\xi)| < 2^{N-\alpha-1}$, *then* $\phi \in C^\alpha$.

Proof.

1. Since $m_0(0) = 1$, $\mathcal{L}(0) = 1$; hence $|\mathcal{L}(\xi)| \le 1 + C|\xi|$. Consequently,

$$\sup_{|\xi|\le 1} \prod_{j=1}^{\infty} |\mathcal{L}(2^{-j}\xi)| \le \sup_{|\xi|\le 1} \prod_{j=1}^{\infty} \exp\left[C2^{-j}|\xi|\right] \le e^C .$$

2. Now take any ξ, with $|\xi| \ge 1$. There exists $J \ge 1$ so that $2^{J-1} \le |\xi| < 2^J$. Hence

$$\begin{aligned}\prod_{j=1}^{\infty} |\mathcal{L}(2^{-j}\xi)| &= \prod_{j=1}^{J} |\mathcal{L}(2^{-j}\xi)| \prod_{j=1}^{\infty} |\mathcal{L}(2^{-j}2^{-J}\xi)| \\ &\le q^J \cdot e^C \le C'\ 2^{J(N-\alpha-1-\epsilon)} \\ &\le C''\ (1+|\xi|)^{N-\alpha-1-\epsilon} .\end{aligned}$$

Consequently, $|\hat{\phi}(\xi)| \le C''' \, (1+|\xi|)^{-\alpha-1-\epsilon}$, and $\phi \in C^\alpha$. ∎

Grouping together several $\mathcal{L}$ leads to better estimates, as follows.

LEMMA 7.1.2. *Define*

$$
\begin{aligned}
q_j &= \sup_\xi \left| \prod_{k=0}^{j-1} \mathcal{L}(2^{-k}\xi) \right| , & (7.1.5)\\
\mathcal{K}_j &= \frac{\log q_j}{j \log 2} , & (7.1.6)\\
\mathcal{K} &= \inf_{j\in\mathbb{N}} \mathcal{K}_j .
\end{aligned}
$$

Then $\mathcal{K} = \lim_{j\to\infty} \mathcal{K}_j$*; if* $\mathcal{K} < N-1-\alpha$*, then* $\phi \in C^\alpha$.

Proof.

1. Take $j_2 > j_1$. Then $j_2 = nj_1 + r$ with $0 \le r < j_1$, and

$$q_{j_2} \le (q_{j_1})^n \; q_1^r .$$

Consequently,

$$\mathcal{K}_{j_2} \le \frac{n \log q_{j_1} + r \log q_1}{j_2 \log 2} \le \mathcal{K}_{j_1} \; + \; C \; j_1/j_2 .$$

2. For any $\epsilon > 0$, there exists j_0 so that $\mathcal{K} = \inf_j \; \mathcal{K}_j > \mathcal{K}_{j_0} - \epsilon$. For $j \ge j_0$ we then have $\mathcal{K}_j \le \mathcal{K} + \epsilon + C \; j_0/j \underset{j\to\infty}{\longrightarrow} \mathcal{K} + \epsilon$. Since ϵ was arbitrary, it follows that $\mathcal{K} = \lim_{j\to\infty} \; \mathcal{K}_j$.

3. If $\mathcal{K} < N-1-\alpha$, then $\mathcal{K}_\ell < N-1-\alpha$ for some $\ell \in \mathbb{N}$. We can then repeat the argument in the proof of Lemma 7.1.1, applying it to

$$\prod_{j=1}^{\infty} \mathcal{L}(2^{-j}\xi) \;=\; \prod_{j=0}^{\infty} \; \mathcal{L}_\ell(2^{-\ell j-1}\xi) ,$$

with $\mathcal{L}_\ell(\xi) \;=\; \prod_{j=0}^{\ell-1} \; \mathcal{L}(2^{-j}\xi)$, and with 2^ℓ playing the role of 2 in Lemma 7.1.1. This leads to $|\hat{\phi}(\xi)| \le C(1+|\xi|)^{-N+\mathcal{K}_\ell} \le C(1+|\xi|)^{-\alpha-1-\epsilon}$, hence $\phi \in C^\alpha$. ∎

The following lemma shows that in most cases, we will not be able to obtain much better by the brute force method.

LEMMA 7.1.3. *There exists a sequence* $(\xi_\ell)_{\ell\in\mathbb{N}}$ *so that*

$$(1+|\xi_\ell|)^{-\mathcal{K}} \left| \prod_{j=1}^{\infty} \mathcal{L}(2^{-j}\xi_\ell) \right| \ge C > 0 .$$

Proof.

1. By Theorem 6.3.1, the orthonormality of the $\phi(\cdot - n)$ implies the existence of a compact set K congruent to $[-\pi, \pi]$ modulo 2π, such that $|\hat{\phi}(\xi)| \geq C > 0$ for $\xi \in K$. Since K is congruent to $[-\pi, \pi]$ and $\mathcal{L}_\ell$ is periodic with period $2^{\ell+1}\pi$, we have

$$q_\ell = \sup_{|\xi| \leq 2^\ell \pi} |\mathcal{L}_\ell(\xi)| = \sup_{\xi \in 2^\ell K} |\mathcal{L}_\ell(\xi)| ,$$

i.e., there exists $\zeta_\ell \in 2^\ell K$ so that $|\mathcal{L}_\ell(\zeta_\ell)| = q_\ell$. Since K is compact, the $2^{-\ell}\ \zeta_\ell \in K$ are uniformly bounded. We therefore have

$$|\zeta_\ell| \leq 2^\ell\ C' \tag{7.1.7}$$

for $0 < C'$.

2. Moreover, since $\left|\frac{1+e^{i\xi}}{2}\right| = |\cos\xi/2| \leq 1$, we have for all $\xi \in 2^\ell K$,

$$\left|\prod_{j=\ell+1}^{\infty} \mathcal{L}(2^{-j}\xi)\right| \geq \left|\prod_{j=\ell+1}^{\infty} m_0(2^{-j}\xi)\right| = |\hat{\phi}(2^{-\ell}\xi)| \geq C > 0 .$$

Putting it all together we find for $\xi_\ell = 2\zeta_\ell$

$$\begin{aligned} \left|\prod_{j=1}^{\infty} \mathcal{L}(2^{-j}\xi_\ell)\right| &= |\mathcal{L}_\ell(\zeta_\ell)| \left|\prod_{j=\ell+1}^{\infty} \mathcal{L}(2^{-j}\zeta_\ell)\right| \\ &\geq C\ q_\ell = C\, 2^{\ell \mathcal{K}_\ell} . \end{aligned}$$

By (7.1.7),

$$(1 + |\xi_\ell|)^{-\mathcal{K}} \left|\prod_{j=1}^{\infty} \mathcal{L}(2^{-j}\xi_\ell)\right| \geq C\ 2^{\ell\mathcal{K}_\ell}\ C''\ 2^{-\ell\mathcal{K}} .$$

Since $\mathcal{K} = \inf_\ell\ \mathcal{K}_\ell$, this is bounded below by a strictly positive constant. ∎

Let us now turn to the particular family of ${}_N\phi$ constructed in §6.4, and see how these estimates perform. We have

$${}_N m_0(\xi) = \left(\frac{1 + e^{-i\xi}}{2}\right)^N \mathcal{L}_N(\xi) ,$$

with

$$|\mathcal{L}_N(\xi)|^2 = P_N(\sin^2 \xi/2) = \sum_{n=0}^{N-1} \binom{N-1+n}{n} (\sin^2 \xi/2)^n .$$

We start by establishing a few elementary properties of P_N.

LEMMA 7.1.4. *The polynomial* $P_N(x) = \sum_{n=0}^{N-1} \binom{N-1+n}{n} x^n$ *satisfies the following properties:*

$$0 \le x \le y \quad \Rightarrow \quad x^{-N+1}\, P_N(x) \ge y^{-N+1} P_N(y)\;, \tag{7.1.8}$$

$$0 \le x \le 1 \quad \Rightarrow \quad P_N(x) \le 2^{N-1} \max\,(1,\, 2x)^{N-1}\;. \tag{7.1.9}$$

Proof.

1. If $0 \le x \le y$, then

$$\begin{aligned} x^{-(N-1)}\, P_N(x) &= \sum_{n=0}^{N-1} \binom{N-1+n}{n} x^{-(N-1-n)} \\ &\ge \sum_{n=0}^{N-1} \binom{N-1+n}{n} y^{-(N-1-n)} \quad = y^{-(N-1)}\, P_N(y)\;. \end{aligned}$$

2. Recall (see §6.1) that P_N is the solution to

$$x^N P_N(1-x) \;+\; (1-x)^N\; P_N(x) \;=\; 1\;.$$

On substituting $x = \frac{1}{2}$, it follows that $P_N(1/2) = 2^{N-1}$. For $x \le \frac{1}{2}$, we have $P_N(x) \le P_N(\frac{1}{2}) = 2^{N-1}$ because P_N is increasing. For $x \ge \frac{1}{2}$, applying (7.1.8) leads to $P_N(x) \le x^{N-1} 2^{N-1} P_N(\frac{1}{2}) = 2^{N-1}(2x)^{N-1}$. This proves (7.1.9). ∎

It is now easy to apply Lemmas 7.1.1 and 7.1.2. We have

$$\begin{aligned} \sup_\xi |\mathcal{L}_N(\xi)| &= \left[\sum_{n=0}^{N-1} \binom{N-1+n}{n}\right]^{1/2} \\ &< \left[2^{N-1} \sum_{n=0}^{N-1} \binom{N-1+n}{n} 2^{-n}\right]^{1/2} \\ &= \left[2^{N-1} P_N(1/2)\right]^{1/2} \;=\; 2^{N-1}\;. \end{aligned}$$

Lemma 7.1.1 allows us to conclude that the ${}_N\phi$ are continuous. In view of the graphs in Figure 6.3, which show the ${}_N\phi$ to have increasing degree of regularity as N increases, this is clearly not optimal! Using $\mathcal{K}_j$ for $j > 1$ immediately leads to sharper results. We have, for instance,

$$\begin{aligned} q_2 &= \sup_\xi\; |\mathcal{L}_N(\xi)\; \mathcal{L}_N(2\xi)| \\ &= \sup_{0 \le y \le 1}\; [P_N(y)\; P_N(4y(1-y))]^{1/2} \\ &\qquad\qquad (\text{because } \sin^2\xi = 4\sin^2\xi/2\;\; (1-\sin^2\xi/2))\;. \end{aligned}$$

If either $y \leq 1/2$ or $y \geq \frac{1}{2} + \frac{\sqrt{2}}{4}$ (implying $4y(1-y) \leq \frac{1}{2}$), then $[P_N(y)P_N(4y(1-y))] \leq 2^{3(N-1)}$ by (7.1.9). In the remaining window $\frac{1}{2} + \frac{\sqrt{2}}{4} \geq y \geq \frac{1}{2}$, we have

$$\begin{aligned} P_N(y)\ P_N(4y(1-y)) &\leq 2^{2(N-1)}\ [16y^2(1-y)]^{N-1} \\ &\leq 2^{6(N-1)} \left(\frac{4}{27}\right)^{N-1} \\ &\qquad \left(\text{because } y^2(1-y) \leq \frac{4}{27} \text{ for } 0 \leq y \leq 1\right) . \end{aligned}$$

Consequently $q_2 \leq 2^{4(N-1)}\ 3^{-3(N-1)/2}$, and $\mathcal{K}_2 \leq (N-1)[2 - \frac{3}{4}\frac{\log 3}{\log 2}]$. It follows that asymptotically, for large N, ${}_N\phi \in C^{\mu N}$ with $\mu = \frac{3}{4}\frac{\log 3}{\log 2} - 1 \simeq .1887$. A slightly better value can be obtained by estimating q_4 rather than q_2; one finds then $\mu \simeq .1936$.

Note that $y = \frac{3}{4}$ is a fixed point for the map $y \mapsto 4y(1-y)$, so that $q_k \geq [P_N(3/4)]^k$ for any k, leading to a lower bound on $\mathcal{K}$ and an upper bound on the regularity of ϕ. In terms of ξ, $y = \sin^2\frac{\xi}{2} = \frac{3}{4}$ corresponds to $\xi = \frac{2\pi}{3}$; we already saw earlier that $\pm\frac{2\pi}{3}$ play a special role because $\{\frac{2\pi}{3}, \frac{-2\pi}{3}\}$ is an invariant cycle for multiplication by 2 modulo 2π. In the next subsection, we will see how these invariant cycles can be used to derive decay estimates for $\hat{\phi}$.

7.1.2. Decay estimates from invariant cycles. The values of $\mathcal{L}$ at an invariant cycle give rise to lower bounds for the decay of $\hat{\phi}$.

LEMMA 7.1.5. *If $\{\xi_0, \xi_1, \cdots, \xi_{M-1}\} \subset [-\pi, \pi]$ is any non-trivial invariant cycle (i.e., $\xi_0 \neq 0$) for the map $\tau\xi = 2\xi$ (modulo 2π), with $\xi_m = \tau\xi_{m-1}$, $m = 1, \cdots, N-1$, $\tau\xi_{M-1} = \xi_0$, then, for all $k \in \mathbb{N}$,*

$$|\hat{\phi}(2^{kM+1}\ \xi_0)| \geq C\ (1 + |2^{kM+1}\ \xi_0|)^{-N+\tilde{\mathcal{K}}},$$

where $\tilde{\mathcal{K}} = \sum_{m=0}^{M-1} \log|\mathcal{L}(\xi_m)|/(M\log 2)$, and $C > 0$ is independent of k.

Proof.

1. First note that there exists $C_1 > 0$ so that, for all $k \in \mathbb{N}$,

$$|\sin(2^{kM}\ \xi_0)| \geq C_1 \ . \tag{7.1.10}$$

Indeed, $2^{kM}\ \xi_0 = \xi_0 \pmod{2\pi}$, so that (7.1.10) follows if $\xi_0 \neq 0$ or $\pm\pi$. We already know that $\xi_0 \neq 0$; if $\xi_0 = \pm\pi$, then $\xi_1 = 0 \pmod{2\pi}$ and hence $\xi_0 = 2^{M-1}\xi_1 = 0 \pmod{2\pi}$, which is impossible.

2. Now

$$|\hat{\phi}(2^{kM+1}\ \xi_0)| = \left|\frac{\sin 2^{kM}\ \xi_0}{2^{kM}\ \xi_0}\right|^N \left|\prod_{j=0}^{\infty} \mathcal{L}(2^{kM-j}\ \xi_0)\right| .$$

Since $\mathcal{L}$ is a trigonometric polynomial and $\mathcal{L}(0) = 1$, there exists C_2 so that $|\mathcal{L}(\xi)| \geq 1 - C_2\ |\xi| \geq e^{-2\ C_2\ |\xi|}$ for $|\xi|$ small enough. Hence, for r large enough,

$$\prod_{j=rM}^{\infty} |\mathcal{L}(2^{-j}\ \xi_0)| \ \geq \ \prod_{j=rM}^{\infty} \exp\left[-2\ C_2\ 2^{-j}\ |\xi_0|\ \right]$$
$$\geq \ \exp\left[-2^{-rM+2}\ C_2\ |\xi_0|\ \right] \geq e^{-4C_2\ |\xi_0|} = C_3\ .$$

Hence

$$\begin{aligned} |\hat{\phi}(2^{kM+1}\ \xi_0)| \ &\geq \ C_1^N\ (2^{kM}\ |\xi_0|\ )^{-N}\ C_3 \left|\prod_{\ell=0}^{(r+k)M-1} \mathcal{L}(2^{kM-\ell}\ \xi_0)\right| \\ &\geq \ C_4\ |\mathcal{L}(\xi_0)\ \mathcal{L}(\xi_1)\cdots\mathcal{L}(\xi_{M-1})|^{r+k+1}\ (1+|2^{kM}\ \xi_0|)^{-N} \\ &\geq \ C_5\ 2^{\tilde{\mathcal{K}}Mk}\ (1+|2^{kM}\ \xi_0|)^{-N} \\ &\geq \ C\ (1+|2^{kM+1}\ \xi_0|)^{-N+\tilde{\mathcal{K}}}\ . \ \blacksquare \end{aligned}$$

We can apply this to the example at the end of the last subsection: Lemma 7.1.5 implies $|\hat{\phi}(2^n\frac{2\pi}{3})| \geq C(1+|2^n\frac{2\pi}{3}|)^{-N+\tilde{\mathcal{K}}}$, with $\tilde{\mathcal{K}} = \log|\mathcal{L}(\frac{2\pi}{3})\ \mathcal{L}(-\frac{2\pi}{3})|/2\log 2$. If $\mathcal{L}$ has only real coefficients (as is the case in most applications of practical interest), then $|\mathcal{L}(-\frac{2\pi}{3})| = |\mathcal{L}(\frac{2\pi}{3})|$, and $\tilde{\mathcal{K}} = \log|\mathcal{L}(\frac{2\pi}{3})|/\log 2$. The next short invariant cycles are $\{\frac{2\pi}{5}, \frac{4\pi}{5}, -\frac{2\pi}{5}, -\frac{4\pi}{5}\}$, $\{\frac{2\pi}{7}, \frac{4\pi}{7}, -\frac{6\pi}{7}\}$, etc.; each of them gives an upper bound for the decay exponent of $\hat{\phi}$.

In some cases one of these upper bounds on α can be proved to be a lower bound as well. We first prove the following lemma.

LEMMA 7.1.6. *Suppose that* $[-\pi,\pi] = D_1 \cup D_2 \cdots \cup D_M$, *and that there exists* $q > 0$ *so that*

$$\begin{array}{ll} |\mathcal{L}(\xi)| \leq q & \xi \in D_1 \\ |\mathcal{L}(\xi)\ \mathcal{L}(2\xi)| \leq q^2 & \xi \in D_2 \\ \vdots & \vdots \\ |\mathcal{L}(\xi)\ \mathcal{L}(2\xi)\cdots\mathcal{L}(2^{M-1}\xi)| \leq q^M & \xi \in D_M\ . \end{array}$$

Then $|\hat{\phi}(\xi)| \leq C(1+|\xi|)^{-N+\mathcal{K}}$, *with* $\mathcal{K} = \log q/\log 2$.

Proof.

1. Let us estimate $\left|\prod_{k=0}^{j-1}\mathcal{L}(2^{-k}\xi)\right|$, for some large but arbitrary j. Since $\zeta = 2^{-j+1}\xi \in D_m$ for some $m \in \{1,\ 2,\cdots,M\}$, we have

$$\left|\prod_{k=0}^{j-1}\mathcal{L}(2^{-k}\xi)\right| = \left|\prod_{\ell=0}^{j-1}\mathcal{L}(2^{\ell}\zeta)\right| \leq q^m \left|\prod_{\ell=m}^{j-1}\mathcal{L}(2^{\ell}\zeta)\right|\ .$$

We can now apply the same trick to $2^m\zeta$, and keep doing so until we cannot go on. At that point we have

$$\left|\prod_{k=0}^{j-1} \mathcal{L}(2^{-k}\xi)\right| \le q^{j-r} \left|\prod_{k=0}^{r} \mathcal{L}(2^{-k}\xi)\right| ,$$

with at most $M-1$ different $\mathcal{L}$-factors remaining (i.e., $r \le M-1$). Hence

$$\left|\prod_{k=0}^{j-1} \mathcal{L}(2^{-k}\xi)\right| \le q^{j-M+1}\, q_1^{M-1} ,$$

with q_1 defined as in (7.1.5). Consequently, with the definition (7.1.6),

$$\mathcal{K}_j \le \frac{1}{j \log 2}\, [C + j \log q] ,$$

and $\mathcal{K} = \lim_{j\to\infty} \mathcal{K}_j \le \log q / \log 2$. The bound on $\hat{\phi}$ now follows from Lemma 7.1.2. ∎

In particular, one has the following lemma.

LEMMA 7.1.7. *Suppose that*

$$\begin{aligned} &|\mathcal{L}(\xi)| \le |\mathcal{L}(\tfrac{2\pi}{3})| && \text{for } |\xi| \le \tfrac{2\pi}{3} , \\ &|\mathcal{L}(\xi)\ \mathcal{L}(2\xi)| \le |\mathcal{L}(\tfrac{2\pi}{3})|^2 && \text{for } \tfrac{2\pi}{3} \le |\xi| \le \pi . \end{aligned} \tag{7.1.11}$$

Then $|\hat{\phi}(\xi)| \le C(1+|\xi|)^{-N+\mathcal{K}}$, *with* $\mathcal{K} = \log |\mathcal{L}(\frac{2\pi}{3})| / \log 2$, *and this decay is optimal.*

Proof. The proof is a straightforward consequence of Lemmas 7.1.5 and 7.1.6. ∎

Of course, Lemma 7.1.7 is only applicable to very special $\mathcal{L}$; in most cases (7.1.11) will not be satisfied: there even exist $\mathcal{L}$ for which $\mathcal{L}(\frac{2\pi}{3}) = 0$. Similar optimal bounds can be derived by using other invariant cycles as breakpoints for a partition of $[-\pi, \pi]$, and applying Lemma 7.1.6. Let us return to our "standard" example ${}_N\phi$. In this case Cohen and Conze (1992) proved that $\mathcal{L}_N(\xi)$ does indeed satisfy (7.1.11), as follows.

LEMMA 7.1.8. *For all* $N \in \mathbb{N}$, $N \ge 1$, $P_N(y) = \sum_{n=0}^{N-1} \binom{N-1+n}{n} y^n$ *satisfies*

$$P_N(y) \le P_N\left(\tfrac{3}{4}\right) \quad \text{if } \ 0 \le y \le \tfrac{3}{4} \tag{7.1.12}$$

$$P_N(y)\ P_N(4y(1-y)) \le \left[P_N\left(\tfrac{3}{4}\right)\right]^2 \quad \text{if } \ \tfrac{3}{4} \le y \le 1 . \tag{7.1.13}$$

We start by proving yet another property of P_N.

LEMMA 7.1.9.

$$P_N'(x) = \frac{N}{1-x}\left[P_N(x) - P_N(1)x^{N-1}\right] . \qquad (7.1.14)$$

Proof.

1.

$$\begin{aligned} P_N'(x) &= \sum_{n=1}^{N-1} \binom{N-1+n}{n} n\, x^{n-1} = N \sum_{n=0}^{N-2} \binom{N+n}{n} x^n \\ &= N\left[P_{N+1}(x) - \binom{2N}{N} x^N - \binom{2N-1}{N-1} x^{N-1}\right] . \end{aligned} \qquad (7.1.15)$$

2.

$$\begin{aligned} (1-x)P_{N+1}(x) &= 1 + \sum_{n=1}^{N}\left[\binom{N+n}{n} - \binom{N+n-1}{n-1}\right] x^n \\ &\quad - \binom{2N}{N} x^{N+1} \\ &= 1 + \sum_{n=1}^{N} \binom{N-1+n}{n} x^n - \binom{2N}{N} x^{N+1} \\ &= P_N(x) + \binom{2N-1}{N} x^N (1-2x) . \end{aligned} \qquad (7.1.16)$$

3. Combining (7.1.15) and (7.1.16) gives

$$(1-x)P_N'(x) = N\left[P_N(x) - \binom{2N-1}{N} x^{N-1}\right] .$$

Since $P_N(1) = \sum_{n=0}^{N-1} \binom{N-1+n}{n} = \binom{2N-1}{N}$, (7.1.14) follows. ∎

We now tackle the proof of Lemma 7.1.8.

Proof of Lemma 7.1.8.

1. Since $P_N(y)$ is increasing on [0,1], we only need to prove (7.1.13).
2. Define $f(y) = P_N(y)\, P_N(4y(1-y))$. Applying Lemma 7.1.9 leads to

$$f'(y) = \frac{N}{(1-y)(2y-1)}\, g(y) ,$$

with

$$g(y) = P_N(y)\, P_N(4y(1-y))(6y-5)$$
$$-y^{N-1}(2y-1)P_N(1)P_N(4y(1-y))$$
$$+4(1-y)[4y(1-y)]^{N-1}P_N(1)P_N(y)\,. \qquad (7.1.17)$$

3. Since $4y(1-y) \le y$ for $y \ge 3/4$, we can apply (7.1.8) to derive

$$P_N(y)\, y^{-N+1} \le [4y(1-y)]^{-N+1} P_N(4y(1-y))$$

or

$$[4(1-y)]^{N-1} P_N(y) \le P_N(4y(1-y))\,.$$

Substituting this into (7.1.17) leads to

$$g(y) \le (6y-5)\; P_N(4y(1-y))\; [P_N(y) - y^{N-1}\; P_N(1)]\,.$$

The quantity in square brackets equals $\frac{1}{N}(1-y)P'_N(y) \ge 0$ for $y \le 1$, so that $g(y) \le 0$ for $\frac{3}{4} \le y \le \frac{5}{6}$. It follows that $P_N(y)\; P_N(4y(1-y))$ is decreasing on $[\frac{3}{4}, \frac{5}{6}]$, which proves (7.1.13) for $y \le \frac{5}{6}$.

4. For $\frac{5}{6} \le y \le 1$ we follow a different strategy. Since $P_N(y) \le \left(\frac{4y}{3}\right)^{N-1} P_N(\frac{3}{4})$ by Lemma 7.1.4, it suffices to prove

$$\left(\frac{4y}{3}\right)^{N-1} P_N(4y(1-y)) \le P_N\left(\frac{3}{4}\right)\,. \qquad (7.1.18)$$

But $P_N(4y(1-y)) \le [1-4y(1-y)]^{-N} = (2y-1)^{-2N}$ (because $(1-x)^N$ $P_N(x) = 1 - x^N\; P_N(1-x) \le 1$), and

$$P_N\left(\frac{3}{4}\right) \ge \left(\frac{3}{4}\right)^{N-1} P_N(1) \ge \frac{1}{\sqrt{N}}\; 3^{N-1}\,, \qquad (7.1.19)$$

where we have used Lemma 7.1.4 again, as well as

$$P_N(1) = \binom{2N-1}{N} = \frac{1}{2}\binom{2N}{N} \ge \frac{1}{\sqrt{N}}\; 4^{N-1}\,.$$

To prove (7.1.18) it is therefore sufficient to prove

$$\left[\frac{y}{(2y-1)^2}\right]^{N-1} (2y-1)^{-2} \le \frac{1}{\sqrt{N}}\left(\frac{9}{4}\right)^{N-1}\,. \qquad (7.1.20)$$

Since both $(2y-1)^{-2}$ and $y(2y-1)^{-2}$ are decreasing on $[\frac{5}{6}, 1]$, it suffices to verify that (7.1.20) holds for $y = \frac{5}{6}$, i.e., that

$$\left(\frac{5}{6}\right)^{N-1} \le \frac{4}{9\sqrt{N}}\,.$$

This is true for $N \ge 13$.

5. It remains to prove (7.1.13) for $\frac{5}{6} \le y \le 1$ and $1 \le N \le 12$. We do this in two steps: $y \le y_0 = \frac{2+\sqrt{2}}{4}$, and $y \ge y_0$. For $y \le \frac{2+\sqrt{2}}{4}$, $4y(1-y) \ge \frac{1}{2}$, hence, again by Lemma 7.1.4,

$$P_N(4y(1-y)) \le [8y(1-y)]^{N-1} P_N(\tfrac{1}{2}) = [16y(1-y)]^{N-1} .$$

Similarly $P_N(y) \le \left(\frac{6y}{5}\right)^{N-1} P_N(\frac{5}{6})$, so that

$$\begin{aligned} P_N(y)\, P_N(4y(1-y)) &\le \left(\frac{6}{5}\right)^{N-1} P_N\left(\frac{5}{6}\right) [16y^2(1-y)]^{N-1} \\ &\le \left(\frac{20}{9}\right)^{N-1} P_N\left(\frac{5}{6}\right) , \end{aligned} \tag{7.1.21}$$

because $y^2(1-y)$ is decreasing on $[\frac{5}{6}, \frac{2+\sqrt{2}}{4}]$. One checks by numerical computation that (7.1.21) is indeed smaller than $[P_N(\frac{3}{4})]^2$ for $1 \le N \le 12$.

6. For $\frac{2+\sqrt{2}}{4} = y_0 \le y \le 1$ we use the bounds $P_N(4y(1-y)) \le (2y-1)^{-2N}$ and $P_N(y) \le (\frac{y}{y_0})^{N-1} P_N(y_0)$ to derive

$$\begin{aligned} P_N(4y(1-y))P_N(y) &\le y_0^{-N+1} P_N(y_0)(2y-1)^{-2} \left[\frac{y}{(2y-1)^2}\right]^{N-1} \\ &\le 2^N\ P_N(y_0) , \end{aligned} \tag{7.1.22}$$

where the last inequality uses that $(2y-1)^{-2}$ and $y(2y-1)^{-2}$ are both decreasing in $[y_0, 1]$. One checks by numerical computation that (7.1.22) is smaller than $[P_N(\frac{3}{4})]^2$ for $5 \le N \le 12$.

7. It remains to prove (7.1.13) for $1 \le N \le 4$ and $\frac{2+\sqrt{2}}{4} \le y \le 1$. For these small values of N the polynomial $P_N(y)\, P_N(4y(1-y)) - P_N(\frac{3}{4})^2$ has degree at most 9, and its roots can be computed easily (numerically). One checks that there are none in $]\frac{3}{4}, 1]$, which finishes the proof, because (7.1.13) is satisfied in $y = 1$. ∎

It follows from Lemmas 7.1.8 and 7.1.7 that we know the exact asymptotic decay of ${}_N\hat{\phi}(\xi)$:

$$|\,{}_N\hat{\phi}(\xi)| \le C(1+|\xi|)^{-N+\log|P_N(3/4)|/2\log 2} . \tag{7.1.23}$$

For the first few values of N this translates into ${}_N\phi \in C^{\alpha-\epsilon}$ with the following estimates for α:

N	α
2	.339
3	.636
4	.913
5	1.177
6	1.432
7	1.682
8	1.927
9	2.168
10	2.406

We can also use Lemma 7.1.7 to estimate the smoothness of ${}_N\phi$ as $N\to\infty$. Since

$$\frac{1}{\sqrt{N}}\, 3^{N-1} \le P_N(\tfrac{3}{4}) \le 3^{N-1}$$

(use Lemma 7.1.4 for the upper bound, (7.1.19) for the lower bound),

$$\frac{\log|P_N(3/4)|}{2\log 2} = \frac{\log 3}{2\log 2}\, N[1 - O(N^{-1}\log N)]\ ,$$

implying that asymptotically, for large N, ${}_N\phi \in C^{\mu N}$ with $\mu = 1 - \frac{\log 3}{2\log 2} \simeq .2075$.[1] One does not, in fact, need the full force of Lemma 7.1.8 to prove this asymptotic result: it is sufficient to prove that

$$P_N(y) \le C\, 3^{N-1} \quad \text{for } y \le \tfrac{3}{4} \tag{7.1.24}$$

$$P_N(y)\, P_N(4y(1-y)) \le C^2\, 3^{2(N-1)} \quad \text{for } \tfrac{3}{4} \le y \le 1\ , \tag{7.1.25}$$

with C independent of N. The asymptotic result then follows immediately from Lemma 7.1.6. The estimate (7.1.24) is immediate from $P_N(y) \le P_N(\frac{3}{4}) \le 3^{N-1}$ for $y \le \frac{3}{4}$; the estimate (7.1.25) follows easily from Lemma 7.1.4 as follows. If $\frac{3}{4} \le y \le \frac{2+\sqrt{2}}{4}$, then $P_N(y)\, P_N(4y(1-y)) \le (4y)^{N-1}\, (16y(1-y))^{N-1} = [64y^2(1-y)]^{N-1} \le 3^{2(N-1)}$ because $y^2(1-y)$ is decreasing on $[\frac{3}{4}, 1]$; if $\frac{2+\sqrt{2}}{4} \le y \le 1$, then $P_N(y)\, P_N(4y(1-y)) \le (4y)^{N-1} P_N(\frac{1}{2}) = (8y)^{N-1} < 3^{2(N-1)}$. This much simpler argument for the exact asymptotic decay of $\hat{\phi}$ is due to Volkmer (1991), who derived it independently of Cohen and Conze's work.

7.1.3. Littlewood–Paley type estimates. The estimates in this subsection are L^1 or L^2-estimates for $(1+|\xi|)^\alpha \hat{\phi}$ rather than pointwise decay estimates for $\hat{\phi}$ itself. The basic idea is the usual Littlewood–Paley technique: the Fourier transform of the function gets broken up into dyadic pieces (i.e., roughly $2^jC \le |\xi| \le 2^{j+1}C$) and the integral of each piece estimated. If $\int_{2^j \le |\xi| \le 2^{j+1}} d\xi |\hat{\phi}(\xi)| \le C\lambda^j$ for $j \in \mathbb{N}$, then $\int d\xi (1+|\xi|)^\alpha |\hat{\phi}(\xi)| \le$

$C[1+\sum_{j=1}^{\infty} 2^{j\alpha}\lambda^j] < \infty$ if $\alpha < -\log\lambda/\log 2$, implying $\phi \in C^\alpha$. To obtain estimates of this nature, we exploit the special structure of $\hat{\phi}$ as the infinite product of $m_0(2^{-j}\xi)$; the operator P_0 defined in §6.3 will be a basic tool in this derivation.

We will first restrict ourselves to positive trigonometric polynomials $M_0(\xi)$. (Later on, we will take $M_0(\xi) = |m_0(\xi)|^2$ to extend our results to non-positive m_0.) As in §6.3 we define the operator P_0 acting on 2π-periodic functions by

$$(P_0 f)(\xi) = M_0\left(\frac{\xi}{2}\right) f\left(\frac{\xi}{2}\right) + M_0\left(\frac{\xi}{2}+\pi\right) f\left(\frac{\xi}{2}+\pi\right) .$$

This operator was studied by Conze and Raugi and several of the results in this subsection come from their work (Conze and Raugi (1990), Conze (1991)). Similar ideas were also developed independently by Eirola (1991) and Villemoes (1992). A first useful lemma is the following.

LEMMA 7.1.10. *For all $m > 0$ and all 2π-periodic functions f,*

$$\int_{-\pi}^{\pi} d\xi (P_0^m f)(\xi) = \int_{-2^m\pi}^{2^m\pi} d\xi \ f(2^{-m}\xi) \prod_{j=1}^{m} M_0(2^{-j}\xi) . \qquad (7.1.26)$$

Proof.

1. By induction. For $m = 1$,

$$\begin{aligned} \int_{-\pi}^{\pi} d\xi (P_0 f)(\xi) &= \int_{-\pi}^{\pi} d\xi \left[M_0\left(\frac{\xi}{2}\right) f\left(\frac{\xi}{2}\right) + M_0\left(\frac{\xi}{2}+\pi\right) f\left(\frac{\xi}{2}+\pi\right)\right] \\ &= 2\int_{-\pi/2}^{\pi/2} d\zeta \ [M_0(\zeta)f(\zeta) + M_0(\zeta+\pi)f(\zeta+\pi)] \\ &= 2\int_{-\pi/2}^{3\pi/2} d\zeta \ M_0(\zeta)f(\zeta) = \int_{-2\pi}^{2\pi} d\xi \ f\left(\frac{\xi}{2}\right) M_0\left(\frac{\xi}{2}\right) . \end{aligned}$$

2. Suppose (7.1.26) holds for $m = n$. Then it holds for $m = n+1$:

$$\begin{aligned} &\int_{-\pi}^{\pi} d\xi \ (P_0^{n+1} f)(\xi) = \int_{-\pi}^{\pi} d\xi \ (P_0^n P_0 f)(\xi) \\ &= \int_{-2^n\pi}^{2^n\pi} d\xi \ [M_0(2^{-n-1}\xi)f(2^{-n-1}\xi) \\ &\qquad + M_0(2^{-n-1}\xi+\pi)f(2^{-n-1}\xi+\pi)] \prod_{j=1}^{n} M_0(2^{-j}\xi) \\ &= 2^{n+1}\int_{-\pi/2}^{\pi/2} d\zeta \left[\prod_{j=1}^{n} M_0(2^j\zeta)\right] [M_0(\zeta)f(\zeta) + M_0(\zeta+\pi)f(\zeta+\pi)] \end{aligned}$$

$$= 2^{n+1} \int_{-\pi/2}^{3\pi/2} d\zeta \left[\prod_{j=0}^{n} M_0(2^j \zeta) \right] f(\zeta)$$

$$= 2^{n+1} \int_{-\pi}^{\pi} d\zeta \left[\prod_{j=0}^{n} M_0(2^j \zeta) \right] f(\zeta)$$

$$= \int_{-2^{n+1}\pi}^{2^{n+1}\pi} d\xi \left[\prod_{j=1}^{n} M_0(2^{-j} \xi) \right] f(2^{-n-1}\xi) \ . \quad \blacksquare$$

Since M_0 is a positive trigonometric polynomial, it can be written as

$$\begin{aligned} M_0(\xi) &= \sum_{j=-J}^{J} a_j \, e^{-ij\xi} \ , \quad \text{with } a_j = a_{-j} \in \mathbb{R} \\ &= \sum_{j=0}^{J} b_j \cos(j\xi) \ . \end{aligned}$$

One then finds that the $(2J+1)$-dimensional vector space of trigonometric polynomials defined by

$$V_J = \left\{ f(\xi); \quad f = \sum_{j=-J}^{J} f_j \, e^{-ij\xi} \right\}$$

is invariant for P_0. The action of P_0 in V_J can be represented by a $(2J+1) \times (2J+1)$ matrix which we will also denote by P_0,

$$(P_0)_{k\ell} = 2a_{2k-\ell} \ , \qquad -J \le k, \ \ell \le J \ , \tag{7.1.27}$$

with the convention $a_r = 0$ if $|r| > J$. For M_0 of the type

$$M_0(\xi) = \left(\cos \frac{\xi}{2} \right)^{2K} L(\xi) \ , \tag{7.1.28}$$

where L is a trigonometric polynomial such that $L(\pi) \neq 0$, the matrix P_0 has very special spectral properties.

LEMMA 7.1.11. *The values* $1, \frac{1}{2}, \cdots, 2^{-2K+1}$ *are eigenvalues for* P_0. *The row vectors* $e_k = (j^k)_{j=-J,\cdots,J}$, $k = 0, \cdots, 2K-1$ *generate a subspace which is left invariant for* P_0. *More precisely,*

$$e_k P_0 = 2^{-k} e_k + \textit{linear combination of the } e_n, \quad n < k \ .$$

Proof.

1. The factorization (7.1.28) is equivalent to

$$\sum_{j=-J}^{J} a_j \, j^k (-1)^j \;=\; 0 \quad \text{for } k = 0, \cdots, 2K-1 \;. \tag{7.1.29}$$

Moreover, since $M_0(0) = 1$, $\sum a_{2j} = \sum a_{2j+1} = \frac{1}{2}$. This means that the sum of each column in the matrix (7.1.27) is equal to 1; e_0 is thus a left eigenvector of P_0 with eigenvalue 1.

2. For $0 < k \leq 2K - 1$, define $g_k = e_k P_0$, i.e.,

$$(g_k)_m \;=\; 2 \sum_j j^k \, a_{2j-m} \;.$$

For m even, $m = 2\ell$,

$$(g_k)_{2\ell} \;=\; 2 \sum_j (j+\ell)^k a_{2j} \;=\; 2^{-k+1} \sum_m (2\ell)^m \sum_j \binom{k}{m} \, (2j)^{k-m} a_{2j} \;.$$

For m odd, $m = 2\ell + 1$,

$$\begin{aligned}(g_k)_{2\ell+1} \;&=\; 2 \sum_j (j+\ell+1)^k a_{2j+1} \\ &=\; 2^{-k+1} \sum_m (2\ell+1)^m \sum_j \binom{k}{m} (2j+1)^{k-m} a_{2j+1} \;.\end{aligned}$$

Hence

$$e_k P_0 \;=\; g_k \;=\; 2^{-k+1} \sum_{m=0}^{k} \binom{k}{m} A_{k-m} \, e_m \;,$$

where

$$A_m \;=\; \sum_j a_{2j} (2j)^m \;=\; \sum_j a_{2j+1} (2j+1)^m$$

by (7.1.29). ∎

A consequence of Lemma 7.1.11 is that the spaces E_k,

$$E_k \;=\; \left\{ f \in V_J; \;\; \sum_{j=-J}^{J} j^n f_j = 0 \;\text{ for }\; n \;=\; 0, \cdots, k-1 \right\} \;,$$

with $1 \leq k \leq 2K$, are all right invariant for P_0. The main result of this subsection is then the following.

THEOREM 7.1.12. *Let λ be the eigenvalue of $P_0|_{E_{2K}}$ with the largest absolute value. Define F, α by*

$$\hat{F}(\xi) = (2\pi)^{-1/2} \prod_{j=1}^{\infty} M_0(2^{-j}\xi) ,$$

$$\alpha = -\log|\lambda|/\log 2 .$$

If $|\lambda| < 1$, then $F \in C^{\alpha-\epsilon}$ for all $\epsilon > 0$.

Proof.

1. Define $f(\xi) = (1-\cos\xi)^K$. Since $\frac{d^k}{d\xi^k} f|_{\xi=0} = 0$ for $k \le 2K-1$, $f \in E_{2K}$.

2. The spectral radius $\rho(P_0|_{E_{2K}})$ equals $|\lambda|$. Since, for any $\delta > 0$, there exists $C > 0$ so that $\|A^n\| \le C(\rho(A)+\delta)^n$ for all $n \in \mathbb{N}$, it follows that

$$\int_{-\pi}^{\pi} d\xi\ (P_0^m f)(\xi) \le C(|\lambda|+\delta)^m . \tag{7.1.30}$$

3. On the other hand, $f(\xi) \ge 1$ for $\frac{\pi}{2} \le |\xi| \le \pi$. Together with the boundedness of $\prod_{j=1}^{\infty} M_0(2^{-j}\xi)$ for $|\xi| \le \pi$ (derived as usual from $|M_0(\xi)| \le 1 + C|\xi|$), this implies

$$\begin{aligned} \int_{2^{n-1}\pi \le |\xi| \le 2^n\pi} d\xi\ \hat{F}(\xi) &\le C \int_{2^{n-1}\pi \le |\xi| \le 2^n\pi} d\xi \prod_{j=1}^{n} M_0(2^{-j}\xi) \\ &\le C \int_{2^{n-1}\pi \le |\xi| \le 2^n\pi} d\xi\ f(2^{-n}\xi) \prod_{j=1}^{n} M_0(2^{-j}\xi) = C\int_{-\pi}^{\pi} d\xi\ (P_0^n f)(\xi) \\ &\qquad \text{(use Lemma 7.1.10)} \\ &\le C'(|\lambda|+\delta)^n . \end{aligned}$$

By the argument at the start of this subsection, this implies $F \in C^{\alpha-\epsilon}$. ∎

In fact, a slightly stronger result can be proved. If we extend the definition of C^n (n integer) to include all the functions for which the $(n-1)$th derivative is in the Zygmund class

$$\mathcal{F} = \{f;\ |f(x+y) + f(x-y) - 2f(x)| \le C|y| \ \text{ for all } x, y\} ,$$

then it is also true that $F \in C^\alpha$ if $P_0|_{E_{2K}}$ is diagonal (i.e., we can drop the ϵ in this case). Moreover, both this smoothness estimate and the estimate $F \in C^{\alpha-\epsilon}$ for all $\epsilon > 0$ in Theorem 7.1.12 are optimal if $\hat{F}$ has no zero on $[-\pi, \pi]$. For a proof, see Theorem 2.7 in Cohen and Daubechies (1991).

REMARK. The same result can be derived via an equivalent technique, which uses an operator P_0^L defined in the same way as P_0, but with $M_0(\xi)$ replaced by the factor $L(\xi)$ in (7.1.28). In this case we define $\lambda^L = \rho(P_0^L)$, and we factorize $\hat{F}(\xi) = [2(\sin\xi/2)/\xi]^{2K}(2\pi)^{-1/2}\prod_{j=1}^{\infty} L(2^{-j}\xi)$ to obtain

$$\int_{2^{n-1}\pi\le|\xi|\le 2^n\pi} d\xi\ \hat{F}(\xi) \le C \int_{2^{n-1}\pi\le|\xi|\le 2^n\pi} d\xi\ |\xi|^{-2K}\prod_{j=1}^{n} L(2^{-j}\xi)$$
$$\le C\ 2^{-2nK}\int_{-\pi}^{\pi} d\xi\ \left[(P_0^L)^n 1\right](\xi)$$
$$\le C\ 2^{-2nK}(\lambda^L+\epsilon)^n ,$$

so that $F \in C^{\alpha-\epsilon}$ with $\alpha = 2K - \frac{\log\lambda}{\log 2}$. This method has the advantage that we start directly with a smaller matrix P_0^L, so that the computation of the spectral radius is simpler. The two methods are completely equivalent, as shown by the following argument. If μ is an eigenvalue of P_0, with eigenfunction $f_\mu \in E_{2K}$, then f_μ can be written as

$$f_\mu(\xi) = \left(\sin^2\frac{\xi}{2}\right)^K g_\mu(\xi) .$$

Replacing $M_0(\xi)$ by its factorized form in

$$\mu f_\mu(\xi) = M_0\left(\frac{\xi}{2}\right) f_\mu\left(\frac{\xi}{2}\right) + M_0\left(\frac{\xi}{2}+\pi\right) f_\mu\left(\frac{\xi}{2}+\pi\right) ,$$

we obtain after dividing by $[\sin^2\frac{\xi}{2}\cos^2\frac{\xi}{2}]^N$,

$$\mu 2^{2K} g_\mu(\xi) = L\left(\frac{\xi}{2}\right) g_\mu\left(\frac{\xi}{2}\right) + L\left(\frac{\xi}{2}+\pi\right) g_\mu\left(\frac{\xi}{2}+\pi\right) ,$$

so that the eigenvalues of P_0^L are exactly given by $\mu^L = 2^{2K}\mu$. □

In general, m_0 will not be positive. (Indeed, in the framework of orthonormal wavelet bases, m_0 is never positive, except for the Haar basis; see Janssen (1992).) However, we can then define $M_0 = |m_0|^2$; the same techniques lead to

$$\int_{2^{n-1}\pi\le|\xi|\le 2^n\pi} d\xi\ |\hat{\phi}(\xi)|^2 \le C\ 2^{-2nN}(\lambda^L+\epsilon)^n ,$$

where λ^L is the spectral radius for P_0^L, with $L(\xi) = |\mathcal{L}(\xi)|^2$. Hence

$$\int d\xi\ (1+|\xi|)^\alpha\ |\hat{\phi}(\xi)|$$
$$\le C\left[1+\sum_{n=0}^{\infty} 2^{n\alpha}\ 2^{n/2}\left(\int_{2^{n-1}\pi\le|\xi|\le 2^n\pi} d\xi\ |\hat{\phi}(\xi)|^2\right)^{1/2}\right] < \infty$$

if $\alpha + \frac{1}{2} < N + \frac{\log \lambda^L}{2 \log 2}$. Consequently $\phi \in C^{\alpha - \epsilon}$ for $\alpha \leq N + \frac{\log \lambda^L}{2 \log 2} - \frac{1}{2}$.[2]

For the special ${}_N\phi$ of §6.4, the resulting α, for the first few values of N, are:

N	α
2	.5
3	.915
4	1.275
5	1.596
6	1.888
7	2.158
8	2.415
9	2.661
10	2.902

These are much better than the values obtained from the pointwise decay of $\hat{\phi}$ (see §7.1.2). The size of the matrix P_0^L increases with N (linearly), and I do not know of any way to determine the asymptotics of its spectral radius as $N \to \infty$; for asymptotic estimates, pointwise decay of $\hat{\phi}$ is the best method.

7.2. A direct method.

The smoothness results obtained at the end of §7.1.3 for ${}_N\phi$ with small N are still not optimal. Moreover, Fourier-based methods can only give information on the global Hölder exponent, whereas it is clear from Figure 6.3 that ${}_2\phi$, for instance, is smoother in some points than in others. In fact, we will see that there exists a whole hierarchy of (fractal) sets in which ${}_2\phi$ has different Hölder exponents, ranging from .55 to 1. Results such as these can be obtained by direct methods, not involving $\hat{\phi}$. For the sake of simplicity, I will explain the setup in the general case, but expose the method in detail for the example ${}_2\phi$ only, and then state the general theorems on global and local regularity without proof. Proofs can be found in Daubechies and Lagarias (1991, 1992). Similar results about the global regularity were also proved (independently, and in fact before Daubechies and Lagarias) in Micchelli and Prautzsch (1989), in the framework of subdivision schemes.

The method is completely independent of wavelet theory. The starting point is the equation

$$F(x) = \sum_{k=0}^{K} c_k \, F(2x - k) , \tag{7.2.1}$$

with $\sum_{k=0}^{K} c_k = 2$, and we are interested in the compactly supported L^1-solution F, which, if it exists, is uniquely determined[3] (up to its normalization): since $F \in L^1$, $\hat{F}$ is continuous, and (7.2.1) implies

$$\hat{F}(\xi) = (2\pi)^{-1/2} \left[\int_{-\pi}^{\pi} dx \; F(x) \right] \prod_{j=1}^{\infty} m(2^{-j}\xi) ,$$

with $m(\xi) = \frac{1}{2}\sum_{k=0}^{K} c_k\, e^{-ik\xi}$, and Lemma 6.2.2 then tells us that support $F = [0, K]$. Equation (7.2.1) can be considered as a fixed point equation. Define, for functions g supported on $[0, K]$, Tg by

$$(Tg)(x) \;=\; \sum_{k=0}^{K} c_k\, g(2x-k)\;.$$

Then F solves (7.2.1) if $TF = F$. We will try to find this fixed point by the usual method: find a suitable F_0,[4] define $F_j = T^j F_0$, and prove that F_j has a limit. To determine F_0, note first that (7.2.1) imposes a constraint on the values $F(n)$, $n \in \mathbb{Z}$ if F is continuous. Since support $F = [0, K]$, we only need to determine the $F(k)$, $1 \le k \le K-1$; the other $F(n)$ are zero. Substituting $x = k$, $1 \le k \le K-1$ into (7.2.1) leads to $K-1$ linear equations for the $K-1$ unknowns $F(k)$; the system of equations can also be read as the requirement that the vector $(F(1), \cdots, F(K-1))$ be an eigenvector with eigenvalue 1 of a $(K-1)\times(K-1)$-dimensional matrix derived from the c_k. It turns out that, modulo some technical conditions (see below), this matrix does indeed have 1 as a nondegenerate eigenvalue, so that $(F(1), \cdots, F(K-1))$ can be fixed up to an overall multiplicative constant. Let us suppose this is done. One can prove that $\sum_{k=1}^{K-1} F(k) \neq 0$, so we can fix the normalization so that $\sum_{k=1}^{K-1} F(k) = 1$. (All this will be illustrated by an example below.) Define now $F_0(x)$ to be the piecewise linear function which takes exactly the values $F(k)$ at the integers, i.e.,

$$F_0(x) \;=\; F(k)(k+1-x) \;+\; F(k+1)(x-k) \quad \text{for } k \le x \le k+1\;. \tag{7.2.2}$$

Successive applications of T define the $F_j = T^j F_0$, i.e.,

$$F_{j+1}(x) = (TF_j)(x) = \sum_{k=0}^{K} c_k\, F_j(2x-k)\;; \tag{7.2.3}$$

it easily follows that the F_j are piecewise linear with nodes at the $2^{-j}n \in [0, K]$, $n \in \mathbb{N}$. To discuss whether the F_j have a limit as $j\to\infty$, and to study the regularity of this limit, it is convenient to recast (7.2.3) in another form.

The key idea is to study the $F_j(x)$, $F_j(x+1) \cdots F_j(x+K-1)$ simultaneously, for $x \in [0,1]$. We define $v_j(x) \in \mathbb{R}^K$[5] by

$$[v_j(x)]_k \;=\; F_j(x+k-1)\;, \qquad k \;=\; 1, \cdots, K,\;\; x \in [0,1]\;. \tag{7.2.4}$$

For $0 \le x \le \frac{1}{2}$, (7.2.3), together with support $F_j \subset [0, K]$, implies that $F_{j+1}(x)$, $F_{j+1}(x+1), \cdots, F_{j+1}(x+K-1)$ are all linear combinations of $F_j(2x)$, $F_j(2x+1), \cdots, F_j(2x+K-1)$. More precisely, in terms of $v_j(x)$,

$$v_{j+1}(x) \;=\; T_0\; v_j(2x) \;\; \text{for } 0 \le x \le \tfrac{1}{2}\;, \tag{7.2.5}$$

where T_0 is the $K \times K$ matrix defined by

$$(T_0)_{mn} \;=\; c_{2m-n-1}\;, \qquad 1 \le m, n \le K\;, \tag{7.2.6}$$

with the convention $c_k = 0$ for $k < 0$ or $k > K$. Similarly,

$$v_{j+1}(x) \;=\; T_1 \; v_j(2x-1) \quad \text{for } \tfrac{1}{2} \le x \le 1 \;, \tag{7.2.7}$$

where

$$(T_1)_{mn} \;=\; c_{2m-n} \;, \qquad 1 \le m, n \le K \;. \tag{7.2.8}$$

Equations (7.2.5), (7.2.7) both apply for $x = \frac{1}{2}$: because of the special structure of T_0, T_1 and v_j (in particular, $(T_0)_{mn} = (T_1)_{m\,n+1}$, $[v_j(0)]_n = [v_j(1)]_n$ for $n = 2, \cdots, K$), the two equations are identical for $x = \frac{1}{2}$. We can combine (7.2.5) and (7.2.7) into a single vector equation as follows. Every $x \in [0,1]$ can be represented by a binary sequence,

$$x \;=\; \sum_{n=1}^{\infty} d_n(x)\; 2^{-n} \;,$$

with $d_n(x) = 1$ or 0 for all n. Strictly speaking, two possibilities exist for every dyadic rational x, i.e. every x of the type $k\,2^{-j}$: we can replace the last digit 1 followed by all zeros by a digit 0 followed by all ones. This will not cause a problem, but to be clear we distinguish these two sequences by a subscript: $d_n^+(x)$ for the sequence ending in zeros (the expansion "from above," i.e., the expansion that will start by the same $J-1$ digits as $x + 2^{-J}$ for $J \to \infty$), $d_n^-(x)$ for the sequence ending in ones (expansion "from below"). For instance,

$$d_1^+(\tfrac{1}{2}) \;=\; 1 \;, \quad d_n^+(\tfrac{1}{2}) \;=\; 0 \;, \quad n \ge 2 \;,$$

$$d_1^-(\tfrac{1}{2}) \;=\; 0 \;, \quad d_n^-(\tfrac{1}{2}) = 1 \;, \qquad n \ge 2 \;.$$

The two definition regions $0 \le x < \frac{1}{2}$ and $\frac{1}{2} < x \le 1$ of (7.2.5) and (7.2.7) are completely characterized by $d_1(x)$: $d_1(x) = 0$ if $x < \frac{1}{2}$, 1 if $x > \frac{1}{2}$.

For every binary sequence $d = (d_n)_{n \in \mathbb{N} \setminus \{0\}}$ we also define its right shift τd by

$$(\tau d)_n \;=\; d_{n+1} \;, \qquad n \;=\; 1, 2, \cdots \;.$$

It is then clear that $\tau d(x) \;=\; d(2x)$ if $0 \le x < \frac{1}{2}$, $\tau d(x) = d(2x-1)$ if $\frac{1}{2} < x \le 1$. (For $x = \frac{1}{2}$, we have two possibilities: $\tau d^+(\frac{1}{2}) = d(0)$, $\tau d^-(\frac{1}{2}) = d(1)$.) Although τ is really defined on binary sequences, we will make a slight abuse of notation and write $\tau x = y$ rather than $\tau d(x) = d(y)$. With this new notation, we can rewrite (7.2.5), (7.2.7) as the single equation

$$v_{j+1}(x) \;=\; T_{d_1(x)} \; v_j(\tau x) \;. \tag{7.2.9}$$

If the v_j have a limit v, then this vector-valued function v will therefore be a fixed point of the linear operator $\mathbf{T}$ defined by

$$(\mathbf{T}w)(x) \;=\; T_{d_1(x)} \; w(\tau x) \;;$$

$\mathbf{T}$ acts on all the vector-valued functions $w : \; [0,1] \longrightarrow \mathbb{R}^K$ that satisfy the requirements

$$[w(0)]_1 = 0, \;\; [w(1)]_K = 0, \;\; [w(0)]_k = [w(1)]_{k-1}, \;\; k = 2, \cdots, N \;. \tag{7.2.10}$$

(As a result of these conditions $\mathbf{T}w$ is defined unambiguously at the dyadic rationals: the two expansions lead to the same result.)

What has all this recasting the equations into different forms done for us? Well, it follows from (7.2.9) that

$$v_j(x) = T_{d_1(x)} T_{d_2(x)} \cdots T_{d_j(x)} v_0(\tau^j x) ,$$

which implies

$$v_j(x) - v_{j+\ell}(x) = T_{d_1(x)} \cdots T_{d_j(x)} [v_0(\tau^j x) - v_\ell(\tau^j x)] . \qquad (7.2.11)$$

In other words, information on the spectral properties of products of the T_d-matrices will help us to control the difference $v_j - v_{j+\ell}$, so that we can prove $v_j \longrightarrow v$, and derive smoothness for v. But let us turn to an example.

For the function ${}_2\phi$ (7.2.1) reads

$${}_2\phi(x) = \sum_{k=0}^{3} c_k \; {}_2\phi(2x-k) , \qquad (7.2.12)$$

with

$$c_0 = \frac{1+\sqrt{3}}{4}, \quad c_1 = \frac{3+\sqrt{3}}{4}, \quad c_2 = \frac{3-\sqrt{3}}{4}, \quad c_3 = \frac{1-\sqrt{3}}{4} .$$

Note that

$$c_0 + c_2 = c_1 + c_3 = 1 \qquad (7.2.13)$$

and

$$2c_2 = c_1 + 3c_3 , \qquad (7.2.14)$$

both of which are consequences of the divisibility of $m_0(\xi) = \frac{1}{2} \sum_{k=0}^{3} c_k e^{-ik\xi}$ by $(1+e^{-i\xi})^2$. The values ${}_2\phi(1)$, ${}_2\phi(2)$ are determined by the system

$$\begin{pmatrix} {}_2\phi(1) \\ {}_2\phi(2) \end{pmatrix} = M \begin{pmatrix} {}_2\phi(1) \\ {}_2\phi(2) \end{pmatrix} , \quad \text{with} \quad M = \begin{pmatrix} c_1 & c_0 \\ c_3 & c_2 \end{pmatrix} .$$

Because of (7.2.13), the columns of M all sum to 1, ensuring that $(1,\ 1)$ is a left eigenvector of M with eigenvalue 1. This eigenvalue is nondegenerate; the right eigenvector for the same eigenvalue is therefore not orthonormal to $(1,\ 1)$, which means it can be normalized so that the sum of its entries is 1. This choice of $({}_2\phi(1),\ {}_2\phi(2))$ leads to

$${}_2\phi(1) = \frac{1+\sqrt{3}}{2} , \quad {}_2\phi(2) = \frac{1-\sqrt{3}}{2} .$$

The matrices T_0, T_1 are 3×3 matrices given by

$$T_0 = \begin{pmatrix} c_0 & 0 & 0 \\ c_2 & c_1 & c_0 \\ 0 & c_3 & c_2 \end{pmatrix} , \quad T_1 = \begin{pmatrix} c_1 & c_0 & 0 \\ c_3 & c_2 & c_1 \\ 0 & 0 & c_3 \end{pmatrix} .$$

Because of (7.2.13), T_0 and T_1 have a common left eigenvector $e_1 = (1,\ 1,\ 1)$ with eigenvalue 1. Moreover, for all $x \in [0, 1]$,

$$\begin{aligned} e_1 \cdot v_0(x) &= e_1 \cdot [(1-x)\, v_0(0) \; + \; x\, v_0(1)] \\ &= (1-x)\, [\, {}_2\phi(1) + {}_2\phi(2)] \; + \; x\, [\, {}_2\phi(1) + {}_2\phi(2)] \\ &\qquad\qquad\qquad\qquad\qquad\qquad \text{(use (7.2.2))} \\ &= 1\,. \end{aligned}$$

It follows that, for all $x \in [0, 1]$, all $j \in \mathbb{N}$,

$$\begin{aligned} e_1 \cdot v_j(x) &= e_1 \cdot T_{d_1(x)} \cdots T_{d_j(x)}\, v_0(\tau^j x) \\ &= e_1 \cdot v_0(\tau^j x) \qquad\qquad \text{(because } e_1 T_d = e_1 \text{ for } d = 0,\, 1) \\ &= 1\,. \end{aligned}$$

Consequently, $v_0(y) - v_\ell(y) \in E_1 = \{w;\ e_1 \cdot w = w_1 + w_2 + w_3 = 0\}$, the space orthogonal to e_1. In view of (7.2.11), we therefore only need to study products of T_d-matrices restricted to E_1 in order to control the convergence of the v_j. But more is true! Define $e_2 = (1,\ 2,\ 3)$. Then (7.2.14) implies

$$\begin{aligned} e_2 T_0 &= \tfrac{1}{2} e_2 \; + \; \alpha_0 e_1\,, \\ e_2 T_1 &= \tfrac{1}{2} e_2 \; + \; \alpha_1 e_1\,, \end{aligned} \tag{7.2.15}$$

with $\alpha_0 = c_0 + 2c_2 - \frac{1}{2} = \frac{5-\sqrt{3}}{4}$, $\alpha_1 = c_1 + 2c_3 - \frac{1}{2} = \frac{3-\sqrt{3}}{4}$. If we define $e_2^0 = e_2 - 2\alpha_0\, e_1$, then (7.2.15) becomes

$$e_2^0\, T_0 = \tfrac{1}{2}\, e_2^0 \;\text{ and }\; e_2^0\, T_1 = \tfrac{1}{2}\, e_2^0 - \tfrac{1}{2}\, e_1, \;\text{ or }\; e_2^0\, T_d = \tfrac{1}{2}\, e_2^0 - \tfrac{1}{2}\, d e_1\,.$$

On the other hand,

$$e_2^0 \cdot v_0(x) = (1-x)\, e_2^0 \cdot v_0(0) \; + \; x\, e_2^0 \cdot v_0(1) = -x\,;$$

consequently,

$$\begin{aligned} e_2^0 \cdot v_j(x) &= e_2^0 \cdot T_{d_1(x)}\, v_{j-1}(\tau x) \\ &= -\frac{1}{2}\, d_1(x) \; + \; \frac{1}{2}\, e_2^0 \cdot v_{j-1}(\tau x) \\ &= -\sum_{m=1}^{j} 2^{-m}\, d_m(x) \; + \; 2^{-j}\, e_2^0 \cdot v_0(\tau^j x) \\ &= -\sum_{m=1}^{j} 2^{-m}\, d_m(x) \; - \; 2^{-j}\, \tau^j x = -x\,. \end{aligned}$$

It follows that $e_2^0 \cdot [v_0(x) - v_\ell(x)] = 0$. This means that we only need to study products of T_d-matrices restricted to E_2, the space spanned by e_1 and e_2^0, in order to control $v_j - v_{j+\ell}$. But, because this is a simple example, E_2 is one-dimensional, and $T_d|_{E_2}$ is simply multiplication by some constant, namely the third eigenvalue of T_d, which is $\frac{1+\sqrt{3}}{4}$ for T_0, $\frac{1-\sqrt{3}}{4}$ for T_1. Consequently,

$$\|v_j(x) - v_{j+\ell}(x)\| \le \left[\frac{1+\sqrt{3}}{4}\right]^j \left|\frac{1-\sqrt{3}}{1+\sqrt{3}}\right|^{\sum_{n=1}^j d_j(x)} C , \tag{7.2.16}$$

where we have used that the v_ℓ are uniformly bounded.[6] Since $\left|\frac{1-\sqrt{3}}{1+\sqrt{3}}\right| < 1$, (7.2.16) implies

$$\|v_j(x) - v_{j+\ell}(x)\| \le C\, 2^{-\alpha j} ,$$

with $\alpha = |\log((1+\sqrt{3})/4)|/\log 2 = .550$. It follows that the v_j have a limit function v, which is continuous since all the v_j are and since the convergence is uniform. Moreover v automatically satisfies (7.2.10), since all the v_j do, so that it can be "unfolded" into a continuous function F on $[0,3]$. This function solves (7.2.1), so that ${}_2\phi = F$, and it is uniformly approachable by piecewise linear spline functions F_j with nodes at the $k2^{-j}$,

$$\|\,{}_2\phi - F_j\|_{L^\infty} \le C\, 2^{-\alpha j} . \tag{7.2.17}$$

It follows from standard spline theory (see, e.g., Schumaker (1981))[7] that ${}_2\phi$ is Hölder continuous with exponent $\alpha = .550$. Note that this is better than the best result in §7.1 (we found $\alpha = .5 - \epsilon$ at the end of §7.1.3). This Hölder exponent is optimal: from (7.2.12) we have

$${}_2\phi(2^{-j}) = \left(\frac{1+\sqrt{3}}{4}\right) {}_2\phi(2^{-j+1}) = \cdots = \left(\frac{1+\sqrt{3}}{4}\right)^j {}_2\phi(1) = C\, 2^{-\alpha j} ,$$

hence

$$|\,{}_2\phi(2^{-j}) - {}_2\phi(0)| = C(2^{-j})^\alpha .$$

But this matrix method can do even better than determine the optimal Hölder exponent. Since $v(x) = T_{d_1(x)} v(\tau x)$, we have, for t small enough so that x and $x+t$ have the same first j digits in their binary expansion,

$$v(x) - v(x+t) = T_{d_1(x)} \cdots T_{d_j(x)} \left[v(\tau^j x) - v(\tau^j(x+t))\right] .$$

This can be studied in exactly the same way as $v_j(x) - v_{j+\ell}(x)$ above; we find

$$e_1 \cdot [v(x) - v(x+t)] = 0 ,$$

$$e_2^0 \cdot [v(x) - v(x+t)] = t .$$

For the remainder, only the $T_d|_{E_2}$ matter, and we find

$$\|v(x) - v(x+t)\| \le C|t| + C\, 2^{-\alpha j} \left|\frac{1-\sqrt{3}}{1+\sqrt{3}}\right|^{\sum_{n=1}^j d_n(x)} , \tag{7.2.18}$$

where t itself is of order 2^{-j}. With the notation $r_j(x) = \frac{1}{j}\sum_{n=1}^{j} d_n(x)$, (7.2.18) can be rewritten as

$$\|v(x) - v(x+t)\| \leq C|t| \; + \; C\, 2^{-(\alpha+\beta r_j(x))j} \;, \tag{7.2.19}$$

where $\beta = |\log|(1-\sqrt{3})/(1+\sqrt{3})||/\log 2$. Suppose $r_j(x)$ tends to a limit $r(x)$ for $j\to\infty$. If $r(x) < \frac{1-\alpha}{\beta} = .2368$, then the second term in (7.2.19) dominates the first, and v, hence ${}_2\phi$, is Hölder continuous with exponent $\alpha + \beta r(x)$. If $r(x) > \frac{1-\alpha}{\beta}$, then the first term, of order 2^{-j}, dominates, and ${}_2\phi$ is Lipschitz. In fact, one can even prove that ${}_2\phi$ is differentiable in these points, which constitute a set of full measure. This establishes a whole hierarchy of fractal sets (the sets on which $r(x)$ takes some preassigned value) on which ${}_2\phi$ has different Hölder exponents. And what happens at dyadic rationals? Well, there you can define $r_\pm(x)$, depending on whether you come "from above" (associated with $d^+(x)$) or "from below" ($d^-(x)$); $r_+(x) = 0$, $r_-(x) = 1$. As a consequence, ${}_2\phi$ is left differentiable at dyadic rationals, x, but has Hölder exponent .550 when x is approached from the right. This is illustrated by Figure 7.1, which shows blowups of ${}_2\phi$, exhibiting the characteristic lopsided peaks at even very fine scales.

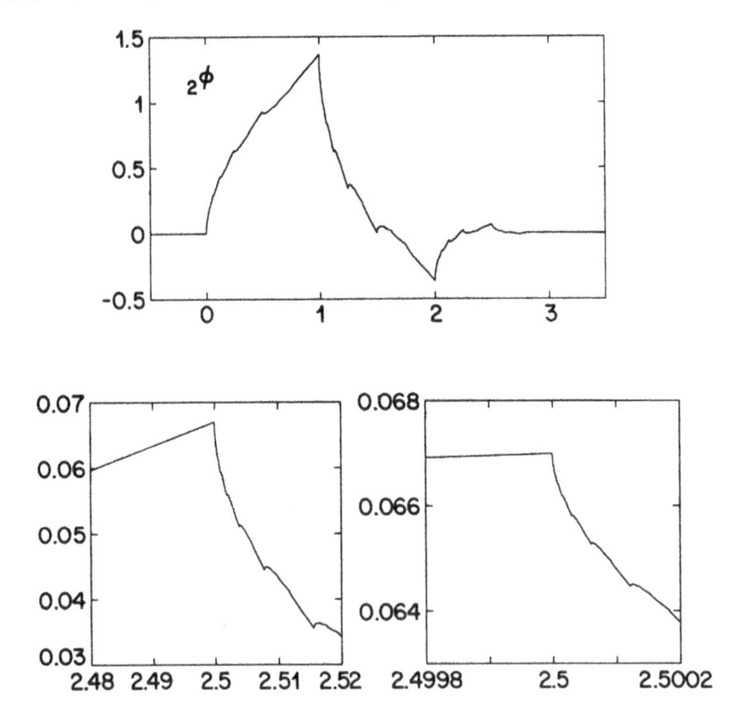

FIG. 7.1. *The function ${}_2\phi(x)$ and two successive blowups near $x = 2.5$.*

In this example, we had two "sum rules" (7.2.13), (7.2.14), reflecting that $m_0(\xi) = \frac{1}{2}\sum_k c_k\, e^{-ik\xi}$ was divisible by $((1+e^{-i\xi})/2)^2$. In general, m_0 is divisible by $((1+e^{-i\xi})/2)^N$, and we have N sum rules. The subspace E_N

will, however, be more than one-dimensional, which complicates estimates. The general theorem about global regularity is as follows.

THEOREM 7.2.1. *Assume that the* c_k, $k = 0, \cdots, K$, *satisfy* $\sum_{k=0}^{K} c_k = 2$ *and*

$$\sum_{k=0}^{K} (-1)^k \, k^\ell \, c_k = 0 \quad for\ \ell = 0,\ 1, \cdots, L\ . \tag{7.2.20}$$

For every $m = 1, \cdots, L+1$, *define* E_m *to be the subspace of* $\mathbb{R}^N$ *orthogonal to* $U_M = Span\ \{e_1, \cdots, e_m\}$, *where* $e_j = (1^{j-1},\ 2^{j-1}, \cdots, N^{j-1})$. *Assume that there exist* $1/2 \le \lambda < 1$, $0 \le \ell \le L\,(\ell \in \mathbb{N})$ *and* $C > 0$ *such that, for all binary sequences* $(d_j)_{j \in \mathbb{N}}$, *and all* $m \in \mathbb{N}$,

$$\|T_{d_1} \cdots T_{d_m}|_{E_{L+1}}\| \le C\ \lambda^m 2^{-m\ell}\ . \tag{7.2.21}$$

Then

1. *there exists a non-trivial continuous* L^1*-solution* F *for the two-scale equation* (7.2.1) *associated with the* c_n,
2. *this solution* F *is* ℓ *times continuously differentiable, and*
3. *if* $\lambda > \frac{1}{2}$, *then the* ℓ*th derivative* $F^{(\ell)}$ *of* f *is Hölder continuous, with exponent at least* $|\ln \lambda|/\ln 2$; *if* $\lambda = 1/2$, *then the* ℓ*th derivative* $F^{(\ell)}$ *of* F *is almost Lipschitz: it satisfies*

$$|F^{(\ell)}\ (x+t) - F^{(\ell)}\ (x)| \le C|t|\ |\ln|t||\ .$$

REMARK. The restriction $\lambda \ge \frac{1}{2}$ means only that we pick the largest possible integer $\ell \le L$ for which (7.2.21) holds with $\lambda < 1$. If $\ell = L$, then necessarily $\lambda \ge \frac{1}{2}$ (see Daubechies and Lagarias (1992)); if $\ell < L$ and $\lambda < \frac{1}{2}$, then we could replace ℓ by $\ell + 1$ and λ by 2λ, and (7.2.21) would hold for a larger integer ℓ. □

A similar general theorem can be formulated for the local regularity fluctuations exhibited by the example of ${}_2\phi$. For a precise statement, more details and proofs, I refer to Daubechies and Lagarias (1991, 1992).

When applied to the ${}_N\phi$, these methods lead to the following optimal Hölder exponents:

N	α
2	.5500
3	1.0878
4	1.6179

These are clearly better than what was obtained in §7.1.3; moreover, we see to our surprise that ${}_3\phi$ is continuously differentiable, even though its graph seems to have a "peak" at $x = 1$. Blowups show that this is deceptive: the true maximum lies a little to the right of $x = 1$, and everything is indeed smooth (see Figure 7.2). The derivative of ${}_3\phi$ is continuous, but has a very small Hölder exponent, as illustrated by Figure 7.3.

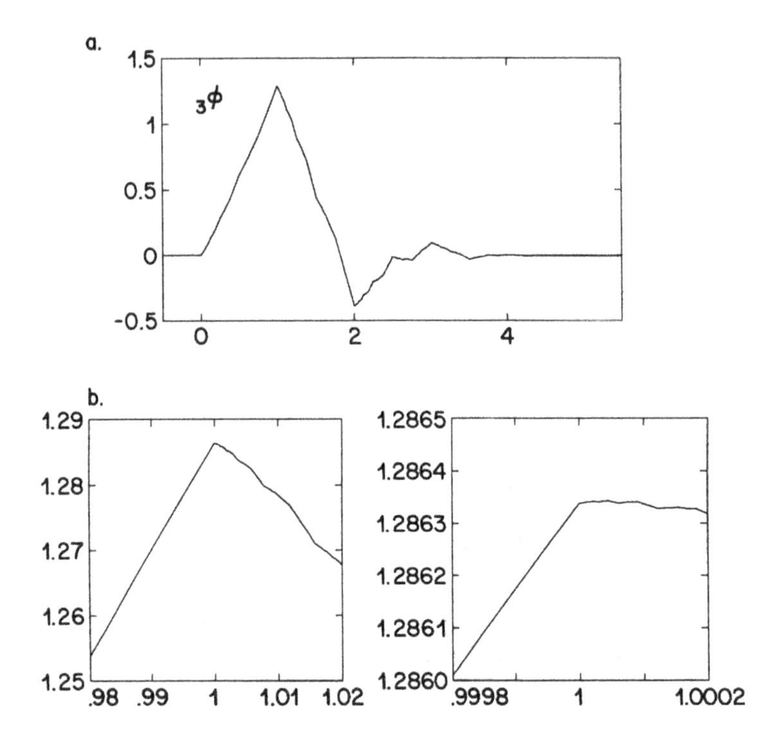

FIG. 7.2. *The function* $_3\phi(x)$ *and successive blowups around* $x = 1$.

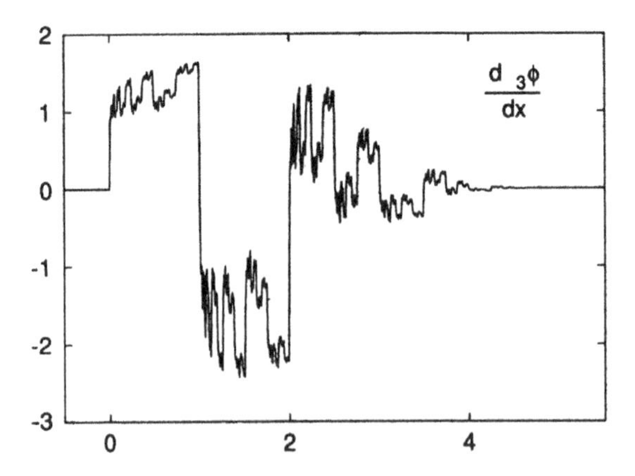

FIG. 7.3. *The derivative of* $_3\phi(x)$.

Unfortunately, these matrix methods are too cumbersome to treat large examples. Another, more recent "direct method" has been developed in Dyn and Levin (1990) and Rioul (1992); when applied to the ${}_N\phi$ with $N = 2, 3, 4$ it reproduces the α-values above; since it is computationally less heavy, it can also tackle larger values of N with better results than in §7.1.3 (see Rioul (1991)).

REMARKS.

1. Note the similarity of the matrices T_0, T_1 and P_0 in §7.1.3 (see (7.1.27))! Even the spectral analysis, with the nested invariant subspaces, is the same. This shows that the result in Theorem 7.1.12 is indeed optimal: if $\tilde{\lambda}$ is the spectral radius of $P_0|_{E_{2K}} = T_1|_{E_{2K}}$, then

$$\|(T_1|_{E_{2K}})^m\| \geq C(\tilde{\lambda} - \epsilon)^m$$

so that λ in (7.2.21) must be at least $\tilde{\lambda}2^\ell$, and the Hölder exponent is at most $\ell + |\log \lambda| / \log 2 \leq |\log \tilde{\lambda}| / \log 2$. The difference between the two approaches is that the present method also gives optimal estimates if $M_0(\xi)$ is not positive, unlike §7.1.3.

2. The condition (7.2.21) suggests that infinitely many conditions on the T_0, T_1 have to be checked before Theorem 7.2.1 can be applied. Fortunately, (7.2.21) can be reduced to equivalent conditions which can be checked in a finite-time computer search. For details, see Daubechies and Lagarias (1992).

3. In practice, it is not necessary to work with T_0, T_1 and restrict them to E_{2K}. One can also define directly the matrices $\tilde{T}_0$, $\tilde{T}_1$ corresponding to the coefficients of $m_0(\xi)/((1 + e^{-i\xi})/2)^K$; it turns out that bounds on $\|T_{d_1} \cdots T_{d_m}|_{E_{2K}}\|$ are equivalent to bounds on $\|\tilde{T}_{d_1} \cdots \tilde{T}_{d_m}\| \cdot 2^{-Lm}$ (see Daubechies and Lagarias (1992), §5). The matrices $\tilde{T}_d$ are much smaller than T_d ($(N - K) \times (N - K)$ instead of $N \times N$). □

Since this method works for any function satisfying an equation of the type (7.2.1), we can apply it to the basic functions in subdivision schemes. For the Lagrangian interpolation function corresponding to (6.5.14), a detailed analysis shows that F is "almost" C^2: it is C^1, and F' satisfies

$$|F'(x) - F'(x + t)| \leq C|t| \, |\log |t|| \, .$$

This had already been obtained previously by Dubuc (1986). But our matrix methods can do more! They can prove that F' is almost everywhere differentiable, and they can even compute F'' where it is well defined. For details, see again Daubechies and Lagarias (1992).

7.3. Compactly supported wavelets with more regularity.

By Corollary 5.5.2, an orthonormal basis of wavelets can consist of C^{N-1} wavelets only if the basic wavelet ψ has N vanishing moments. (We implicitly assume that ψ stems from a multiresolution analysis and that ϕ, ψ have sufficient decay; both conditions are trivially satisfied for the compactly supported wavelet bases as constructed in Chapter 6.) This was our motivation to construct the ${}_N\phi$, which lead to ${}_N\psi$ with N vanishing moments. The asymptotic results in §7.1.2 show however that the ${}_N\phi$, ${}_N\psi \in C^{\mu N}$ with $\mu \simeq .2$. This means that 80% of the zero moments are "wasted," i.e., the same regularity could be achieved with only $N/5$ vanishing moments.

Something similar happens for small values of N. For instance, ${}_2\phi$ is continuous but not C^1, ${}_3\phi$ is C^1 but not C^2, even though ${}_2\psi$, ${}_3\psi$ have, respectively, two and three vanishing moments. We can therefore "sacrifice" in each of these two cases one of the vanishing moments and use the additional degree of freedom to obtain ϕ with a better Hölder exponent than ${}_2\phi$ or ${}_3\phi$ have, with the same support width. This amounts to replacing $|m_0(\xi)|^2 = (\cos^2 \frac{\xi}{2})^N P_N(\sin^2 \frac{\xi}{2})$ by $|m_0(\xi)|^2 = (\cos^2 \frac{\xi}{2})^{N-1}[P_{N-1}(\sin^2 \frac{\xi}{2}) + a(\sin^2 \frac{\xi}{2})^N \cos\xi]$ (see (6.1.11)), and to choose a so that the regularity of ϕ is improved. Examples for $N = 2$, 3 are shown in Figures 7.4 and 7.5; the corresponding h_n are as follows:

$$
\begin{aligned}
N = 2 \qquad h_0 &= \frac{3}{5\sqrt{2}} \\
h_1 &= \frac{6}{5\sqrt{2}} \\
h_2 &= \frac{2}{5\sqrt{2}} \\
h_3 &= \frac{-1}{5\sqrt{2}}
\end{aligned}
$$

$$
\begin{aligned}
N = 3 \qquad h_0 &= .37432841633/\sqrt{2} \\
h_1 &= .109093396059/\sqrt{2} \\
h_2 &= .786941229301/\sqrt{2} \\
h_3 &= -.146269859213/\sqrt{2} \\
h_4 &= -.161269645631/\sqrt{2} \\
h_5 &= .0553358986263/\sqrt{2}
\end{aligned}
$$

These examples correspond to a choice of a such that $\max[\rho(T_0|_{E_\ell}), \rho(T_1|_{E_\ell})]$ is minimized; the eigenvalues of T_0, T_1 are then degenerate.[8] One can prove that the Hölder exponents of these two functions are at least .5864, 1.40198 respectively, and at most .60017, 1.4176; these last values are probably the true Hölder exponents. For more details, see Daubechies (1993).

7.4. Regularity or vanishing moments?

The examples in the previous section show that for fixed support width of ϕ, ψ, or equivalently, for fixed length of the filters in the associated subband coding scheme, the choice of the h_n that leads to maximum regularity is different from the choice with maximum number N of vanishing moments for ψ. The question

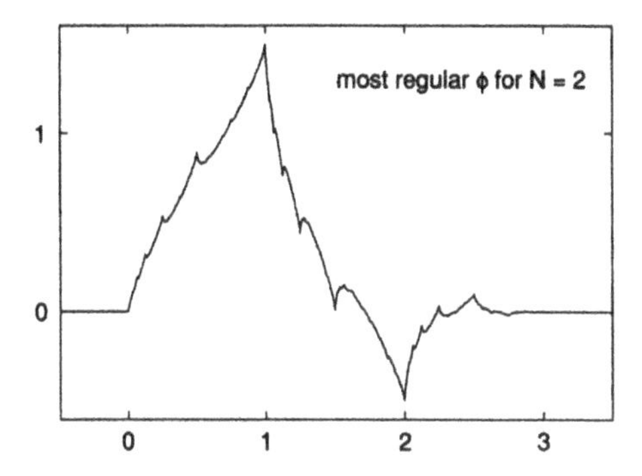

FIG. 7.4. *The scaling function ϕ for the most regular wavelet construction with support width* 3.

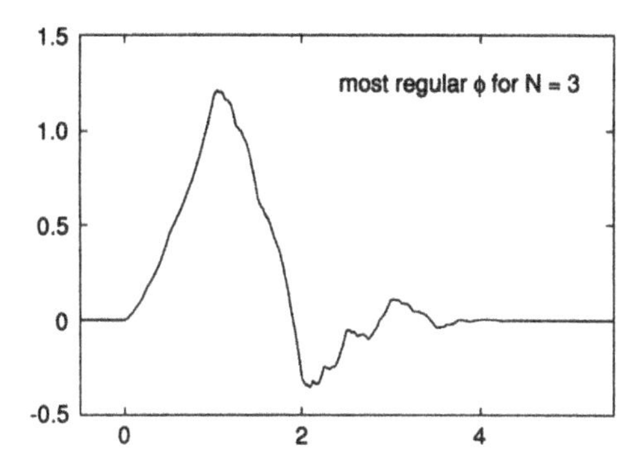

FIG. 7.5. *The scaling function ϕ for the most regular wavelet construction with support width* 5.

then arises: what is more important, vanishing moments or regularity? The answer depends on the application, and is not always clear. Beylkin, Coifman, and Rokhlin (1991) use compactly supported orthonormal wavelets to compress large matrices, i.e., to reduce them to a sparse form. For the details of this application, the reader should consult the original paper, or the chapter by Beylkin in Ruskai et al. (1991); one of the things that make their method work is the number of vanishing moments. Suppose you want to decompose a function $F(x)$ into wavelets (strictly speaking, matrices should be modelled by a function of two variables, but the point is illustrated just as well, and in a simpler way, with one variable). You compute all the wavelet coefficients $\langle F,\ \psi_{j,k}\rangle$, and to compress all that information, you throw away all the coefficients smaller than some threshold ϵ. Let us see what this means at some fine scale; $j = -J$, $J \in \mathbb{N}$ and J "large." If F is C^{L-1} and ψ has L vanishing moments, then, for x near

$2^{-J}k$, we have

$$\begin{aligned} F(x) &= F(2^{-J}k) + F'(2^{-J}k)(x-2^{-J}k) \\ &\quad + \cdots + \frac{1}{(L-1)!} F^{(L-1)}(2^{-J}k)(x-2^{-J}k)^{L-1} + (x-2^{-J}k)^L R(x), \end{aligned}$$

where R is bounded. If we multiply this by $\psi(2^J x - k)$ and integrate, then the first L terms will not contribute because $\int dx\ x^\ell \psi(x) = 0,\ \ell = 0, \cdots, L-1$. Consequently,

$$\begin{aligned} |\langle F,\ \psi_{-J,k}\rangle| &= \left| \int dx\ (x-2^{-J}k)^L\ R(x)\ 2^{J/2}\ \psi(2^J x - k) \right| \\ &\le C\ 2^{-J(L-1/2)} \int dy\ |y|^L\ |\psi(y)|\ . \end{aligned}$$

For J large, this will be negligibly small, unless R is very large near $k2^{-J}$. After thresholding, we will therefore only retain fine-scale wavelet coefficients near singularities of F or its derivatives. The effect will be all the more pronounced if the number L of vanishing moments of ψ is large.[9] Note that the regularity of ψ does not play a role at all in this argument; it seems that for Beylkin, Coifman, and Rokhlin-type applications the number of vanishing moments is far more important than the regularity of ψ.

For other applications, regularity may be more relevant. Suppose you want to compress the information in an image. Again, you decompose into wavelets (two-dimensional wavelets, e.g., associated with a tensor product multiresolution analysis), and you throw away all the small coefficients. (This is a rather primitive procedure. In practice, one chooses to allocate more precision to some coefficients than to others, by means of a quantization rule.) You end up with a representation of the type

$$\tilde{I} = \sum_{j,k \in S} \langle I,\ \psi_{j,k}\rangle\ \psi_{j,k}\ ,$$

where S is only a (small) subset of all the possible values, chosen in function of I. The mistakes you have made will consist of multiples of the deleted $\psi_{j,k}$. If these are very wild objects, then the difference between I and $\tilde{I}$ might well be much more perceptible than if ψ is smoother. This is admittedly very much a hand-waving argument, but it suggests that at least some regularity might be required. Some first experiments reported in Antonini et al. (1992) seem to confirm this, but more experiments are required for a convincing answer.

The sum rules (7.2.20), equivalent to the divisibility of $m_0(\xi)$ by $(1+e^{-i\xi})^{L+1}$, have another interesting consequence. In the example studied in detail, ${}_2\phi$, we saw that (7.2.13) and (7.2.14) implied that

$$e_1 \cdot v(x) = 1, \qquad e_2^0 \cdot v(x) = -x$$

(we proved both for v_j, so that they also hold for $v = \lim_{j\to\infty} v_j$), or, in terms of ϕ rather than v,

$$\phi(x) + \phi(x+1) + \phi(x+2) = 1\ ,$$

$$(1 - 2\alpha_0)\ \phi(x) + (2 - 2\alpha_0)\ \phi(x+1) + (3 - 2\alpha_0)\ \phi(x+2) = -x$$

for all $x \in [0, 1]$. Because support$(\phi) = [0, 3]$, one easily checks that this implies, for all $y \in \mathbb{R}$,

$$\sum_{n \in \mathbb{Z}} \phi(y+n) = 1\ ,$$

$$\sum_{n \in \mathbb{Z}} (n + 1 - 2\alpha_0)\ \phi(y+n) = -y\ .$$

All polynomials of degree less than or equal to 1 can therefore be written as linear combinations of the $\phi(x - n)$. Something similar happens in general: the conditions (7.2.20) ensure that all the polynomials of degree less than or equal to L can be generated by linear combinations of the $\phi(x - n)$. (See Fix and Strang (1969), Cavaretta, Dahmen, and Micchelli (1991).) This can again be used to explain why the condition $\frac{d^\ell}{d\xi^\ell} m_0|_{\xi=\pi} = 0,\ \ell = 0, \cdots, L$ is useful in subband filtering schemes. Ideally, one wants the low frequency channel, after the filtering, to contain all the slowly changing features, and to find only true "high frequency" features in the other channel. Polynomials of low degree are essentially slowly changing features, and the sum rules (7.2.20) ensure that they (or their restrictions to a large interval, to keep it all in $L^2(\mathbb{R})$; we disregard border effects here) are in every V_J, i.e., they are given completely by the low frequency channel.

In the design of FIR filters for subband coding, it is not customary to pay much attention to the number of vanishing moments of m_0, which is reflected by the "flatness" of the filter at $\xi = \pi$.[1]0 What follows is yet another argument showing that in implementations where filters are cascaded, it is nevertheless important to have at least some zero moments. Suppose that we apply three successive low pass filtering + decimation steps to a signal. If we call the original signal f^0, with Fourier transform $\hat{f}^0(\xi) = \sum_n f_n^0\ e^{-in\xi}$, then the result of one filtering + decimation is the sequence f_n^1, where $\hat{f}^1(\xi) = \sum_n f_n^1\ e^{-in\xi}$ satisfies

$$\hat{f}^1(\xi) = \frac{1}{\sqrt{2}} \left[\hat{f}^0\left(\frac{\xi}{2}\right) m_0\left(\frac{\xi}{2}\right) + \hat{f}^0\left(\frac{\xi}{2} + \pi\right) m_0\left(\frac{\xi}{2} + \pi\right) \right]. \qquad (7.4.1)$$

The second term can be viewed as the result of aliasing, due to the lower sampling rate in the f^1. Similarly, three such operations lead to

$$\hat{f}^3(\xi) = 2^{-3/2} \left[\hat{f}^0\left(\frac{\xi}{8}\right) m_0\left(\frac{\xi}{8}\right) m_0\left(\frac{\xi}{4}\right) m_0\left(\frac{\xi}{2}\right) + 7 \text{ "folding" terms} \right]. \qquad (7.4.2)$$

It follows that the product $m_0(\xi)\ m_0(2\xi)\ m_0(4\xi)$ plays an important role. Figure 7.6 shows what this product looks like for the ideal low pass filter, $m_0(\xi) = 1$ for $|\xi| \le \pi/2$, 0 for $\pi/2 \le |\xi| \le \pi$. If the low pass filter is not ideal, then it will "leak" a little into the high-pass region $\pi/2 \le |\xi| \le \pi$. It is then important

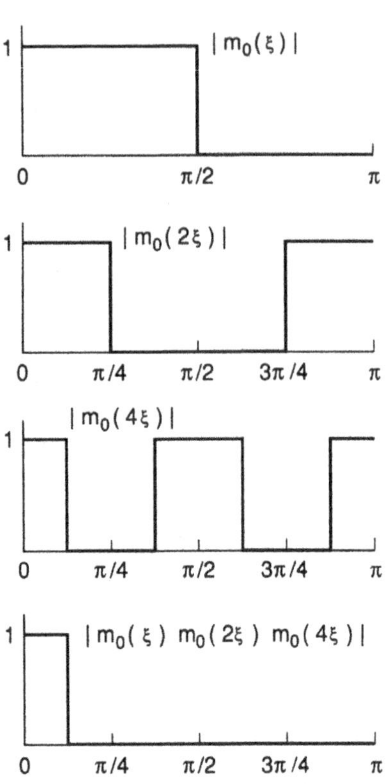

FIG. 7.6. *Plots of $m_0(\xi)$, $m_0(2\pi)$, $m_0(4\xi)$ and of their product for the ideal low-pass filter.*

to contain this leakage, especially when the filters are cascaded: it contributes to the "folding" terms in (7.4.1) and can lead to audible or visible aliasing once quantization is introduced and perfect reconstruction is no longer attained. In the ideal case in Figure 7.6, the "bump" of $m_0(2\xi)$ for $\xi \in [3\pi/4,\ \pi]$ is killed in the product $m_0(\xi)\ m_0(2\xi)\ m_0(4\xi)$ because $m_0(\xi) = 0$ in this interval. The same happens for extra "bumps" of $m_0(4\xi)$, resulting in $m_0(\xi)\ m_0(2\xi)\ m_0(4\xi) = 1$ if $\xi \in [0, \pi/8[, = 0$ if $\xi \in]\pi/8, \pi]$. In the non-ideal case, a similar effect can be achieved by imposing that m_0 have a zero of reasonable multiplicity at $\xi = \pi$, which "kills" the maximum of $m_0(2\xi)$ in a concatenation. This phenomenon is illustrated in Figure 7.7, where a wavelet filter is compared with a non-wavelet perfect reconstruction filter. In Figure 7.7a we see plots of $|m_0(\xi)|$ for two orthonormal perfect reconstruction filters (i.e., $|m_0(\xi)|^2 + |m_0(\xi+\pi)|^2 = 1$), each with eight taps; the filter on the left corresponds to the example constructed in §6.4, with two vanishing moments (i.e., m_0 has a zero with double multiplicity at $\xi = \pi$), and an extra zero at $\xi = 7\pi/9$. The filter on the right is not a wavelet filter, since $m_0(\pi) \neq 0$ and hence $m_0(0) \neq 1$; it is constructed more according to standard wisdom, with an "equi-ripple" design: in this case the location of the nodes is chosen so that the amplitude of the two ripples is the same as in the one ripple in the wavelet filter on the left, while keeping the transition band as narrow as possible within this constraint. The resulting filter is a little steeper than the wavelet filter (its first zero is at $\xi = .76\pi$ rather than $.78\pi$ for the wavelet example), and seems therefore closer to the ideal filter. (Of course, both are quite far from the ideal case, but remember that we have used only eight taps!) Figure 7.7b plots $| m_0(\xi)\ m_0(2\xi)\ m_0(4\xi)|$ for these two examples, and Figure 7.7c blows up these plots in the region $\pi/2 \leq \xi \leq \pi$. It is clear that in the second (non-wavelet) case the leakage into this high frequency region is more important than in the wavelet case; this is true in L^2-sense as well as in amplitude (the highest peak at right is about 3 dB higher than at left). This effect can become even more pronounced when larger filters are considered.[11]

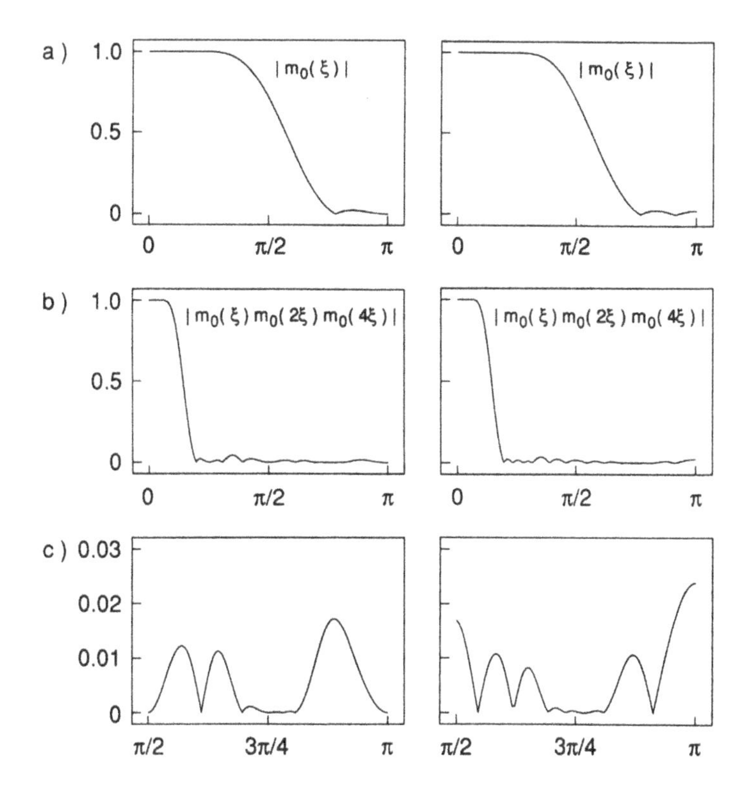

FIG. 7.7. *Comparison of three concatenations of two 8-tap low-pass filters with the perfect reconstruction property:* (a) *plots of* $|m_0(\xi)|$, (b) *plots of* $|m_0(\xi)\ m_0(2\xi)\ m_0(4\xi)|$, (c) *blowups of b for* $\pi/2 \leq \xi \leq \pi$.

Notes.

1. Incidentally, this proves that the statement in Remark 3 on p. 983 of Daubechies (1988b) is *wrong*; I made a mistake in the extraction of a numerical value for μ from Meyer's proof.

2. This is essentially what was done in the Appendix of Daubechies (1988b). Beware of the typos in that Appendix, however!

3. If the restriction that F is a compactly supported L^1-function is removed, then many other solutions are possible. On the other hand, if we insist on compact support and $F \in L^1$, then necessarily $\sum_{k=0}^{K} c_k = 2^{m+1}$ for some $m \in \mathbb{N}$, and F is the mth derivative of the compactly supported L^1-solution to the equation obtained after replacing the c_k by $2^{-m}c_k$; no generality is lost by the restriction $\sum c_k = 2$. For proofs, see Daubechies and Lagarias (1991).

4. In all the examples we will consider, it is not really necessary to choose the particular F_0 constructed later: the algorithm would work with any F_0 with integral one.

5. We implicitly assume that the c_k are real. Everything still carries through for complex c_k, but then $v(x) \in \mathbb{C}^K$.

6. Equipping E_1 in this special case with the norm $|||(a, -a-b, b)|||^2 = a^2 + b^2$ (equivalent with the standard Euclidian norm on E_1), one finds $\sup_{u \in E_1} |||T_0 u|||/|||u||| \simeq .728, \sup_{u \in E_1} |||T_1 u|||/|||u||| \simeq .859$, so that $|||v_{j+1}(x) - v_j(x)||| \leq A^j |||v_1(x) - v_0(x)|||$, with $A = .859$, by (7.2.11). It immediately follows that

$$\begin{aligned} \|v_j(x)\| &\leq \|v_0(x)\| + C \sum_{k=1}^{j} |||v_k(x) - v_{k-1}(x)||| \\ &\leq \|v_0(x)\| + (1-A)^{-1}\, C|||v_1(x) - v_0(x)|||\ , \end{aligned}$$

which is uniformly bounded in x and j.

7. The following argument also gives a direct proof. Suppose that $2^{-(j+1)} \leq y - x \leq 2^{-j}$. Then there exists $\ell \in \mathbb{N}$ so that one of the two following alternatives holds: $(\ell-1)2^{-j} \leq x \leq y \leq \ell 2^{-j}$ or $(\ell-1)2^{-j} \leq x \leq \ell 2^{-j} \leq y \leq (\ell+1)2^{-j}$. We will only discuss the second case; the first is similar. We then have

$$\begin{aligned} |f(x) - f(y)| &\leq |f(x) - f_j(x)| + |f_j(x) - f_j(\ell 2^{-j})| \\ &\quad + |f_j(\ell 2^{-j}) - f_j(y)| + |f_j(y) - f(y)| \\ &\leq 2C\, 2^{-\alpha j} + |f_j(x) - f_j(\ell 2^{-j})| + |f_j(y) - f_j(\ell 2^{-j})|\ , \end{aligned}$$

by (7.2.17). Because of the choice of ℓ, there exists $k \in \mathbb{N}$ so that $x' = x - k$ and $\ell' 2^{-j} = \ell 2^{-j} - k$ are both in $[0,1]$. We can moreover choose binary expansions for x' and $\ell' 2^{-j}$ with coinciding first j digits (choose the expansion ending in ones for $\ell' 2^{-j}$, and if x' is dyadic, the expansion ending in zeros for x'). It follows that

$$\begin{aligned} |f_j(x) - f_j(\ell 2^{-j})| &\le \|v_j(x') - v_j(\ell' 2^{-j})\| \\ &= \|T_{d_1(x')} \cdots T_{d_j(x')} \left[v_0(\tau^j x') - v_0(\tau^j(\ell' 2^{-j}))\right] \| \\ &\le C\, 2^{-\alpha j} , \end{aligned}$$

where we have used $\|T_{d_1} \cdots T_{d_m}|_{E_2}\| \le C\, 2^{-\alpha j}$, the boundedness of v_0, and $v_0(u) - v_0(u') \in E_1$ for all u, u'. We can similarly bound $|f_j(y) - f_j(\ell 2^{-j})|$; putting it all together leads to

$$|f(x) - f(y)| \le C'\, 2^{-\alpha j} \le C''|x - y|^\alpha ,$$

which proves the Hölder continuity with exponent α.

8. This corrects a mistake in the first printing, where a too large value was given for the Hölder exponent for $N = 2$. I thank L. Villemoes and C. Heil for pointing out this mistake to me. Incidentally, the $N = 2$ example is an instance where the best possible λ in (7.2.21) is strictly larger than $\max[\rho(T_0|_{E_1}), \rho(T_1|_{E_1})]$. In this case $\rho(T_0|_{E_1}) = \rho(T_1|_{E_1}) = \frac{3}{5}$, and $\frac{5}{3}[\rho(T_0 T_1^{12})]^{1/13} \simeq 1.09946\ldots > 1$.

9. Of course Beylkin, Coifman, and Rokhlin (1991) contains much more than this! For a large class of matrices it turns out that after an orthonormal basis transform using wavelets, dense $N \times N$ matrices reduce to sparse structures with only $O(N)$ entries larger than the threshold ϵ. The total L^2-error made by throwing away all the entries below ϵ turns out to be $O(\epsilon)$, which is a much deeper result than the "compression" explained here; it is essentially the $T(1)$ theorem of David and Journé, the proof of which uses "hard" analysis.

10. The argument below also holds for the biorthogonal case (see Chapter 8), where the flatness of $|m_0|$ at $\xi = 0$ and at $\xi = \pi$ need not be the same; it is the multiplicity of the zero at $\xi = \pi$ that counts.

11. In Cohen and Johnston (1992), filters are constructed which optimize criteria that are a mixture of "standard" wisdom and wavelet desiderata.

CHAPTER 8

Symmetry for Compactly Supported Wavelet Bases

All the examples we have seen so far of compactly supported orthonormal wavelet bases are conspicuously non-symmetric, in contrast to the infinitely supported wavelet bases we saw before, such as the Meyer and Battle–Lemarié bases. In this chapter we discuss why this asymmetry occurs, what can be done about it, and whether anything should be done about it.

8.1. Absence of symmetry for compactly supported orthonormal wavelets.

In Chapter 5 we already saw that a multiresolution analysis does not determine ϕ, ψ uniquely. This is again borne out by the following lemma.

LEMMA 8.1.1. *If $f_n(x) = f(x-n)$ and $g_n(x) = g(x-n)$, $n \in \mathbb{Z}$, constitute orthonormal bases of the same subspace E of $L^2(\mathbb{R})$, then there exists a 2π-periodic function $\alpha(\xi)$, with $|\alpha(\xi)| = 1$, so that $\hat{g}(\xi) = \alpha(\xi)\hat{f}(\xi)$.*

Proof.

1. Since the f_n are an orthonormal basis for $E \ni g$, $g = \sum_n \alpha_n f_n$, with $\sum_n |\alpha_n|^2 = \|g\|^2 = 1$. Consequently, $\hat{g}(\xi) = \alpha(\xi)\ \hat{f}(\xi)$, with $\alpha(\xi) = \sum_n \alpha_n e^{-in\xi}$.

2. As shown in Chapter 5, orthonormality of the $f(\cdot - n)$ is equivalent with $\sum_m |\hat{f}(\xi - 2\pi m)|^2 = (2\pi)^{-1}$ a.e. Similarly $\sum_m |\hat{g}(\xi - 2\pi m)|^2 = (2\pi)^{-1}$. It follows that $|\alpha(\xi)| = 1$. ∎

However, we also have the following lemma.

LEMMA 8.1.2. *If $(\alpha_n)_{n\in\mathbb{Z}}$ is a finite sequence (all but finitely many α_n equal 0), and if $|\alpha(\xi)| = 1$, then $\alpha_n = \alpha\delta_{n,n_0}$ for some $n_0 \in \mathbb{Z}$.*

Proof.

1. Since $|\alpha(\xi)|^2 = 1$, $\sum_n \alpha_n\ \overline{\alpha_{n+\ell}} = \delta_{\ell,0}$. (8.1.1)

2. Define n_1, n_2 so that $\alpha_{n_1} \neq 0 \neq \alpha_{n_2}$, and $\alpha_n = 0$ if $n < n_1$ or $n > n_2$.

3. By (8.1.1), $\sum_n \alpha_n \overline{\alpha_{n+n_2-n_1}} = \delta_{n_2-n_1,0}$. But, by the definition of n_1, n_2, the sum consists of the single term $\alpha_{n_1}\overline{\alpha_{n_2}}$, which is nonzero by definition. Hence $n_1 = n_2$. ■

Together, these two lemmas imply that *compactly supported* ϕ, ψ are unique, for a given multiresolution analysis, up to a shift.

COROLLARY 8.1.3. *If f, g are both compactly supported, and the $f_n = f(\cdot - n)$, $g_n = g(\cdot - n)$, $n \in \mathbb{Z}$ are both orthonormal bases for the same space E, then $g(x) = \alpha f(x - n_0)$ for some $\alpha \in \mathbb{C}$, $|\alpha| = 1$ and $n_0 \in \mathbb{Z}$.*

Proof. By Lemma 8.1.1, $\hat{g}(\xi) = \alpha(\xi)\hat{f}(\xi)$, with $\alpha_n = \int dx\ g(x)\overline{f(x-n)}$. Because f, g have compact support, only finitely many $\alpha_n \neq 0$. Consequently, by Lemma 8.1.2, $\alpha(\xi) = \alpha e^{-in_0\xi}$, hence $g(x) = \alpha f(x - n_0)$. ■

In particular, if ϕ_1, ϕ_2 are both compactly supported, and "orthonormalized"[1] scaling functions for the same multiresolution analysis, then ϕ_2 is a shifted version of ϕ_1: the constant α is necessarily 1, because by convention $\int dx\ \phi_2(x) = 1 = \int dx\ \phi_1(x)$ (see Chapter 5). This uniqueness result can be used to prove that, except for the Haar basis, *all* real orthonormal wavelet bases with compact support are asymmetric.

THEOREM 8.1.4. *Suppose that ϕ and ψ, the scaling function and wavelet associated with a multiresolution analysis, are both real and compactly supported. If ψ has either a symmetry or an antisymmetry axis, then ψ is the Haar function.*

Proof.

1. We can always shift ϕ so that $h_n = \int dx\ \phi(x)\ \phi(x-n) = 0$ for $n < 0$, $h_0 \neq 0$. Since ϕ is real, so are the h_n. Let N be the largest index for which h_n does not vanish: $h_N \neq 0$, $h_n = 0$ for $n > N$. Then N is odd, because N even, $N = 2n_0$ together with

$$\sum_n h_n\ h_{n+2\ell} = \delta_{\ell,0}\ ,$$

would lead to a contradiction if $\ell = n_0$.

2. Since $h_n = 0$ for $n < 0$, $n > N$, support $\phi = [0, N]$, by Lemma 6.2.2.[2] The standard definition (5.1.34) then leads to support $\psi = [-n_0, n_0 + 1]$, where $n_0 = \frac{N-1}{2}$. The symmetry axis is therefore necessarily at $\frac{1}{2}$; we have either $\psi(1-x) = \psi(x)$ or $\psi(1-x) = -\psi(x)$.

3. Consequently,

$$\begin{aligned}\psi_{j,k}(-x) &= \pm 2^{-j/2}\ \psi(2^{-j}x + k + 1) \\ &= \pm \psi_{j,-(k+1)}(x)\ ,\end{aligned}$$

which means that the W_j-spaces are invariant under the map $x \mapsto -x$. Since $V_j = \overline{\bigoplus_{k>j} W_k}$, V_j is invariant as well.

4. Define now $\tilde{\phi}(x) = \phi(N - x)$. Then the $\tilde{\phi}(\cdot - n)$ generate an orthonormal basis of V_0 (since V_0 is invariant for $x \mapsto -x$), $\int dx\, \tilde{\phi}(x) = \int dx\, \phi(x) = 1$, and support $\tilde{\phi}$ = support ϕ. It follows from Corollary 8.1.3 that $\tilde{\phi} = \phi$, i.e., $\phi(N - x) = \phi(x)$. Consequently,

$$\begin{aligned} h_n &= \sqrt{2} \int dx\, \phi(x)\, \phi(2x - n) \\ &= \sqrt{2} \int dx\, \phi(N - x)\, \phi(N - 2x + n) \\ &= \sqrt{2} \int dy\, \phi(y)\, \phi(2y - N + n) = h_{N-n}\,. \end{aligned} \tag{8.1.2}$$

5. On the other hand,

$$\begin{aligned} \delta_{\ell,0} &= \sum_n h_n\, h_{n+2\ell} \\ &= \sum_m h_{2m}\, h_{2m+2\ell} + \sum_m h_{2m+1}\, h_{2m+2\ell+1} \\ &= \sum_m h_{2m}\, h_{2m+2\ell} + \sum_m h_{2n_0-2m} h_{2n_0-2m-2\ell} \\ &\qquad\qquad \text{(use (8.1.2) on the second term)} \\ &= 2 \sum_m h_{2m}\, h_{2m+2\ell}\,. \end{aligned}$$

By Lemma 8.1.2, this implies $h_{2m} = \delta_{m,m_0}\alpha$ for some $m_0 \in \mathbb{Z}$, $|\alpha| = 2^{-1/2}$. Since we assumed $h_0 \neq 0$, this means that $h_{2m} = \delta_{m,0}\, \alpha$. By (8.1.2), $h_N = h_0 = \alpha$ as well, and $h_{2m+1} = \alpha\, \delta_{m,n_0}$ in general. The normalization $\Sigma h_n = \sqrt{2}$ (see Chapter 5) fixes the value of α, $\alpha = \frac{1}{\sqrt{2}}$.

6. We have thus $h_{2m} = \frac{1}{\sqrt{2}}\, \delta_{m,0}$, $h_{2m+1} = \frac{1}{\sqrt{2}}\, \delta_{m,n_0}$, or $m_0(\xi) = \frac{1}{2}(1 + e^{-iN\xi})$. It follows that $\hat{\phi}(\xi) = (2\pi)^{-1/2}\, ((1 - e^{-iN\xi})/iN\xi)$, or $\phi(x) = N^{-1}$ for $0 \le x \le N$, $\phi(x) = 0$ otherwise. If $N = 1$, then this gives exactly the Haar basis; if $N > 1$, then the $\phi(\cdot - n)$ are not orthonormal, which contradicts the assumptions in the theorem. ∎

REMARKS.

1. The nonexistence of symmetric or antisymmetric real compactly supported wavelets should be no surprise to anybody familiar with subband coding: it had already been noted by Smith and Barnwell (1986) that symmetry is not compatible with the the exact reconstruction property in subband filtering. The only extra result of Theorem 8.1.4 is that symmetry for ψ necessarily implies symmetry for the h_n, but that is a rather intuitively true result anyway.

2. If the restriction that ϕ be real is lifted, then symmetry is possible, even if ϕ is compactly supported (Lawton, private communication, 1990). □

The asymmetry of all the examples plotted in §6.4 is therefore unavoidable. But why should we care? Symmetry is nice, but can't we do without? For some applications it does not really matter at all. The numerical analysis applications in Beylkin, Coifman, and Rokhlin (1991), for instance, work very well with very asymmetric wavelets. For other applications, the asymmetry can be a nuisance. In image coding, for example, quantization errors will often be most prominent around edges in the images; it is a property of our visual system that we are more tolerant of symmetric errors than asymmetric ones. In other words, less asymmetry would result in greater compressibility for the same perceptual error.[3] Moreover, symmetric filters make it easier to deal with the boundaries of the image (see also Chapter 10), another reason why the subband coding engineering literature often sticks to symmetry. The following subsections discuss what we can do to make orthonormal wavelets less asymmetric, or how we can recover symmetry if we give up orthonormality.

8.1.1. Closer to linear phase. Symmetric filters are often called *linear phase* filters by engineers; if a filter is not symmetric, then its deviation from symmetry is judged by how much its phase deviates from a linear function. More precisely, a filter with filter coefficients a_n is called *linear phase* if the phase of the function $a(\xi) = \sum_n a_n e^{-in\xi}$ is a linear function of ξ, i.e., if, for some $\ell \in \mathbb{Z}$,

$$a(\xi) = e^{-i\ell\xi} \, |a(\xi)| \, .$$

This means that the a_n are symmetric around ℓ, $a_n = a_{2\ell-n}$. Note that according to this definition, the Haar filter $m_0(\xi) = (1 + e^{-i\xi})/2$ is not linear phase, although the filter coefficients are clearly symmetric. This is because the h_n^{Haar} are symmetric around $\frac{1}{2} \notin \mathbb{Z}$; in this case

$$m_0(\xi) = \begin{cases} e^{-i\xi/2} \, |m_0(\xi)| & \text{if } \; 0 \le \xi \le \pi \, , \\ -e^{-i\xi/2} \, |m_0(\xi)| & \text{if } \; \pi \le \xi \le 2\pi \, . \end{cases}$$

The phase has a discontinuity at π, where $|m_0| = 0$. If we extend the definition of linear phase to include also the filters for which the phase of $a(\xi)$ is piecewise linear, with constant slope, and has discontinuities only where $|a(\xi)|$ is zero, then filters with the same symmetry as the Haar filter are also included. To make a filter "close" to symmetric, the idea is then to juggle with its phase so that it is "almost" linear. Let us apply this to the "standard" construction of the ${}_N\phi$, ${}_N\psi$, as given in §6.4. In that case we have

$$|\,{}_N m_0(\xi)|^2 = (\cos\,\xi/2)^{2N} \, P_N(\sin^2\,\xi/2) \, ,$$

and the coefficients ${}_N h_n$ were determined by taking the "square root" of P_N via spectral factorization. Typically this means writing the polynomial $L(z)$, defined by $L(e^{i\xi}) = P_N\,(\sin^2 \xi/2)$, as a product of $(z - z_\ell)(z - \overline{z}_\ell)(z - z_\ell^{-1})(z - \overline{z}_\ell^{-1})$ or $(z - r_\ell)(z - r_\ell^{-1})$, where z_ℓ, r_ℓ are the complex, respectively, real roots of L, and selecting one pair $\{z_\ell, \, \overline{z}_\ell\}$ out of each quadruple of complex roots, and one value r_ℓ out of each pair of real roots. Up to normalization, the resulting m_0 is then

$$ {}_N m_0(\xi) = \left(\frac{1+e^{-i\xi}}{2}\right)^N \prod_\ell (e^{-i\xi} - z_\ell)(e^{-i\xi} - \overline{z}_\ell) \prod_k (e^{-i\xi} - r_k) . $$

The phase of ${}_N m_0$ can therefore be computed from the phase of each contribution. Since

$$ (e^{-i\xi} - R_\ell \, e^{-i\alpha_\ell})(e^{-i\xi} - R_\ell \, e^{i\alpha_\ell}) = e^{-i\xi}(e^{-i\xi} - 2R_\ell \, \cos\alpha_\ell + R_\ell^2 \, e^{i\xi}) $$

and

$$ (e^{-i\xi} - r_\ell) = e^{-i\xi/2} \, (e^{-i\xi/2} - r_\ell \, e^{i\xi/2}) , $$

the corresponding phase contributions are

$$ \Phi_\ell(\xi) = \mathrm{arctg} \left(\frac{(R_\ell^2 - 1) \, \sin\xi}{(1+R_\ell^2) \, \cos\xi - 2R_\ell \, \cos\alpha_\ell} \right) $$

and

$$ \Phi_\ell(\xi) = \mathrm{arctg} \left(\frac{r_\ell + 1}{r_\ell - 1} \, \mathrm{tg} \, \frac{\xi}{2} \right) . $$

Let us choose the valuation of arctg so that Φ_ℓ is continuous in $[0, 2\pi]$, and $\Phi_\ell(0) = 0$; as shown by the example of the Haar basis, this may not be the "true" phase: we have ironed out possible discontinuities. To see how linear the phase is, this ironing out is exactly what we want to do, however. Moreover, we would like to extract only the nonlinear part of Φ_ℓ; we therefore define

$$ \Psi_\ell(\xi) = \Phi_\ell(\xi) - \frac{\xi}{2\pi} \, \Phi_\ell(2\pi) . $$

In §6.4 we systematically chose all the z_ℓ, r_ℓ with absolute value less than 1 when we constructed ${}_N\phi$. This is a so-called "extremal phase" choice; it results in a total phase $\Psi_{\mathrm{tot}}(\xi) = \sum_\ell \, \Psi_\ell(\xi)$ which is very nonlinear (see Figure 8.1). In order to obtain m_0 as close to linear phase as possible, we have to choose the zeros to retain from every quadruplet or duplet in such a way that $\Psi_{\mathrm{tot}}(\xi)$ is as close to zero as possible. In practice, we have $2^{\lfloor N/2 \rfloor}$ choices. This number can be reduced by another factor of 2: for every choice, the complementary choice (choosing all the other zeros) leads to the complex conjugate m_0 (up to a phase shift), and therefore to the mirror image of ϕ. For $N = 2$ or 3, there is therefore effectively only one pair ϕ_N, ψ_N. For $N \geq 4$, one can compare the $2^{\lfloor N/2 \rfloor - 1}$ different graphs for Ψ_{tot} in order to find the closest to linear phase. The net effect of a change of choice from z_ℓ, $\overline{z}_\ell$ to z_ℓ^{-1}, $\overline{z}_\ell^{-1}$ will be most significant if R_ℓ is close to 1, and if α_ℓ is close to either 0 or π. In Figure 8.1 we show the graphs for $\Psi_{\mathrm{tot}}^{(\xi)}$ for $N = 4$, 6, 8, 10, both for the original construction in §6.4, and for the case with flattest Ψ_{tot}. Incidentally, in all cases the original construction corresponded to the least flat Ψ_{tot}, i.e., to the most asymmetric ϕ. The "least asymmetric" ϕ and ψ, associated with the flattest possible Ψ_{tot}, were plotted in Figure 6.4 for $N = 4$, 6, 8, 10; the corresponding filter coefficients were given in Table 6.3, for all N from 4 to 10.

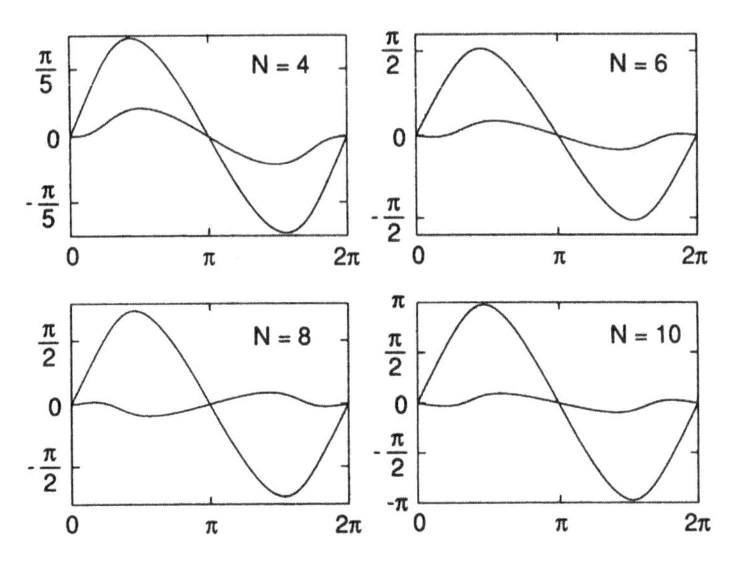

FIG. 8.1. *The non-linear part* $\Psi_{\mathrm{tot}}(\xi)$ *of the phase of* $m_0(\xi)$ *for* $N = 4, 6, 8,$ *and* 10, *for the extremal phase choice (largest amplitude) and for the "closest to linear phase" choice (flattest curve).*

REMARKS.

1. In this discussion we have restricted ourselves to the case where m_0 and $|\mathcal{L}|^2$ are given by (6.1.10) and (6.1.12), respectively. This means that the ϕ in Figure 6.4 are the least asymmetric possible, given that N moments of ψ are zero, and that ϕ has support width $2N - 1$. (This is the minimum width for N vanishing moments.) If ϕ may have larger support width, then it can be made even more symmetric. These wider solutions correspond to a choice $R \not\equiv 0$ in (6.1.11). The functions ϕ in the next subsection, for instance, are more symmetric than those in Figure 6.4, but they have larger support width.

2. One can achieve even more symmetry by going a little beyond the "standard" multiresolution scheme explained in Chapter 5. Suppose h_n are the coefficients associated to a "standard" multiresolution analysis and the corresponding orthonormal basis (compactly supported or not). Define functions ϕ^1, ϕ^2, ψ^1, ψ^2 by

$$\begin{aligned}
\phi^1(x) &= \frac{1}{\sqrt{2}} \sum_n h_n \, \phi^2(2x - n) \; , \\
\phi^2(x) &= \frac{1}{\sqrt{2}} \sum_n h_{-n} \, \phi^1(2x - n) \; , \\
\psi^1(x) &= \frac{1}{\sqrt{2}} \sum_n (-1)^n \, h_{-n+1} \, \phi^2(2x - n) \; ,
\end{aligned}$$

$$\psi^2(x) = \frac{1}{\sqrt{2}} \sum_n (-1)^n \, h_{n-1} \, \phi^1(2x-n) \, .$$

Then the same calculations as in Chapter 5 show that the functions $\psi^1_{2j,k}(x) = 2^{-j}\psi^1(2^{-2j}x - k)$, $\psi^2_{2j+1,k}(x) = 2^{-j-1/2}\psi^2(2^{-2j-1}x - k)$ $(j, k \in \mathbb{Z})$ constitute an orthonormal basis for $L^2(\mathbb{R})$. Since the recursions above correspond to

$$\begin{aligned} \hat{\phi}^1(\xi) &= m_0(\xi/2)\, \overline{m_0(\xi/4)}\, m_0(\xi/8)\, \overline{m_0(\xi/16)} \cdots \\ &= \prod_{j=1}^{\infty} \left[m_0(2^{-2j-1}\xi)\, \overline{m_0(2^{-2j-2}\xi)} \right] , \end{aligned}$$

the phase of $\hat{\phi}^1$ can be expected to be closer to linear phase than that of $\hat{\phi}(\xi) = \prod_{j=1}^{\infty} m_0(2^{-j}\xi)$. Note also that $\hat{\phi}_2(\xi) = \overline{\hat{\phi}_1(\xi)}$, $\hat{\psi}_2(\xi) = \overline{\hat{\psi}_1(\xi)}$; hence $\phi_2(x) = \phi_1(-x)$, $\psi_2(x) = \psi_1(-x)$. Figure 8.2 shows ϕ_1, ψ_1 computed from the h_n for $N = 2$, i.e., $h_0 = \frac{1+\sqrt{3}}{2\sqrt{2}}$, $h_1 = \frac{3+\sqrt{3}}{2\sqrt{2}}$, $h_2 = \frac{3-\sqrt{3}}{2\sqrt{2}}$, $h_3 = \frac{1-\sqrt{3}}{2\sqrt{2}}$. (Unlike the previous construction, this "switching" makes a difference even for $N = 2$.) For the "least asymmetric" h_n given in Table 6.3, this switching technique leads to slightly "better" ϕ, but seems to have little effect on ψ. $\square$

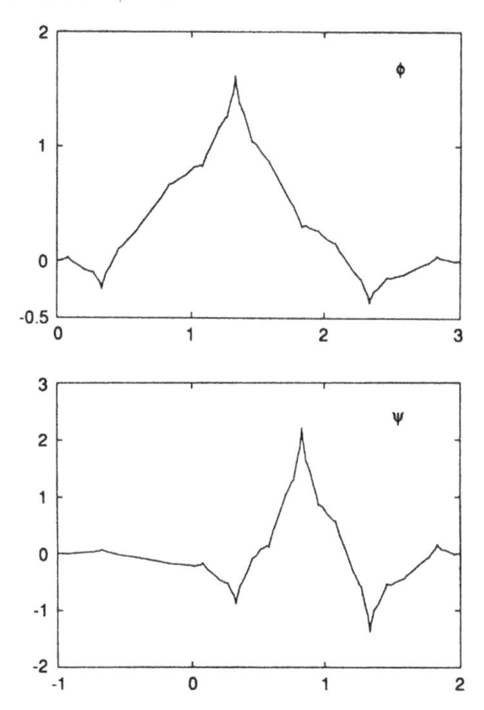

FIG. 8.2. *Scaling function ϕ_1 and wavelet ψ_1 obtained by applying the "switching trick" to the 4-tap wavelet filters of §6.4.*

8.2. Coiflets.

In §7.4 we saw one advantage of having a high number of vanishing moments for ψ: it led to high compressibility because the fine scale wavelet coefficients of a function would be essentially zero where the function was smooth. Since $\int dx\ \phi(x) = 1$, the same thing can never happen for the $\langle f, \phi_{j,k}\rangle$. Still, if $\int dx\ x^\ell\ \phi(x) = 0$ for $\ell = 1, \cdots, L$, then we can apply the same Taylor expansion argument and conclude that for J large, $\langle f,\ \phi_{-J,k}\rangle \simeq 2^{J/2}\ f(2^{-J}k)$, with an error that is negligibly small where f is smooth. This means that we have a particularly simple quadrature rule to go from the samples of f to its fine scale coefficients $\langle f,\ \phi_{-J,k}\rangle$. For this reason, R. Coifman suggested in the spring of 1989 that it might be worthwhile to construct orthonormal wavelet bases with vanishing moments not only for ψ, but also for ϕ.[4] In this section I give a brief account of how this can be done; more details are given in Daubechies (1993). Because they were first requested by Coifman (with a view to applying them for the algorithms in Beylkin, Coifman, and Rokhlin), I have named the resulting wavelets "coiflets."

The goal is to find ψ, ϕ so that

$$\int dx\ x^\ell \psi(x) = 0, \qquad \ell = 0, \cdots, L-1 \tag{8.2.1}$$

and

$$\int dx\ \phi(x) = 1, \quad \int dx\ x^\ell \phi(x) = 0, \quad \ell = 1, \cdots, L-1; \tag{8.2.2}$$

L is then called the *order* of the coiflet. We already know how to express (8.2.1) in terms of m_0; it is equivalent with

$$m_0(\xi) \;=\; \left(\frac{1+e^{-i\xi}}{2}\right)^L\ \mathcal{L}(\xi)\ . \tag{8.2.3}$$

What does (8.2.2) correspond to? It is equivalent to the condition $\frac{d^\ell}{d\xi^\ell}\hat{\phi}\Big|_{\xi=0} = 0$, $\ell = 1, \cdots, L-1$. Let us check what $\hat{\phi}'(0) = 0$ means for m_0. Because $\hat{\phi}(\xi) = m_0(\xi/2)\ \hat{\phi}(\xi/2)$, we have

$$\hat{\phi}'(\xi) \;=\; \tfrac{1}{2}\ m_0'(\xi/2)\ \hat{\phi}(\xi/2) \;+\; \tfrac{1}{2}\ m_0(\xi/2)\ \hat{\phi}'(\xi/2)\ ;$$

hence

$$\hat{\phi}'(0) \;=\; \tfrac{1}{2}\ m_0'(0)\ (2\pi)^{-1/2} \;+\; \tfrac{1}{2}\ \hat{\phi}'(0)\ ,$$

or

$$m_0'(0) \;=\; (2\pi)^{1/2}\ \hat{\phi}'(0)\ .$$

Consequently, $\int dx\ x\phi(x) = 0$ is equivalent with $m_0'(0) = 0$. Similarly, one sees that (8.2.2) is equivalent with $\left(\frac{d^\ell}{d\xi^\ell}\hat{\phi}\Big|_{\xi=0}\right) = 0$, $\ell = 1, \cdots, L-1$, or with

$$m_0(\xi) \;=\; 1 \;+\; (1-e^{-i\xi})^L\ \tilde{\mathcal{L}}(\xi)\ , \tag{8.2.4}$$

where $\tilde{\mathcal{L}}$ is a trigonometric polynomial. In addition to (8.2.3) and (8.2.4), m_0 will of course also have to satisfy $|m_0(\xi)|^2 + |m_0(\xi+\pi)|^2 = 1$. Let us specialize to L even (the easiest case, although odd L are not much harder), $L = 2K$. Then (8.2.3), (8.2.4) imply that we have to find two trigonometric polynomials $\mathcal{P}_1, \mathcal{P}_2$ so that

$$\left(\cos^2 \frac{\xi}{2}\right)^K \mathcal{P}_1(\xi) = 1 + \left(\sin^2 \frac{\xi}{2}\right)^K \mathcal{P}_2(\xi) . \tag{8.2.5}$$

$$\left(\text{Because} \left(\frac{1+e^{-i\xi}}{2}\right)^{2K} = e^{-i\xi K}\left(\cos^2 \frac{\xi}{2}\right)^K, (1-e^{-i\xi})^{2K} = e^{-iK\xi}\left(2i \sin \frac{\xi}{2}\right)^{2K}.\right)$$

But we already know what the general form of such $\mathcal{P}_1$, $\mathcal{P}_2$ are: (8.2.5) is nothing other than the Bezout equation which we already solved in §6.1. In particular, $\mathcal{P}_1$ has the form

$$\mathcal{P}_1(\xi) = \sum_{k=0}^{K-1} \binom{K-1+k}{k} \left(\sin^2 \frac{\xi}{2}\right)^k + \left(\sin^2 \frac{\xi}{2}\right)^K f(\xi) ,$$

where f is an arbitrary trigonometric polynomial. It then remains to taylor f in $m_0(\xi) = ((1+e^{-i\xi})/2)^{2K} \mathcal{P}_1(\xi)$ so that $|m_0(\xi)|^2 + |m_0(\xi+\pi)|^2 = 1$ is satisfied. With the ansatz $f(\xi) = \sum_{n=0}^{2K-1} f_n \, e^{-in\xi}$, it is shown in Daubechies (1990) how to reduce this "tayloring" to the solution of a system of K quadratic equations for K unknowns. A heuristic, perturbative argument suggests that this system will have a solution for large K, and explicit numerical solutions are computed for $K = 1, \cdots, 5$. Figure 8.3 shows the plots of the resulting ϕ, ψ; the corresponding coefficients are listed in Table 8.1. It is clear from the figure that ϕ, ψ are much more symmetric than the ${}_N\phi$, ${}_N\psi$ of §6.4, or even than the ϕ, ψ in §8.1, but there is of course a price to pay: a coiflet with $2K$ vanishing moments typically has support width $6K-1$, as compared to $4K-1$ for ${}_{2K}\phi$.

REMARK. The ansatz $f(\xi) = \sum_{n=0}^{2K-1} f_n \, e^{-in\xi}$ is not the only possible one, but it makes the computations easier. For small values of K ($K = 1, 2, 3$), different ansatzes are also tried out in Daubechies (1993). It turns out that the smoothest coiflets (at least at these small values for K) are not the most symmetric ones; for $K = 1$, for instance, there exists a (very asymmetric) coiflet with Hölder exponent 1.191814, whereas the coiflet of order 2 in Figure 8.3 is not C^1; both have support width 5. Similar gains of regularity can be found for $K = 2, 3$. For graphs, coefficients and more details, see Daubechies (1990b). □

8.3. Symmetric biorthogonal wavelet bases.

As mentioned above, it is well known in the subband filtering community that symmetry and exact reconstruction are incompatible, if the same FIR filters are used for reconstruction and decomposition. As soon as this last requirement is given up, symmetry is possible. This means that we replace the block diagram of Figure 5.11 by Figure 8.4. Several questions naturally arise: what does Figure 8.4

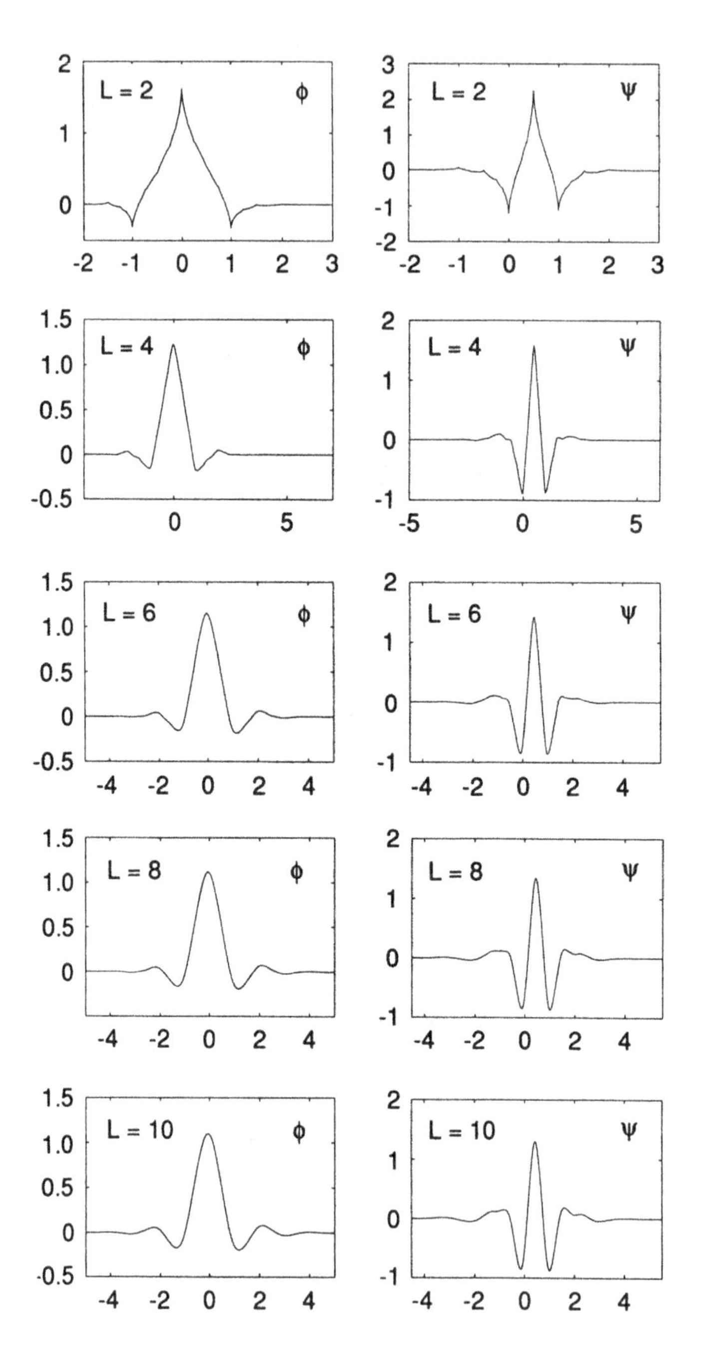

FIG. 8.3. *Coiflets ψ and their corresponding scaling functions ϕ for $L = 2, 4, 6, 8$, and 10. The support width of ϕ and ψ is $3L - 1$ in all cases.*

TABLE 8.1

The coefficients for coiflets of order $L = 2K$, $K = 1$ to 5. (The coefficients listed are normalized so that their sum is 1; they are equal to the $2^{-1/2}h_n$.)

	n	$h_n/\sqrt{2}$
$K = 1$	-2	$-.051429728471$
	-1	.238929728471
	0	.602859456942
	1	.272140543058
	2	$-.051429972847$
	3	$-.011070271529$
$K = 2$	-4	.011587596739
	-3	$-.029320137980$
	-2	$-.047639590310$
	-1	.273021046535
	0	.574682393857
	1	.294867193696
	2	$-.054085607092$
	3	$-.042026480461$
	4	.016744410163
	5	.003967883613
	6	$-.001289203356$
	7	$-.000509505399$
$K = 3$	-6	$-.002682418671$
	-5	.005503126709
	-4	.016583560479
	-3	$-.046507764479$
	-2	$-.043220763560$
	-1	.286503335274
	0	.561285256870
	1	.302983571773
	2	$-.050770140755$
	3	$-.058196250762$
	4	.024434094321
	5	.011229240962
	6	$-.006369601011$
	7	$-.001820458916$
	8	.000790205101
	9	.000329665174
	10	$-.000050192775$
	11	$-.000024465734$
$K = 4$	-8	.000630961046
	-7	$-.001152224852$
	-6	$-.005194524026$
	-5	.011362459244
	-4	.018867235378
	-3	$-.057464234429$
	-2	$-.039652648517$
	-1	.293667390895
$K = 4$	0	.553126452562
	1	.307157326198
	2	$-.047112738865$
	3	$-.068038127051$
	4	.027813640153
	5	.017735837438
	6	$-.010756318517$
	7	$-.004001012886$
	8	.002652665946
	9	.000895594529
	10	$-.000416500571$
	11	$-.000183829769$
	12	.000044080354
	13	.000022082857
	14	$-.000002304942$
	15	$-.000001262175$
$K = 5$	-10	$-.0001499638$
	-9	.0002535612
	-8	.0015402457
	-7	$-.0029411108$
	-6	$-.0071637819$
	-5	.0165520664
	-4	.0199178043
	-3	$-.0649972628$
	-2	$-.0368000736$
	-1	.2980923235
	0	.5475054294
	1	.3097068490
	2	$-.0438660508$
	3	$-.0746522389$
	4	.0291958795
	5	.0231107770
	6	$-.0139736879$
	7	$-.0064800900$
	8	.0047830014
	9	.0017206547
	10	$-.0011758222$
	11	$-.0004512270$
	12	.0002137298
	13	.0000993776
	14	$-.0000292321$
	15	$-.0000150720$
	16	.0000026408
	17	.0000014593
	18	$-.0000001184$
	19	$-.0000000673$

mean in terms of multiresolution analysis? What do c^j and d^j now stand for? (They were coefficients of orthogonal projections in Chapter 5.) Is there an associated wavelet basis? How does it differ from the bases constructed earlier? The answer is that, provided the filters satisfy certain technical conditions, such a scheme corresponds to *two* dual wavelet bases, associated with two different multiresolution ladders. In this section we will see how to prove all this, and give several families of (symmetric!) examples. Except for an improved argument due to Cohen and Daubechies (1992), all these results are from Cohen, Daubechies, and Feauveau (1992). Many of the same examples are also derived independently in Vetterli and Herley (1990), who present a treatment from the "filter design" point of view.

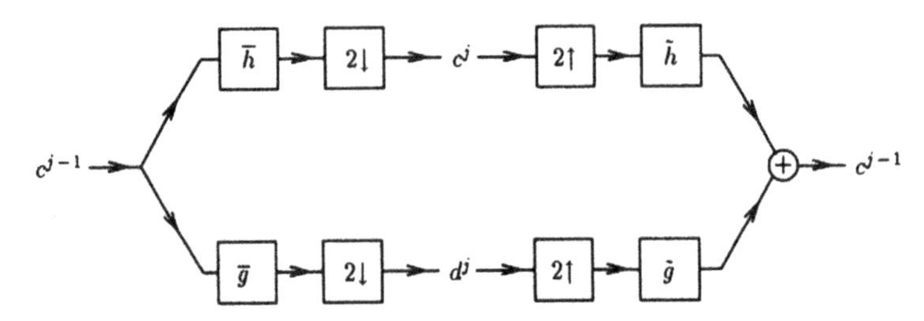

FIG. 8.4. *Subband filtering scheme with exact reconstruction but reconstruction filters different from the decomposition filters.*

8.3.1. Exact reconstruction. Since we have now *four* filters instead of two, we have to rewrite (5.6.5), (5.6.6) as

$$c_n^1 = \sum_k h_{k-2n}\, c_k^0 \,, \qquad d_n^1 = \sum_k g_{k-2n}\, c_k^0$$

and

$$c_\ell^0 = \sum_n \left[\tilde{h}_{\ell-2n}\, c_n^1 + \tilde{g}_{\ell-2n}\, d_n^1\right] .$$

In the z-notation introduced in §5.6, this can be rewritten as

$$\begin{aligned} c^0(z) &= \frac{1}{2}\left[\tilde{h}(z)\, \overline{h}(z) + \tilde{g}(z)\, \overline{g}(z)\right] c^0(z) \\ &\quad + \frac{1}{2}\left[\tilde{h}(z)\, \overline{h}(-z) + \tilde{g}(z)\, \overline{g}(-z)\right] c^0(-z) \,. \end{aligned}$$

Consequently, we require

$$\tilde{h}(z)\, \overline{h}(z) \;+\; \tilde{g}(z)\, \overline{g}(z) \;=\; 2 \,, \tag{8.3.1}$$

$$\tilde{h}(z)\, \overline{h}(-z) + \tilde{g}(z)\, \overline{g}(-z) \;=\; 0 \,, \tag{8.3.2}$$

where we assume $\tilde{h}, \tilde{g}, \overline{h}, \overline{g}$ to be polynomials since the filters are all FIR. (For simplicity, we use the term "polynomial" in a slightly wider sense than usual:

we also allow negative powers. In other words, $\sum_{n=-N_1}^{N_2} a_n\, z^n$ is a polynomial in this terminology.) From (8.3.1) it follows that $\overline{h}$ and $\overline{g}$ have no common zeros; consequently, (8.3.2) implies that

$$\tilde{g}(z) \;=\; \overline{h}(-z)p(z), \qquad \tilde{h}(z) = -\overline{g}(-z)p(z) \tag{8.3.3}$$

for some polynomial p. Substitution into (8.3.1) leads to

$$\overline{p}(z)\,\left[h(-z)g(z) - h(z)g(-z)\right] \;=\; 2\;.$$

The only polynomials that divide constants are monomials; hence

$$p(z) \;=\; \alpha z^k$$

for some $\alpha \in \mathbb{C}$, $k \in \mathbb{Z}$, and (8.3.3) becomes

$$\tilde{g}(z) \;=\; \alpha z^k\, \overline{h}(-z), \qquad g(z) \;=\; -\alpha^{-1}(-1)^k z^k\, \overline{\tilde{h}}(-z)\;. \tag{8.3.4}$$

Any choice for α and k will do; we choose $\alpha = 1$, $k = 1$, which makes the equations (8.3.4) for g and $\tilde{g}$ symmetric. Substitution into (8.3.1) gives

$$h(z)\,\overline{\tilde{h}}(z) \;+\; h(-z)\,\overline{\tilde{h}}(-z) \;=\; 2\;. \tag{8.3.5}$$

In terms of the filter coefficients, all this becomes

$$\sum_n h_n\, \tilde{h}_{n+2k} \;=\; \delta_{k,0}\;, \tag{8.3.6}$$

$$g_n \;=\; (-1)^{n+1}\, \tilde{h}_{-n+1}, \qquad \tilde{g}_n \;=\; (-1)^{n+1}\, h_{-n+1}\;, \tag{8.3.7}$$

where we have implicitly assumed that all the coefficients are real. These equations are obvious generalizations of (5.1.39), (5.1.34).

8.3.2. Scaling functions and wavelets. Because we have two pairs of filters, we also have two pairs of scaling function + wavelet: ϕ, ψ and $\tilde{\phi}, \tilde{\psi}$. They are defined by

$$\hat{\phi}(\xi) \;=\; m_0(\xi/2)\,\hat{\phi}(\xi/2)\;, \qquad \widehat{\tilde{\phi}}(\xi) \;=\; \tilde{m}_0(\xi/2)\,\widehat{\tilde{\phi}}(\xi/2)\;, \tag{8.3.8}$$

$$\hat{\psi}(\xi) \;=\; m_1(\xi/2)\,\hat{\phi}(\xi/2)\;, \qquad \widehat{\tilde{\psi}}(\xi) \;=\; \tilde{m}_1(\xi/2)\,\widehat{\tilde{\phi}}(\xi/2)\;, \tag{8.3.9}$$

where $m_0(\xi) = \frac{1}{\sqrt{2}} \sum_n h_n\, e^{-in\xi}$, $m_1(\xi) \;=\; \frac{1}{\sqrt{2}} \sum_n g_n\, e^{-in\xi}$; $\tilde{m}_0$, $\tilde{m}_1$ are defined analogously. Note that (8.3.7) implies

$$m_1(\xi) \;=\; e^{-i\xi}\, \overline{\tilde{m}_0(\xi+\pi)}, \qquad \tilde{m}_1(\xi) \;=\; e^{-i\xi}\, \overline{m_0(\xi+\pi)}\;. \tag{8.3.10}$$

We saw in Chapter 3 that in order to generate wavelet Riesz bases, ψ and $\tilde{\psi}$ have to satisfy $\hat{\psi}(0) = 0 = \widehat{\tilde{\psi}}(0)$. A necessary condition is therefore $m_1(0) = 0 =$

$\tilde{m}_1(0)$; in terms of the polynomials $h(z)$, $\tilde{h}(z)$ this is equivalent to $h(-1) = 0 = \tilde{h}(-1)$. Substitution into (8.3.5) then leads to $h(1)\,\overline{\tilde{h}(1)} = 2$, or

$$\left(\sum_n h_n\right)\left(\sum_n \tilde{h}_n\right) = 2 .$$

This implies that we can normalize both h and $\tilde{h}$ so that $\sum_n h_n = \sqrt{2} = \sum_n \tilde{h}_n$. Consequently, $m_0(0) = 1 = \tilde{m}_0(0)$, and we can solve (8.3.8) by defining

$$\hat{\phi}(\xi) = (2\pi)^{-1/2} \prod_{j=1}^{\infty} m_0(2^{-j}\xi) ,$$

$$\widehat{\tilde{\phi}}(\xi) = (2\pi)^{-1/2} \prod_{j=1}^{\infty} \tilde{m}_0(2^{-j}\xi) .$$

The same arguments as in Chapter 6 show that these infinite products converge uniformly on compact sets, and that ϕ and $\tilde{\phi}$ have compact support, with support width given by the filter lengths. As finite linear combinations of ϕ and $\tilde{\phi}$, ψ and $\tilde{\psi}$ also have compact support. This is by no means sufficient to guarantee that the $\psi_{j,k} = 2^{-j/2}\,\psi(2^{-j}x - k)$ and $\tilde{\psi}_{j,k}$ are dual Riesz bases of wavelets, however. Indeed, even in the orthogonal case (reconstruction filters = decomposition filters), it was possible for ψ to fail to generate an orthonormal basis (see §6.2, §6.3). In this nonorthogonal case we have to be even more careful. Let us summarize the different steps in the argument proving that we have dual wavelet bases (with certain restrictions).

First of all, if $\phi,\ \tilde{\phi} \in L^2(\mathbb{R})$ (which will have to be proved too! See below.) then we can define bounded operators T_j by

$$\langle T_j f,\ g\rangle = \sum_k \langle f,\ \phi_{j,k}\rangle\langle\tilde{\phi}_{j,k},\ g\rangle ,$$

where

$$\phi_{j,k} = 2^{-j/2}\,\phi(2^{-j}x - k), \qquad \tilde{\phi}_{j,k} = 2^{-j/2}\,\tilde{\phi}(2^{-j}x - k) ,$$

as usual.[5] A consequence of the definitions (8.3.8), (8.3.9) is

$$\phi_{1,n}(x) = \sum_k h_{k-2n}\,\phi_{0,k} , \qquad \tilde{\phi}_{1,n} = \sum_k \tilde{h}_{k-2n}\,\tilde{\phi}_{0,k} ,$$

$$\psi_{1,n}(x) = \sum_k g_{k-2n}\,\phi_{0,k} , \qquad \tilde{\psi}_{1,n} = \sum_k \tilde{h}_{k-2n}\,\tilde{\phi}_{0,k} ;$$

together with the properties of the filter coefficients imposed in §8.1, this implies (as can easily be checked by substitution)

$$\sum_k \langle f, \phi_{0,k}\rangle\langle\tilde{\phi}_{0,k}, g\rangle = \sum_n \left[\langle f, \phi_{1,n}\rangle\langle\tilde{\phi}_{1,n}, g\rangle + \langle f, \psi_{1,n}\rangle\langle\tilde{\psi}_{1,n}, g\rangle\right] .$$

The same trick can be applied for other values of j; "telescoping" all the identities together leads to

$$\sum_{j=-J}^{J} \sum_{\ell} \langle f, \psi_{j,\ell}\rangle \langle \tilde{\psi}_{j,\ell}, g\rangle = \langle T_{-J-1} f, g\rangle - \langle T_J f, g\rangle$$
$$= \sum_k \langle f, \phi_{-J-1,k}\rangle \langle \tilde{\phi}_{-J-1,k}, g\rangle - \sum_k \langle f, \phi_{J,k}\rangle \langle \tilde{\phi}_{J,k}, g\rangle$$

Exactly the same arguments as in Chapter 5, used in the estimation of (5.3.9), (5.3.13), respectively, show that $\langle T_J f, g\rangle \longrightarrow 0$, $\langle T_{-J} f, g\rangle \rightarrow \langle f, g\rangle$, for $J \rightarrow \infty$. Consequently,

$$\lim_{J\rightarrow\infty} \sum_{j=-J}^{J} \sum_{\ell} \langle f, \psi_{j,\ell}\rangle \langle \tilde{\psi}_{j,\ell}, g\rangle = \langle f, g\rangle , \tag{8.3.11}$$

or, in a weak sense,

$$f = \lim_{J\rightarrow\infty} \sum_{j=-J}^{J} \sum_{\ell} \langle f, \psi_{j,\ell}\rangle \, \tilde{\psi}_{j,\ell} .$$

This is not sufficient to establish that the $\psi_{j,\ell}$, $\tilde{\psi}_{j,\ell}$ constitute dual Riesz bases. For one thing, the $\psi_{j,\ell}$ or $\tilde{\psi}_{j,\ell}$ may fail to constitute frames; in this case the convergence in (8.3.11) could depend crucially on the order of summation. To avoid this, we need to impose that

$$\sum_{j,k} |\langle f, \psi_{j,k}\rangle|^2 \text{ and } \sum_{j,k} |\langle f, \tilde{\psi}_{j,k}\rangle|^2$$

converge for all $f \in L^2(\mathbb{R})$, or equivalently,

$$\sum_{j,k} |\langle f, \psi_{j,k}\rangle|^2 \le A \, \|f\|^2, \qquad \sum_{j,k} |\langle f, \tilde{\psi}_{j,k}\rangle|^2 \le \tilde{A} \, \|f\|^2 . \tag{8.3.12}$$

If these upper bounds hold, then it follows from (8.3.11) that[6]

$$\sum_{j,k} |\langle f, \psi_{j,k}\rangle|^2 \ge \tilde{A}^{-1} \, \|f\|^2, \qquad \sum_{j,k} |\langle f, \tilde{\psi}_{j,k}\rangle|^2 \ge A^{-1} \, \|f\|^2 ,$$

so that we automatically have frames. But even then the $\psi_{j,k}$, $\tilde{\psi}_{j,k}$ may merely be (redundant) dual frames and not dual Riesz bases; this redundancy is eliminated by the requirement

$$\langle \psi_{j,k}, \tilde{\psi}_{j',k'}\rangle = \delta_{j,j'} \, \delta_{k,k'} , \tag{8.3.13}$$

which, exactly like in the orthonormal case (see §6.2), can be shown to be equivalent with

$$\langle \phi_{0,k}, \tilde{\phi}_{0,k'}\rangle = \delta_{k,k'} . \tag{8.3.14}$$

If the conditions (8.3.12) and (8.3.14) are satisfied (we will come back to them shortly), then we do indeed have two multiresolution analysis ladders.

$$\cdots V_2 \subset V_1 \subset V_0 \subset V_{-1} \subset V_{-2} \subset \cdots ,$$
$$\cdots \tilde{V}_2 \subset \tilde{V}_1 \subset \tilde{V}_0 \subset \tilde{V}_{-1} \subset \tilde{V}_{-2} \subset \cdots ,$$

with $V_0 = \overline{\text{Span}\,\{\phi_{0,k};\ k \in \mathbb{Z}\}}$, $\tilde{V}_0 = \overline{\text{Span}\,\{\tilde{\phi}_{0,k};\ k \in \mathbb{Z}\}}$. The spaces $W_j = \overline{\text{Span}\,\{\psi_{j,k};\ k \in \mathbb{Z}\}}$, $\tilde{W}_j = \overline{\text{Span}\,\{\tilde{\psi}_{j,k};\ k \in \mathbb{Z}\}}$ are again complements of V_j, respectively, $\tilde{V}_j$ in V_{j-1}, respectively, $\tilde{V}_{j-1}$, but they are not orthogonal complements: typically the angle[7] between V_j, W_j or $\tilde{V}_j, \tilde{W}_j$ will be smaller than 90°. This is the reason why we have to prove (8.3.12) in this case, whereas it was automatic in the orthonormal case. Another way of seeing this is the following. Because of the non-orthogonality we have

$$\alpha \sum_k \left[|\langle f, \phi_{j,k}\rangle|^2 + |\langle f, \psi_{j,k}\rangle|^2\right] \le \sum_k |\langle f, \phi_{j-1,k}\rangle|^2$$
$$\le \beta \left[\sum_k |\langle f, \phi_{j,k}\rangle|^2 + \sum_k |\langle f, \psi_{j,k}\rangle|^2\right] ,$$

with $\alpha < 1$, $\beta > 1$ (in the orthogonal case, equality holds, with $\alpha = \beta = 1$). Unlike the orthonormal case, we cannot telescope these inequalities to prove that the $\psi_{j,k}$ constitute a Riesz basis: telescoping would lead to a blowup of the constants. We therefore have to follow a different strategy. Note that (8.3.13) implies that $W_j \perp \tilde{V}_j$, $\tilde{W}_j \perp V_j$. The two multiresolution hierarchies and their sequences of complement spaces fit together like a giant zipper, and this is what allows us to control expressions like $\sum_{j,k} |\langle f, \psi_{j,k}\rangle|^2$.

But let us return to the conditions (8.3.12) and (8.3.14). We already saw how to tackle condition (8.3.14) in §6.3, in the simpler orthogonal case. Our strategy here is essentially the same. We again define an operator P_0 acting on 2π-periodic functions,

$$(P_0 f)(\xi) = \left|m_0\left(\frac{\xi}{2}\right)\right|^2 f\left(\frac{\xi}{2}\right) + \left|m_0\left(\frac{\xi}{2}+\pi\right)\right|^2 f\left(\frac{\xi}{2}+\pi\right) ;$$

a second operator $\tilde{P}_0$ is defined analogously. In terms of the Fourier coefficients of f, the action of P_0 is given by

$$(P_0 f)_k = \sum_\ell \left(\sum_m h_m \, \overline{h_{m+\ell-2k}}\right) f_\ell ;$$

we will be mostly interested in invariant trigonometric polynomials for P_0. This means that we can restrict our attention to the $2(N_2 - N_1) + 1$-dimensional subspace of f for which $f_\ell = 0$ if $\ell > N_2 - N_1$ (we assume $h_n = 0$ if $n < N_1$ or $n > N_2$), on which P_0 is represented by a matrix. Theorems 6.3.1 and 6.3.4 have the following analog.

THEOREM 8.3.1. *The following three statements are equivalent:*

1. $\phi, \tilde{\phi} \in L^2(\mathbb{R})$ and $\langle \phi_{0,k}, \tilde{\phi}_{0,\ell} \rangle = \delta_{k,\ell}$.

2. *There exist strictly positive trigonometric polynomials $f_0, \tilde{f}_0$ invariant for P_0, $\tilde{P}_0$; there also exists a compact set K congruent to $[-\pi, \pi]$ modulo 2π so that*

$$\inf_{k \geq 1,\ \xi \in K} \left| m_0(2^{-k}\xi) \right| > 0 , \qquad \inf_{k \geq 1,\ \xi \in K} \left| \tilde{m}_0(2^{-k}\xi) \right| > 0 .$$

3. *There exist strictly positive trigonometric polynomials f_0, $\tilde{f}_0$ invariant for P_0, $\tilde{P}_0$, and these are the only invariant polynomials for P_0, $\tilde{P}_0$ (up to normalization).*

The proof is very similar to the proofs in Chapter 6, but a bit more complicated. In §6.3, the functions f_0, $\tilde{f}_0$ were simply constant; in this case, they are essentially $f_0(\xi) = \sum_\ell |\hat{\phi}(\xi + 2\pi\ell)|^2$, $\tilde{f}_0(\xi) = \sum_\ell |\hat{\tilde{\phi}}(\xi + 2\pi\ell)|^2$. For details on how to adapt the proofs of §6.3 to the present case, see Cohen, Daubechies, and Feauveau (1992).

Condition (8.3.14) therefore simply amounts to checking that two matrices have a nondegenerate eigenvalue 1 and that the entries of the corresponding eigenvectors define a strictly positive trigonometric polynomial. (Note that if the trigonometric polynomial takes *negative* values, then $\phi \notin L^2(\mathbb{R})$. This happens for some exact reconstruction filter quadruplets.) Condition (8.3.12) is something we had not encountered in the orthogonal case. It turns out that this condition is satisfied if any of the three conditions in Theorem 8.3.1 holds. The proof of this surprising fact is in the following steps[8]:

- First, one shows that the existence of an eigenvalue λ of P_0 with $|\lambda| \geq 1$, $\lambda \neq 1$ would contradict the square integrability of ϕ. It follows therefore from Theorem 8.3.1 that all the other eigenvalues of P_0 have absolute value strictly smaller than 1 if the eigenvalue 1 is nondegenerate and the associated eigenvector corresponds to a strictly positive trigonometric polynomial. The proof of this step uses Lemma 7.1.10.

- Since $m_0(\pi) = 0 = \tilde{m}_0(\pi)$, we have obviously $M_0(\pi) = |m_0(\pi)|^2 = 0 = |\tilde{m}_0(\pi)|^2 = \tilde{M}_0(\pi)$. We saw in Chapter 7 that this means that the columns of the matrix representing P_0 all sum to 1, so that the row vector (of the appropriate dimension) with all entries 1 is a left eigenvector for P_0 with eigenvalue 1. It follows from the first point that ρ, the spectral radius of $P_0|_{E_1}$, with $E_1 = \{f;\ \sum_n f_n = 0\}$, is strictly smaller than 1. One then uses that $f(\xi) = 1 - \cos\xi$ is in E_1 to prove (the estimates are analogous to those in the proof of Theorem 7.1.12) that $\int_{2^{n-1}\pi \leq |\xi| \leq 2^n\pi} d\xi\ |\hat{\phi}(\xi)|^2 \leq C\left(\frac{1+\rho}{2}\right)^n$.

- Via Hölder's inequality this implies $\int d\xi\ |\hat{\phi}(\xi)|^{2(1-\delta)} < \infty$ for sufficiently small δ. This can then be used to prove a "discretized" version, i.e., $\sum_{m\in\mathbb{Z}} |\hat{\phi}(\xi + \pi m)|^{2(1-\delta')} \leq C < \infty$ for all $\xi \in \mathbb{R}$, again for sufficiently small δ'. Because m_1 is bounded, $\hat{\psi}$ satisfies a similar bound,

$$\sum_{m\in\mathbb{Z}} |\hat{\psi}(\xi + 2\pi m)|^{2(1-\delta')} \le C < \infty \,. \tag{8.3.15}$$

- On the other hand, one can also prove that

$$\sup_{\pi\le|\xi|\le 2\pi} \sum_{j\in\mathbb{Z}} |\hat{\psi}(2^j\xi)|^{2\delta'} < \infty \,. \tag{8.3.16}$$

Since $\hat{\psi}$ is entire and $\hat{\psi}(0) = 0$, $|\hat{\psi}(\xi)| \le C|\xi|$ for sufficiently small $|\xi|$, so that $\sum_{j=-\infty}^{0} |\hat{\psi}(2^j\xi)|^{2\delta'}$ is uniformly bounded for $|\xi| \le 2\pi$, and we only need to concentrate on $j \ge 0$ in (8.3.16). But

$$\begin{aligned}
\sup_{2^j\pi\le|\zeta|\le 2^{j+1}\pi} |\hat{\psi}(\zeta)|^2 &\le \int_{2^j\pi\le|\xi|\le 2^{j+1}\pi} d\xi \, \frac{d}{d\xi}|\hat{\psi}(\xi)|^2 \\
&\le 2 \int_{2^j\pi\le|\xi|\le 2^{j+1}\pi} d\xi \, |\hat{\psi}(\xi)| \, \left|\frac{d}{d\xi}\hat{\psi}(\xi)\right| \\
&\le C\left[\int_{2^{j-1}\pi\le|\zeta|\le 2^j\pi} d\zeta \, |\hat{\phi}(\zeta)|^2\right]^{1/2} \cdot \left[\int dx \, |x\psi(x)|^2\right]^{1/2} .
\end{aligned}$$

The second factor is finite because ψ is compactly supported and in $L^2(\mathbb{R})$; the first factor is bounded by $C\lambda^j$, with $|\lambda| < 1$, as shown above. This establishes (8.3.16), which is also equivalent to

$$\sup_{|\xi|\ne 0} \sum_{j\in\mathbb{Z}} |\hat{\psi}(2^j\xi)|^{2\delta'} < \infty \,.$$

- Finally, a combination of the Poisson summation formula and the Cauchy–Schwarz inequality leads to

$$\sum_k |\langle f, \psi_{j,k}\rangle|^2 \le 2\pi \int d\xi \, |\hat{f}(\xi)|^2 \, |\hat{\psi}(2^j\xi)|^{2\delta'} \sum_m |\hat{\psi}(2^j\xi + 2\pi m)|^{2(1-\delta')} .$$

It therefore follows from (8.3.15) and (8.3.16) that

$$\sum_{j,k} |\langle f, \psi_{j,k}\rangle|^2 \le A \, \|f\|^2 \,.$$

For more details concerning this argument, see Cohen and Daubechies (1992). In order to ensure that we have indeed two dual Riesz bases of wavelets, we therefore only have to check that 1 is a nondegenerate eigenvalue of $P_0, \tilde{P}_0$ and that the corresponding trigonometric polynomial is strictly positive.

8.3.3. Regularity and vanishing moments. If the $\psi_{j,k}$, $\tilde{\psi}_{j,k}$ constitute dual Riesz bases (of compactly supported wavelets, since we have assumed the filters to be FIR), then we can apply Theorem 5.5.1 to link vanishing moments of one function with regularity for the other: if $\psi \in C^m$, then automatically $\int dx\; x^\ell\; \tilde{\psi}(x) = 0\; \ell = 0, \cdots, m$.[9] This is equivalent with $\frac{d^\ell}{d\xi^\ell}\hat{\tilde{\psi}}\Big|_{\xi=0} = 0$ for $\ell = 0, \cdots, m$. Because of (8.3.9) and $\hat{\tilde{\phi}}(0) = 1$, this implies $\frac{d^\ell}{d\xi^\ell}\tilde{m}_1\Big|_{\xi=0}$ for $\ell = 0, \cdots, m$. By (8.3.10) this implies that m_0 is divisible by $((1+e^{-i\xi})/2)^m$. In order to produce regular ψ, we therefore need to construct filter pairs m_0, $\tilde{m}_0$ such that $m_0(\xi)$ has a multiple zero at $\xi = \pi$.

Note that nothing prevents ψ and $\tilde{\psi}$ from having very different regularity properties, as illustrated by some of the examples below. If $\tilde{\psi}$ is much more regular than ψ, corresponding to many more vanishing moments for ψ than for $\tilde{\psi}$, then the two formulas

$$f = \sum_{j,k} \langle f,\; \psi_{j,k}\rangle\; \tilde{\psi}_{j,k} \tag{8.3.17}$$

$$= \sum_{j,k} \langle f,\; \tilde{\psi}_{j,k}\rangle\; \psi_{j,k}\;, \tag{8.3.18}$$

both equally valid, have very different interpretations (Tchamitchian (1987)). In practice, (8.3.17) is much more useful than (8.3.18): on the one hand, the large number of vanishing moments of ψ leads to much more "compression potential" in the regions where f is reasonably smooth (see §7.4); on the other hand, the "elementary building blocks" $\tilde{\psi}_{j,k}$ are smoother. In Antonini et al. (1992) an experiment was carried out with biorthogonal wavelets of this type: the same filter pair was used twice, the second time with roles of decomposition and reconstruction filters exchanged. The case corresponding to (8.3.17) gave rise to much better results after quantization than (8.3.18). As we already mentioned in §7.4, it is not clear whether the high number of vanishing moments of ψ or the regularity of $\tilde{\psi}$ is the most important factor; it is possible that they are both important.

8.3.4. Symmetry. One advantage of biorthogonal over orthonormal bases is that $m_0, \tilde{m}_0$ can both be symmetric. If the filter corresponding to m_0 has an odd number of taps, and is symmetric, i.e. $m_0(-\xi) = e^{2ik\xi}m_0(\xi)$, then m_0 can be written as

$$m_0(\xi) = e^{-ik\xi}\; p_0(\cos\xi)\;, \tag{8.3.19}$$

where p_0 is a polynomial. It then follows that $\tilde{m}_0$ can be chosen of the same form,

$$\tilde{m}_0(\xi) = e^{-ik\xi}\; \tilde{p}_0(\cos\xi)\;, \tag{8.3.20}$$

where $\tilde{p}_0$ is any polynomial that satisfies

$$p_0(x)\; \overline{\tilde{p}_0(x)} + p_0(-x)\; \overline{\tilde{p}_0(-x)} = 1\;; \tag{8.3.21}$$

we then have indeed

$$m_0(\xi)\ \overline{\tilde{m}_0(\xi)}\ +\ m_0(\xi+\pi)\ \overline{\tilde{m}_0(\xi+\pi)}\ =\ 1\ , \tag{8.3.22}$$

which is the same as (8.3.5). Polynomials $\tilde{p}_0$ solving (8.3.21) can only be found if $p_0(x)$ and $p_0(-x)$ have no common zeros; once this is the case, there always exist solutions by Bezout's theorem (see §6.1). Note that this also means that biorthogonal bases are much easier to construct than orthonormal bases: we need to solve only linear equations to find $\tilde{p}_0$ satisfying (8.3.21) once p_0 is fixed, instead of the spectral factorization needed in §6.1.

If the filter corresponding to m_0 has an even number of taps and is symmetric (such as, e.g., the Haar filter), then m_0 satisfies $m_0(-\xi) = e^{2ik\xi+i\xi}\ m_0(\xi)$; hence

$$m_0(\xi)\ =\ e^{-ik\xi-i\xi/2}\ \cos\frac{\xi}{2}\ p_0(\cos\xi)\ . \tag{8.3.23}$$

One can then again choose $\tilde{m}_0$ of the same type,

$$\tilde{m}_0(\xi)\ =\ e^{-ik\xi-i\xi/2}\ \cos\frac{\xi}{2}\ \tilde{p}_0(\cos\xi)\ ; \tag{8.3.24}$$

equation (8.3.22) becomes

$$\cos^2\frac{\xi}{2}\ p_0(\cos\xi)\ \overline{\tilde{p}_0(\cos\xi)}\ +\ \sin^2\frac{\xi}{2}\ p_0(-\cos\xi)\ \overline{\tilde{p}_0(-\cos\xi)}\ =\ 1\ ,$$

which means that $\tilde{p}_0$ solves the Bezout problem

$$p_0^{\#}(x)\ \overline{\tilde{p}_0(x)}\ +\ p_0^{\#}(-x)\ \overline{\tilde{p}_0(-x)}\ =\ 1\ ,$$

with $p_0^{\#}(x) = \frac{1+x}{2}\ p_0(x)$.

Examples. All the examples we give here have both symmetry and some regularity. The trigonometric polynomials m_0 and $\tilde{m}_0$ are therefore of type (8.3.19), (8.3.20) or (8.3.23), (8.3.24), with $p_0(\cos\xi)$, $\tilde{p}_0(\cos\xi)$ divisible by $(1+e^{-i\xi})^\ell$ for some $\ell > 0$. Since we are dealing with polynomials in $\cos\xi$, ℓ will automatically be even; $(1+e^{-i\xi})^2\ =\ 4e^{-i\xi}\ \cos^2\frac{\xi}{2}\ =\ 2e^{-i\xi}(1+\cos\xi)$. Consequently, we are looking for m_0, $\tilde{m}_0$ of type

$$\left(\cos\frac{\xi}{2}\right)^{2\ell}\ q_0(\cos\xi)$$

if they have an even number of taps (we have assumed that $k=0$, i.e., that the h_n, $\tilde{h}_n$ are symmetric around 0), or of type

$$e^{-i\xi/2}\ \left(\cos\frac{\xi}{2}\right)^{2\ell+1}\ q_0(\cos\xi)$$

if the number of taps is odd (again we have taken $k\ =\ 0$, corresponding to $h_{1-n} = h_n$, $\tilde{h}_{1-n} = \tilde{h}_n$). In both cases, substitution into (8.3.22) gives

$$\left(\cos\frac{\xi}{2}\right)^{2L}\ q_0(\cos\xi)\overline{\tilde{q}_0(\cos\xi)} + \left(\sin\frac{\xi}{2}\right)^{2L}\ q_0(-\cos\xi)\overline{\tilde{q}_0(-\cos\xi)} = 1, \tag{8.3.25}$$

with $L = \ell + \tilde{\ell}$ in the first case, $L = \ell + \tilde{\ell} + 1$ in the second case. If we define $q_0(\cos\xi)\ \overline{\tilde{q}_0(\cos\xi)} = P\left(\sin^2 \frac{\xi}{2}\right)$, then (8.3.25) reduces to

$$(1-x)^L\ P(x)\ +\ x^L P(1-x)\ =\ 1\ , \tag{8.3.26}$$

an equation we already encountered in §6.1. The solutions to (8.3.26) are all given by

$$P(x) = \sum_{m=0}^{L-1} \binom{L-1+m}{m}\ x^m\ +\ x^L\ R(1-2x)\ ,$$

where R is an odd polynomial (see Proposition 6.1.2). We now present three families of examples, based on different choices for R and different factorizations of P into q_0 and $\tilde{q}_0$.

Spline examples. Here we take $R \equiv 0$, and $\tilde{q}_0 \equiv 1$. It follows that $\tilde{m}_0(\xi) = (\cos\frac{\xi}{2})^{\tilde{N}}$, $\tilde{N} = 2\tilde{\ell}$, or $\tilde{m}_0(\xi) = e^{-i\xi/2}\ (\cos\frac{\xi}{2})^{\tilde{N}}$, $\tilde{N} = 2\tilde{\ell}+1$, so that $\tilde{\phi}$ is a B-spline centered around 0, respectively $\frac{1}{2}$. In the first case, we then have, with $N = 2\ell$,

$$m_0(\xi) = \left(\cos\frac{\xi}{2}\right)^N \sum_{m=0}^{\ell+\tilde{\ell}-1} \binom{\ell+\tilde{\ell}-1+m}{m} \left(\sin^2\frac{\xi}{2}\right)^m ,$$

in the second case, with $N = 2\ell+1$,

$$m_0(\xi) = e^{-i\xi/2} \left(\cos\frac{\xi}{2}\right)^N \sum_{m=0}^{\ell+\tilde{\ell}} \binom{\ell+\tilde{\ell}+m}{m} \left(\sin^2\frac{\xi}{2}\right)^m .$$

In both cases we can choose ℓ freely, subject to the constraint that the eigenvalue 1 of P_0 is nondegenerate and that the associated eigenvector corresponds to a strictly positive trigonometric polynomial (see §8.4.2). The result is a family of biorthogonal bases in which $\tilde{\psi}$ is a spline function of compact support; for every preassigned order of this spline function (i.e., fixed $\tilde{\ell}$), there exists an infinity of choices for ℓ, corresponding to different ψ (with increasing support widths) and different $\tilde{\psi}$, with increasing number of vanishing moments. Note that $\tilde{\phi}$ is completely fixed by $\tilde{N}$ alone, while m_0, hence ϕ depends on both $N, \tilde{N}$. We have plotted the functions ${}_{\tilde{N}}\tilde{\phi}$, ${}_{\tilde{N},N}\tilde{\psi}$, ${}_{\tilde{N},N}\phi$, and ${}_{\tilde{N},N}\psi$, for the first few values of $N, \tilde{N}$, in Figures 8.5–8.7 ($\tilde{N} = 1$ in Figure 8.5, $\tilde{N} = 2$ in Figure 8.6, $\tilde{N} = 3$ in Figure 8.7); the corresponding filters are given in Table 8.2. In all these cases, the conditions derived in §8.4.2 are satisfied. A striking feature in Figures 8.5–8.7 is that from some point on, increasing N (for fixed $\tilde{N}$) does not alter the shape of ${}_{\tilde{N},N}\tilde{\psi}$; one sees the "wrinkles" in the corresponding ${}_{\tilde{N},N}\phi$ and ${}_{\tilde{N},N}\psi$ get ironed out as N increases.

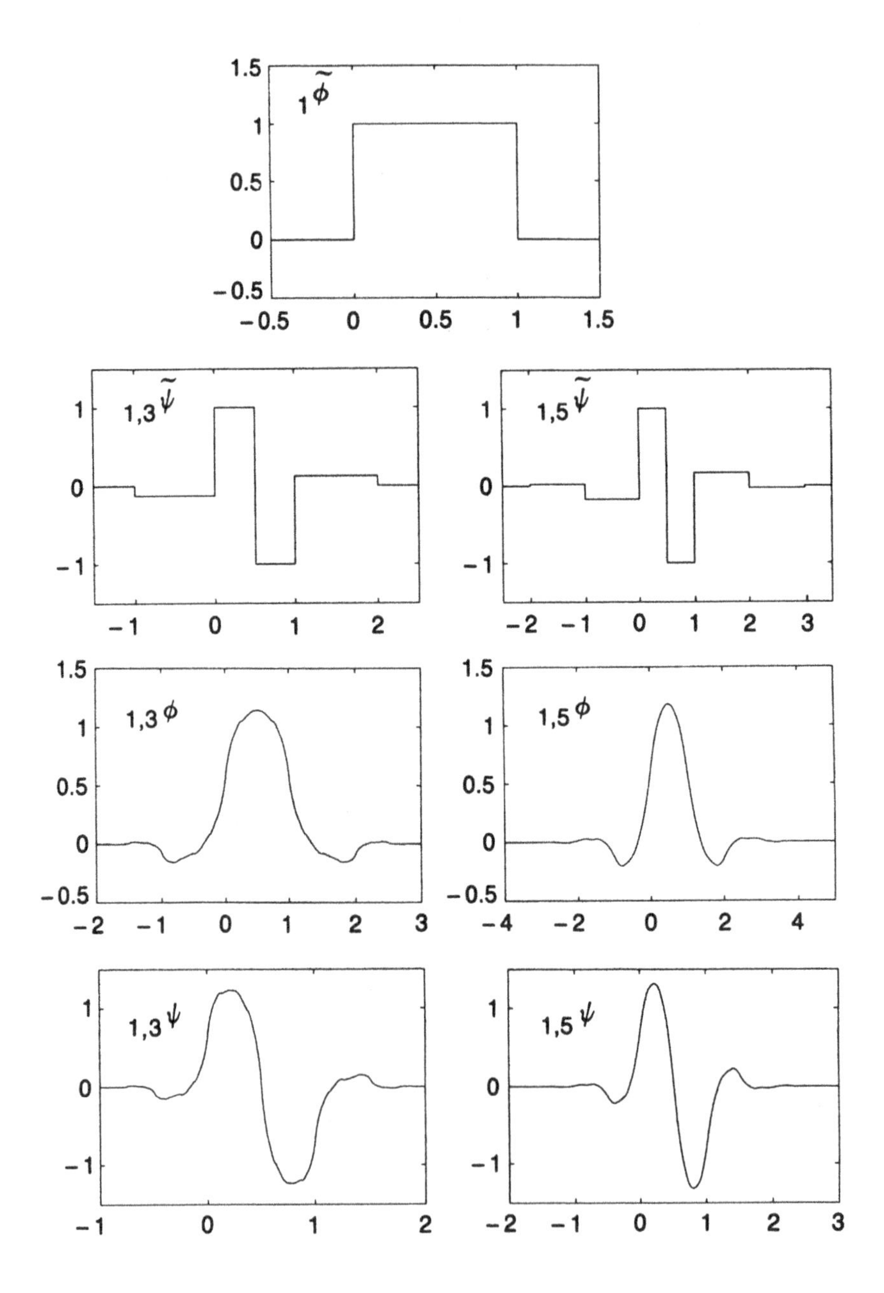

FIG. 8.5. *Spline examples with* $\tilde{N} = 2$, $N = 2, 4, 6$, *and* 8. *Here support* ${}_{2,N}\phi = [-N, N]$, *support* ${}_{2,N}\psi =$ *support* ${}_{2,N}\tilde{\psi} = \left[-\frac{N}{2}, \frac{N}{2}+1\right]$. *As always, the plots of* ϕ, ψ *are in fact plots of approximations obtained by the cascade algorithm (see pp. 205–206), with* 8 *or* 9 *iterations.*[10]

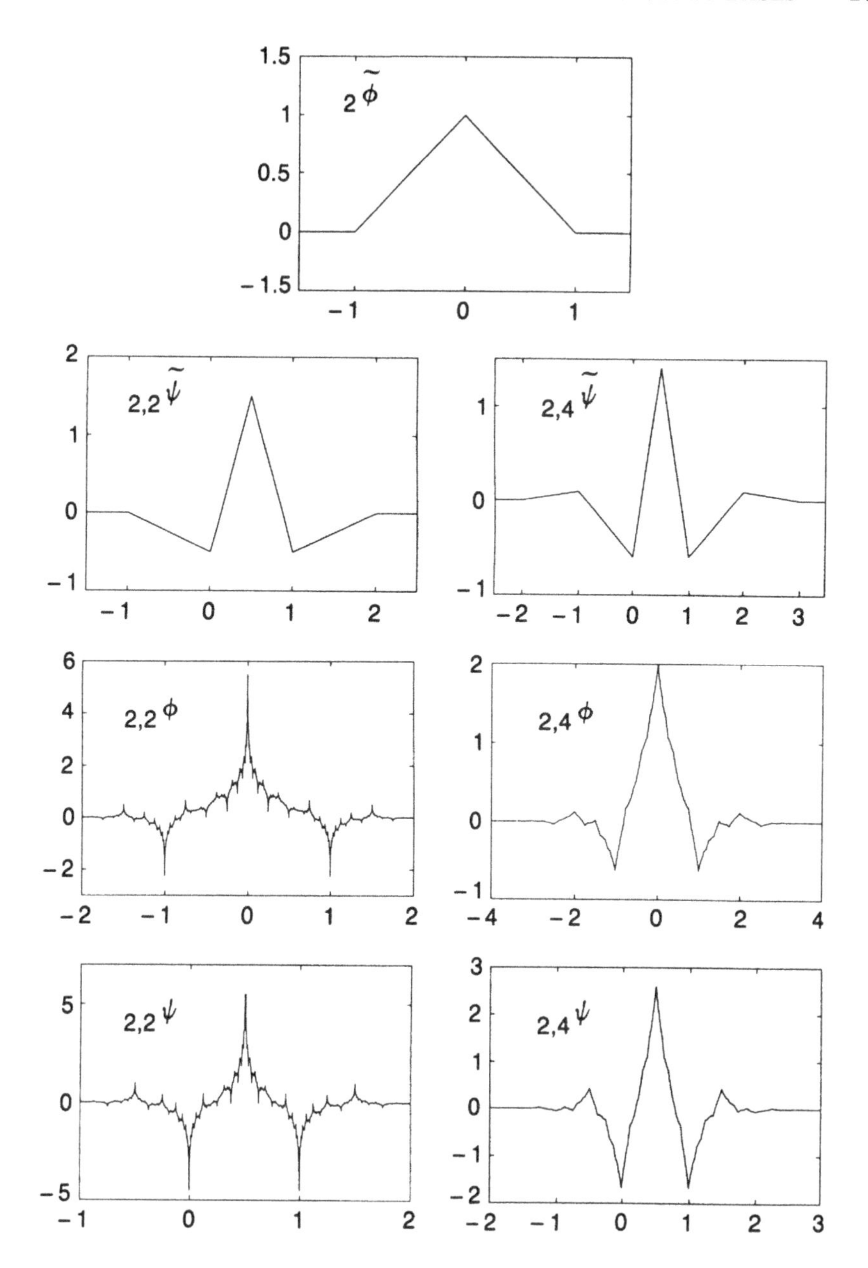

FIG. 8.6. *Spline examples with* $\tilde{N} = 3$, $N = 3, 5, 7$, *and* 9. *For* $N = 1$ *(not plotted),* ${}_{3,1}\phi$ *is not square integrable. Here support* ${}_{3,N}\phi = [-N, N+1]$, *support* ${}_{3,N}\psi =$ *support* ${}_{3,N}\tilde{\psi} = \left[-\frac{N+1}{2}, \frac{N+3}{2}\right]$. *The functions* ${}_{3,3}\phi$ *and* ${}_{3,3}\psi$ *are examples of the fact that the cascade algorithm may diverge while the direct algorithm still converges (see Note* 11 *at the end of Chapter* 6*).*

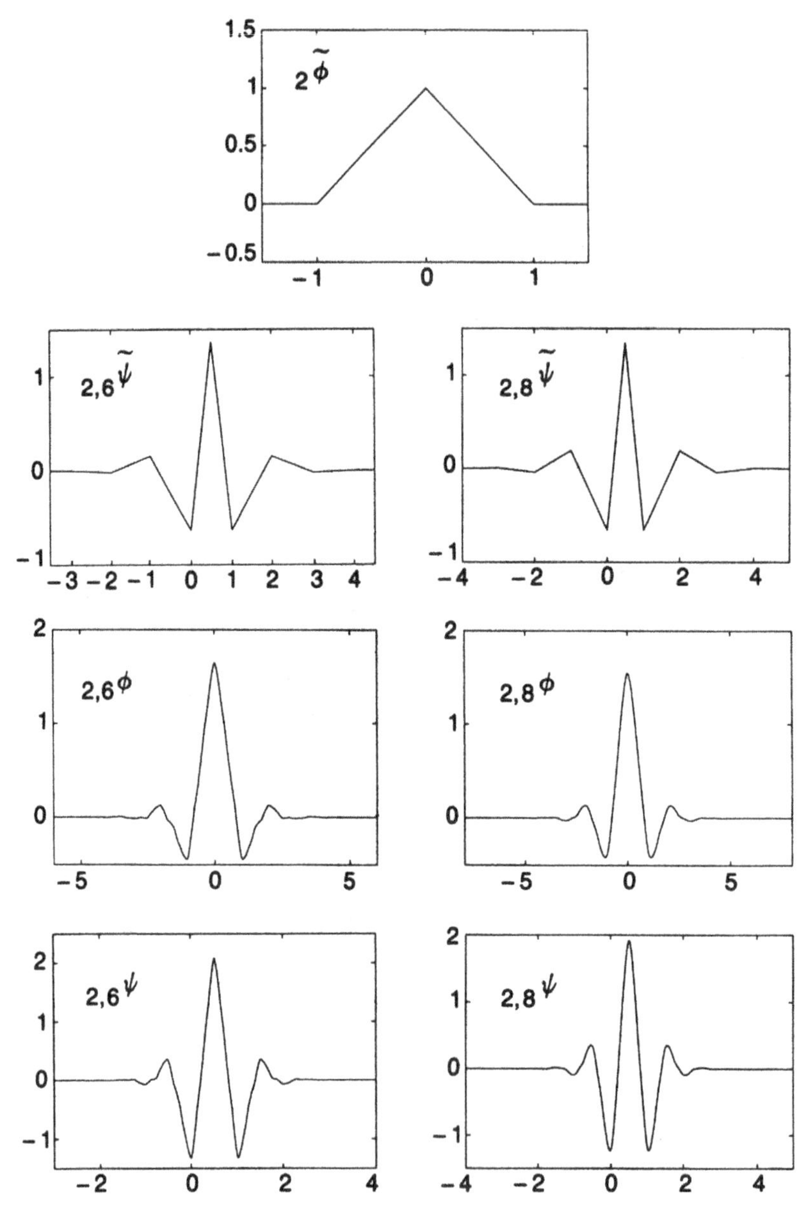

FIG. 8.6. (*continued*).

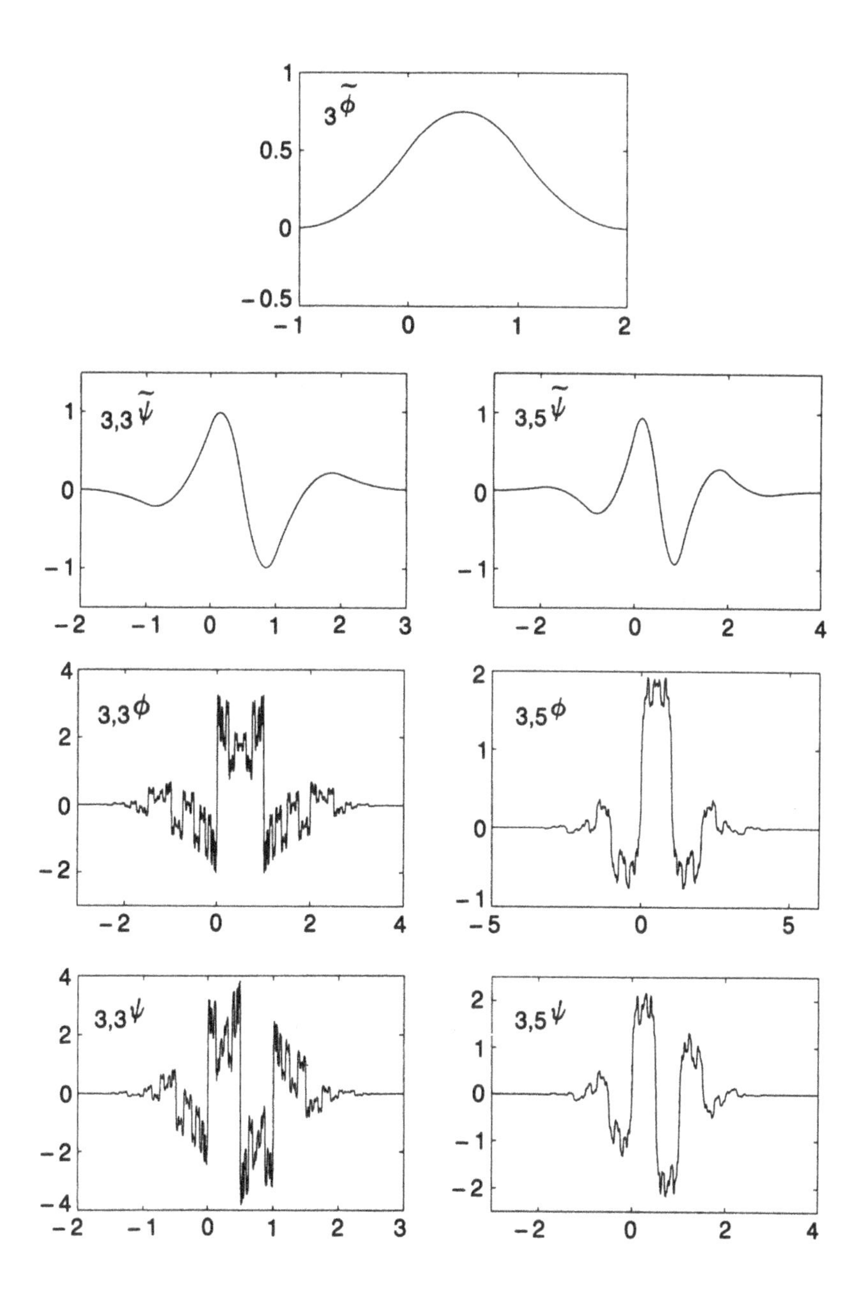

FIG. 8.7. *Spline examples with* $\tilde{N} = 3$, $N = 3, 5, 7$, *and* 9. *For* $N = 1$ *(not plotted),* $_{3,1}\phi$ *is not square integrable. Here support* $_{3,N}\phi = [-N, N+1]$, *support* $_{3,N}\psi =$ *support* $_{3,N}\tilde{\psi} = \left[-\frac{N+1}{2}, \frac{N+3}{2}\right]$.

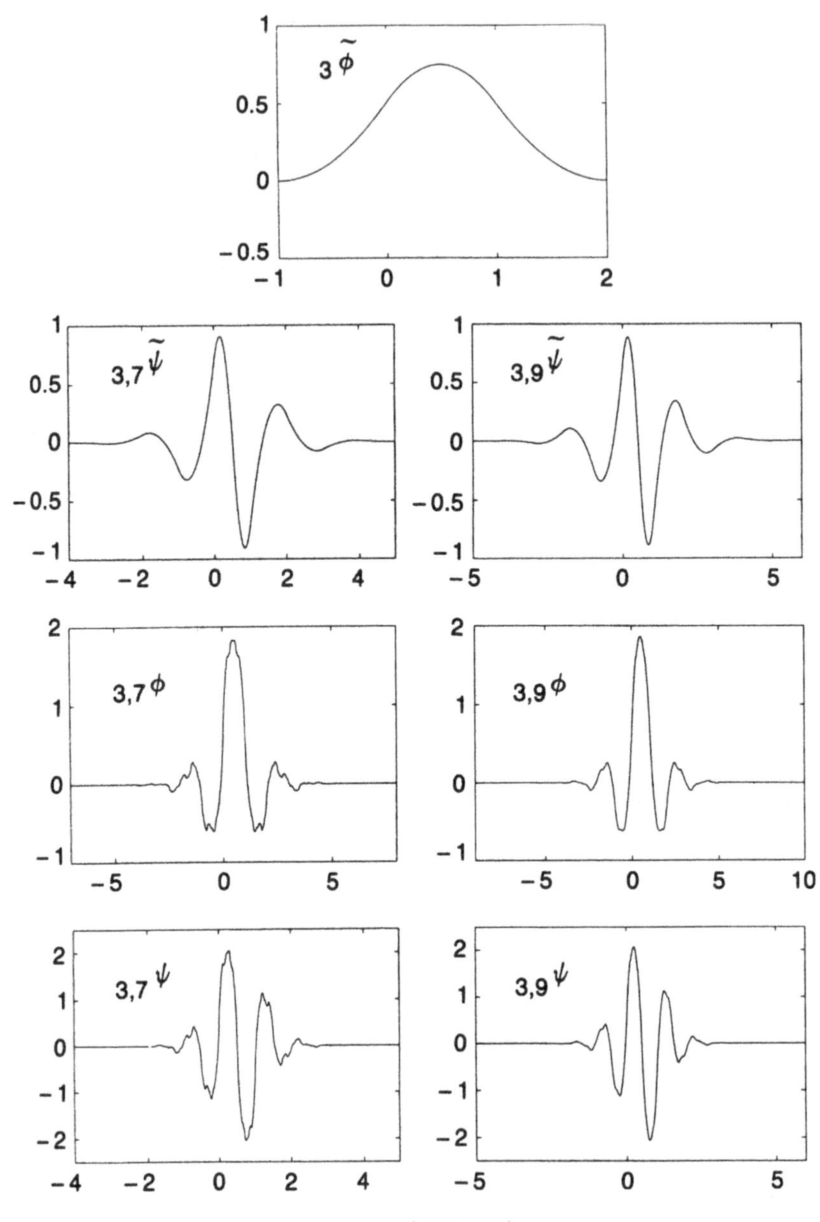

FIG. 8.7. (*continued*).

TABLE 8.2

List of ${}_{\tilde N}\tilde m_0$, ${}_{\tilde N,N}m_0$ for the first few values of $\tilde N, N$, with $z = e^{-i\xi}$. The corresponding filter coefficients ${}_{\tilde N}\tilde h_k$, ${}_{\tilde N,N}h_k$ are obtained by multiplying $\sqrt{2}$ with the coefficient of z^k in ${}_{\tilde N,N}\tilde m_0$, respectively. Note that the coefficients of ${}_{\tilde N,N}m_0$ are always symmetric; for very long ${}_{\tilde N,N}m_0$ we only list about half the coefficients (the others can be deduced by symmetry).

$\tilde N$	${}_{\tilde N}\tilde m_0$	N	${}_{\tilde N,N}m_0$
1	$\frac{1}{2}(1+z)$	1	$\frac{1}{2}(1+z)$
		3	$-\frac{z^{-2}}{16} + \frac{z^{-1}}{16} + \frac{1}{2} + \frac{z}{2} + \frac{z^2}{16} - \frac{z^3}{16}$
		5	$\frac{3}{256}z^{-4} - \frac{3}{256}z^{-3} - \frac{11}{128}z^{-2} + \frac{11}{128}z^{-1} + \frac{1}{2} + \frac{z}{2} + \frac{11}{128}z^2 - \frac{11}{128}z^3 - \frac{3}{256}z^4 + \frac{3}{256}z^5$
2	$\frac{1}{4}(z^{-1}+2+z)$	2	$-\frac{1}{8}z^{-2} + \frac{1}{4}z^{-1} + \frac{3}{4} + \frac{1}{4}z - \frac{1}{8}z^2$
		4	$\frac{3}{128}z^{-4} - \frac{3}{64}z^{-3} - \frac{1}{8}z^{-2} + \frac{19}{64}z^{-1} + \frac{45}{64} + \frac{19}{64}z - \frac{1}{8}z^2 - \frac{3}{64}z^3 + \frac{3}{128}z^4$
		6	$-\frac{5}{1024}z^{-6} + \frac{5}{512}z^{-5} + \frac{17}{512}z^{-4} - \frac{39}{512}z^{-3} - \frac{123}{1024}z^{-2} + \frac{81}{256}z^{-1} + \frac{175}{256} + \frac{81}{256}z - \frac{123}{1024}z^2 \cdots$
		8	$2^{-15}(35z^{-8} - 70z^{-7} - 300z^{-6} + 670z^{-5} + 1228z^{-4} - 3126z^{-3} - 3796z^{-2} + 10718z^{-1} + 22050 + 10718z - 3796z^2 \cdots)$
3	$\frac{1}{8}(z^{-1}+3+3z+z^2)$	1	$-\frac{1}{4}z^{-1} + \frac{3}{4} + \frac{3}{4}z - \frac{1}{4}z^2$
		3	$\frac{3}{64}z^{-3} - \frac{9}{64}z^{-2} - \frac{7}{64}z^{-1} + \frac{45}{64} + \frac{45}{64}z - \frac{7}{64}z^2 - \frac{9}{64}z^3 + \frac{3}{64}z^4$
		5	$-\frac{5}{512}z^{-5} + \frac{15}{512}z^{-4} + \frac{19}{512}z^{-3} - \frac{97}{512}z^{-2} - \frac{13}{256}z^{-1} + \frac{175}{256} + \frac{175}{256}z - \frac{13}{256}z^2 \cdots$
		7	$2^{-14}(35z^{-7} - 105z^{-6} - 195z^{-5} + 865z^{-4} + 336z^{-3} - 3489z^{-2} - 307z^{-1} + 11025 + 11025z \cdots)$
		9	$2^{-17}(-63z^{-9} + 189z^{-8} + 469z^{-7} - 1911z^{-6} - 1308z^{-5} + 9188z^{-4} + 1140z^{-3} - 29676z^{-2} + 190z^{-1} + 87318 + 87318z \cdots)$

The functions ${}_{1,3}\psi$ and ${}_{1,3}\tilde{\psi}$ were first constructed in Tchamitchian (1987) as an example of two dual wavelet bases with very different regularity properties. Here they constitute the first non-orthonormal example of the family ($\tilde{N} = 1 = N$ gives the Haar basis). As in the orthonormal case, arbitrarily high regularity can be attained with these examples, for both ψ and $\tilde{\psi}$. As a spline function, ${}_{\tilde{N},N}\tilde{\psi}$ is piecewise polynomial of degree $\tilde{N} - 1$ and is $C^{\tilde{N}-2}$ at the knots; the regularity of ${}_{\tilde{N},N}\psi$ can be assessed with any of the techniques in Chapter 7. Asymptotically, for large $\tilde{N}$, one finds that ${}_{\tilde{N},N}\psi \in C^m$ if $N > 4.165\,\tilde{N} + 5.165\,(m+1)$. These spline examples have several remarkable features. For one thing, all the filter coefficients are dyadic rationals; since division by 2 can be done very fast on a computer, this makes them very suitable for fast computations. Another attractive property is that the functions ${}_{\tilde{N},N}\psi(x)$ are known exactly and explicitly for all x, unlike the orthonormal compactly supported wavelets we saw before.[11] One disadvantage they have is that m_0 and $\tilde{m}_0$ are very unequal in length, as is apparent from Table 8.2. This is reflected in very different support widths for ϕ and $\tilde{\phi}$; because they are determined by both m_0 and $\tilde{m}_0$, ψ and $\tilde{\psi}$ always have the same support width, given by the average of the filter lengths of m_0, $\tilde{m}_0$, minus 1. The large difference in filter lengths for m_0, $\tilde{m}_0$ can be a nuisance in some applications, such as image analysis.

Examples with less disparate filter lengths. Even if we still take $R \equiv 0$, it is possible to find m_0 and $\tilde{m}_0$ with closer filter lengths by choosing an appropriate factorization of $P(\sin^2 \frac{\xi}{2})$ into $q_0(\cos\xi)$ and $\tilde{q}_0(\cos\xi)$. For fixed $\ell + \tilde{\ell}$ there is a limited number of factorizations. One way to find them is to use spectral factorization again: we determine all the zeros (real and pairs of conjugated complex zeros) of P, so that we can write this polynomial as a product of real first and second order polynomials,

$$P\,(x) \;=\; A \prod_{j=1}^{j_1} (x - x_j)\, \prod_{i=1}^{j_2} (x^2 - 2\mathrm{Re} z_i x + |z_i|^2)\;.$$

Regrouping of these factors leads to all the possibilities for q_0 and $\tilde{q}_0$. Table 8.3 gives the coefficients for m_0, $\tilde{m}_0$ for three examples of this kind, for $\ell + \tilde{\ell} = 4$ and 5. (Note that $\ell + \tilde{\ell} = 4$ is the smallest value for which a non-trivial factorization of this type is possible, with q_0, $\tilde{q}_0$ both real.) For $\ell + \tilde{\ell} = 4$, the factorization is unique, for $\ell + \tilde{\ell} = 5$ there are two possibilities. In both cases we have chosen $\ell, \tilde{\ell}$ so as to make the length difference of $m_0, \tilde{m}_0$ as small as possible. The corresponding wavelets and scaling functions are given in Figures 8.8 and 8.9. In all cases the conditions of §8.4.2 are satisfied.

8.3.5. Biorthogonal bases close to an orthonormal basis. This first example of this family was suggested by M. Barlaud, whose research group in vision analysis tried out the filters in §6A, 6B for image coding (see Antonini et al. (1992)). Because of the popularity of the Laplacian pyramid scheme (Burt and Adelson (1983)), Barlaud wondered whether dual systems of wavelets could

TABLE 8.3

The coefficients of $m_0, \tilde{m}_0$ for three cases of "variations on the spline case" with filters of similar length, corresponding to $\ell + \tilde{\ell} = 4$ and 5 (see text). For each filter we have also given the number of $(\cos \xi/2)$ factors (denoted $N, \tilde{N}$). As in Table 8.2, multiplying the entries below with $\sqrt{2}$ gives the filter coefficients $h_n, \tilde{h}_n$.

$N, \tilde{N}$	n	coefficient of $e^{-in\xi}$ in m_0	coefficient of $e^{-in\xi}$ in $\tilde{m}_0$
	0	.557543526229	.602949018236
$N = 4$	1, −1	.295635881557	.266864118443
$\tilde{N} = 4$	2, −2	−.028771763114	−.078223266529
	3, −3	−.045635881557	−.016864118443
	4, −4	0	.026748757411
	0	.636046869922	.520897409718
$N = 5$	1, −1	.337150822538	.244379838485
$\tilde{N} = 5$	2, −2	−.066117805605	−.038511714155
	3, −3	−.096666153049	.005620161515
	4, −4	−.001905629356	.028063009296
	5, −5	.009515330511	0
	0	.382638624101	.938348578330
$N = 5$	1, −1	.242786343133	.333745161515
$\tilde{N} = 5$	2, −2	.043244142922	−.257235611210
	3, −3	.000197904543	−.083745161515
	4, −4	.015436545027	.038061322045
	5, −5	.007015752324	0

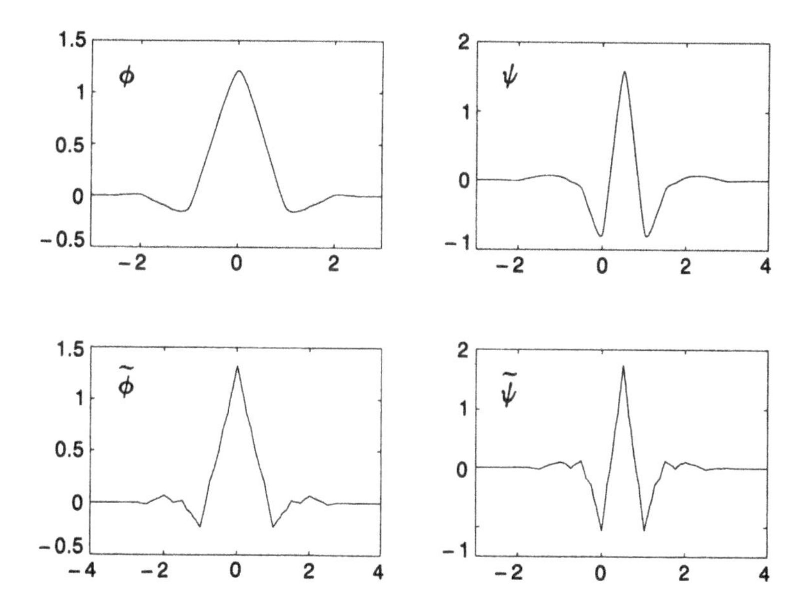

FIG. 8.8. *The functions $\phi, \tilde{\phi}, \psi, \tilde{\psi}$ corresponding to the case $N = 4 = \tilde{N}$ in Table 8.3.*

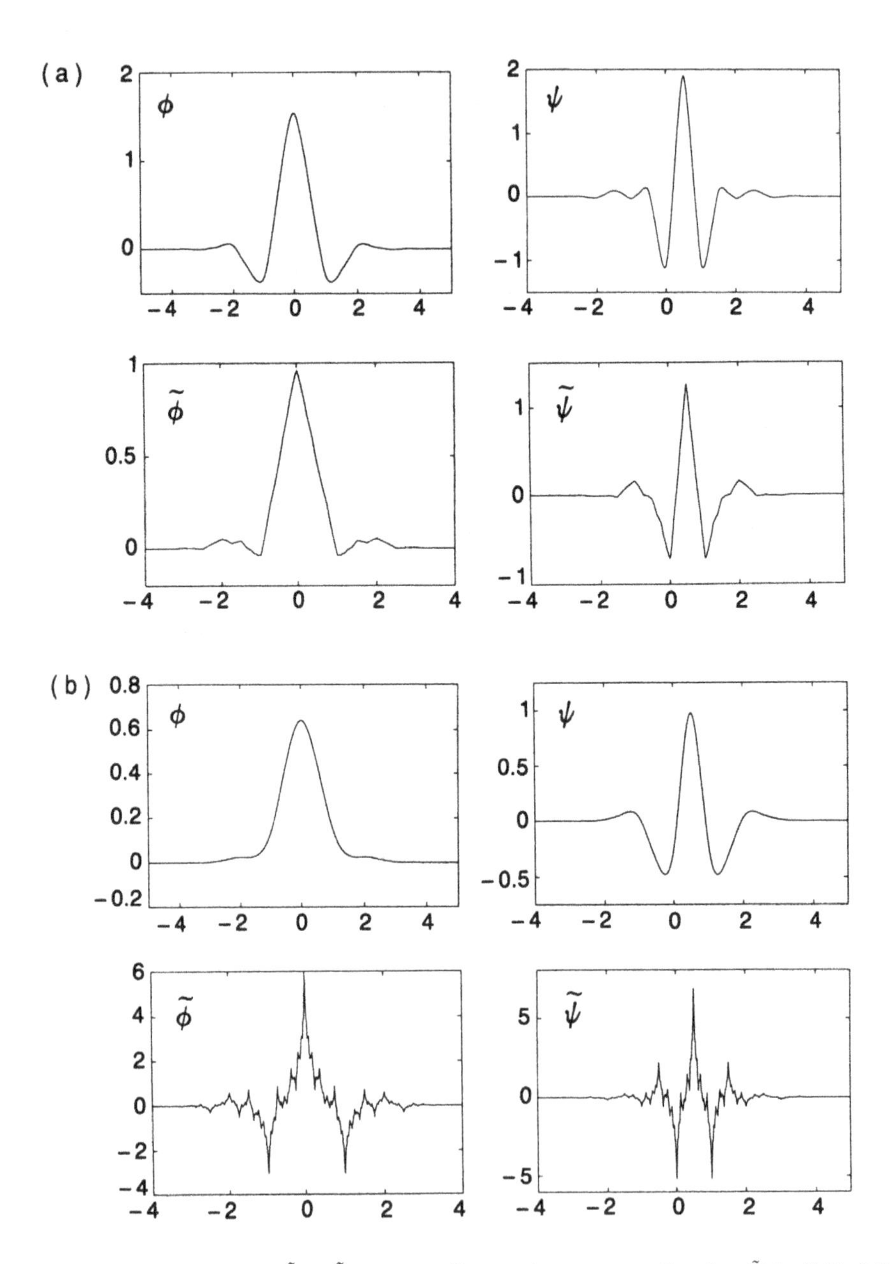

FIG. 8.9. *The functions $\phi, \tilde{\phi}, \psi, \tilde{\psi}$ corresponding to the two cases $N = 5 = \tilde{N}$ in Table 8.3.*

be constructed, using the Laplacian pyramid filter as either m_0 or $\tilde{m}_0$. These filters are given explicitly by

$$- a\, e^{-2i\xi} \;+\; .25e^{-i\xi} \;+\; (.5+2a) \;+\; .25e^{i\xi} \;-\; a\, e^{2i\xi} \,. \tag{8.3.27}$$

For $a = -1/16$, this reduces to the spline filter ${}_4\tilde{m}_0$ as described under the "spline examples" above. For applications in vision, the choice $a = .05$ is especially popular: even though the corresponding $\tilde{\phi}$ has less regularity than ${}_4\tilde{\phi}$, it seems to lead to results that are better from the point of view of visual perception. Following Barlaud's suggestion, we chose therefore $a = .05$ in (8.3.27), or

$$\begin{aligned} m_0(\xi) &= .6 \;+\; .5 \cos\xi \;-\; .1 \cos 2\xi \\ &= \left(\cos\frac{\xi}{2}\right)^2 \left(1 \;+\; \frac{4}{5}\sin^2\frac{\xi}{2}\right) . \end{aligned} \tag{8.3.28}$$

Candidates for $\tilde{m}_0$ dual to this m_0 have to satisfy

$$m_0(\xi)\,\overline{\tilde{m}_0(\xi)} \;+\; m_0(\xi+\pi)\,\overline{\tilde{m}_0(\xi+\pi)} \;=\; 1 \,.$$

As shown in §8.4.4, such $\tilde{m}_0$ can be chosen to be symmetric (since m_0 is symmetric); we also opt for $\tilde{m}_0$ divisible by $(\cos\xi/2)^2$ (so that the corresponding ψ, $\tilde{\psi}$ both have two zero moments). In other words,

$$\tilde{m}_0(\xi) \;=\; \left(\cos\frac{\xi}{2}\right)^2 P\left(\sin^2\frac{\xi}{2}\right) ,$$

where

$$(1-x)^2 \left(1 \;+\; \frac{4}{5}x\right) P(x) \;+\; x^2\left(\frac{9}{5} \;-\; \frac{4}{5}x\right) P(1-x) \;=\; 1 \,.$$

By Theorem 6.1.1, together with the symmetry of this equation for substitution of x by $1-x$, this equation has a unique solution P of degree 2, which is easily found to be

$$P(x) \;=\; 1 \;+\; \frac{6}{5}x \;-\; \frac{24}{35}x^2 \,.$$

This leads to

$$\tilde{m}_0(\xi) \;=\; \left(\cos\frac{\xi}{2}\right)^2 \left(1 + \frac{6}{5}\sin^2\frac{\xi}{2} \;-\; \frac{24}{35}\sin^4\frac{\xi}{2}\right) \tag{8.3.29}$$

$$\begin{aligned} &= -\frac{3}{280}e^{-3i\xi} \;-\; \frac{3}{56}e^{-2i\xi} \;+\; \frac{73}{280}e^{-i\xi} \;+\; \frac{17}{28} \;+\; \frac{73}{280}e^{i\xi} \\ &\quad -\frac{3}{56}e^{2i\xi} \;-\; \frac{3}{280}e^{3i\xi} \,. \end{aligned} \tag{8.3.30}$$

One can check that both (8.3.28) and (8.3.29) satisfy all the conditions in §8.4.2. It follows that these m_0 and $\tilde{m}_0$ do indeed correspond to a pair of biorthogonal wavelet bases. Figure 8.10 shows graphs of the corresponding ϕ, $\tilde{\phi}$, ψ and $\tilde{\psi}$. All four functions are continuous but not differentiable. It is very striking how similar $\tilde{\phi}$ and ϕ are, or ψ and $\tilde{\psi}$. This can be traced back to a similarity of m_0 and $\tilde{m}_0$, which is not immediately obvious from (8.3.27) and (8.3.30), but becomes apparent by comparison of the explicit numerical values of the filter coefficients, as in Table 8.4. In fact, both filters are very close to the (necessarily nonsymmetric) filter corresponding to one of the orthonormal coiflets (see §8.3), which we list again, for comparison, in the third column in Table 8.4. This proximity of m_0 to an orthonormal wavelet filter explains why the $\tilde{m}_0$ dual to m_0 is so close to m_0 itself. A first application to image analysis of these biorthogonal bases associated to the Laplacian pyramid is given in Antonini et al. (1992).

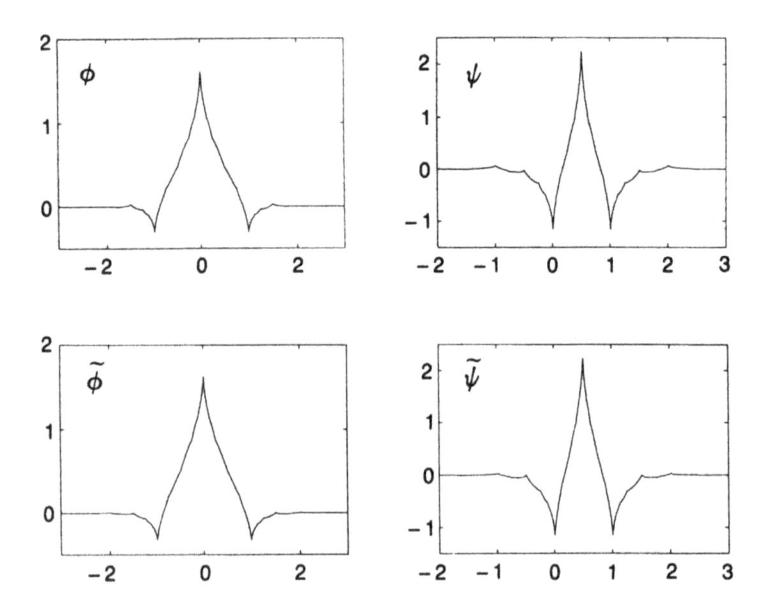

FIG. 8.10. *Graphs of $\phi, \psi, \tilde{\phi}, \tilde{\psi}$ for the biorthogonal pair constructed from the Burt–Adelson low-pass filter.*

M. Barlaud's suggestion led to the accidental discovery that the Burt filter is very close to an orthonormal wavelet filter. (One wonders whether this closeness makes the filter so effective in applications?) This example suggested that maybe other biorthogonal bases, with symmetric filters and rational filter coefficients, can be constructed by approximating and "symmetrizing" existing orthonormal wavelet filters, and computing the corresponding dual filter. The coiflet coefficients listed in §8.3 were obtained via a construction method that naturally led to close to symmetric filters; it is natural, therefore, to expect that symmetric biorthogonal filters close to an orthonormal basis will in fact be close to these

TABLE 8.4

Filter coefficients for $(m_0)_{\text{Burt}}$, *for the dual filter* $(\tilde{m}_0)_{\text{Burt}}$ *computed in this section, and for a very close filter* $(m_0)_{\text{coiflet}}$ *corresponding to an orthonormal basis of coiflets (see the entries for* $K = 1$ *in Table* 8.1).

n	$(m_0)_{\text{Burt}}$	$(\tilde{m}_0)_{\text{Burt}}$	$(m_0)_{\text{coiflet}}$
−3	0.	−.010714285714	0.
−2	−.05	−.053571428571	−.051429728471
−1	.25	.260714285714	.238929728471
0	.6	.607142857143	.602859456942
1	.25	.260714285714	.272140543058
2	−.05	−.053571428571	−.051429972847
3	0.	−.010714285714	−.011070271529

coiflet bases. The analysis in §8.3 suggests, therefore,

$$m_0(\xi) \; = \; (\cos\ \xi/2)^{2K} \left[\sum_{k=1}^{K-1} \binom{K-1+k}{k} (\sin\ \xi/2)^{2k} \; + \; O((\sin\ \xi/2)^{2K}) \right] .$$

In the examples below we have chosen in particular

$$m_0(\xi) \; = \; (\cos\ \xi/2)^{2K} \left[\sum_{k=0}^{K-1} \binom{K-1+k}{k} (\sin\ \xi/2)^{2k} \; + \; a(\sin\ \xi/2)^{2K} \right]$$

and we have then followed the following procedure:

1. Find a such that $\left| \int_{-\pi}^{\pi} d\xi \; [1 - |m_0(\xi)|^2 - |m_0(\xi+\pi)|^2] \right|$ is minimal (zero in the examples below). This optimization criterium can of course be replaced by other criteria (e.g., least sum of squares of *all* the Fourier coefficients of $1 - |m_0(\xi)|^2 - |m_0(\xi+\pi)|^2$ instead of only the coefficient of $e^{i\ell\xi}$ with $\ell = 0$). For the cases $K = 1, 2, 3$, the smallest root for a is .861001748086, 3.328450120793, 13.113494845221, respectively.

2. Replace this (irrational) "optimal" value for a by a close value expressible as a simple fraction.[12] For our examples $a = .8 = 4/5$ was chosen for $K = 1$, $a = 3.2 = 16/5$ for $K = 2$ and $a = 13$ for $K = 3$. For $K = 1$, this reduces then to the example above.

3. Since m_0 is now fixed, we can compute $\tilde{m}_0$. If we require that $\tilde{m}_0$ be also divisible by $(\cos\ \xi/2)^{2K}$, then

$$\tilde{m}_0(\xi) \; = \; (\cos\ \xi/2)^{2K} \; P_K((\sin\ \xi/2)^2) \; , \qquad (8.3.31)$$

where P_K is a polynomial of degree $3K - 1$. The same analysis as in Daubechies (1990) shows that

$$P_K(x) \; = \; \sum_{k=0}^{K-1} \binom{K-1+k}{k} x^k \; + \; O(x^K) \; ,$$

thereby determining already K of the $3K$ coefficients of P_K. The others can be computed easily. For $K = 2$ and 3 we find

$$P_2(x) = 1 + 2x + \frac{14}{5}x^2 + 8x^3 - \frac{8024}{455}x^4 + \frac{3776}{455}x^5 , \qquad (8.3.32)$$

$$P_3(x) = 1 + 3x + 6x^2 + 7x^3 + 30x^4 + 42x^5 - \frac{1721516}{6075}x^6 + \frac{1921766}{6075}x^7 - \frac{648908}{6075}x^8 . \qquad (8.3.33)$$

In Table 8.5 we list the explicit numerical values of the filter coefficients for m_0, $\tilde{m}_0$ and the closest coiflet, for $K = 2$ and 3. We have graphed ϕ, $\tilde{\phi}$, ψ, and $\tilde{\psi}$ for both cases in Figure 8.11. It is worthwhile to note that the computation of the biorthogonal filters m_0, $\tilde{m}_0$, as explained by the above procedure, is much simpler than the computation in Daubechies (1990) of the orthonormal coiflet filters! This illustrates the greater flexibility of the construction of biorthogonal wavelet bases versus orthonormal wavelet bases.

TABLE 8.5

Numerical values for the filters $m_0, \tilde{m}_0$ for biorthogonal bases close to coiflets, for the cases $K = 2$ and 3 (see text). The third column lists the coefficients of the orthonormal coiflet filter to which m_0 and $\tilde{m}_0$ are very close. In order to compare the different coefficients more easily, we have expressed everything in decimal notation; in fact, the coefficients of m_0 and $\tilde{m}_0$ are rational.

K	n	coefficients of m_0	coefficients of $\tilde{m}_0$	coefficients of $(m_0)_{\text{coiflet}}$ $n \leq 0$	$n \geq 0$
2	0	.575	.575291895604	.574682393857	
	± 1	.28125	.286392513736	.273021046535	.294867193696
	± 2	−.05	−.052305116758	−.047639590310	−.054085607092
	± 3	−.03125	−.039723557692	−.029320137980	−.042026480461
	± 4	.0125	.015925480769	.011587596739	.016744410163
	± 5	0	.003837568681	0	.003967883613
	± 6	0	−.001266311813	0	−.001289203356
	± 7	0	−.000506524725	0	−.000509505399
3	0	.5634765625	.560116167736	.561285256870	
	± 1	.29296875	.296144908701	.286503335274	.302983571773
	± 2	−.047607421875	−.047005100329	−.043220763560	−.050770140755
	± 3	−.048828125	−.055220135661	−.046507764479	−.058196250762
	± 4	.01904296875	.021983637555	.016583560479	.024434094321
	± 5	.005859375	.010536373594	.005503126709	.011229240962
	± 6	−.003173828125	−.005725661541	−.002682418671	−.006369601011
	± 7	0	−.001774953991	0	−.001820458916
	± 8	0	.000736056355	0	.000790205101
	± 9	0	.000339274308	0	.000329665174
	± 10	0	−.000047015908	0	−.000050192775
	± 11	0	−.000025466950	0	−.000024465734

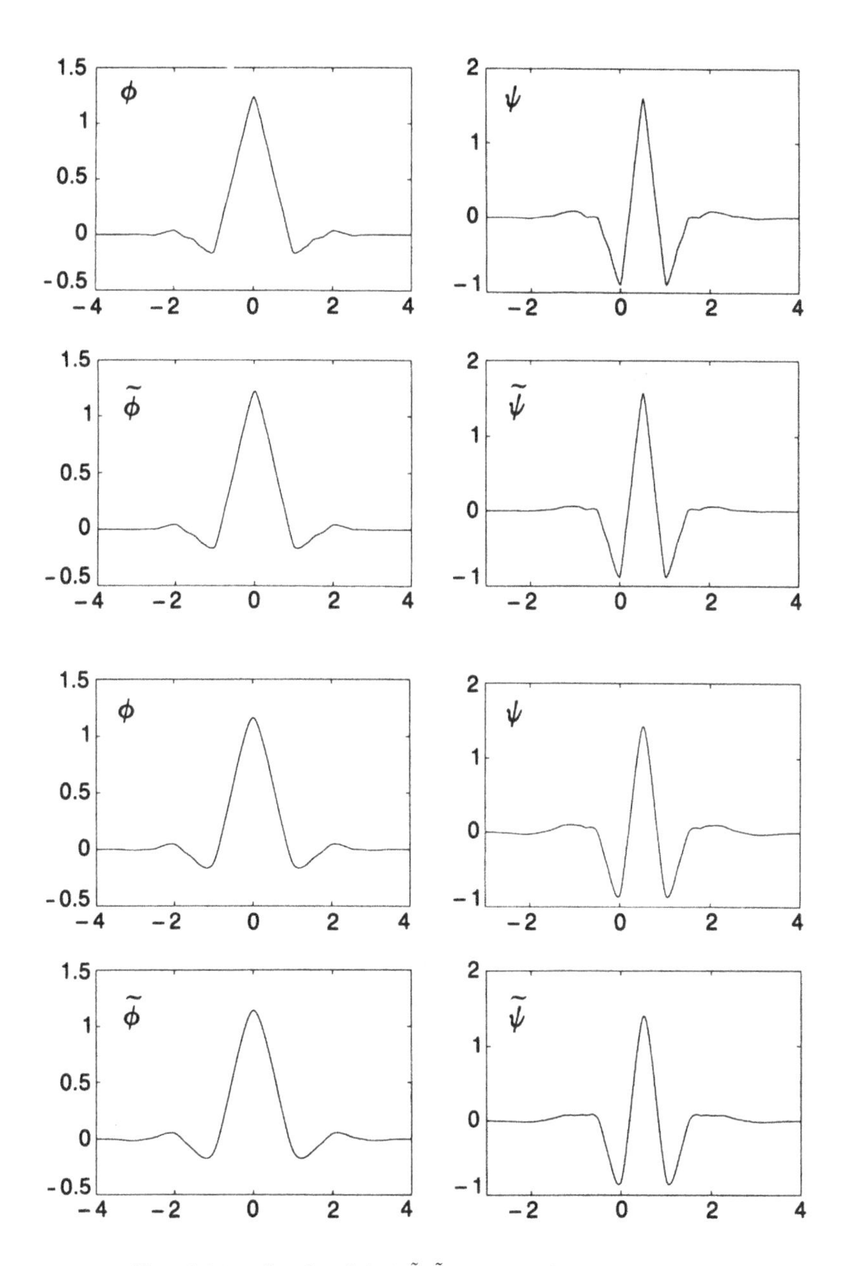

FIG. 8.11. *Graphs of* $\phi, \psi, \tilde{\phi}, \tilde{\psi}$ *corresponding to Table* 8.5.

Notes.

1. In the sense that the $\phi_1(\cdot - n)$ are orthonormal, as are the $\phi_2(\cdot - n)$.

2. Strictly speaking, Lemma 6.2.2 only proves support$(\phi) \subset [0, N]$. A recent paper by Lemarié and Malgouyres (1991) shows that support(ϕ) is necessarily an interval, which in this case then has to be $[0, N]$.

3. Nevertheless, AWARE, Inc. uses the asymmetric filters from §6.4 with excellent results in image and video coding. Note also that "perceptually" small or large errors are difficult to quantify mathematically; the norm most often used to measure "distance" is the ℓ^2-norm, but that is more because this is the easiest norm to handle than for any other reason. All experts agree that the ℓ^2-norm is not a good candidate for a "perceptual" norm, but as far as I know, there is no agreement on a better candidate.

4. The ${}_N\phi$ from §6.4 do not have this property. The graph of $|{}_N\hat{\phi}(\xi)|$ is very flat near $\xi = 0$, showing that $\frac{d^\ell}{d\xi^\ell} \left. |{}_N\hat{\phi}| \right|_{\xi=0} = 0$ for $\ell = 1 \cdots N$, but the phase of ${}_N\hat{\phi}(\xi)$ does not share this property.

5. The proof that T_0 is a bounded operator is easy: if support $\phi = [-N_1, N_2]$, then

$$
\begin{aligned}
|\langle f, \phi_{0,k}\rangle|^2 &= \left| \int dx \; f(x) \; \overline{\phi(x-k)} \right|^2 \\
&\le \left(\int_{-N_1-k}^{N_2-k} dx \; f(x)|^2 \right)^{1/2} \|\phi\|^2 \qquad \text{(by Cauchy–Schwarz);}
\end{aligned}
$$

hence

$$
\begin{aligned}
\sum_k |\langle f, \phi_{0,k}\rangle|^2 &\le \|\phi\|^2 \sum_k \cdot \int_{-N_1-k}^{N_2-k} dx \; |f(x)|^2 \\
&\le \|\phi\|^2 \, (N_2 + N_1) \, \|f\|^2 \, .
\end{aligned}
$$

Similarly, one proves that all the T_j are bounded.

6. We have

$$
\|f\|^2 = \inf_{\|g\| \le 1} |\langle f, g\rangle|^2 \le \inf_{\|g\| \le 1} \left[\lim_{J \to \infty} \sum_{j=-J}^{J} \sum_{\ell} |\langle f, \psi_{j\ell}\rangle| |\langle \tilde{\psi}_{j,\ell}, g\rangle| \right]^2
$$

$$
\le \inf_{\|g\| \le 1} \left(\sum_{j,\ell} |\langle f, \psi_{j,\ell}\rangle|^2 \right) \left(\sum_{j,\ell} |\langle \tilde{\psi}_{j,\ell}, g\rangle|^2 \right)
$$

$$\leq \inf_{\|g\|\leq 1} \left(\sum_{j,\ell} |\langle f, \psi_{j,\ell}\rangle|^2 \right) \tilde{A} \, \|g\|^2$$

$$\leq \tilde{A} \sum_{j,\ell} |\langle f, \psi_{j,\ell}\rangle|^2 \, .$$

7. The angle between two subspaces is defined as the minimum angle between elements,

$$\text{angle } (E, F) = \inf_{e\in E,\ f\in F} \cos^{-1} \frac{|\langle e, f\rangle|}{\|e\| \ \|f\|} \, .$$

8. The proof in Cohen, Daubechies, and Feauveau (1992) imposes a much stronger decay condition on $\hat{\phi}$, namely $|\hat{\phi}(\xi)| \leq C(1 + |\xi|)^{-1/2-\epsilon}$ (which is known not to be satisfied in even some orthonormal cases) in order to derive (8.3.15) and (8.3.16). The argument sketched here comes from Cohen and Daubechies (1992).

9. The derivatives $\psi^{(\ell)}$, $\ell = 0 \cdots m$, are automatically bounded because ψ has compact support.

10. In the case $N = 2 = \tilde{N}$, a curious phenomenon happens. The function ${}_{2,2}\phi$, although an element of $L^2([-2, 2])$ (hence also of $L^1([-2, 2])$, has in fact a singularity at every dyadic rational. The true graph of ${}_{2,2}\phi$ (or ${}_{2,2}\psi$) would therefore consist of a black rectangle (since our lines have some thickness), but the graph in Fig. 8.6 is nevertheless a close approximation in L^2 or L^1, although not in L^∞. I would like to thank Win Sweldens for pointing this out to me.

11. Auscher (1989) and Chui and Wang (1991) contain another construction of non-orthonormal wavelet bases where one of the two wavelets, say ψ, is a compactly supported spline function, and is therefore also known exactly and explicitly everywhere. In this construction the W_j-spaces are orthogonal, unlike here, and $\tilde{W}_j = W_j$. As a result, the dual wavelet $\tilde{\psi}$ has infinite support (compact support for both ψ, $\tilde{\psi}$ can only be achieved by giving up the orthogonality of the W_j), with exponential decay. The associated multiresolution analysis is the same as for the Battle–Lemarié wavelets; ψ is chosen so that it is orthogonal to the B-spline of the right order and all its integer translates, and $\tilde{\psi}$ is then given by $\hat{\tilde{\psi}}(\xi) = \hat{\psi}(\xi)/[\sum_k |\hat{\psi}(\xi + 2\pi k)|^2]$.

12. Choosing a rational leads to m_0, $\tilde{m}_0$ with rational coefficients. Note that there is nothing sacred about the original irrational values of a: changing criterion in point 1 will lead to slightly different values of a.

CHAPTER 9

Characterization of Functional Spaces by Means of Wavelets

The major message of this chapter is that the orthonormal bases we have discussed for the last four chapters also give good (i.e., unconditional) bases for many other spaces than L^2, out-performing the Fourier basis functions in this respect. Almost all the material in this chapter is borrowed from Meyer (1990), but it is here presented in (I believe) a more pedestrian way, accessible to readers with a lower level of mathematical sophistication. (Meyer's book also contains much more on this subject than is explained in this chapter.) In §9.1 I start by reviewing a classic theorem of pure harmonic analysis, the Calderón–Zygmund decomposition. It can be found also in many textbooks (such as Stein (1970)); I include a detailed proof here as an illustration of techniques using different (dyadic) scales, practiced in pure harmonic analysis long before wavelets came along. Together with some other classic theorems, it leads to the proof that wavelets are an unconditional bases for L^p, $1 < p < \infty$. Section 9.2 lists the characterizations, by means of wavelets, of other functional spaces, without proof. Also included is a short discussion on the detection of singularities with orthonormal wavelet bases. Section 9.3 treats expansions of L^1-functions by means of wavelets; since L^1 has no unconditional bases, wavelets cannot do the impossible, but they still do a better job than Fourier expansions. Finally, §9.4 points out an amusing difference in emphasis between wavelet and Fourier expansions.

9.1. Wavelets: Unconditional bases for $\mathbf{L}^p(\mathbb{R})$, $1 < \mathbf{p} < \infty$.

We start by proving the Calderón–Zygmund decomposition theorem.

THEOREM 9.1.1. *Suppose f is a positive function in $L^1(\mathbb{R})$. Fix $\alpha > 0$. Then $\mathbb{R}$ can be decomposed as follows:*

1. $\mathbb{R} = G \cup B$, with $G \cap B = \emptyset$.

2. *On the "good" set G, $f(x) \le \alpha$ a.e.*

3. *The "bad" set B can be written as*

$$B = \bigcup_{k \in \mathbb{N}} Q_k, \quad \textit{where the } Q_k \textit{ are non-overlapping intervals,}$$

$$\text{and} \quad \alpha \leq |Q_k|^{-1} \int_{Q_k} dx\ f(x) \leq 2\alpha, \quad \text{for all} \quad k \in \mathbb{N}\ .$$

Proof.

1. Choose $L = 2^\ell$ so that $2^{-\ell} \int_{\mathbb{R}} dx\ f(x) \leq \alpha$. It follows that $L^{-1} \int_{kL}^{(k+1)L} dx\ f(x) \leq \alpha$ for all $k \in \mathbb{Z}$. This defines a first partition of $\mathbb{R}$.

2. Take a fixed interval $Q = [kL,\ (k+1)L[$ in this first partition. Split it into two halves, $[kL,\ (k+\frac{1}{2})L[$ and $[(k+\frac{1}{2})L,\ (k+1)L[$. Take either of the halves, call it Q', and compute $I_{Q'} = |Q'|^{-1} \int_{Q'} dx\ f(x)$. If $I_{Q'} > \alpha$, then put Q' in the bag of intervals that will make up B. We have indeed
$$\alpha < I_{Q'} \leq |Q'|^{-1} \int_Q dx\ f(x) = 2|Q|^{-1} \int_Q dx\ f(x) \leq 2\alpha\ .$$
If $I_{Q'} < \alpha$, keep going (split into halves, etc.), if necessary, ad infinitum. Do the same for the other half of Q, and also for all the other intervals $[kL,\ (k+1)L[$. At the end we have a countable bag of "bad" intervals which all satisfy the equation at the top of this page; call their union B and the complement set G.

3. By the construction of B, we find that for any $x \notin B$, there exists an infinite sequence of smaller and smaller intervals $Q_1, Q_2, Q_3, \cdots$ so that $x \in Q_n$ for every n, and $|Q_n|^{-1} \int_{Q_n} dy\ f(y) \leq \alpha$. In fact, $|Q_j| = \frac{1}{2}|Q_{j-1}|$ for every j, and $Q_j \subset Q_{j-1}$. Because the Q_n "shrink to" x,
$$|Q_n|^{-1} \int_{Q_n} dy\ f(y) \longrightarrow f(x) \quad \text{almost surely}\ .$$
Since the left side is $\leq \alpha$ by construction, it follows that $f(x) \leq \alpha$ a.e. in G. ■

Note that the choice $L = 2^\ell$ implies that all the intervals occurring in this proof are automatically dyadic intervals, i.e., of the form $[k2^{-j},\ (k+1)2^{-j}[$ for some $k, j \in \mathbb{Z}$.

Next we define Calderón–Zygmund operators and prove a classical property.

DEFINITION.[1] *A Calderón–Zygmund operator T on $\mathbb{R}$ is an integral operator*

$$(Tf)(x) = \int dy\ K(x,y)\ f(y) \tag{9.1.1}$$

for which the integral kernel satisfies

$$|K(x,y)| \leq \frac{C}{|x-y|}\ , \tag{9.1.2}$$

$$\left|\frac{\partial}{\partial x}\ K(x,y)\right| + \left|\frac{\partial}{\partial y}\ K(x,y)\right| \leq \frac{C}{|x-y|^2}\ , \tag{9.1.3}$$

and which defines a bounded operator on $L^2(\mathbb{R})$.

THEOREM 9.1.2. *A Calderón–Zygmund operator is also a bounded operator from $L^1(\mathbb{R})$ to $L^1_{\text{weak}}(\mathbb{R})$.*

The space $L^1_{\text{weak}}(\mathbb{R})$ in this theorem is defined as follows.

DEFINITION. $f \in L^1_{\text{weak}}(\mathbb{R})$ *if there exists* $C > 0$ *so that, for all* $\alpha > 0$,

$$\Big|\{x;\ |f(x)| \geq \alpha\}\Big| \leq \frac{C}{\alpha} \,. \tag{9.1.4}$$

The infinum of all C for which (9.1.4) holds (for all $\alpha > 0$) is sometimes called $\|f\|_{L^1_{\text{weak}}}$.[2]

EXAMPLES.

1. If $f \in L^1(\mathbb{R})$, then (9.1.4) is automatically satisfied. Indeed, if $S_\alpha = \{x;\ |f(x)| \geq \alpha\}$, then

$$\alpha \cdot |S_\alpha| \leq \int_{S_\alpha} dx\ |f(x)| \leq \int_{\mathbb{R}} dx\ |f(x)| \ = \ \|f\|_{L^1} \,;$$

   hence

$$\|f\|_{L^1_{\text{weak}}} \leq \|f\|_{L^1} \,.$$

2. $f(x) = |x|^{-1}$ is in L^1_{weak}, since $|\{x;\ |x|^{-1} \geq \alpha\}| = \frac{2}{\alpha}$. However, $f(x) = |x|^{-\beta}$ is not in L^1_{weak} if $\beta > 1$.

The name L^1_{weak} is justified by these examples: L^1_{weak} extends L^1, and contains the functions f for which $\int |f|$ "just" misses to be finite because of logarithmic singularities in the primitive of $|f|$.

We are now ready for the proof of the theorem.

Proof of Theorem 9.1.2.

1. We want to estimate $|\{x;\ |Tf(x)| \geq \alpha\}|$. We start by making a Calderón–Zygmund decomposition of $\mathbb{R}$ for the function $|f|$, with threshold α. Define now

$$g(x) = \begin{cases} f(x) & \text{if } \ x \in G\,, \\ |Q_k|^{-1} \displaystyle\int_{Q_k} dy\ f(y) & \text{if } \ x \in \text{ interior of } Q_k\,, \end{cases}$$

$$b(x) = \begin{cases} 0 & \text{if } \ x \in G\,, \\ f(x) - |Q_k|^{-1} \displaystyle\int_{Q_k} dy\ f(y) & \text{if } \ x \in \text{ interior of } Q_k\,. \end{cases}$$

   Then $f(x) = g(x) + b(x)$ a.e.; hence $Tf = Tg + Tb$. It follows that $|Tf(x)| \geq \alpha$ is only possible if either $|Tg(x)| \geq \alpha/2$ or $|Tb(x)| \geq \alpha/2$ (or both); consequently,

$$
\begin{aligned}
&|\{x;\ |Tf(x)| \geq \alpha\}| \\
&\quad \leq \left|\left\{x;\ |Tg(x)| \geq \frac{\alpha}{2}\right\}\right| + \left|\left\{x;\ |Tb(x)| \geq \frac{\alpha}{2}\right\}\right| .
\end{aligned}
\tag{9.1.5}
$$

The theorem will therefore be proved if each of the terms in the right-hand side of (9.1.5) is bounded by $\frac{C}{\alpha}\ \|f\|_{L^1}$.

2. We have

$$
\begin{aligned}
\left(\frac{\alpha}{2}\right)^2 \left|\left\{x;\ |Tg(x)| \geq \frac{\alpha}{2}\right\}\right| &\leq \int\limits_{\{x;\ |Tg(x)| \geq \frac{\alpha}{2}\}} dx \quad |Tg(x)|^2 \\
&\leq \int_{\mathbb{R}} dx\ |Tg(x)|^2 \ = \ \|Tg\|_{L^2}^2 \leq C\|g\|_{L^2}^2 \ ,
\end{aligned}
\tag{9.1.6}
$$

because T is a bounded operator on L^2. Moreover,

$$
\begin{aligned}
\|g\|_{L^2}^2 \ &= \ \int_G dx\ |g(x)|^2 \ + \ \int_B dx\ |g(x)|^2 \\
&\leq \ \alpha \int_G dx\ |f(x)| \ + \ \sum_k |Q_k| \left| \frac{1}{|Q_k|} \int_{Q_k} dy\ f(y) \right|^2 \\
&\qquad \text{(use the definition of } g, \text{ and } |f(x)| \leq \alpha \text{ on } G) \\
&\leq \ \alpha \int_G dx\ |f(x)| \ + \ \sum_k 2\alpha \int_{Q_k} dy\ |f(y)| \\
&\qquad \left(\text{use } |Q_k|^{-1} \int_{Q_k} dy\ |f(y)| \leq 2\alpha\right) \\
&\leq \ 2\alpha \int_{\mathbb{R}} dx\ |f(x)| \ = \ 2\alpha\ \|f\|_{L^1} \ .
\end{aligned}
$$

Combining this with (9.1.6), we obtain

$$
\left|\left\{x;\ |Tg(x)| \geq \frac{\alpha}{2}\right\}\right| \leq \frac{8}{\alpha}\ C\ \|f\|_{L^1} \ . \tag{9.1.7}
$$

3. We now concentrate on b. For each k, we define new intervals Q_k^* by "stretching" the Q_k: Q_k^* has the same center y_k as Q_k, but twice its length. We define then $B^* \ = \ \cup_k Q_k^*$, and $G^* \ = \ \mathbb{R} \backslash B^*$. Now

$$
\begin{aligned}
|B^*| \ &\leq \ \sum_k |Q_k^*| \ = \ 2 \sum_k |Q_k| \\
&\leq \ \frac{2}{\alpha} \sum_k \int_{Q_k} dx\ |f(x)| \\
&\qquad \left(\text{because } |Q_k|^{-1} \int_{Q_k} dx\ |f(x)| \geq \alpha\right) \\
&\leq \ \frac{2}{\alpha}\ \|f\|_{L^1} \ ,
\end{aligned}
$$

so that

$$\left|\left\{x \in B^*;\ |Tb(x)| \geq \frac{\alpha}{2}\right\}\right| \leq |B^*| \leq \frac{2}{\alpha}\ \|f\|_{L^1}\ . \tag{9.1.8}$$

4. It remains to estimate $|\{x \in G^*;\ |Tb(x)| \geq \frac{\alpha}{2}\}|$. We have

$$\begin{aligned} \frac{\alpha}{2}\ \left|\left\{x \in G^*;\ |Tb(x)| \geq \frac{\alpha}{2}\right\}\right| &\leq \int_{\{x\in G^*;\ |Tb(x)|\geq\frac{\alpha}{2}\}} dx \qquad |Tb(x)| \\ &\leq \int_{G^*} dx\ |Tb(x)|\ . \end{aligned} \tag{9.1.9}$$

5. To estimate this last integral, we separate the different contributions to b. Define $b_k(x)$ by

$$b_k(x) = \begin{cases} 0 & \text{if } x \notin Q_k\ , \\ f(x) - \frac{1}{|Q_k|} \int dy\ f(y) & \text{if } x \in \text{interior of } Q_k\ . \end{cases}$$

Then $b(x) = \sum_k b_k(x)$ a.e., since the Q_k do not overlap. Consequently, $Tb = \sum_k Tb_k$, and

$$\begin{aligned} \int_{G^*} dx\ |Tb(x)| &\leq \sum_k \int_{G^*} dx\ |Tb_k(x)| \leq \sum_k \int_{\mathbb{R}\setminus Q_k^*} dx\ |Tb_k(x)| \\ &= \sum_k \int_{\mathbb{R}\setminus Q_k^*} dx\ \left|\int_{Q_k} dy\ K(x,y)\ b_k(y)\right| \\ &= \sum_k \int_{\mathbb{R}\setminus Q_k^*} dx\ \left|\int_{Q_k} dy\ [K(x,y) - K(x,y_k)]\ b_k(y)\right| \\ &\qquad \text{(y_k is the center of Q_k; we can insert this} \\ &\qquad \text{extra term because } \textstyle\int_{Q_k} dy\ b_k(y) = 0) \\ &\leq \sum_k \int_{\mathbb{R}\setminus Q_k^*} dx \int_{Q_k} dy\ |K(x,y) - K(x,y_k)|\ |b_k(y)|\ . \end{aligned} \tag{9.1.10}$$

The difference $K(x,y) - K(x,y_k)$ can be estimated by using the bound on the partial derivative $\partial_2 K$ of K with respect to its second variable:

$$\begin{aligned} &\int_{\mathbb{R}\setminus Q_k^*} dx\ |K(x,y) - K(x,y_k)| \\ &\quad \leq \int_{\mathbb{R}\setminus Q_k^*} dx \int_0^1 dt\ |\partial_2 K(x, y_k + t(y - y_k))| \cdot |y - y_k| \\ &\quad \leq \int_{|x-y_k|\geq 2R_k} dx \int_0^1 dt\ C|y - y_k|\ |(x - y_k) - t(y - y_k)|^{-2} \end{aligned}$$

$$\begin{aligned}
&\quad \text{(where we write } Q_k = [y_k - R_k,\ y_k + R_k[,\\
&\quad Q_k^* = [y_k - 2R_k,\ y_k + 2R_k[\ )\\
&= R_k^2\ |v| \int_{|u|>2} du \int_0^1 dt\ \frac{C}{R_k^2\ |u - tv|^2}\\
&\quad \text{(after the substitution } x = y_k + R_k u,\quad y = y_k + R_k v,\\
&\quad \text{where } |u| \geq 2,\quad |v| \leq 1)\\
&\leq C' \quad \text{(independent of } k) \ .
\end{aligned}$$

Substituting this into (9.1.10) yields

$$\begin{aligned}
\int_{G^*} dx\ |Tb(x)| &\leq C' \sum_k \int_{Q_k} dy\ |b_k(y)|\\
&\leq C' \sum_k \int_{Q_k} dy\ \left[|f(y)| + \frac{1}{|Q_k|} \int_{Q_k} dx\ |f(x)|\right]\\
&\leq 2C' \sum_k \int_{Q_k} dy\ |f(y)| \leq 2C'\ \|f\|_{L^1} \ .
\end{aligned}$$

Together with (9.1.7), (9.1.8), and (9.1.9), this proves the theorem. ∎

Once we know that T maps L^2 to L^2 and L^1 to L^1_{weak}, we can extend T to other L^p-spaces by interpolation theorem of Marcinkiewicz.

THEOREM 9.1.3. *If an operator T satisfies*

$$\|Tf\|_{L^{q_1}_{\text{weak}}} \leq C_1\ \|f\|_{L^{p_1}}\ , \tag{9.1.11}$$

$$\|Tf\|_{L^{q_2}_{\text{weak}}} \leq C_2\ \|f\|_{L^{p_2}}\ , \tag{9.1.12}$$

where $q_1 \leq p_1$, $q_2 \leq p_2$, then for $\frac{1}{p} = \frac{t}{p_1} + \frac{1-t}{p_2}$, $\frac{1}{q} = \frac{t}{q_1} + \frac{1-t}{q_2}$, with $0 < t < 1$, there exists a constant K, depending on p_1, q_1, p_2, q_2, and t, so that

$$\|Tf\|_{L^q} \leq K\ \|f\|_{L^p}\ .$$

Here L^q_{weak} stands for the space of all functions f for which

$$\|f\|_{L^q_{\text{weak}}} = [\inf\ \{C;\ |\{x;\ |f(x)| \geq \alpha\}| \leq C\ \alpha^{-q}\ \text{ for all }\ \alpha > 0\}]^{1/q}$$

is finite.

This theorem is remarkable in that it only needs weaker bounds at the two extrema, and nevertheless derives bounds on L^q-norms (not L^q_{weak}) for intermediate values q.[3] The proof of this theorem is outside the scope of this chapter; a proof of a more general version can be found in Stein and Weiss (1971). The Marcinkiewicz interpolation theorem implies that the $L^1 \to L^1_{\text{weak}}$-boundedness proved in Theorem 9.1.2 is sufficient to derive $L^p \to L^p$ boundedness for $1 < p < \infty$, as follows.

THEOREM 9.1.4. *If T is an integral operator with integral kernel K satisfying (9.1.2) and (9.1.3), and if T is bounded from $L^2(\mathbb{R})$ to $L^2(\mathbb{R})$, then T extends to a bounded operator from $L^p(\mathbb{R})$ to $L^p(\mathbb{R})$ for all p with $1 < p < \infty$.*

Proof.

1. Theorem 9.1.2 proves that T is bounded from L^1 to L^1_{weak}; by Marcinkiewicz' theorem, T extends to a bounded operator from L^p to L^p for $1 < p \leq 2$.

2. For the range $2 \leq p < \infty$, we use the adjoint $\tilde{T}$ of T, defined by

$$\int dx \; (\tilde{T}f)(x) \; \overline{g(x)} \;=\; \int dx \; f(x) \; \overline{(Tg)(x)} \; .$$

It is associated to the integral kernel $\tilde{K}(x,y) \;=\; \overline{K(y,x)}$, which also satisfies the conditions (9.1.2) and (9.1.3). On $L^2(\mathbb{R})$, it is exactly the adjoint in L^2-sense T^*, so that it is bounded. It follows then from Theorem 9.1.2 that $\tilde{T}$ is bounded from L^1 to L^1_{weak}, and hence by Theorem 9.1.3, that it is bounded from L^p to L^p for $1 < p \leq 2$. Since for $\frac{1}{p} + \frac{1}{q} = 1$, $\tilde{T}: \; L^p \to L^p$ is the adjoint of $T: \; L^q \to L^q$, it follows that T is bounded for $2 \leq q < \infty$. More explicitly, for readers unfamiliar with adjoints on Banach spaces,

$$\begin{aligned} \|Tf\|_q &= \sup_{\substack{g \in L^p \\ \|g\|_{L^p}=1}} \left| \int dx \; (Tf)(x) \; \overline{g(x)} \right| \qquad \left(\text{if } \frac{1}{p} + \frac{1}{q} = 1 \right) \\ &= \sup_{\substack{g \in L^p \\ \|g\|_{L^p}=1}} \left| \int dx \int dy \; f(y) \; K(x,y) \; \overline{g(x)} \right| \\ &= \sup_{\substack{g \in L^p \\ \|g\|_{L^p}=1}} \left| \int dy \; f(y) \; (\tilde{T}g)(y) \right| \\ &\leq \sup_{\substack{g \in L^p \\ \|g\|_{L^p}=1}} \|f\|_{L^q} \; \|\tilde{T}g\|_{L^p} \leq C \; \|f\|_{L^q} \; . \end{aligned}$$

(Strictly speaking, changing the order of integrations in the third equality is not allowed for all f, g, but we can restrict to a dense subspace where there is no such problem.) ∎

We can now apply this to prove that if ψ has some decay and some regularity and if the $\psi_{j,k}(x) \;=\; 2^{-j/2} \; \psi(2^{-j}x - k)$ constitute an orthonormal basis for $L^2(\mathbb{R})$, then the $\psi_{j,k}$ also provide unconditional bases for $L^p(\mathbb{R})$, $1 < p < \infty$. What we need to prove (see Preliminaries) is that if

$$f \;=\; \sum_{j,k} c_{j,k} \; \psi_{j,k} \in L^p \; ,$$

then

$$\sum_{j,k} \omega_{j,k} \; c_{j,k} \; \psi_{j,k} \in L^p$$

for any choice of the $\omega_{j,k} = \pm 1$.

We will assume that ψ is continuously differentiable, and that both ψ and ψ' decay faster than $(1 + |x|)^{-1}$:

$$|\psi(x)|, \; |\psi'(x)| \leq C(1 + |x|)^{-1-\epsilon} \; . \tag{9.1.13}$$

Then $\psi \in L^p$ for $1 < p < \infty$, and $f = \sum_{j,k} c_{j,k}\psi_{j,k}$ implies $c_{j,k} = \int dx f(x) \; \psi_{j,k}(x)$, because of the orthonormality of the $\psi_{j,k}$. We therefore want to show that, for any choice of the $\omega_{j,k} = \pm 1$, T_ω defined by

$$T_\omega f \; = \; \sum_{j,k} \omega_{j,k} \; \langle f, \; \psi_{j,k} \rangle \; \psi_{j,k}$$

is a bounded operator from L^p to L^p. We already know that T_ω is bounded from L^2 to L^2, since

$$\|T_\omega f\|^2_{L^2} \; = \; \sum_{j,k} |\omega_{j,k} \; \langle f, \; \psi_{j,k} \rangle|^2 \; = \; \sum_{j,k} |\langle f, \; \psi_{j,k} \rangle|^2 \; = \; \|f\|^2 \; ,$$

so the L^p-boundedness will follow by Theorem 9.1.3 if we can prove that T_ω is an integral operator with kernel satisfying (9.1.2), (9.1.3). This is the content of the following lemma.

LEMMA 9.1.5.

Choose $\omega_{j,k} = \pm 1$, *and define* $K(x,y) = \sum_{j,k} \omega_{j,k}\psi_{j,k}(x)\overline{\psi_{j,k}(y)}$. *Then there exists* $C < \infty$ *so that*

$$|K(x,y)| \leq \frac{C}{|x-y|}$$

and

$$\left| \frac{\partial}{\partial x} \; K(x,y) \right| \; + \; \left| \frac{\partial}{\partial y} \; K(x,y) \right| \leq \frac{C}{|x-y|^2} \; .$$

Proof.

1.

$$\begin{aligned} |K(x,y)| \; &\leq \; \sum_{j,k} \; |\psi_{j,k}(x)| \; |\psi_{j,k}(y)| \\ &\leq \; C \sum_{j,k} \; 2^{-j}(1 + |2^{-j}x - k|)^{-1-\epsilon}(1 + |2^{-j}y - k|)^{-1-\epsilon} \\ &\qquad \text{(by (9.1.13))} \; . \end{aligned}$$

Find $j_0 \in \mathbb{Z}$ so that $2^{j_0} \leq |x - y| \leq 2^{j_0+1}$. We split the sum over j into two parts: $j < j_0$ and $j \geq j_0$.

2. Since $\sum_k (1+|a-k|)^{-1-\epsilon}(1+|b-k|)^{-1-\epsilon}$ is uniformly bounded for all values of a, b,[4] we have

$$\sum_{j=j_0}^{\infty} \sum_k 2^{-j}(1+|2^{-j}x-k|)^{-1-\epsilon}(1+|2^{-j}y-k|)^{-1-\epsilon}$$
$$\leq C \sum_{j=j_0}^{\infty} 2^{-j} \leq C\, 2^{-j_0+1} \leq \frac{4C}{|x-y|} .$$

3. The part $j < j_0$ is a little less easy.

$$\sum_{j=-\infty}^{j_0-1} 2^{-j} \sum_k [(1+|2^{-j}x-k|)(1+|2^{-j}y-k|)]^{-1-\epsilon}$$
$$= \sum_{j=-j_0+1}^{\infty} 2^{j} \sum_k [(1+|2^{j}x-k|)(1+|2^{j}y-k|)]^{-1-\epsilon}$$
$$\leq 4^{1+\epsilon} \sum_{j=-j_0+1}^{\infty} 2^{j} \sum_k [(2+|2^{j}x-k|)(2+|2^{j}y-k|)]^{-1-\epsilon} . \quad (9.1.14)$$

Find $k_0 \in \mathbb{Z}$ so that $k_0 \leq 2^j \frac{x+y}{2} \leq k_0+1$, and define $\ell = k - k_0$. Then

$$\begin{aligned} 2+|2^j\, x-k| &= 2 + \left|2^j\, \frac{x-y}{2} - \ell + \left(2^j\, \frac{x+y}{2} - k_0\right)\right| \\ &\geq 1 + \left|2^j\, \frac{x-y}{2} - \ell\right| ; \end{aligned}$$

similarly,

$$2 + |2^j\, y-k| \geq 1 + \left|2^j \frac{y-x}{2} - \ell\right| .$$

Consequently, with $a = 2^j \frac{x-y}{2}$,

$$\sum_k [(2+|2^jx-k|)(2+|2^jy-k|)]^{-1-\epsilon}$$
$$\leq \sum_\ell [(1+|a+\ell|)(1+|a-\ell|)]^{-1-\epsilon} \leq C(1+|a|)^{-1-\epsilon} ,$$[5]

so that

$$\begin{aligned} (9.1.14) \quad &\leq C \sum_{j=-j_0+1}^{\infty} 2^j \left(1+2^j\left|\frac{x-y}{2}\right|\right)^{-1-\epsilon} \\ &\leq C \sum_{j'=1}^{\infty} 2^{j'-j_0} \left(1+2^{j'-j_0}\frac{1}{2}\, 2^{j_0}\right)^{-1-\epsilon} \end{aligned}$$

(because $|x-y| \geq 2^{j_0+1}$)

$$\leq \; C\, 2^{-j_0} \sum_{j'=1}^{\infty} 2^{j'} (1 + 2^{j'-1})^{-1-\epsilon}$$
$$\leq \; C'\, 2^{-j_0} \leq 2C'\, |x-y|^{-1} \,.$$

It therefore follows that $|K(x,y)| \leq C|x-y|^{-1}$.

4. For the estimates on $\partial_x K,\ \partial_y K$, we write

$$|\partial_x K(x,y)| \;\leq\; \sum_{j,k} 2^{-j} |\psi'(2^{-j}x - k)|\; |\psi(2^{-j}y - k)|$$
$$\leq\; C \sum_{j,k} 2^{-2j}\; \left[(1 + |2^{-j}x - k|)(1 + |2^{-j}y - k|)\right]^{-1-\epsilon}$$

and we follow the same technique; we obtain without difficulty

$$|\partial_x K(x,y)|,\;\; |\partial_y K(x,y)| \leq C|x-y|^{-2} \,. \quad \blacksquare$$

From the discussion preceding the lemma it therefore follows that we have proved the following theorem.

THEOREM 9.1.6. *If ψ is C^1 and $|\psi(x)|,\ |\psi'(x)| \leq C(1+|x|)^{-1-\epsilon}$, and if the $\psi_{j,k}(x) = 2^{-j/2}\ \psi(2^{-j}x - k)$ constitute an orthonormal basis for $L^2(\mathbb{R})$, then the $\{\psi_{j,k};\ j,k \in \mathbb{Z}\}$ also constitute an unconditional basis for all the L^p-spaces, $1 < p < \infty$.*

9.2. Characterization of function spaces by means of wavelets.

Since the $\psi_{j,k}$ constitute an unconditional basis for $L^p(\mathbb{R})$, there exists a characterization for functions $f \in L^p(\mathbb{R})$ using only the absolute values of the wavelet coefficients of f. In other words, given f, then we can decide whether $f \in L^p$ by looking only at the $|\langle f,\ \psi_{j,k}\rangle|$. The explicit criterion is, again for $1 < p < \infty$,

$$f \in L^p(\mathbb{R}) \iff \left[\sum_{j,k} |\langle f,\ \psi_{j,k}\rangle|^2\ |\psi_{j,k}(x)|^2\right]^{1/2} \in L^p(\mathbb{R})$$
$$\iff \left[\sum_{j,k} |\langle f,\ \psi_{j,k}\rangle|^2 2^{-j}\ \chi_{[2^j k,\, 2^j(k+1)]}(x)\right]^{1/2} \in L^p(\mathbb{R}) \,.$$

For a proof that these are indeed equivalent characterizations of $L^p(\mathbb{R})$, see Meyer (1990).

Similarly, wavelets provide unconditional bases and characterizations for many other functional spaces. We list a few here, without proofs.

The Sobolev spaces $\mathbf{W}^s(\mathbb{R})$. The Sobolev spaces are defined by

$$W^s(\mathbb{R}) \;=\; \left\{f;\ \int d\xi\ (1 + |\xi|^2)^s\ |\hat{f}(\xi)|^2 < \infty\right\} .$$

Their characterization by means of wavelet coefficients is

$$f \in W^s(\mathbb{R}) \Leftrightarrow \sum_{j,k} |\langle f, \psi_{j,k}\rangle|^2 \, (1 + 2^{-2js}) < \infty \, .$$

The Hölder spaces $\mathbf{C}^s(\mathbb{R})$. For $0 < s < 1$, we define

$$C^s(\mathbb{R}) = \left\{ f \in L^\infty(\mathbb{R}); \ \sup_{x,h} \frac{|f(x+h) - f(x)|}{|h|^s} < \infty \right\} .$$

For $s = n + s'$, $0 < s' < 1$, we define

$$C^s(\mathbb{R}) = \left\{ f \in L^\infty(\mathbb{R}) \cap C^n(\mathbb{R}); \ \frac{d^n}{dx^n} f \in C^{s'} \right\} .$$

For integer values of s, the appropriate spaces in this ladder are not the traditional C^n-spaces (consisting of the functions that are n times continuously differentiable), nor even the Lipschitz-spaces, but slightly larger spaces defined by

$$\begin{aligned} \Lambda_* &= \text{“Zygmund class”} \\ &= \left\{ f \in L^\infty(\mathbb{R}) \cap C^0(\mathbb{R}); \ \sup_{x,h} \frac{|f(x+h) + f(x-h) - 2f(x)|}{|h|} < \infty \right\} , \end{aligned}$$

which takes the place of $C^1(\mathbb{R})$, and

$$\Lambda_*^n = \left\{ f \in L^\infty(\mathbb{R}) \cap C^{n-1}(\mathbb{R}); \ \frac{d^{n-1}}{dx^{n-1}} f \in \Lambda_* \right\} .$$

For this ladder of Hölder spaces one has the following characterization:

> A locally integrable f is in $C^s(\mathbb{R})$ (s noninteger) or Λ_*^n ($s = n$ integer) if and only if there exists $C < \infty$ so that
>
> $$\begin{aligned} &\cdot \ |\langle f, \phi_{0,k}\rangle| \le C && \text{for all } k \in \mathbb{Z} \, , \\ &\cdot \ |\langle f, \psi_{-j,k}\rangle| \le C \, 2^{-j(s+1/2)} && \text{for all } j \ge 0, \quad k \in \mathbb{Z} \, . \end{aligned} \tag{9.2.1}$$

We have implicitly assumed here that $\psi \in C^r$, with $r > s$.

For proofs and more examples, see Meyer (1990). Of the examples given here, the only spaces that can be completely characterized (with “if and only if” conditions) by Fourier transforms are the Sobolev spaces.

The conditions (9.2.1) characterize *global* regularity. Local regularity can also be studied by means of coefficients with respect to an orthonormal wavelet basis. The most general theorem is the following, due to Jaffard (1989b). For simplicity, we assume that ψ has compact support and is C^1 (the formulation of the theorem is slightly different for more general ψ).

THEOREM 9.2.1. *If f is Hölder continuous with exponent α, $0 < \alpha < 1$, at x_0, i.e.,*

$$|f(x) - f(x_0)| \leq C|x - x_0|^\alpha , \tag{9.2.2}$$

then

$$\max_k [|\langle f, \psi_{-j,k}\rangle| \; dist(x_0, \; support(\psi_{-j,k}))^{-\alpha}] = O\left(2^{-(\frac{1}{2}+\alpha)j}\right) \tag{9.2.3}$$

for $j \to \infty$. Conversely, if (9.2.3) holds and if f is known to be C^ϵ for some $\epsilon > 0$, then

$$|f(x) - f(x_0)| \leq C|x - x_0|^\alpha \; \log \frac{2}{|x - x_0|} . \tag{9.2.4}$$

We do not have exact equivalence between (9.2.3) and (9.2.2) here. The estimate (9.2.4) is in fact optimal, as is the condition $f \in C^\epsilon$: if f is merely continuous, or if the logarithm in (9.2.4) is omitted, then counterexamples can be found (Jaffard (1989b)). Non-equivalence of (9.2.2) and (9.2.3) can be caused by the existence of less regular points near x_0, or by wild oscillations of $f(x)$ near x_0 (see, e.g., Mallat and Hwang (1992)). If we modify condition (9.2.3) slightly, then these problems are circumvented. More precisely (again with compactly supported $\psi \in C^1$), we have the following.

THEOREM 9.2.2. *Define, for $\epsilon > 0$,*

$$S(x_0, j; \; \epsilon) \; = \; \{k \in \mathbb{Z}; \; \text{support}(\psi_{j,k}) \cap]x_0 - \epsilon, \; x_0 + \epsilon[\neq \phi\}.$$

If, for some $\epsilon > 0$, and some α, $0 < \alpha < 1$,

$$\max_{k \in S(x_0, -j; \epsilon)} |\langle f, \; \psi_{-j,k}\rangle| \; = \; O\left(2^{-j(\frac{1}{2}+\alpha)}\right) , \tag{9.2.5}$$

then f is Hölder continuous with exponent α in x_0.

Proof.

1. Choose any x in $]x_0 - \epsilon, \; x_0 + \epsilon[$. Since either $\psi_{j,k}(x) \neq 0$ or $\psi_{j,k}(x_0) \neq 0$ implies $k \in S(x_0, j; \epsilon)$, we have

$$\begin{aligned} f(x) \; - \; f(x_0) \; &= \; \sum_{j,k} \langle f, \; \psi_{j,k}\rangle \; [\psi_{j,k}(x) - \psi_{j,k}(x_0)] \\ &= \; \sum_j \; \sum_{k \in S(x_0, j; \epsilon)} \langle f, \; \psi_{j,k}\rangle \; [\psi_{j,k}(x) - \psi_{j,k}(x_0)] \; . \end{aligned}$$

It follows that

$$|f(x) - f(x_0)| \leq \sum_j \; C_1 \; 2^{j(\frac{1}{2}+\alpha)} \; \sum_{k \in S(x_0, j; \epsilon)} |\psi_{j,k}(x) - \psi_{j,k}(x_0)| \; .$$

2. Since ψ has compact support, the number of k for which $\psi_{j,k}(x) \neq 0$ or $\psi_{j,k}(x_0) \neq 0$ is bounded, uniformly in j, by 2 $|\text{support}(\psi)|$. Consequently,

$$\begin{aligned}\sum_{k \in S(x_0, j; \epsilon)} & |\psi_{j,k}(x) - \psi_{j,k}(x_0)| \\ & \leq C_2 \max_k |\psi_{j,k}(x) - \psi_{j,k}(x_0)| \\ & \leq C_2\, 2^{-j/2} \max_k |\psi(2^{-j}x - k) - \psi(2^{-j}x_0 - k)| \ .\end{aligned}$$

Since ψ is bounded and C^1,

$$|\psi(2^{-j}x - k) - \psi(2^{-j}x_0 - k)| \leq C_3 \min\,(1,\ 2^{-j}|x - x_0|) \ .$$

3. Now choose j_0 so that $2^{j_0} \leq |x - x_0| \leq 2^{j_0+1}$. Then

$$\begin{aligned}|f(x) - f(x_0)| & \leq C_1 C_2 C_3 \left[\sum_{j=-\infty}^{j_0} 2^{\alpha j} + \sum_{j=j_0+1}^{\infty} 2^{\alpha j - j}|x - x_0|\right] \\ & \leq C_4 \left[2^{\alpha j_0} + 2^{(\alpha-1)j_0}|x - x_0|\right] \leq C_5 |x - x_0|^\alpha \ . \quad \blacksquare\end{aligned}$$

REMARKS.

1. Similar theorems can, of course, be proved for C^α-spaces with $\alpha > 1$.

2. If $\alpha = 1$ (or more generally, $\alpha \in \mathbb{N}$), then the very last step of the proof does not work any more, because the second series will not converge. One can avoid this divergence by using $| < f, \psi_{j,k} > | \leq C$ for $j \geq 0$, but the sum over j between $j_0 < 0$ and 0 then still leads to a term in $|x - x_0|\,\big|\ln|x - x_0|\big|$. That is why one has to be more circumspect for integer α, and why the Zygmund class enters.

3. Theorems 9.2.1 and 9.2.2 are also true if ψ has infinite support, and ψ and ψ' have good decay at ∞ (see Jaffard (1989b)). Compact support for ψ makes the estimates easier. □

Local regularity can therefore be studied by means of wavelet coefficients. For practical purposes, one should beware, however: it may be that very large values of j are needed to determine α in (9.2.5) reliably. This is illustrated by the following example. Take

$$f(x - a) = \begin{cases} 2\, e^{-|x-a|} & \text{if } x \leq a - 1 \ , \\ e^{-|x-a|} & \text{if } a - 1 \leq x \leq a + 1 \ , \\ e^{-(x-a)}[(x - a - 1)^2 + 1] & \text{if } x \geq a + 1 \ ; \end{cases}$$

this function is graphed in Figure 9.1 (with $a = 0$). This function has Hölder exponents 0, 1, 2 at $x = a - 1$, a, $a + 1$, respectively, and is C^∞ elsewhere. One can then, for each of the three points $x_0 = a - 1$, a, or $a + 1$, compute

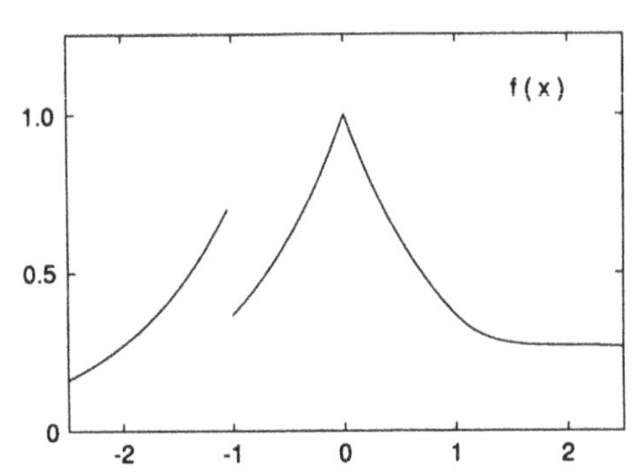

FIG. 9.1. *This function is C^∞ except at $x = -1, 0$ and 1, where, respectively, f, f', and f'' are discontinuous.*

$A_j = \max\,\{|\langle f,\ \psi_{-j,k}\rangle|;\ \ x_0 \in \text{support}\ (\psi_{-jk})\}$, and plot $\log A_j/\log 2$. If $a = 0$, then these plots line up on straight lines, with slope $1/2$, $3/2$ and $5/2$, with pretty good accuracy, leading to good estimates for α. A decomposition in orthonormal wavelets is not translation invariant, however, and dyadic rationals, particularly 0, play a very special role with respect to the dyadic grid $\{2^{-j}k;\ j,k \in \mathbb{Z}\}$ of localization centers for our wavelet basis. Choosing different values for a illustrates this: for $a = 1/128$, we have very different $\langle f,\ \psi_{j,k}\rangle$, but still a reasonable line-up in the plots of $\log A_j/\log 2$, with good estimates for α; for irrational a, the line-up is much less impressive, and determining α becomes correspondingly less precise. All this is illustrated in Figure 9.2, showing the plots of $\log A_j/\log 2$ as a function of j, for $x_0 = a-1,\ a,\ a+1$ and for the three choices $a = 0,\ 1/128$ and $\sqrt{2} - 11/8$ (we subtract $11/8$ to obtain a close to zero, for programming convenience). To make the figure, $|\langle f, \psi_{-j,k}\rangle|$ was computed for the relevant values of k and for j ranging from 3 to 10. (Note that this means that f itself had to be sampled with a resolution 2^{-17}, in order to have a reasonable accuracy for the $j = 10$ integrals.) For $a = 0$, the eight points line up beautifully and the estimate for $\alpha + \frac{1}{2}$ is accurate to less than 1.5% at all three locations. For $a = 1/128$, the points at the coarser resolution scales do not align as well, but if $\alpha + \frac{1}{2}$ is estimated from only the finest four resolution points, then the estimates are still within 2%. For the irrational choice $a = \sqrt{2} - 11/8$ no alignment can be seen at the discontinuity at $a - 1$ (one probably needs even smaller scales), and the estimate for $\alpha + \frac{1}{2}$ at a, where f is Lipschitz, is off by about 13% (interestingly enough, the estimate would be much better if the scale-10 point were deleted); at $a + 1$, where f' is Lipschitz, the estimate is within 2.5%. This illustrates that to determine the local regularity of a function, it is more useful to use very redundant wavelet families, where this translational non-invariance is much less pronounced (discrete case) or absent (continuous case). (See Holschneider and Tchamitchian (1990), Mallat and Hwang (1992).) Another reason for using very redundant wavelet families for the characterization of local regularity is that then only the number of vanishing moments of ψ limits

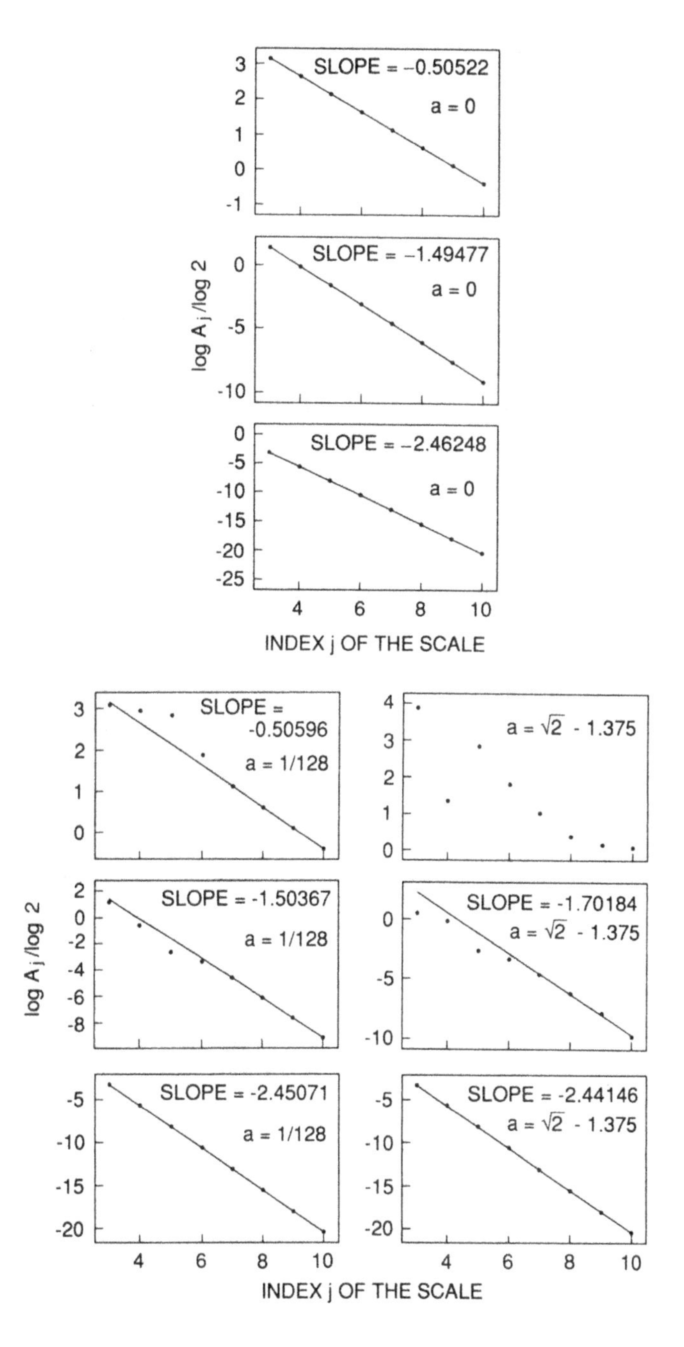

FIG. 9.2. *Estimates of the Hölder exponents of $f(x-a)$ (see Figure 9.1) at $a-1$ (top), a (middle), $a+1$ (bottom), computed from $\log A_j/\log 2$, for different values of a. (This figure was contributed by M. Nitzsche, whom I would like to thank for her help.)*

the maximum regularity that can be characterized; the regularity of ψ plays no role (see §2.9). If orthonormal bases are used, then we are necessarily limited by the regularity of ψ itself, as is illustrated by choosing $f = \psi$. For this choice we have indeed $\langle f, \psi_{-j,k}\rangle = 0$ for all $j > 0$, all k; it follows that with orthonormal wavelets we can hope to characterize only regularity up to $C^{r-\epsilon}$ if $\psi \in C^r$.

9.3. Wavelets for $\mathbf{L}^1([\mathbf{0}, \mathbf{1}])$.

Since L^1-spaces do not have unconditional bases, wavelets cannot provide one. Nevertheless, they still outperform Fourier analysis in some sense. We will illustrate this by a comparison of expansions in wavelets versus Fourier series of $L^1([0,1])$-functions. But first we must introduce "periodized wavelets."

Given a multiresolution analysis with scaling function ϕ and wavelet ψ, both with reasonable decay (say, $|\phi(x)|$, $|\psi(x)| \le C(1+|x|)^{-1-\epsilon}$), we define

$$\phi^{\rm per}_{j,k}(x) = \sum_{\ell\in\mathbb{Z}} \phi_{j,k}(x+\ell), \qquad \psi^{\rm per}_{j,k} = \sum_{\ell\in\mathbb{Z}} \psi_{j,k}(x+\ell)\ ;$$

and

$$V_j^{\rm per} = \overline{\text{Span}\ \{\phi^{\rm per}_{\rm j,k};\ \ {\rm k}\in\mathbb{Z}\}}, \qquad W_j^{\rm per} = \overline{\text{Span}\ \{\psi^{\rm per}_{\rm j,k};\ \ {\rm k}\in\mathbb{Z}\}}\ .$$

Since $\sum_{\ell\in\mathbb{Z}} \phi(x+\ell) = 1$,[6] we have, for $j \ge 0$, $\phi^{\rm per}_{j,k}(x) = 2^{-j/2} \sum_\ell \phi(2^{-j}x - k + 2^{-j}\ell) = 2^{j/2}$, so that the $V_j^{\rm per}$, for $j \ge 0$, are all identical one-dimensional spaces, containing only the constant functions. Similarly, because $\sum_\ell \psi(x+\ell/2) = 0$,[7] $W_j^{\rm per} = \{0\}$ for $j \ge 1$. We therefore restrict our attention to the $V_j^{\rm per}$, $W_j^{\rm per}$ with $j \le 0$. Obviously $V_j^{\rm per}$, $W_j^{\rm per} \subset V_{j-1}^{\rm per}$, a property inherited from the non-periodized spaces. Moreover, $W_j^{\rm per}$ is still orthogonal to $V_j^{\rm per}$, because

$$\begin{aligned}
&\int_0^1 dx\ \psi^{\rm per}_{j,k}(x)\ \phi^{\rm per}_{j,k'}(x)\\
&\quad = \sum_{\ell,\ell'\in\mathbb{Z}} 2^{-j} \int_0^1 dx\ \psi(2^{-j}x + 2^{-j}\ell - k)\ \overline{\phi(2^{-j}x + 2^{-j}\ell' - k')}\\
&\quad = \sum_{\ell,\ell'\in\mathbb{Z}} 2^{|j|} \int_{\ell'}^{\ell'+1} dy\ \psi(2^{|j|}y + 2^{|j|}(\ell-\ell') - k)\ \overline{\phi(2^{|j|}y - k')}\\
&\qquad\qquad\qquad\qquad\qquad\qquad (\text{because } j \le 0)\\
&\quad = \sum_{r\in\mathbb{Z}} \langle \psi_{j,k+2^{|j|}r},\ \phi_{j,k'}\rangle = 0\ .
\end{aligned}$$

It follows that, as in the non-periodized case, $V_{j-1}^{\rm per} = V_j^{\rm per} \oplus W_j^{\rm per}$. The spaces $V_j^{\rm per}$, $W_j^{\rm per}$ are all finite-dimensional: since $\phi^{\rm per}_{j,k+m2^{|j|}} = \phi^{\rm per}_{j,k}$ for $m \in \mathbb{Z}$, and the same is true for $\psi^{\rm per}$, both $V_j^{\rm per}$ and $W_j^{\rm per}$ are spanned by the $2^{|j|}$ functions obtained from $k = 0, 1, \cdots, 2^{|j|}-1$. These $2^{|j|}$ functions are moreover orthonormal; in e.g., W_j^{per} we have, for $0 \le k, k' \le 2^{|j|} - 1$,

$$\langle \psi_{j,k}^{\mathrm{per}}, \psi_{j,k'}^{\mathrm{per}} \rangle = \sum_{r \in \mathbb{Z}} \langle \psi_{j,k+2^{|j|}r}, \psi_{j,k'} \rangle = \delta_{k,k'} .$$

We have therefore a ladder of multiresolution spaces,

$$V_0^{\mathrm{per}} \subset V_{-1}^{\mathrm{per}} \subset V_{-2}^{\mathrm{per}} \subset \cdots ,$$

with successive orthogonal complements W_0^{per} (of V_0^{per} in V_{-1}^{per}), $W_1^{\mathrm{per}}, \cdots$, and orthonormal bases $\{\phi_{j,k};\ k = 0, \cdots 2^{|j|} - 1\}$ in V_j^{per}, $\{\psi_{j,k};\ k = 0, \cdots, 2^{|j|} - 1\}$ in W_j^{per}. Since $\overline{\cup_{j \in -\mathbb{N}} V_j^{\mathrm{per}}} = L^2([0,1])$ (this follows again from the corresponding non-periodized version), the functions in $\{\phi_{0,0}^{\mathrm{per}}\} \cup \{\psi_{j,k}^{\mathrm{per}};\ -j \in \mathbb{N},\ k = 0, \cdots, 2^{|j|} - 1\}$ constitute an orthonormal basis in $L^2([0,1])$. We will relabel this basis as follows:

$$\begin{aligned}
g_0(x) &= 1 = \phi_{0,0}^{\mathrm{per}}(x) \\
g_1(x) &= \psi_{0,0}^{\mathrm{per}}(x) \\
g_2(x) &= \psi_{-1,0}^{\mathrm{per}}(x) \\
g_3(x) &= \psi_{-1,1}^{\mathrm{per}}(x) = \psi_{-1,0}^{\mathrm{per}}(x - \tfrac{1}{2}) = g_2(x - \tfrac{1}{2}) \\
g_4(x) &= \psi_{-2,0}^{\mathrm{per}}(x) \\
&\ \vdots \\
g_{2^j}(x) &= \psi_{-j,0}^{\mathrm{per}}(x) \\
&\ \vdots \\
g_{2^j+k}(x) &= \psi_{-j,k}^{\mathrm{per}}(x) = g_{2^j}(x - k2^{-j}) \qquad \text{for } 0 \le k \le 2^j - 1 \\
&\ \vdots
\end{aligned}$$

Then this basis has the following remarkable property.

THEOREM 9.3.1. *If f is a continuous periodic function with period 1, then there exist $\alpha_n \in \mathbb{C}$ so that*

$$\left\| f - \sum_{n=0}^{N} \alpha_n \, g_n \right\|_{L^\infty} \longrightarrow 0 \quad as \quad N \to \infty . \tag{9.3.1}$$

Proof.

1. Since the g_n are orthonormal, we necessarily have $\alpha_n = \langle f, g_n \rangle$. Define S_N by

$$S_N f = \sum_{n=0}^{N-1} \langle f, g_n \rangle \, g_n .$$

In a first step we prove that the S_N are uniformly bounded, i.e.,

$$\|S_N f\|_{L^\infty} \le C\ \|f\|_{L^\infty}\ , \tag{9.3.2}$$

with C independent of f or N.

2. If $N = 2^j$, then $S_{2^j} = \text{Proj}_{V_{-j}^{\text{per}}}$; hence

$$(S_{2^j} f)(x) = \sum_{k=0}^{2^{|j|}-1} \langle f,\ \phi_{-j,k}^{\text{per}}\rangle\ \phi_{-j,k}^{\text{per}}(x) = \int_0^1 dy\ K_j(x,y)\ f(y)\ ,$$

with

$$K_j\ (x,y) = \sum_{k=0}^{2^{|j|}-1} \phi_{-j,k}^{\text{per}}(x)\ \overline{\phi_{-j,k}^{\text{per}}\ (y)}\ .$$

Consequently,

$$\|S_{2^j} f\|_{L^\infty} \le \left[\sup_{x\in[0,1]} \int_0^1 dy\ |K_j(x,y)|\right]\ \|f\|_{L^\infty}\ .$$

Now

$$\begin{aligned}
&\sup_{x\in[0,1]} \int_0^1 dy\ |K_j(x,y)| \\
&\quad\le \sup_{x\in[0,1]} \int_0^1 dy \sum_{k=0}^{2^{|j|}-1} \sum_{\ell,\ell'\in\mathbb{Z}} |\phi_{-j,k}(x+\ell)|\ |\phi_{-j,k}(y+\ell')| \\
&\quad\le \sup_x \int_{-\infty}^{\infty} dy \sum_{k=0}^{2^{|j|}-1} \sum_{\ell\in\mathbb{Z}} 2^j\ |\phi(2^j(x+\ell)-k)|\ |\phi(2^j y-k)| \\
&\quad\le C\ \sup_{x'} \sum_{k=0}^{2^{|j|}-1} \sum_{\ell\in\mathbb{Z}} |\phi(x'+2^j\ell-k)| \\
&\quad\le C\ \sup_{x'} \sum_{m\in\mathbb{Z}} |\phi(x'+m)|\ ,
\end{aligned}$$

and this is uniformly bounded if $|\phi(x)| \le C(1+|x|)^{-1-\epsilon}$. This establishes (9.3.2) for $N = 2^j$.

3. If $N = 2^j + m$, $0 \le m \le 2^j - 1$, then

$$(S_N f)(x) = (S_{2^j} f)(x) + \sum_{k=0}^{m} \langle f,\ \psi_{-j,k}^{\text{per}}\rangle\ \psi_{-j,k}^{\text{per}}(x)\ .$$

Estimates exactly similar to those in point 2 show that the L^∞-norm of the second sum is also bounded by $C\ \|f\|_{L^\infty}$, uniformly in j, which proves (9.3.2) for all N.

4. Take now $f \in E = \cup_{j \in -\mathbb{N}} V_j^{\text{per}}$. Then $f \in V_{-J}^{\text{per}}$ for some $J > 0$, so that $\langle f, \psi_{-j',k}^{\text{per}} \rangle = 0$ for $j' \geq J$, i.e., $\langle f, g_\ell \rangle = 0$ for $\ell \geq 2^J$. Consequently, $f = S_N f$ if $N \geq 2^J$, so that (9.3.1) clearly holds. Since E is dense in $C(\mathbb{T})$, the space of continuous periodic functions equipped with the $\| \quad \|_\infty$-norm, the theorem follows. ∎

By duality, we obtain a similar theorem for $L^1([0,1])$.

THEOREM 9.3.2. *If $f \in L^1([0,1])$, then*

$$\left\| f - \sum_{n=0}^{N} \langle f, g_n \rangle g_n \right\|_{L^1} \longrightarrow 0 \quad as \quad N \to \infty .$$

Proof. We exploit that $L^1([0,1])$ is contained in the dual of $C(\mathbb{T})$, i.e.,

$$\|f\|_{L^1} = \sup \{ |\langle f, g \rangle|; \ g \text{ continuous, periodic with period 1, } \|g\|_{L^\infty} \leq 1 \} .$$

This immediately leads to

$$\begin{aligned} \|S_N f\|_{L^1} &= \sup \{ |\langle S_N f, g \rangle|; \ g \text{ continuous, 1-periodic, } \|g\|_{L^\infty} \leq 1 \} \\ &= \sup \{ |\langle f, S_N g \rangle|; \ g \text{ continuous, 1-periodic, } \|g\|_{L^\infty} \leq 1 \} \\ &\leq C \|f\|_{L^1} \qquad (9.3.3) \\ &\qquad \text{(by the uniform bound (9.3.2) and} \\ &\qquad \text{because } |\langle f, h \rangle| \leq \|f\|_{L^1} \|h\|_{L^\infty}) . \end{aligned}$$

Since $E = \cup_{j \in -\mathbb{N}} V_j^{\text{per}}$ is also dense in $L^1([0,1])$, the uniform bound (9.3.3) is sufficient to prove the theorem. ∎

What makes Theorems 9.3.1 and 9.3.2 remarkable is that no such property holds for Fourier series: in order to obtain, for example, uniform convergence of the Fourier series of f to f itself, one needs to impose more conditions than just continuity (e.g., $f \in C^1$).

Note that the *ordering* of the g_n is important in Theorems 9.3.1 and 9.3.2: we have a Schauder basis, but not an unconditional basis!

9.4. An amusing contrast between wavelet expansions and Fourier series.

The amusing contrast lies in the different behavior of "full" versus "lacunary" series with respect to the two expansion methods, Fourier series or wavelets. We start with a simple lemma, borrowed, as is this whole section, from Meyer (1990).

LEMMA 9.4.1. *Suppose f is a function on $[0,1]$, differentiable in $x_0 \in]0,1[$. Let g_m be the orthonormal basis for $L^2([0,1])$ as introduced above, and assume that the corresponding wavelet ψ satisfies $\int dx \ x \ \psi(x) = 0$. Then the α_m in $f = \sum_{m=0}^\infty \alpha_m g_m$, with m restricted to the set $m = 2^j + k$ with $|2^{-j}k - x_0| \leq 2^{-j}$, satisfy $\alpha_m = o(m^{-3/2})$ for $m \to \infty$.*

Proof.

1. Suppose, for simplicity, that ψ is compactly supported, with support $\psi \subset [-L, L]$. For sufficiently large j, this means that $\psi^{\text{per}}_{-j,k}(x) = \psi_{-j,k}(x)$ if $|2^{-j}k - x_0| \leq 2^{-j}$. (Again, this is not crucial. For non-compactly supported ψ, one only has to be a little more careful in the estimates below.[8])

2. For $m = 2^j + k$, $\alpha_m = \int dx\, f(x)\, \overline{\psi_{-j,k}(x)}$. Here

$$\begin{aligned}\text{support } \psi_{-j,k} &\subset [2^{-j}(k-L),\ 2^{-j}(k+L)] \\ &\subset [x_0 - 2^{-j}(L+1), x_0 + 2^{-j}(L+1)] \\ &\qquad\qquad \text{(because } |2^{-j}k - x_0| \leq 2^{-j}) ;\end{aligned}$$

hence

$$\begin{aligned}\alpha_m &= \int_{|x-x_0|\leq 2^{-j}(L+1)} dx\ f(x)\, \overline{\psi_{-j,k}(x)} \\ &= \int_{|x-x_0|\leq 2^{-j}(L+1)} dx\ [f(x) - f(x_0) - (x-x_0)f'(x_0)]\ 2^{j/2}\overline{\psi(2^j x - k)} \\ &= o(2^{j/2}\, 2^{-2j}) \\ &\qquad \text{(use } f(x) - f(x_0) - (x-x_0)f'(x_0) = o(x-x_0), \\ &\qquad \text{and change variables: } y = 2^j(x-x_0)) \\ &= o(2^{-3j/2}) = o(m^{-3/2}) \\ &\qquad \text{(because } 2^j \leq m \leq 2^{j+1}) . \quad \blacksquare\end{aligned}$$

This has the following corollary.

COROLLARY 9.4.2. *If, for all m, $C_1\ m^{-3/2} \leq |\alpha_m| \leq C_2\ m^{-3/2}$, with $C_1 > 0$, $C_2 < \infty$, then $\sum_{m=0}^{\infty} \alpha_m\, g_m$ is in C^α for all $\alpha < 1$, but is nowhere differentiable.*

Proof. Immediate from Theorem 9.2.2 and Lemma 9.4.1. $\blacksquare$

Let us now construct a very particular function. Take $\alpha_m = \alpha_{2^j+k} = \beta_j$, independently of k. Then

$$\begin{aligned}\sum_{m=0}^{\infty} \alpha_m\, g_m &= \sum_{j=0}^{\infty} \beta_j \sum_{k=0}^{2^j-1} g_{2^j+k} \\ &= \sum_{j=0}^{\infty} \beta_j \sum_{k=0}^{2^j-1} \sum_{\ell\in\mathbb{Z}} 2^{j/2}\ \psi(2^j x + 2^j \ell - k) \\ &= \sum_{j=0}^{\infty} 2^{j/2}\ \beta_j \sum_m \psi(2^j x - m) = \sum_{j=0}^{\infty} 2^{j/2}\ \beta_j F(2^j x),\end{aligned}$$

where $F(x) = \sum_m \psi(x-m)$ is a periodic function. We have

$$F(x) = \sum_n F_n \, e^{-2\pi i n x} ,$$

with

$$F_n = \frac{1}{2\pi} \int_0^1 dx \, F(x) \, e^{2\pi i n x} = \sqrt{2\pi} \, \hat{\psi}(-2\pi n) .$$

In the special case where $\psi = \psi_{\text{Meyer}}$ (see Chapters 4 and 5), support $\hat{\psi} = \{\xi; \frac{2\pi}{3} \le |\xi| \le \frac{8\pi}{3}\}$, so that $\hat{\psi}(2\pi n) \neq 0$ only if $n = \pm 1$. Moreover, $\hat{\psi}(-2\pi) = \hat{\psi}(2\pi)$. Consequently, $F(x) = A \cos(2\pi x)$, and

$$\sum_{m=0}^{\infty} \alpha_m \, g_m(x) = A \sum_{j=0}^{\infty} \beta_j \, 2^{j/2} \cos(2^j 2\pi x) .$$

The "full" wavelet series of the left-hand side has a lacunary Fourier expansion! If now the β_j are chosen so that $C_1 2^{-j} \le 2^{j/2} \beta_j \le C_2 \, 2^{-j}$, then we can apply Corollary 9.4.2,[9] and conclude that the function is nowhere differentiable. For this special case, this is in fact a well-known result about lacunary Fourier series: $\sum_{j=0}^{\infty} \gamma_j \cos(\lambda_j x)$, with $\sum_{j=0}^{\infty} |\gamma_j| < \infty$ but $\gamma_j \lambda_j \not\to 0$, defines a continuous, nowhere differentiable function.

On the other hand, if we take a function with a localized singularity, but which is C^∞ elsewhere, such as, e.g., $f(x) = |\sin \pi x|^{-\alpha}$, with $0 < \alpha < 1$, then its wavelet expansion will be more or less lacunary (all the coefficients decay very fast as $-j \to \infty$, except the few for which $2^{-|j|}k$ is close to the singularity), while the Fourier series is "full": $f_n = \gamma_\alpha n^{-1+\alpha} + O(n^{-3+\alpha})$, with $\gamma_\alpha \neq 0$; the effects of the singularity are felt in *all* the Fourier coefficients.

Notes.

1. There exist many different definitions of Calderón–Zygmund operators. A discussion of these different definitions and their evolution is given at the start of Meyer (1990, vol. 2). Note that the bounds are infinite on the diagonal $x = y$; in general K will be singular on the diagonal. Strictly speaking, we should be more careful about what happens on the diagonal. One way to make sure everything is well defined is to require that T is bounded from $\mathcal{D}$ to $\mathcal{D}'$ ($\mathcal{D}$ is the set of all compactly supported C^∞ functions, $\mathcal{D}'$ its dual, the space of (non-tempered) distributions), and that if $x \notin$ support (f), then $(Tf)(x) = \int dy \, K(x,y) f(y)$. It then follows that K does not completely determine T: the operator $(T_1 f)(x) = (Tf)(x) + m(x) f(x)$, with $m \in L^\infty(\mathbb{R})$, has the same integral kernel. See Meyer (1990, vol. 2) for a clear and extensive discussion.

2. Note that $\| \cdot \|_{L^1_{\text{weak}}}$ constitutes a (very convenient) abuse of notation. As shown, for example, by $\| \, |x-1|^{-1} + |x+1|^{-1} \|_{L^1_{\text{weak}}} \ge \|(x-1)^{-1}\|_{L^1_{\text{weak}}} + \|(x+1)^{-1}\|_{L^1_{\text{weak}}}$, the triangle inequality is not satisfied, so that $\|.\|_{L^1_{\text{weak}}}$ is not a "true" norm.

3. If the "weak" is dropped, then the theorem is known as the Riesz–Thorin theorem; in this case $K = C_1^t C_2^{1-t}$, and the restriction $q_1 \le p_1$, $q_2 \le p_2$ is not necessary.

4.

$$\sum_k (1+|a-k|)^{-1-\epsilon}\,(1+|b-k|)^{-1-\epsilon} \le \sum_k (1+|a-k|)^{-1-\epsilon}$$
$$\le \sup_{0\le a'\le 1} \sum_k (1+|a'-k|)^{-1-\epsilon} \le 2\sum_{\ell=0}^{\infty} (1+\ell)^{-1-\epsilon} < \infty\,.$$

5. We can suppose without loss of generality that $a \ge 0$. Find k so that $k \le a \le k+1$. Then

$$\sum_\ell \left[(1+|a-\ell|)\,(1+|a+\ell|)\right]^{-1-\epsilon}$$
$$\le \sum_{\ell=-\infty}^{-k-1} \left[(1+(k+|\ell|))\,(1+(|\ell|-k-1))\right]^{-1-\epsilon}$$
$$+ \sum_{\ell=-k}^{k} \left[(1+(k-\ell))\,(1+(k+\ell))\right]^{-1-\epsilon}$$
$$+ \sum_{\ell=k+1}^{\infty} \left[(1+(\ell-k-1))\,(1+(\ell+k))\right]^{-1-\epsilon}$$
$$\le 2\sum_{\ell=0}^{k} \left[1+(k-\ell)\right]^{-1-\epsilon}\,(1+k)^{-1-\epsilon}$$
$$+2\sum_{\ell=k+1}^{\infty} \left[1+(\ell-k-1)\right]^{-1-\epsilon}(1+2k)^{-1-\epsilon} \le C(1+|a|)^{-1-\epsilon}\,.$$

6. In Note 9 of Chapter 5 we saw that $\sum_\ell \phi(x+\ell) = \text{constant}$. Since $\int_{-\infty}^{\infty} dx\ \phi(x) = 1$, this constant is necessarily equal to 1.

7.

$$\begin{aligned}
\sum_\ell \psi(x+\ell/2) &= \sum_\ell \sum_n (-1)^n\, h_{-n+1}\, \phi(2x+\ell-n) \\
&= \sum_{k,m} (-1)^{m+1}\, h_m\, \phi(2x+k) \\
&\qquad (k=\ell-n,\ m=-n+1) \\
&= 0 \qquad \left(\text{because } \textstyle\sum h_{2m} = \sum h_{2m+1}\right).
\end{aligned}$$

8. By now, the reader has seen so many instances of this type of estimate that I leave the proof of Lemma 9.4.1 for non-compactly supported but well-decaying ψ as an exercise.

9. Yes, Meyer's wavelet does not have compact support, and the proof of Lemma 9.4.1 uses that ψ is compactly supported. See, however, Note 8 above.

CHAPTER 10

Generalizations and Tricks for Orthonormal Wavelet Bases

This chapter consists of several generalizations and extensions of earlier constructions. These are not treated with the same detail as in the previous chapters. Some of the topics are still developing, and I expect that any write-up on them would look very different even two years from now. The sections cover multidimensional wavelets with dilation factor 2, via tensor product multiresolution analysis, or via nonseparable schemes; orthonormal wavelet bases with dilation factor different from 2, integer or non-integer; the "splitting trick" for better frequency resolution (in fact, merely a special case of the "wavelet packets" of Coifman and Meyer); wavelet bases on an interval.

10.1. Multidimensional wavelet bases with dilation factor 2.

For simplicity, we will consider only the two-dimensional case; higher dimensions are analogous. One trivial way of constructing an orthonormal basis for $L^2(\mathbb{R}^2)$, starting from an orthonormal wavelet basis $\psi_{j,k}(x) = 2^{-j/2}\ \psi(2^{-j}x - k)$ for $L^2(\mathbb{R})$, is simply to take the tensor product functions generated by two one-dimensional bases:

$$\Psi_{j_1,k_1;\ j_2,k_2}(x_1, x_2) = \psi_{j_1,k_1}(x_1)\ \psi_{j_2,k_2}(x_2)\ .$$

The resulting functions are indeed wavelets, and $\{\Psi_{j_1,k_1;\ j_2,k_2};\ j_1, j_2, k_1, k_2 \in \mathbb{Z}\}$ is an orthonormal basis for $L^2(\mathbb{R}^2)$. In this basis the two variables x_1 and x_2 are dilated separately.

There exists another construction, more interesting for many applications, in which the dilations of the resulting orthonormal wavelet basis control both variables simultaneously. In this construction, one considers the tensor product of two one-dimensional multiresolution analyses rather than of the corresponding wavelet bases. More precisely, define spaces $\mathbf{V}_j,\ j \in \mathbb{Z}$, by

$$\begin{aligned} \mathbf{V}_0 &= V_0 \otimes V_0 = \overline{\text{Span}\{F(x,y) = f(x)g(y); f, g \in V_0\}}\ , \\ F \in \mathbf{V}_j &\Leftrightarrow F(2^j\cdot, 2^j\cdot) \in \mathbf{V}_0\ . \end{aligned}$$

Then the $\mathbf{V}_j$ form a multiresolution ladder in $L^2(\mathbb{R}^2)$ satisfying

$$\cdots \mathbf{V}_2 \subset \mathbf{V}_1 \subset \mathbf{V}_0 \subset \mathbf{V}_{-1} \subset \mathbf{V}_{-2} \cdots\ ,$$

$$\bigcap_{j\in\mathbb{Z}} \mathbf{V}_j = \{0\}, \quad \overline{\bigcup_{j\in\mathbb{Z}} \mathbf{V}_j} = L^2(\mathbb{R}^2) .$$

Since the $\phi(\cdot - n)$, $n \in \mathbb{Z}$, constitute an orthonormal basis for V_0, the product functions

$$\Phi_{0;n_1,n_2}(x,y) = \phi(x-n_1)\,\phi(x-n_2) , \qquad n_1, n_2 \in \mathbb{Z} ,$$

constitute an orthonormal basis for $\mathbf{V}_0$, generated by the $\mathbb{Z}^2$-translations of a single function Φ. Similarly, the

$$\begin{aligned}\Phi_{j;n_1,n_2}(x,y) &= \phi_{j,n_1}(x)\,\phi_{j,n_2}(y)\\ &= 2^{-j}\Phi(2^{-j}x-n_1,\ 2^{-j}y-n_2) , \qquad n_1, n_2 \in \mathbb{Z} ,\end{aligned}$$

constitute an orthonormal basis for $\mathbf{V}_j$. As in the one-dimensional case, we define, for each $j \in \mathbb{Z}$, the complement space $\mathbf{W}_j$ to be the orthogonal complement in $\mathbf{V}_{j-1}$ of $\mathbf{V}_j$. We have

$$\begin{aligned}\mathbf{V}_{j-1} &= V_{j-1}\otimes V_{j-1} = (V_j\oplus W_j)\otimes(V_j\oplus W_j)\\ &= V_j\otimes V_j \oplus [(W_j\otimes V_j)\oplus(V_j\otimes W_j)\oplus(W_j\otimes W_j)]\\ &= \mathbf{V}_j\oplus\mathbf{W}_j .\end{aligned}$$

It follows that $\mathbf{W}_j$ consists of three pieces, with orthonormal bases given by the $\psi_{j,n_1}(x)\,\phi_{j,n_2}(y)$ (for $W_j\otimes V_j$), $\phi_{j,n_1}(x)\,\psi_{j,n_2}(y)$ (for $V_j\otimes W_j$), and $\psi_{j,n_1}(x)\,\psi_{j,n_2}(y)$ (for $W_j\otimes W_j$). This leads us to define *three* wavelets,

$$\begin{aligned}\Psi^h(x,y) &= \phi(x)\,\psi(y)\\ \Psi^v(x,y) &= \psi(x)\,\phi(y)\\ \Psi^d(x,y) &= \psi(x)\,\psi(y)\end{aligned}$$

(h, v, d stand for "horizontal," "vertical," "diagonal," respectively; see below). Then

$$\{\Psi^{\lambda}_{j;n_1,n_2};\ n_1,n_2\in\mathbb{Z},\ \lambda=h,\ v \text{ or } d\}$$

is an orthonormal basis for $\mathbf{W}_j$, and

$$\{\Psi^{\lambda}_{j;\mathbf{n}};\ j\in\mathbb{Z},\ \ \mathbf{n}\in\mathbb{Z}^2,\ \ \lambda=h,\ v \text{ or } d\}$$

is an orthonormal basis for $\overline{\bigoplus_{j\in\mathbb{Z}} \mathbf{W}_j} = L^2(\mathbb{R}^2)$.

If, in this construction, the original one-dimensional ϕ and ψ have compact support, then obviously so have Φ and the Ψ^{λ}. Moreover, the interpretation in terms of subband filtering of a decomposition with respect to such an orthonormal basis of compactly supported wavelets, as explained in §5.6, carries over to two dimensions. The filtering can be done on "rows" and "columns" in the two-dimensional array, corresponding to horizontal and vertical directions in images, for example. Figure 5.8 becomes, in two dimensions, the schematic representation in Figure 10.1.

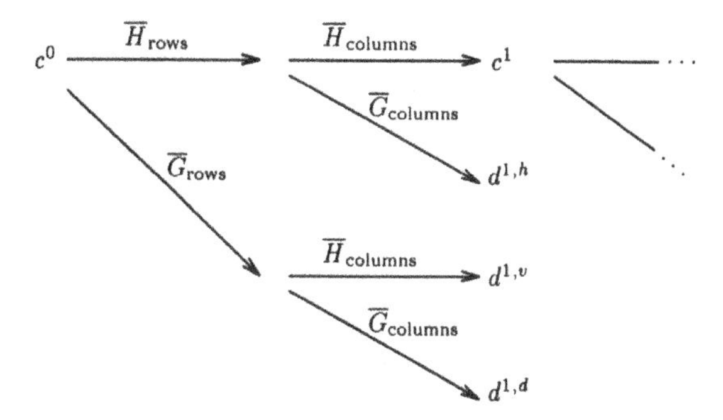

FIG. 10.1. *Schematic representation of repeated low- and high-pass filtering, on rows and columns, for a two-dimensional wavelet decomposition.*

The $d^{1,\lambda}$ correspond exactly to the wavelet coefficients $\langle F, \Psi^{\lambda}_{1;\mathbf{n}}\rangle$, with $F = \sum_{\mathbf{n}} c^0_{\mathbf{n}} \Phi_{0;\mathbf{n}}$. In an image, horizontal edges will show up in $d^{1,h}$, vertical edges in $d^{1,v}$, diagonal edges in $d^{1,d}$, as illustrated in the image example below. (This justifies the h, v, d superscripts.) Note that if the original image (c^0) consists of an $N \times N$ array, then (apart from border effects; see also §10.6), every array $d^{1,\lambda}$ consists of $\frac{N}{2} \times \frac{N}{2}$ elements, and can therefore be represented by an image (magnitudes of the coefficients corresponding to grey levels) of one quarter the size of the original. The whole scheme can therefore be represented as in Figure 10.2. Of course, one can decompose c^2 even further if more multiresolution layers are wanted. Figure 10.3 shows this decomposition scheme on a real image, with three multiresolution layers.

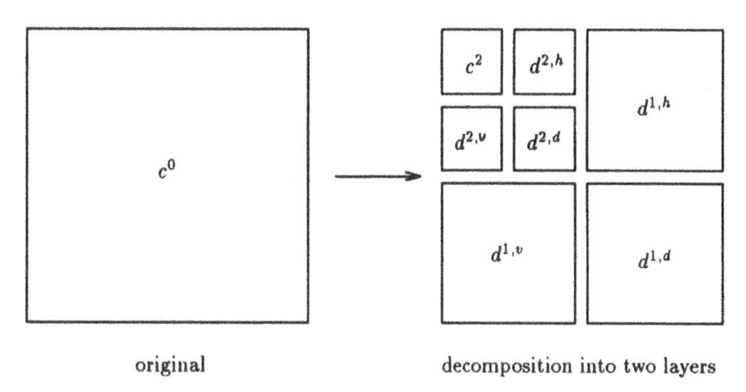

FIG. 10.2. *Schematic representation of the visualization of the two-dimensional wavelet transform of Figure* 10.3.

All this concerned two-dimensional schemes which have a tensor product structure. One can also consider the case in which one starts from a two-dimensional multiresolution analysis (with the $\mathbf{V}_j$ satisfying all the obvious gen-

FIG. 10.3. *A real image, and its wavelet decomposition into three multiresolution layers. On the wavelet components one clearly sees that the* $d^{j,v}$, $d^{j,h}$, $d^{j,d}$ *emphasize, respectively, vertical, horizontal, and diagonal edges. In this picture,* c^3 *has been overexposed to make details in the* $d^{j,\lambda}$ *more apparent. I would like to thank M. Barlaud for providing this figure.*

eralizations of (5.1.1)–(5.1.6)) in which $\mathbf{V}_0$ is not a tensor product of two one-dimensional V_0-spaces.[1] Some (but not all!) of the constructions done in one dimension can be repeated for this case. More precisely, the multiresolution structure of the $\mathbf{V}_j$ implies that the corresponding scaling function Φ satisfies

$$\Phi(x,y) = \sum_{n_1,n_2} h_{n_1,n_2} \Phi(2x-n_1, 2y-n_2) \tag{10.1.1}$$

for some sequence $(h_{\mathbf{n}})_{\mathbf{n}\in\mathbb{Z}^2}$. Orthonormality of the $\Phi_{0;\mathbf{n}}$ forces the trigonometric polynomial

$$m_0(\xi,\zeta) = \frac{1}{2}\sum_{n_1,n_2} h_{n_1,n_2}\, e^{-i(n_1\xi+n_2\zeta)} \tag{10.1.2}$$

to satisfy

$$|m_0(\xi,\zeta)|^2+|m_0(\xi+\pi,\zeta)|^2+|m_0(\xi,\zeta+\pi)|^2+|m_0(\xi+\pi,\zeta+\pi)|^2=1\,. \tag{10.1.3}$$

To construct an orthonormal basis of wavelets corresponding to this multiresolution analysis, one has to find three wavelets Ψ^1, Ψ^2, Ψ^3 in $\mathbf{V}_{-1}$, orthogonal to $\mathbf{V}_0$ and such that the three spaces spanned by their respective integer translates are orthogonal; moreover the $\Psi^\lambda(\cdot-\mathbf{n})$ should also be orthonormal for each fixed λ. This implies that

$$\hat{\Psi}^\lambda(\xi,\zeta) = m_\lambda\left(\frac{\xi}{2},\frac{\zeta}{2}\right)\hat{\Phi}\left(\frac{\xi}{2},\frac{\zeta}{2}\right),$$

where the m_1, m_2, m_3 are such that the matrix

$$\begin{pmatrix} m_0(\xi,\zeta) & m_1(\xi,\zeta) & m_2(\xi,\zeta) & m_3(\xi,\zeta) \\ m_0(\xi+\pi,\zeta) & m_1(\xi+\pi,\zeta) & m_2(\xi+\pi,\zeta) & m_3(\xi+\pi,\zeta) \\ m_0(\xi,\zeta+\pi) & m_1(\xi,\zeta+\pi) & m_2(\xi,\zeta+\pi) & m_3(\xi,\zeta+\pi) \\ m_0(\xi+\pi,\zeta+\pi) & m_1(\xi+\pi,\zeta+\pi) & m_2(\xi+\pi,\zeta+\pi) & m_3(\xi+\pi,\zeta+\pi) \end{pmatrix} \tag{10.1.4}$$

is unitary. The analysis leading to this condition is entirely similar to the one-dimensional analysis in §5.1; see, e.g., Meyer (1990, §III.4).[2]

Note that the number of wavelets to be constructed can be determined by an easy trick. In two dimensions, for example, $\mathbf{V}_0$ is generated by the translates of *one* function $\Phi(x,y)$, over $\mathbb{Z}^2$; the space $\mathbf{V}_{-1}$ is generated by the translates of $\Phi(2x,2y)$ over $\frac{1}{2}\mathbb{Z}^2$, or equivalently, by the $\mathbb{Z}^2$-translates of the *four* functions $\Phi(2x,2y)$, $\Phi(2x-1,2y)$, $\Phi(2x,2y-1)$, $\Phi(2x-1,2y-1)$. $\mathbf{V}_{-1}$ is therefore "four times as big" as $\mathbf{V}_0$. On the other hand, each of the $\mathbf{W}_0^j$-spaces is generated by the $\mathbb{Z}^2$-translates of a single function $\Psi^j(x,y)$, and is therefore "of the same size" as $\mathbf{V}_0$. It follows that one needs three (= four minus one) spaces $\mathbf{W}_0^j$ (hence three wavelets Ψ^j) to make up the complement of $\mathbf{V}_0$ in $\mathbf{V}_{-1}$. This rule may sound like "hand-waving," but we can also rephrase (and prove) it in more mathematical terms: the number of wavelets is equal to the number of different cosets (different from $\mathbb{Z}^2$ itself) of the subgroup $\mathbb{Z}^2$ in the group $\frac{1}{2}\mathbb{Z}^2$.

In the general n-dimensional case, the same rule shows that there are $2^n - 1$ different functions m_j to determine; they have to be such that the $2^n \times 2^n$-dimensional matrix

$$U_{r,\mathbf{s}}(\xi_1, \cdots, \xi_n) = m_{r-1}(\xi_1 + s_1\pi, \cdots, \xi_n + s_n\pi) \qquad (10.1.5)$$

is unitary, with $r = 1, \cdots, 2^n$, and $\mathbf{s} = (s_1, \cdots, s_n) \in \{0,1\}^n$.[3]

In fact the unitarity requirement of (10.1.4) or (10.1.5) calls for a tricky balance: m_1, m_2, m_3 have to be found so that the first row of (10.1.4) has unit norm, which seems harmless enough, but we also simultaneously need orthogonality with and among the other rows, which are all shifted versions (in ξ or ζ) of the first row. These correlations between the rows may be hard to juggle in practice. It is useful to untangle them first, which can be done via the so-called *polyphase* decomposition. We write, e.g.,

$$\begin{aligned} 2m_0(\xi,\zeta) &= m_{0,0}(2\xi,2\zeta) + e^{-i\xi}m_{0,1}(2\xi,2\zeta) + e^{-i\zeta}m_{0,2}(2\xi,2\zeta) \\ &\quad + e^{-i(\xi+\zeta)}m_{0,3}(2\xi,2\zeta); \end{aligned}$$

$m_{\ell,j}$, $j = 0, \cdots, 3$, are defined similarly from m_ℓ, $\ell = 1, \cdots, 3$. One easily checks that (10.1.3) is equivalent to

$$|m_{0,0}(2\xi,2\zeta)|^2 + |m_{0,1}(2\xi,2\zeta)|^2 + |m_{0,2}(2\xi,2\zeta)|^2 + |m_{0,3}(2\xi,2\zeta)|^2 = 1 .$$

Similarly, all the other conditions ensuring unitarity of (10.1.4) can be recast in terms of the $m_{\ell,j}$; one finds that (10.1.4) is unitary if and only if the *polyphase* matrix

$$\begin{pmatrix} m_{0,0}(\xi,\zeta) & m_{1,0}(\xi,\zeta) & m_{2,0}(\xi,\zeta) & m_{3,0}(\xi,\zeta) \\ m_{0,1}(\xi,\zeta) & m_{1,1}(\xi,\zeta) & m_{2,1}(\xi,\zeta) & m_{3,1}(\xi,\zeta) \\ m_{0,2}(\xi,\zeta) & m_{1,2}(\xi,\zeta) & m_{2,2}(\xi,\zeta) & m_{3,2}(\xi,\zeta) \\ m_{0,3}(\xi,\zeta) & m_{1,3}(\xi,\zeta) & m_{2,3}(\xi,\zeta) & m_{3,3}(\xi,\zeta) \end{pmatrix} \qquad (10.1.6)$$

is unitary.

In n dimensions, one similarly defines

$$2^{n/2}m_r(\xi_1, \cdots, \xi_n) = \sum_{\mathbf{s}\in\{0,1\}^n} e^{-i(s_1\xi_1+\cdots+s_n\xi_n)}\, m_{r,\mathbf{s}}(2\xi_1, \cdots, 2\xi_n) ,$$

and the unitarity of U is equivalent to the unitarity of the polyphase matrix $\tilde{U}$ defined by

$$\tilde{U}_{r,\mathbf{s}}(\xi_1, \cdots, \xi_n) = m_{r-1,\mathbf{s}}(\xi_1, \cdots, \xi_n) . \qquad (10.1.7)$$

The construction therefore boils down to the following question: given m_0 (from (10.1.1), (10.1.2)), can $m_1, \cdots, m_{2^n-1}$ be found such that (10.1.6) is unitary? In the two-dimensional case, and if $m_0(\xi,\zeta)$ happens to be a *real* trigonometric polynomial, then one can even dispense with the polyphase matrix: it is easy to check that the choice $m_1(\xi,\zeta) = e^{-i\xi} m_0(\xi+\pi,\zeta)$, $m_2(\xi,\zeta) = e^{-i(\xi+\zeta)}$ $m_0(\xi,\zeta+\pi)$, $m_3(\xi,\zeta) = e^{-i\zeta} m_0(\xi+\pi,\zeta+\pi)$ makes (10.1.4) unitary. If m_0 is not real, then things are more complicated. At first sight one might even think the task is impossible in general in the n-dimensional situation, where (10.1.7) is a $2^n \times 2^n$-matrix: after all, we need to find unit vectors, depending continuously on the ξ_i (namely the second to last columns of (10.1.7)), orthogonal to a unit vector (the first column of (10.1.7)), i.e., tangent to the unit sphere. But it is well known that "it is impossible to comb a sphere," i.e., there exist no nowhere-vanishing continuous vector fields tangent to the unit sphere, except in real dimensions 2, 4, or 8. The first column in (10.1.7) does not describe the full sphere, however; in fact, because it is a continuous function of n variables (the $\xi_1, \cdots, \xi_n$) in a 2^n-dimensional space, and $2^n > n$, it only describes a compact set of measure zero. This fact saves the day and makes it possible to construct $m_1, \cdots, m_{2^n-1}$, as shown by Gröchenig (1987); see also §III.6 in Meyer (1990). Gröchenig's proof is not constructive; a different, constructive proof is given in Vial (1992). Unfortunately, these constructions can not force compact support for the Ψ^j: even if m_0 is a trigonometric polynomial (only finitely many $h_{\mathbf{n}} \neq 0$), the m_j are not necessarily.

10.2. One-dimensional orthonormal wavelet bases with integer dilation factor larger than 2.

For illustration purposes, let us choose dilation factor 3. A multiresolution analysis for dilation 3 is defined in exactly the same way as for dilation 2, i.e., by (5.1.1)–(5.1.6), except that (5.1.4) is replaced by

$$f \in V_j \Leftrightarrow f(3^j \cdot) \in V_0 \ .$$

We can use the same trick again as above: V_0 is generated by the integer translates of *one* function, i.e., by the $\phi(x-n)$, while V_{-1} is generated by the $\phi(3x-n)$, or, equivalently, by the integer translates of *three* functions, $\phi(3x)$, $\phi(3x-1)$, and $\phi(3x-2)$. V_{-1} is "three times as big" as V_0, and two spaces of the "same size" as V_0 are needed to complement V_0 and constitute V_{-1}: we will need two spaces W_0^1, W_0^2, or two wavelets, ψ^1 and ψ^2.

We can again introduce m_0, m_1, m_2 by

$$\hat{\phi}(\xi) = m_0(\xi/3)\ \hat{\phi}(\xi/3), \qquad \hat{\psi}^\ell(\xi) = m_\ell(\xi/3)\hat{\phi}(\xi/3), \qquad \ell = 1, 2\ .$$

Orthonormality of the whole family $\{\phi_{0,n}, \psi_{0,n}^1, \psi_{0,n}^2;\ n \in \mathbb{Z}\}$, where $\phi_{j,n}$ is now defined by

$$\phi_{j,n}(x) = 3^{-j/2}\ \phi(3^{-j}x - n)$$

($\psi_{j,n}^\ell$ are defined analogously), again forces several orthonormality conditions on the m_ℓ, which can be summarized by the requirement that the matrix

$$\begin{pmatrix} m_0(\xi) & m_1(\xi) & m_2(\xi) \\ m_0\left(\xi + \frac{2\pi}{3}\right) & m_1\left(\xi + \frac{2\pi}{3}\right) & m_2\left(\xi + \frac{2\pi}{3}\right) \\ m_0\left(\xi + \frac{4\pi}{3}\right) & m_1\left(\xi + \frac{4\pi}{3}\right) & m_2\left(\xi + \frac{4\pi}{3}\right) \end{pmatrix} \tag{10.2.1}$$

is unitary. Again, one can restate this in terms of a polyphase matrix, removing the correlations between the rows. Explicit choices of m_0, m_1, m_2 for which (10.2.1) is indeed unitary have been constructed in the ASSP literature (see, e.g., Vaidyanathan (1987)). The question is then again, as in Chapter 6, whether these filters correspond to bona fide L^2-functions ϕ, ψ^1, and ψ^2, whether the $\psi^\ell_{j,k}$ constitute an orthonormal basis, and what the regularity is of all these functions. We know, from Chapter 3, that ψ^1 and ψ^2 must necessarily have integral zero, corresponding to $m_1(0) = 0 = m_2(0)$. Since the first row of (10.2.1) must have norm 1 for all ξ, it follows that $m_0(0) = 1$ (which is necessary anyway for the convergence of the infinite product $\prod_{j=1}^\infty m_0(3^{-j}\xi)$ which defines $\hat{\phi}(\xi)$). The first column of (10.2.1) must also have norm 1 for all ξ, so that $m_0(0) = 1$ implies $m_0(\frac{2\pi}{3}) = 0 = m_0(\frac{4\pi}{3})$, i.e., $m_0(\xi)$ is divisible by $\frac{1+e^{-i\xi}+e^{-2i\xi}}{3}$. If, moreover, any smoothness for ψ^1, ψ^2 is desired, then we need additional vanishing moments of ψ^1, ψ^2, which by exactly the same argument as before, lead to divisibility of $m_0(\xi)$ by $((1 + e^{-i\xi} + e^{-2i\xi})/3)^L$ if $\psi^1, \psi^2 \in C^{L-1}$. One is thus led to looking for m_0 of the type $m_0(\xi) = ((1 + e^{-i\xi} + e^{-2i\xi})/3)^N \mathcal{L}(\xi)$ such that $|m_0(\xi)|^2 + |m_0(\xi + \frac{2\pi}{3})|^2 + |m_0(\xi + \frac{4\pi}{3})|^2 = 1$. If m_0 is a trigonometric polynomial, this means that $L = |\mathcal{L}|^2$ is again the solution to a Bezout problem. The minimal degree solution leads to functions ϕ with arbitrarily high regularity; however, the regularity index only grows logarithmically with N (L. Villemoes, private communication).[4] Once m_0 is fixed, m_1 and m_2 have to be determined. The design scheme explained in Vaidyanathan et al. (1989) gives a way to do this. In this scheme, the matrix (10.2.1) (or rather, its z-notation equivalent) is written as a product of similar matrices the entries of which are much lower degree polynomials, with only a few parameters determining each factor matrix.[5] If one imposes that the first column of a product of such matrices is given by the m_0 we have fixed, then the values of these parameters are fixed likewise, and m_1, m_2 can be read off from the product matrix.[6]

If the compact support constraint is lifted, then other constructions are possible. In Auscher (1989) one can find examples where ϕ and ψ^ℓ are C^∞ functions with fast decay (and infinite support).

One final remark about dilation factor 3. We have seen that m_0 must necessarily be divisible by $(1 + e^{-i\xi} + e^{-2i\xi})/3$. This factor does not vanish for $\xi = \pi$ (unlike the factor $(1 + e^{-i\xi})/2$ for the dilation factor 2 case). However, if we want to interpret m_0 as a low-pass filter, then $m_0(\pi) = 0$ would be a good idea. To ensure this, we need $\mathcal{L}(\pi) = 0$, which means going beyond the lowest degree solution to the Bezout equation for $|\mathcal{L}|^2$.

Similar constructions can be made for larger integer dilation factors. For non-prime dilation factors a, one can generate acceptable m_ℓ from constructions for the factors of a, although not all possible solutions for dilation a can be obtained in this way. For $a = 4$, e.g., one can start from a scheme with dilation 2 and filters m_0 and m_1, and one can define the filters $\tilde{m}_0, \tilde{m}_1, \tilde{m}_2, \tilde{m}_3$ (still orthonormal; the $\tilde{}$ distinguishes them from the dilation factor 2 filters) by

$$\tilde{m}_0(\xi) = m_0(\xi)m_0(\xi/2) \ , \qquad \tilde{m}_2(\xi) = m_1(\xi)m_1(\xi/2) \ ,$$
$$\tilde{m}_1(\xi) = m_0(\xi)m_1(\xi/2) \ , \qquad \tilde{m}_3(\xi) = m_1(\xi)m_0(\xi/2) \ .$$

(It is left to the reader as an exercise to prove that this leads indeed to an orthonormal basis. One easily checks that the 4×4 analogue of (10.2.1) is unitary.) Note that the function ϕ is the same for the factor 4 and the factor 2 constructions! We will come back to this in §10.5.

10.3. Multidimensional wavelet bases with matrix dilations.

This is a generalization of both §10.1 and §10.2: the multiresolution spaces are subspaces of $L^2(\mathbb{R}^n)$, and the basic dilation is a matrix D with integer entries (so that $D\mathbb{Z}^n \subset \mathbb{Z}^n$) such that all its eigenvalues have absolute value strictly larger than 1 (so that we are indeed dilating in all directions). The number of wavelets is again determined by the number of cosets of $D\mathbb{Z}^n$; one introduces again $m_0, m_1, \cdots$, and the orthonormality conditions can again be formulated as a unitarity requirement for a matrix constructed from the $m_0, m_1, \cdots$. The analysis for these matrix dilation cases is quite a bit harder than for the one-dimensional case with dilation 2, and, depending on the matrix chosen, there are a few surprises. One surprise is that generalizing the Haar basis (i.e., choosing m_0 so that all its nonvanishing coefficients are equal) leads in many cases to a function ϕ which is the indicator function of a selfsimilar set with fractal boundary, tiling the plane. For two dimensions, with $D = \begin{pmatrix} 1 & -1 \\ 1 & 1 \end{pmatrix}$, e.g., one finds that ϕ can be the indicator function of the twin dragon set, as shown in Gröchenig and Madych (1992) and Lawton and Resnikoff (1991). Note that such fractal tiles may occur even for $D = 2$ Id if m_0 is chosen "non-canonically" (e.g., $m_0(\xi, \zeta) = \frac{1}{4}(1 + e^{-i\zeta} + e^{-i(\xi+\zeta)} + e^{-i(\xi+2\zeta)})$ in two dimensions—see Gröchenig and Madych (1992)). For more complicated m_0 (not all coefficients are equal), the problem is to control regularity. Zero moments for the ψ_j do not lead to factorization of m_0 in these multidimensional cases (because it is not sufficient to know zeros of a multi-variable polynomial to factorize it), and one has to resort to other tricks to control the decay of $\hat{\phi}$.

A particularly interesting case is given by the "quincunx lattice," i.e., the two-dimensional case where $D\mathbb{Z}^2 = \{(m, n);\ m + n \in 2\mathbb{Z}\}$. In this case there is only one other coset, and therefore only one wavelet to construct, so that the choice for m_1 is as straightforward as it was for dilation 2 in one dimension. The conditions on m_0, m_1 reduce to the requirement that the 2×2 matrix

$$\begin{pmatrix} m_0(\xi,\zeta) & m_1(\xi,\zeta) \\ m_0(\xi+\pi,\zeta+\pi) & m_1(\xi+\pi,\zeta+\pi) \end{pmatrix}$$

be unitary. It is convenient to choose

$$m_1(\xi,\zeta) = e^{-i\xi} m_0(\xi+\pi,\zeta+\pi) \ .$$

Note that any orthonormal basis for dilation factor 2 in one dimension automatically gives rise to a pair of candidates for m_0, m_1 for the quincunx scheme: it suffices to take $m_0(\xi,\zeta) = m_0^{\#}(\xi)$ (where $m_0^{\#}$ is the one-dimensional filter).[7] Different choices for D can be made, however. Two possibilities studied in detail in Cohen and Daubechies (1993b) and Kovačević and Vetterli (1992) are $D_1 = \begin{pmatrix} 1 & -1 \\ 1 & 1 \end{pmatrix}$ and $D_2 = \begin{pmatrix} 1 & 1 \\ 1 & -1 \end{pmatrix}$. The same choice for m_0 leads to very different wavelet bases for these two matrices; in particular, if one derives, via the mechanism explained above, the filter m_0 from the "standard" one-dimensional wavelet filters ${}_N m_0$ in §6.4, then the resulting ϕ are increasingly regular if D_2 is chosen (with regularity index proportional to N), whereas choosing D_1 leads to ϕ which are at most continuous, regardless of N. Other choices for D may lead to yet other families, with different regularity properties again. One can of course also choose to construct two biorthogonal bases rather than one orthonormal basis, as in §8.3; for the choices D_1, D_2 several possibilities are explored in Cohen and Daubechies (1993b) and Kovačević and Vetterli (1992). In this biorthogonal case, one can again derive filters from one-dimensional constructions. If one starts from a symmetric biorthogonal filter pair in one dimension, where all the filters are polynomials in $\cos\xi$, then it suffices to replace $\cos\xi$ by $\frac{1}{2}(\cos\xi + \cos\zeta)$ in every filter to obtain symmetric biorthogonal filter pairs for the quincunx case.[8] Because of the symmetry of these examples, the matrices D_1 and D_2 lead to the same functions $\phi, \tilde{\phi}$ in this case. One finds again that symmetric biorthogonal bases with arbitrarily high regularity are possible (see Cohen and Daubechies (1993b)). The quincunx case is of interest in image processing because it treats the different directions more homogeneously than the separable (tensor-product) two-dimensional scheme: instead of having two favorite directions (horizontals and verticals), the quincunx schemes treat horizontals, verticals, and diagonals on the same footing, without introducing redundancy to achieve this. The first quincunx subband filtering schemes, with aliasing cancellation but without exact reconstruction (which had not been discovered even for one dimension at the time) are given in Vetterli (1984); Feauveau (1990) contains orthonormal and biorthogonal schemes, and links them to wavelet bases; Vetterli, Kovačević, and LeGall (1990) discusses the use of perfect reconstruction quincunx filtering schemes for HDTV applications. In Antonini, Barlaud, and Mathieu (1991) biorthogonal quincunx decompositions combined with vector quantization give spectacular results for image compaction.

10.4. One-dimensional orthonormal wavelet bases with non-integer dilation factors.

In one dimension, we have so far only discussed integer dilation factors ≥ 2.[9] Non-integer dilation factors are also possible, however. Within the framework of a multiresolution analysis, the dilation factor must be rational[10] (for a proof, see Auscher (1989)). It had already been pointed out by G. David in 1985 that the construction of the Meyer wavelet could be generalized to dilation $a = \frac{k+1}{k}$, for $k \in \mathbb{N}, k \geq 1$; Auscher (1989) contains constructions for arbitrary rational a (see also Auscher's paper in Ruskai et al. (1992)). Let us illustrate for $a = \frac{3}{2}$ how the factor 2 scheme has to be adapted. We start again from a multiresolution analysis, defined as in (5.1.1)–(5.1.6), with $\frac{3}{2}$ instead of 2 for the dilation factor. We have again $\phi \in V_0 \subset V_{-1} = \overline{\text{Span}\{\phi(\frac{3}{2}\ \cdot -n)\}}$, so that

$$\phi(x) = \sqrt{\frac{3}{2}} \sum_n h_n^0 \, \phi\left(\frac{3}{2}\, x - n\right) .$$

(The reason for the superscript 0 will soon become clear.) Consequently,

$$\phi(x - 2\ell) = \sqrt{\frac{3}{2}} \sum_n h_n^0 \, \phi\left(\frac{3}{2}\, x - 3\ell - n\right) = \sqrt{\frac{3}{2}} \sum_n h_{n-3\ell}^0 \, \phi\left(\frac{3}{2}\, x - n\right) , \tag{10.4.1}$$

and orthonormality of the $\phi(\cdot - 2\ell)$ implies

$$\sum_n h_n^0 \, \overline{h_{n-3\ell}^0} = \delta_{\ell 0} . \tag{10.4.2}$$

On the other hand, $\phi(\cdot - 1)$ is also in V_0, and can therefore also be written as a (different) linear combination of the $\phi(\frac{3}{2}\, x - n)$,

$$\phi(x - 1) = \sqrt{\frac{3}{2}} \sum_n h_n^1 \, \phi\left(\frac{3}{2}\, x - n\right) . \tag{10.4.3}$$

Orthonormality of the $\phi(x - 2\ell - 1)$, and orthogonality of the $\phi(x - 2\ell - 1)$ with respect to the $\phi(x - 2\ell)$ then implies

$$\sum_n h_n^1 \, \overline{h_{n-3\ell}^1} = \delta_{\ell 0} , \tag{10.4.4}$$

$$\sum_n h_n^1 \, \overline{h_{n-3\ell}^0} = 0 . \tag{10.4.5}$$

All this means that we have in fact *two* m_0-functions,

$$m_0^0(\xi) = \sqrt{\frac{2}{3}} \sum_n h_n^0 e^{-in\xi}, \qquad m_0^1(\xi) = \sqrt{\frac{2}{3}} \sum_n h_n^1 e^{-in\xi} .$$

What about m_1? We define again, for $j \in \mathbb{Z}$, the space W_j to be the orthogonal complement in V_{j-1} of V_j. Note that V_{-1} is generated by the $\phi(\frac{3}{2}\,x - n)$, $n \in \mathbb{Z}$, or, equivalently, by the *even* integer translates of *three* functions, namely

$$\phi\left(\frac{3}{2}\,(x-2\ell)\right),\ \phi\left(\frac{3}{2}\,(x-2\ell)-\frac{1}{2}\right),\ \phi\left(\frac{3}{2}\,(x-2\ell)-1\right),\quad \ell \in \mathbb{Z}\,,$$

corresponding to $n = 3\ell$, $n = 3\ell + 1$, and $n = 3\ell + 2$. The space V_0 is generated by the $2\mathbb{Z}$ translates of *two* functions, $\phi(x-2\ell)$ and $\phi(x-2\ell-1)$, $\ell \in \mathbb{Z}$. It follows that the complement space W_0 is generated by the $2\mathbb{Z}$ translates of a single function, $W_0 = \overline{\text{Span}\ \{\psi(\cdot - 2n);\ n \in \mathbb{Z}\}}$. ("$W_0$ is half the size of V_0.") We expect therefore an orthonormal basis of the type $\psi_{j,k}(x) = (\frac{3}{2})^{-j/2}\ \psi\left((\frac{3}{2})^j x - 2k\right)$, $j, k \in \mathbb{Z}$. This function ψ can also be written as a linear combination of the $\phi(\frac{3}{2}\ x - n)$,

$$\psi(x) = \sqrt{\frac{3}{2}}\ \sum_n g_n \phi\left(\frac{3}{2}\ x - n\right)\ ,$$

and orthonormality of the $\psi(x - 2n)$, plus orthogonality with respect to the $\phi(x - 2n)$, $\phi(x - 2n - 1)$ implies

$$\sum_n g_n\ \overline{g_{n-3\ell}} \;=\; \delta_{\ell 0}\ , \tag{10.4.6}$$

$$\sum_n g_n\ \overline{h^0_{n-3\ell}} \;=\; 0, \quad \sum_n g_n\ \overline{h^1_{n-3\ell}} \;=\; 0\ . \tag{10.4.7}$$

With the definition $m_1(\xi) = \sqrt{\frac{2}{3}}\ \sum_n g_n e^{-in\xi}$, the conditions (10.4.2), (10.4.4)–(10.4.7) are equivalent to the unitarity of the matrix

$$\begin{pmatrix} m^0_0(\xi) & m^1_0(\xi) & m_1(\xi) \\ m^0_0\left(\xi + \frac{2\pi}{3}\right) & m^1_0\left(\xi + \frac{2\pi}{3}\right) & m_1\left(\xi + \frac{2\pi}{3}\right) \\ m^0_0\left(\xi + \frac{4\pi}{3}\right) & m^1_0\left(\xi + \frac{2\pi}{3}\right) & m_1\left(\xi + \frac{4\pi}{3}\right) \end{pmatrix}\ . \tag{10.4.8}$$

This matrix looks identical to (10.2.1), but this similarity is deceptive: in (10.4.8) the first two columns are both given by low-pass filters, because they are both related to the scaling function ϕ ($m^0_0(0) = 1 = m^1_0(0)$), whereas the second column in (10.2.1) corresponds to a high-pass filter. Such m^j_0, m_1 can indeed be constructed (see Auscher (1989) for details and graphs). Note that m^1_0 and m^0_0 are closely related. The Fourier transforms of (10.4.1), (10.4.3) are

$$\hat{\phi}(\xi) = m^0_0\left(\frac{2}{3}\ \xi\right)\ \hat{\phi}\left(\frac{2}{3}\ \xi\right), \qquad \hat{\phi}(\xi)e^{-i\xi} = m^1_0\left(\frac{2}{3}\ \xi\right)\ \hat{\phi}\left(\frac{2}{3}\ \xi\right)\ , \tag{10.4.9}$$

implying

$$m^0_0(\zeta)\ \hat{\phi}(\zeta) \;=\; e^{i3\zeta/2}\ m^1_0(\zeta)\ \hat{\phi}(\zeta)\ ,$$

which should hold for almost all ζ. If $\hat{\phi}$ is continuous, then the following argument shows that $\hat{\phi}$ vanishes on some intervals. Since $\hat{\phi}(0) = (2\pi)^{-1/2}$, there exists α so that for $|\zeta| \le \alpha$, $|\hat{\phi}(\zeta)| \ge (2\pi)^{-1/2}/2$. Consequently, for $|\zeta| \le \alpha$,

$$m_0^0(\zeta) = e^{3i\zeta/2}\, m_0^1(\zeta) \ ,$$

or

$$m_0^0(\zeta + 2\pi) \ = \ -e^{3i\zeta/2}\, m_0^1(\zeta + 2\pi) \ ;$$

since m_0^0, m_0^1 are also 2π-periodic, this implies $m_0^0(\zeta + 2\pi) = 0 = m_0^1(\zeta + 2\pi)$ for $|\zeta| \le \alpha$. It follows that $|\hat{\phi}(\frac{3}{2}\,\zeta + 3\pi)| = 0$ for $|\zeta| \le \alpha$. In particular, this means that ϕ cannot be compactly supported (compact support for ϕ means that $\hat{\phi}$ is entire, and non-trivial entire functions can only have isolated zeros).

Nevertheless, subband filtering schemes with rational noninteger dilation factors, in particular with dilation $\frac{3}{2}$, have been proposed and constructed by Kovačević and Vetterli (1993), with FIR filters. The basic idea is simple: starting from c^0, one can first decompose into three subbands, by means of a scheme as in §10.2, and then regroup the two lowest frequency bands by means of a synthesis filter corresponding to dilation 2; the result of this operation is c^1, while the third, highest frequency band after the first decomposition is d^1. The corresponding block diagram is Figure 10.4. If all the filters are FIR, then the whole scheme is FIR as well. But didn't we just prove that there does not exist a multiresolution analysis for dilation factor $\frac{3}{2}$ with FIR filters? The solution to this paradox is that the block diagram above does not correspond to the construction described earlier. A detailed analysis of Figure 10.4 shows that this scheme uses *two* different functions ϕ^1 and ϕ^2, with V_0 generated by the $\phi^1(x - 2n)$, $\phi^2(x - 2n)$, $n \in \mathbb{Z}$. The argument used to prove that ϕ cannot have compact support then no longer applies, and ϕ^1, ϕ^2 can indeed have compact support. The analog of (10.4.9) is now an equation relating the two-dimensional vectors $(\hat{\phi}^1(\xi), \hat{\phi}^2(\xi))$ and $(\hat{\phi}^1(\frac{2}{3}\,\xi), \hat{\phi}^2(\frac{2}{3}\,\xi))$, however, and it is hard to see how to formulate conditions on the filters that result in regularity of ϕ^1, ϕ^2.

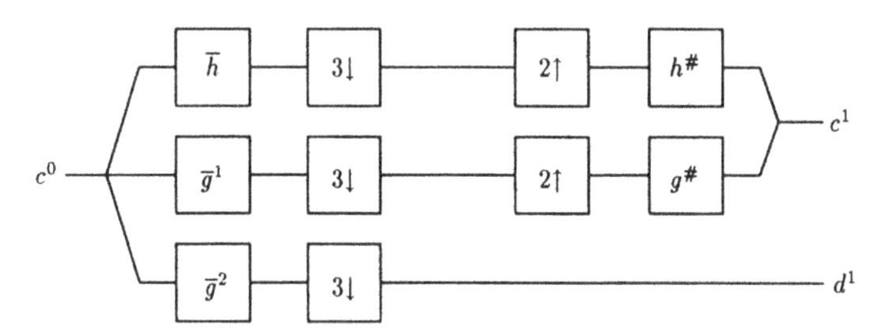

FIG. 10.4. *Block diagram corresponding to a subband filtering with dilation factor* $\frac{3}{2}$, *as constructed in Kovačević and Vetterli* (1993).

One may well wonder what the rationale is for these fractional dilation factors. The answer is that they may provide a sharper frequency localization. If the dilation factor is 2, then $\hat{\psi}$ is essentially localized between π and 2π, as illustrated by the Fourier transform of a "typical" ψ in Figure 10.5. For some applications, it may be useful to have wavelet bases that have a bandwidth narrower than one octave, and fractional dilation wavelet bases are one possible answer. A different answer is given in Cohen and Daubechies (1993a), summarized in the next section.

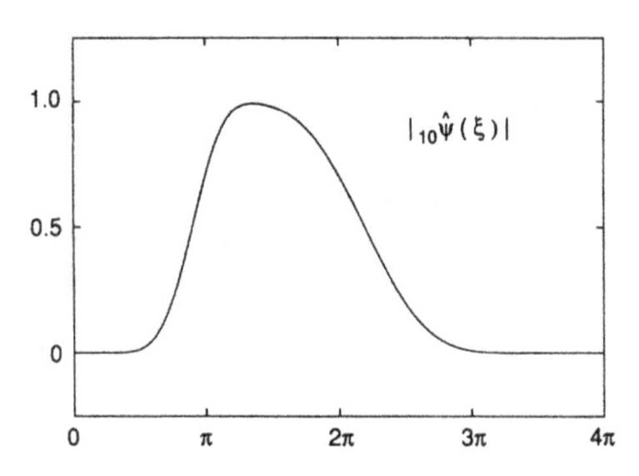

FIG. 10.5. *Modulus of* $|_{10}\hat{\psi}(\xi)|$, *with* $_N\psi$ *as defined in* §6.4.

10.5. Better frequency resolution: The splitting trick

Suppose that h_n, g_n are the filter coefficients associated to an orthonormal wavelet basis with dilation factor 2, i.e.,

$$m_0(\xi) = \frac{1}{\sqrt{2}} \sum_n h_n e^{-in\xi}$$

satisfies

$$|m_0(\xi)|^2 + |m_0(\xi+\pi)|^2 = 1 , \tag{10.5.1}$$

and

$$g_n = (-1)^n \, h_{-n+1} .$$

Then we have the following lemma.

LEMMA 10.5.1. *Take any function f (not necessarily connected to wavelets in any way) so that the $f(\cdot - n)$, $n \in \mathbb{Z}$ are orthonormal. Define*

$$F_1(x) = \sum_n h_n f(x-n)$$

$$F_2(x) = \sum_n g_n f(x-n) .$$

Then $\{F_1(\cdot - 2k), F_2(\cdot - 2k);\ k \in \mathbb{Z}\}$ is an orthonormal basis for $E = \overline{Span\ \{f(\cdot - n);\ n \in \mathbb{Z}\}}$.

Proof.

1. Since $\int dx\ f(x)\ \overline{f(x-n)} = \delta_{n,0}$, we have

$$\int d\xi\ |\hat{f}(\xi)|^2 e^{-in\xi} = \delta_{n,0}, \quad \text{or} \sum_\ell |\hat{f}(\xi + 2\pi\ell)|^2 = (2\pi)^{-1} \text{ a.e.} \tag{10.5.2}$$

2. $$\hat{F}_1(\xi) = \sum_n h_n e^{-in\xi} \hat{f}(\xi) = \sqrt{2}\ m_0(\xi) \hat{f}(\xi)\ . \tag{10.5.3}$$

Consequently,

$$\begin{aligned}
\sum_\ell |\hat{F}_1(\xi + \pi\ell)|^2 &= \sum_k \left[|\hat{F}_1(\xi + 2\pi k)|^2 + |\hat{F}_1(\xi + \pi + 2\pi k)|^2 \right] \\
&= 2\ (2\pi)^{-1} \left[|m_0(\xi)|^2 + |m_0(\xi + \pi)|^2 \right] \\
&\qquad\qquad\qquad\qquad \text{(use (10.5.2) and (10.5.3))} \\
&= \pi^{-1} \qquad \text{(by (10.5.1))}\ .
\end{aligned}$$

This implies

$$\begin{aligned}
\int dx\ F_1(x)\ \overline{F_1(x-2k)} &= \int d\xi\ |\hat{F}_1(\xi)|^2\ e^{-2ik\xi} \\
&= \sum_\ell \int_0^\pi d\xi\ |\hat{F}_1(\xi + \pi\ell)|^2\ e^{-2ik\xi} = \delta_{k0}\ .
\end{aligned}$$

The orthonormality of the $F_2(x - 2k)$ is proved analogously, using $\hat{F}_2(\xi) = \sqrt{2}\ e^{-i\xi}\ \overline{m_0(\xi + \pi)} \hat{f}(\xi)$.

3. Similarly,

$$\int dx F_1(x) \overline{F_2(x-2k)} = \int_0^\pi d\xi \left[\sum_\ell \hat{F}_1(\xi + \pi\ell) \overline{\hat{F}_2(\xi + \pi\ell)} \right] e^{-2ik\xi}\ , \tag{10.5.4}$$

and

$$\begin{aligned}
&\sum_\ell \hat{F}_1(\xi + \pi\ell) \overline{\hat{F}_2(\xi + \pi\ell)} \\
&\quad = \sum_k \left[\hat{F}_1(\xi + 2\pi k) \overline{\hat{F}_2(\xi + 2\pi k)} + \hat{F}_1(\xi + \pi + 2\pi k) \overline{\hat{F}_2(\xi + \pi + 2\pi k)} \right] \\
&\quad = 2\ (2\pi)^{-1} \left[m_0(\xi) m_0(\xi + \pi) e^{i\xi} + m_0(\xi + \pi) m_0(\xi) e^{i(\xi + \pi)} \right] \\
&\quad = 0\ ,
\end{aligned}$$

which proves that the $F_1(x - 2k)$ and the $F_2(x - 2\ell)$ are orthogonal.

4. Finally, the $F_1(\cdot - 2k)$, $F_2(\cdot - 2k)$ span all of E because

$$f(x) = \sum_\ell \left[h_{2\ell}\ F_1(x + 2\ell) + g_{2\ell}\ F_2(x + 2\ell) \right]\ , \tag{10.5.5}$$

and

$$f(x-1) = \sum_\ell \left[h_{2\ell+1}\, F_1(x+2\ell) + g_{2\ell+1}\, F_2(x+2\ell)\right] . \qquad (10.5.6)$$

We have indeed

$$\begin{aligned}
&\sum_\ell h_{2\ell}\, e^{2i\ell\xi} \hat{F}_1(\xi) + \sum_\ell g_{2\ell}\, e^{2i\ell\xi} \hat{F}_2(\xi) \\
&= \left[\overline{m_0(\xi)} + \overline{m_0(\xi+\pi)}\right] m_0(\xi)\hat{f}(\xi) \\
&\quad + \left[\overline{m_1(\xi)} + \overline{m_1(\xi+\pi)}\right] m_1(\xi)\hat{f}(\xi) \\
&= \hat{f}(\xi)\left\{[|m_0(\xi)|^2 + |m_1(\xi)|^2] + [m_0(\xi)\overline{m_0(\xi+\pi)} + m_1(\xi)\overline{m_1(\xi+\pi)}]\right\} \\
&= \hat{f}(\xi) ,
\end{aligned}$$

which proves (10.5.5). Similarly,

$$\begin{aligned}
&\sum_\ell h_{2\ell+1}\, e^{2i\ell\xi} \hat{F}_1(\xi) + \sum_\ell g_{2\ell+1}\, e^{2i\ell\xi} \hat{F}_2(\xi) \\
&\qquad = e^{-i\xi} \left[\overline{m_0(\xi)} - \overline{m_0(\xi+\pi)}\right] m_0(\xi)\hat{f}(\xi) \\
&\qquad\quad + e^{-i\xi} \left[\overline{m_1(\xi)} - \overline{m_1(\xi+\pi)}\right] m_1(\xi)\hat{f}(\xi) \\
&\qquad = e^{-i\xi}\hat{f}(\xi) ,
\end{aligned}$$

which proves (10.5.6). ∎

Lemma 10.5.1 is the "splitting trick": it shows that wavelet filters can be used to split *any* space spanned by orthonormal functions $f(x-n)$ into two parts. Since m_0, m_1 "live" on different frequency ranges (see Figure 10.6), the splitting trick corresponds to cutting the support of $\hat{f}$ into slices, and allotting alternate slices to F_1 and F_2.

We can apply this splitting trick to the (approximately) one-octave bandwidth space W_0 spanned by the $\psi(\cdot - k)$ in a one-dimensional multiresolution analysis with factor 2. We define

$$\psi^1(x) = \sum_n h_n\, \psi(x-n), \qquad \psi^2(x) = \sum_n g_n\, \psi(x-n) ,$$

where the h_n, g_n need not be the same filter coefficients as used in the construction of ψ itself. Then

$$\begin{aligned}
W_0 &= \overline{\text{Span}\ \{\psi(\cdot - k);\ k \in \mathbb{Z}\}} \\
&= \overline{\text{Span}\ \{\psi^1(\cdot - 2\ell);\ \ell \in \mathbb{Z}\}} \oplus \overline{\text{Span}\ \{\psi^2(\cdot - 2\ell);\ \ell \in \mathbb{Z}\}} \\
&= W_0^1 \oplus W_0^2 .
\end{aligned}$$

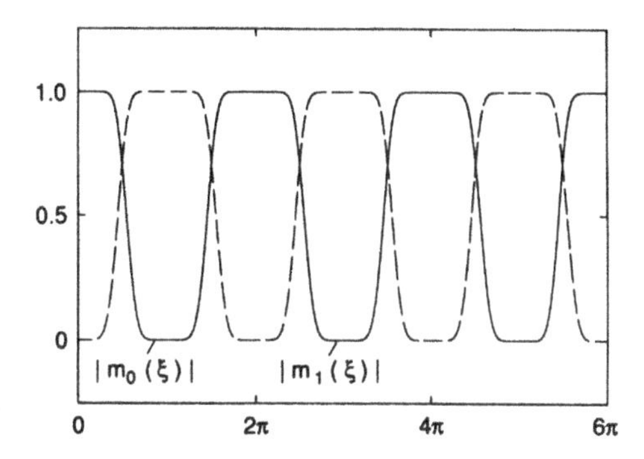

FIG. 10.6. *Graphs of* $|_{10}m_0(\xi)|$ *and* $|_{10}m_1(\xi)|$, *with* $_Nm_0$ *defined as in* §6.4.

Since the W_j-spaces are all dilated copies of W_0, we can construct corresponding orthonormal bases for each W_j, and their union is again a basis for $L^2(\mathbb{R}) = \bigoplus_{j\in\mathbb{Z}} W_j$. We define thus

$$\psi^1_{j,\ell}(x) = 2^{-j/2}\, \psi^1(2^{-j}x - 2\ell), \qquad \psi^2_{j,\ell}(x) = 2^{-j/2}\, \psi^2(2^{-j}x - 2\ell)\ ;$$

the $\{\psi^1_{j,\ell}, \psi^2_{j,\ell};\ j,\ell \in \mathbb{Z}\}$ constitute an orthonormal basis of $L^2(\mathbb{R})$. Since $\hat{\psi}^1, \hat{\psi}^2$ are obtained by "splitting" $\hat{\psi}$, ψ_1 and ψ_2 both have better frequency localization than ψ itself (at the price of having larger supports in x-space!). The splitting up of frequency space corresponding to the W_j on one hand, and the W^1_j, W^2_j on the other hand, is represented schematically in Figure 10.7. Note that frequency is still treated logarithmically, even with the $\psi^1_{j,k}, \psi^2_{j,k}$.

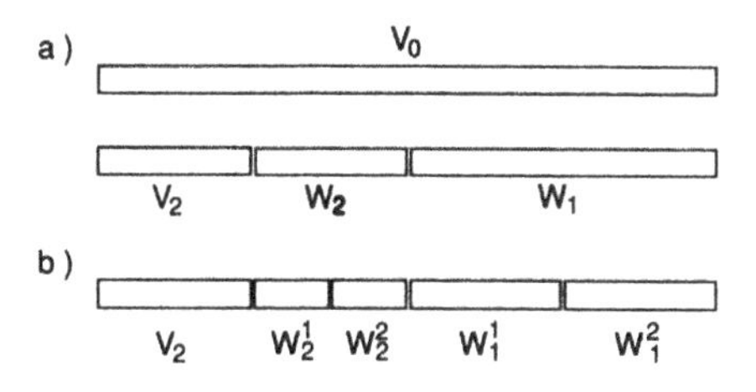

FIG. 10.7. *Schematic representation of the splitting of* V_0 *(corresponding more or less to a bandwidth of* π*) into* (a) W_1, W_2, *and* V_2 *or into* (b) $W^1_1, W^2_1, W^1_2, W^2_2$, *and* V_2.

By construction, $\hat{\psi}^1(\xi) = \sqrt{2}\ m_0(\xi)\ \hat{\psi}(\xi)$, $\hat{\psi}^2(\xi) = \sqrt{2}\ m_1(\xi)\ \hat{\psi}(\xi)$. It follows that $|\hat{\psi}^1(\xi)|^2 + |\hat{\psi}^2(\xi)|^2 = 2|\hat{\psi}(\xi)|^2$, as illustrated by Figure 10.8, which also shows that $\hat{\psi}^1$, $\hat{\psi}^2$ do indeed "split" $\hat{\psi}$ into two pieces, corresponding to its lowest and highest "frequency halves."

Computing the coefficients of a function with respect to the $\psi^1_{j,k}$, $\psi^2_{j,k}$ can again be done via a subband filtering scheme, in which one additional step of

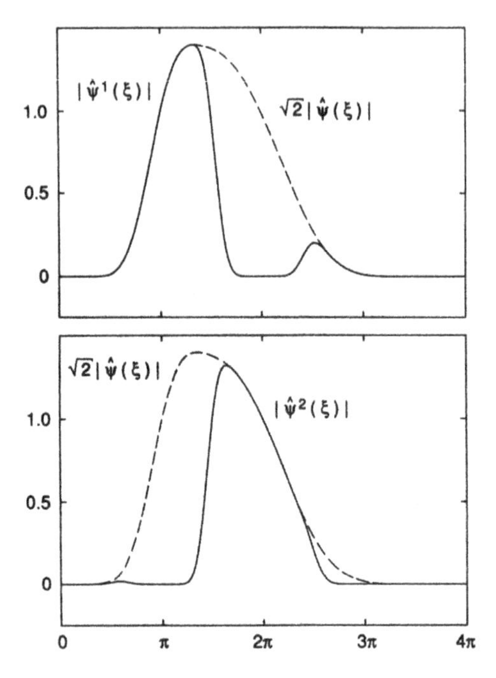

FIG. 10.8. *Graphs of* $|\hat{\psi}^1(\xi)|$, $|\hat{\psi}^2(\xi)|$, *where the low-pass filter is chosen equal to* ${}_{10}m_0(\xi)$, *as defined in* §6.4. *The dotted line is the graph of* $\sqrt{2}\ |\hat{\psi}(\xi)|$.

high- and low-pass splitting is added after the "standard" high-pass filtering. Schematically, we have now Figure 10.9. Note that the dilation factor 4 scheme proposed at the end of §10.2 (derived from a factor 2 scheme) also contains these functions ψ^1, ψ^2. (The wavelets in that scheme are essentially $\psi(x)$, the original wavelet for factor 2, and $\sqrt{2}\ \psi^1(2x)$, $\sqrt{2}\ \psi^2(2x)$, with ψ^1, ψ^2 defined as above.)

If one works with a tensor product multiresolution analysis in higher dimensions, then the splitting trick can be applied in a selective way. Figure 10.10 shows how one can, e.g., use the splitting trick in two dimensions to achieve an orthonormal wavelet basis with better angular resolution in the frequency plane than a "standard" wavelet basis. Figure 10.10a visualizes the construction of §10.1 in the frequency plane: the small central square corresponds to (say) $\mathbf{V}_0$; adding to it the two vertical rectangles corresponding to $\mathbf{W}_0^v = W_0 \otimes V_0$, the two horizontal rectangles for $\mathbf{W}_0^h = V_0 \otimes W_0$ and the four corner squares for $\mathbf{W}_0^d = W_0 \otimes W_0$ leads to the bigger square representing $\mathbf{V}_{-1}$. The structure then repeats in the next annulus, to constitute $\mathbf{V}_{-2}$. The angular resolution in the Fourier plane of this scheme is not very good, as shown by the figure. Figure 10.10b shows what the same two-dimensional construction looks like when one starts from a one-dimensional multiresolution analysis with dilation factor 4, as given at the end of §10.2. In this case the one-dimensional scheme has already three wavelets, so that the two-dimensional product scheme ends up with $2 \times 3 + 3^2 = 15$ wavelets. Figure 10.10b represents one step (with dilation 4)

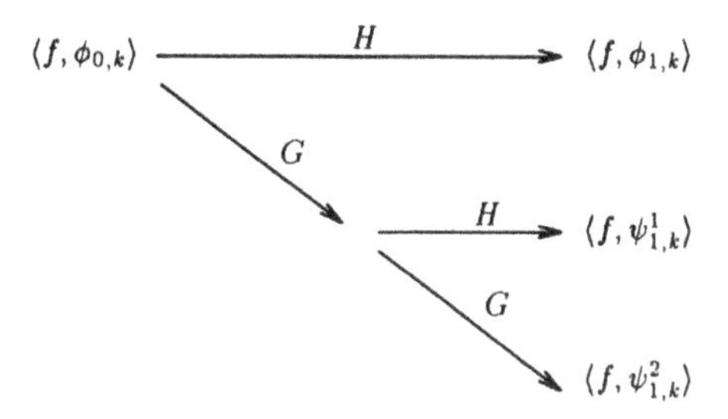

FIG. 10.9. *Schematic representation of the different filtering operations when working with "split" wavelets.*

in the multiresolution scale, as compared to two steps (with dilation factor 2), i.e., two successive annuli in Figure 10.10a. The central part of the two pictures is identical; the only difference between the two pictures is that the outer annulus of Figure 10.10a is split into many pieces to give Figure 10.10b, while the inner annulus is untouched. This corresponds to a "splitting of one level out of two" in terms of the splitting trick lemma, as pointed out above. The result is good angular resolution for some wavelets (corresponding to the outer layer in Figure 10.10b), bad for others (corresponding to the most central rectangles in Figure 10.10b).

Figure 10.10c shows the same picture again, with two steps in a multiresolution analysis with dilation factor 2, but for a product structure starting from the two $\frac{1}{2}$ octave bandwidth wavelets constructed in this section rather than the one octave bandwidth wavelet ψ. The scaling function is the same, but there are now $2 \times 2 + 2^2 = 8$ wavelets (as opposed to 3 for Figure 10.10a, and 15 for Figure 10.10b). Figure 10.10c can be obtained from Figure 10.10a by splitting every annulus (inner as well as outer) into halves by cuts in both horizontal and vertical directions. This improves the angular resolution on the squares in the corners (corresponding to the $\mathbf{W}_j^d$ of Figure 10.10a), but does nothing for the angular resolution of the rectangles (corresponding to $\mathbf{W}_j^h$ or $\mathbf{W}_j^v$ in Figure 2a), which were split better in the outer annulus of Figure 10.10b. The best angular resolution can be obtained by giving up the product structure altogether and just carving up every one of the $\mathbf{W}_j^\lambda$ spaces of Figure 10.10a vertically and/or horizontally, by applying the "splitting trick" in x and/or y, until the desired resolution is achieved. An example is given in Figure 10.10d. This still corresponds to an orthonormal basis, and to a fast algorithm for decomposing and reconstructing functions, although the organization is somewhat more complex. If even better angular resolution is required, then one can repeat the splitting trick where and as many times as necessary.

10.6. Wavelet packet bases.

The better resolution wavelets of the previous section are in fact only special cases in a beautiful construction of Coifman and Meyer, called wavelet packets.

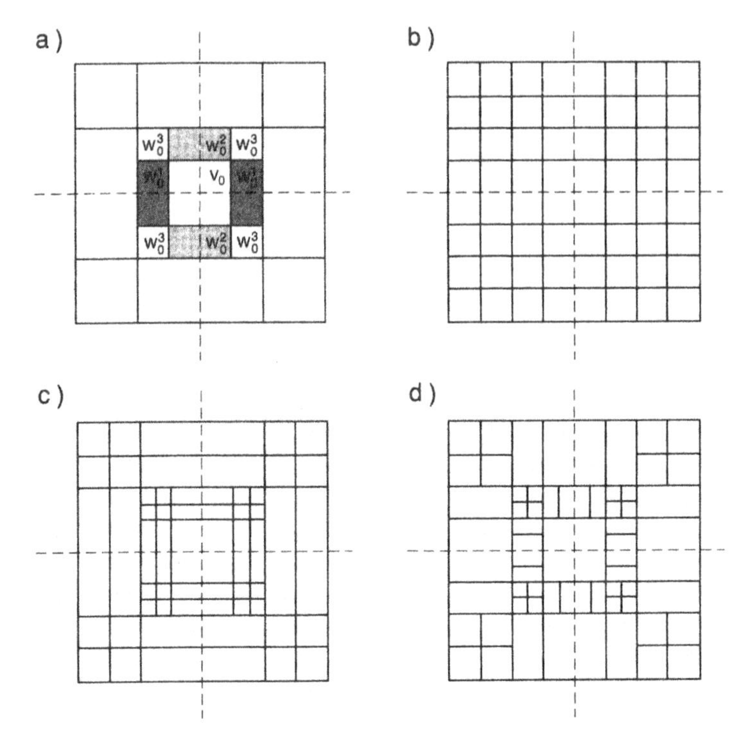

FIG. 10.10. *Visualization of the localization in the Fourier plane achieved by various two-dimensional multiresolution schemes, as explained in the text.*

The present section is only a short description of their construction; more details can be found in, e.g., Coifman, Meyer, and Wickerhauser (1992), and applications to acoustic signals and images are given in Wickerhauser (1990, 1992).

Start with a "usual" multiresolution analysis for factor 2, and consider only the V_j, W_j spaces with $j \leq 0$. The decomposition

$$L^2(\mathbb{R}) = V_0 \oplus \left(\bigoplus_{j \leq 0} W_j \right)$$

corresponds to a frequency splitting as schematically represented in Figure 10.7a. Heuristically, W_{-1} has "twice the size" of V_0 and W_0 (which have the "same size"), W_{-2} has "four times this size," etc. One can imagine reducing all of them, via the splitting trick, to subspaces of the same size; W_{-1} gets split once, W_{-2} twice, etc. This corresponds to defining a host of functions $\psi_{\ell;\, \epsilon_1, \cdots, \epsilon_\ell}$, where ℓ stands for the original $W_{-\ell}$-space (and for the number of splittings this space has undergone), and $\epsilon_j = 0$ or 1 indicates the choice, m_0 or m_1, at the jth splitting. Explicitly,

$$\begin{aligned}\hat{\psi}_{\ell;\,\epsilon_1,\cdots,\epsilon_\ell}(\xi) &= \left[\prod_{j=1}^{\ell} m_{\epsilon_j}(2^{-j}\xi)\right] m_1(2^{-\ell-1}\xi)\ \hat{\phi}(2^{-\ell-1}\xi) \\ &= \left[\prod_{j=1}^{\ell} m_{\epsilon_j}(2^{-j}\xi)\right] \hat{\psi}(2^{-\ell}\xi)\ .\end{aligned}$$

Clearly, the $\psi_{\ell;\epsilon_1,\cdots,\epsilon_\ell}(x)$ are all linear combinations of the $\psi(2^\ell x - k)$, and the splitting trick lemma (applied ℓ times) proves that $\{\psi_{\ell;\epsilon_1,\cdots,\epsilon_\ell}(\cdot - n);\ \epsilon_1,\cdots,\epsilon_\ell = 0 \text{ or } 1,\ n \in \mathbb{Z}\}$ is an orthonormal basis of $\overline{\text{Span}\ \{\psi(2^\ell \cdot -k);\ k \in \mathbb{Z}\}} = W_{-\ell}$. It follows that $\{\psi_{\ell;\epsilon_1,\cdots,\epsilon_\ell}(\cdot - n);\ \ell \in \mathbb{N},\ n \in \mathbb{Z}, \epsilon_1,\cdots,\epsilon_\ell = 0 \text{ or } 1\} \cup \{\phi(\cdot - n);\ n \in \mathbb{Z}\}$ is an orthonormal basis for $L^2(\mathbb{R})$. Note that this basis corresponds to the integer translates of functions which all have more or less equivalent frequency localization (in bands of width more or less π, starting with $|\xi| \le \pi$ for the $\phi(\cdot - n)$, $\pi \le |\xi| \le 2\pi$ for $\psi(\cdot - n), \cdots$).[11] This is very similar in flavor to the windowed Fourier transform and the Wilson bases of §4.2.B, while still having the same ease of computation, via subband filtering schemes, as wavelet bases.

There are of course many intermediate solutions between wavelets on one hand and the wavelet packet basis described above: one may choose to split some $W_{-\ell}$ spaces less often, or to split some of its subspaces more than others. Every such choice corresponds to an orthonormal basis; moreover, there exist efficient algorithms (based on computations of the "entropy of a function" with respect to the different splittings) to determine which of this whole library of choices is the most efficient for given signal; see e.g., Coifman and Wickerhauser (1992).

10.7. Wavelet bases on an interval.

All the one-dimensional wavelet constructions we have discussed so far lead to bases for $L^2(\mathbb{R})$. In many applications one is interested in only part of the real line: numerical analysis computations generally work on an interval, images are concentrated on rectangles, many systems to analyze sound divide it in chunks. All these involve decompositions of functions f supported on an interval, say $[0, 1]$. One could, of course, decide to use standard wavelet bases to analyze f, setting the function equal to zero outside $[0, 1]$, but this introduces an artificial "jump" at the edges, reflected in the wavelet coefficients.[12] It is, moreover, not efficient computationally. It is therefore useful to develop wavelets adapted to "life on an interval."

A first way of achieving this is to use the periodized wavelets described in §9.3. These are computationally efficient, but using them amounts to analyzing the periodized function $\tilde{f}$, defined by $\tilde{f}(x) = f(x - \lfloor x \rfloor)$ (where $\lfloor x \rfloor$ denotes the largest integer not exceeding x), by means of the usual (non-periodized) wavelets. Unless f is already periodic, we have again introduced a "jump" at the boundaries 0, 1, which will be reflected by large fine scale wavelet coefficients near 0 and 1.

There exists another solution which does not present this inconvenience, proposed by Meyer (1992), based on orthonormal wavelet bases with compact support. In this construction, the wavelets with a support contained in $[0, 1]$ that does not reach either 0 or 1 are kept as are, but this family is supplemented at the edges by specially adapted functions. Let us illustrate how the idea works on the half line rather than an interval. This simplification allows us to deal with only one boundary, and to disregard the adaptations needed at the coarsest scales, where both boundaries of $[0, 1]$ have to be dealt with simultaneously. Define

$$\phi_{j,k}^{\text{half}}(x) = \begin{cases} \phi_{j,k}(x) & \text{if } x \geq 0 , \\ 0 & \text{otherwise} ; \end{cases}$$

$$V_j^{\text{half}} = \overline{\text{Span}\ \{\phi_{j,k}^{\text{half}};\ k \in \mathbb{Z}\}} .$$

The space V_j^{half} can also be viewed as the space of all restrictions to $[0, \infty)$ of functions in V_j. If we suppose that the original scaling function ϕ has support $[0, 2N-1]$, then $\phi_{j,k}^{\text{half}}(x) = 0$ if $k \leq -2N+1$; we therefore only consider the $\phi_{j,k}^{\text{half}}$ with $k > -2N+1$. All but $2N-2$ of these are untouched by the restricting procedure: $\phi_{j,k}^{\text{half}}(x) = \phi_{j,k}(x)$ if $k \geq 0$; these functions are therefore still orthonormal. The $2N-2$ functions $\phi_{j,k}^{\text{half}}$, $k = -1, \cdots, -(2N-2)$ are independent of each other and of the $\phi_{j,k}$ with $k \geq 0$. Define now W_j^{half} to be the orthogonal complement in V_{j-1}^{half} of V_j^{half}. If, for convenience, we shift ψ so that it is also supported on $[0, 2N-1]$, then the $\psi_{j,k}^{\text{half}}$ (restrictions of $\psi_{j,k}$ to $[0, \infty)$) are clearly in W_j^{half} if $k \geq 0$, since they are orthogonal to all the $\phi_{j,k}^{\text{half}}$, and lie in V_{j-1}^{half}. What about the $\psi_{j,k}^{\text{half}}$ with $k = -1, \cdots, -(2N-2)$? (If k is even smaller, $k \leq -2N+1$, then $\psi_{j,k}^{\text{half}} \equiv 0$.) It turns out (see Meyer (1992)) that the $\psi_{j,k}^{\text{half}}$ with $k = -N, -(N+1), \cdots, -(2N-2)$ are in V_j^{half}, i.e., they are orthogonal to W_j^{half}. The other $\psi_{j,k}^{\text{half}}$, $k = -1, \cdots, -(N-1)$ contribute to W_j^{half}; in fact, we have that

$$\{\phi_{j,k}^{\text{half}};\ \ k \geq -(2N-2)\} \cup \{\psi_{j,k}^{\text{half}};\ \ k \geq -N(-1)\}$$

is a (non-orthogonal) basis for V_{j-1}^{half}.[13] In order to orthonormalize this basis, one proceeds in the following steps:

(1) Orthonormalize the $\phi_{0,k}^{\text{half}}$, $k = -1, \cdots, -(2N-2)$. The resulting functions $\tilde{\phi}_k$, $k = -1, \cdots, -(2N-2)$ are automatically orthogonal to the $\phi_{0,k}$, $k \geq 0$, and together they provide an orthonormal basis for V_0^{half}. If we define

$$\tilde{\phi}_{j,k}(x) = 2^{-j/2}\ \tilde{\phi}_k(2^{-j}x), \ \ j \in \mathbb{Z},\ k = -1, \cdots, -(2N-2) ,$$

then $\{\phi_{j,k};\ k \geq 0\} \cup \{\tilde{\phi}_{j,k};\ k = -1, \cdots, -(2N-2)\}$ is an orthonormal basis for V_j^{half} for any $j \in \mathbb{Z}$.

(2) Project the $\psi_{0,k}^{\text{half}}$, $k = -1, \cdots, -(N-1)$ onto W_0^{half} by defining

$$\psi_k^{\#} = \psi_{0,k}^{\text{half}} - \sum_{\ell=0}^{2N-2} \langle \psi_{0,k}^{\text{half}}, \tilde{\phi}_\ell \rangle \, \tilde{\phi}_\ell \,.$$

(3) Orthonormalize the $\psi_k^{\#}$. The resulting $\tilde{\psi}_k$, $k = -1, \cdots, -(N-1)$, together with the $\psi_{0,k}^{\text{half}}$, $k \geq 0$, provide an orthonormal basis for W_0^{half}. We can again define

$$\tilde{\psi}_{j,k}(x) = 2^{-j/2} \, \tilde{\psi}_k(2^{-j}x), \qquad j \in \mathbb{Z}, \ k = -1, \cdots, -(N-1) \ ;$$

then $\{\tilde{\psi}_{j,k};\ k = -1, \cdots, -(N-1)\} \cup \{\psi_{j,k};\ k \geq 0\}$ is an orthonormal basis for W_j^{half}. The union of all these bases (j ranging over $\mathbb{Z}$) gives a basis for $L^2([0, \infty))$.

The resulting bases are not only orthonormal bases for $L^2([0,\infty))$, but provide also unconditional bases for the Hölder spaces restricted to the half-line (i.e., they even handle regularity properties at 0 "correctly"), etc.; see Meyer (1992) for proofs. To implement all this in practice, one needs to compute extra filter coefficients at the boundaries, corresponding to the expansion of the $\tilde{\psi}_{0,k}$, $k = -1, \cdots, -(N-1)$, $\tilde{\phi}_{0,k}$, $k = -1, \cdots, -(2N-2)$, in terms of the $\tilde{\phi}_{-1,\ell}$, $\ell = -1, \cdots, -(2N-2)$ and the $\phi_{-1,\ell}$, $\ell = 0, \cdots, 4N-5$. These can be computed from the original h_ℓ; tables are given in Cohen, Daubechies, and Vial (1992); this paper also contains an alternative to Meyer's construction, involving fewer additional functions at the edges (only N instead of $2N-2$), while still handling the regularity properties correctly, even at the edge.

One last remark about wavelet bases on an interval. In image analysis, it is customary to treat border effects by extending the image, beyond the border, by its reflection: this extension avoids the discontinuity that follows from periodization or extending by zero (although there still is a discontinuity in the derivative, however). It is well known that this means that border effects are minimized, and that no extra coefficients (to deal with the borders) have to be introduced, provided that the filters used are symmetric. The same trick can be used to provide biorthogonal wavelet bases on $[0,1]$, with much less effort than the orthonormal wavelet basis on the interval of Meyer (1992) or Cohen, Daubechies, and Vial (1993).

If f is a function on $\mathbb{R}$, then we can define a function on $[0,1]$ by "folding" its graph at 0 and 1. The first "fold," at 0, amounts to replacing $f(x)$ by $f(x) + f(-x)$. Folding back the two tails (one from the original f, the other from the folded over negative part) sticking out beyond 1 leads to $f(x) + f(-x) + f(2-x) + f(x+2)$. If we keep folding like this, then we end up with

$$f^{\text{fold}}(x) = \sum_{\ell \in \mathbb{Z}} f(x - 2\ell) + \sum_{\ell \in \mathbb{Z}} f(2\ell - x) \,. \tag{10.7.1}$$

Note, for later convenience, that[14]

$$\int_0^1 dx \ f^{\text{fold}}(x) \ \overline{g^{\text{fold}}(x)} \ = \ \int_{-\infty}^{\infty} dx \ f(x) \ \overline{g^{\text{fold}}(x)} \,. \tag{10.7.2}$$

Take now $\psi, \tilde{\psi}$, two wavelets which generate biorthogonal wavelet bases of $L^2(\mathbb{R})$, with associated scaling functions $\phi, \tilde{\phi}$, as constructed in §8.3; assume also that $\phi, \tilde{\phi}$ are symmetric around $\frac{1}{2}$, $\phi(1-x) = \phi(x)$, $\tilde{\phi}(1-x) = \tilde{\phi}(x)$, and that ψ, $\tilde{\psi}$ are antisymmetric around $\frac{1}{2}$, $\psi(1-x) = -\psi(x)$, $\tilde{\psi}(1-x) = -\tilde{\psi}(x)$. (Examples were constructed in §8.3.) Apply the "folding" technique to the $\psi_{j,k}$ and $\tilde{\psi}_{j,k}$,

$$\begin{aligned}
\psi_{j,k}^{\text{fold}}(x) &= 2^{-j/2} \sum_{\ell \in \mathbb{Z}} \psi(2^{-j}x - 2^{-j+1}\ell - k) \\
&\quad + 2^{-j/2} \sum_{\ell \in \mathbb{Z}} \psi(2^{-j+1}\ell - 2^{-j}x - k) \\
&= 2^{-j/2} \sum_{\ell} \psi(2^{-j}x - 2^{-j+1}\ell - k) \\
&\quad - 2^{-j/2} \sum_{\ell \in \mathbb{Z}} \psi(2^{-j}x - 2^{-j+1}\ell + 1 + k); \qquad (10.7.3)
\end{aligned}$$

$\tilde{\psi}_{j,k}^{\text{fold}}$ is defined analogously. We will restrict our attention to $j \le 0$, or $j = -J$ with $J \ge 0$, for which (10.7.2) can be rewritten as

$$\psi_{j,k}^{\text{fold}} = \sum_{\ell \in \mathbb{Z}} \left[\psi_{-J,k+2^{J+1}\ell} - \psi_{-J,2^{J+1}\ell-k-1} \right] .$$

For good measure we also define $\phi_{j,k}^{\text{fold}}$, $\tilde{\phi}_{j,k}^{\text{fold}}$; since $\phi(x) = \phi(1-x)$, $\tilde{\phi}(x) = \tilde{\phi}(1-x)$, we find

$$\phi_{j,k}^{\text{fold}} = \sum_{\ell \in \mathbb{Z}} \left[\phi_{-J,k+2^{J+1}\ell} + \phi_{-J,2^{J+1}\ell-k-1} \right] .$$

Obviously, $\phi_{-J,k+2^{J+1}m}^{\text{fold}} = \phi_{-J,k}^{\text{fold}}$ for $m \in \mathbb{Z}$, so that we only need to consider the values $k = 0, \cdots, 2^{J+1} - 1$. Moreover, $\phi_{-J,2^{J+1}-k-1}^{\text{fold}} = \phi_{-J,k}^{\text{fold}}$, which means we can restrict ourselves to only $k = 0, \cdots, 2^J - 1$. A similar argument shows that we only need to consider the $\psi_{-J,k}^{\text{fold}}$ for $k = 0, \cdots, 2^J - 1$. Remarkably, the $\phi_{-J,k}^{\text{fold}}$ and $\tilde{\phi}_{-J,k'}^{\text{fold}}$, with $0 \le k, k' \le 2^J - 1$, are still biorthogonal on $[0,1]$. To prove this, use (10.7.1):

$$\begin{aligned}
&\int_0^1 dx\ \phi_{-J,k}^{\text{fold}}(x)\ \overline{\tilde{\phi}_{-J,k'}^{\text{fold}}(x)} \\
&= \sum_{\ell \in \mathbb{Z}} \left[\langle \phi_{-J,k}, \tilde{\phi}_{-J,k'+2^{J+1}\ell} \rangle + \langle \phi_{-J,k}, \tilde{\phi}_{-J,2^{J+1}\ell-k'} \rangle \right] \\
&= \sum_{\ell \in \mathbb{Z}} \left(\delta_{k,k'+2^{J+1}\ell} + \delta_{k,2^{J+1}\ell-k'} \right) = \delta_{k,k'} \qquad (10.7.4) \\
&\qquad\qquad (\text{because } 0 \le k, k' \le 2^J - 1) .
\end{aligned}$$

This biorthogonality implies, among other things, that the $\phi_{-J,k}^{\text{fold}}$, $k = 0, \cdots, 2^J - 1$ are all independent, and provide a basis for $V_{-J}^{\text{fold}} = \{f^{\text{fold}};\ f \in V_{-J}\}$. (The same is true for the $\tilde{\phi}^{\text{fold}}_{-J,k}$.) We can also define spaces $W^{\text{fold}}_{-J}, \tilde{W}^{\text{fold}}_{-J}$ by $W^{\text{fold}}_{-J} = \{f^{\text{fold}};\ f \in W_{-J}\}$. Obviously, W^{fold}_{-J} is spanned by the $\psi^{\text{fold}}_{-J,k}$, $k = 0, \cdots, 2^J - 1$. Moreover, computations similar to (10.7.4) show that

$$\int_0^1 dx\ \psi^{\text{fold}}_{-J,k}(x)\ \overline{\tilde{\phi}^{\text{fold}}_{-J,k'}(x)} = 0 ,$$

$$\int_0^1 dx\ \psi^{\text{fold}}_{-J,k}(x)\ \overline{\tilde{\psi}^{\text{fold}}_{-J,k'}(x)} = \delta_{k,k'} ,$$

proving that $W^{\text{fold}}_{-J} \perp \tilde{V}^{\text{fold}}_{-J}$ and that the $\psi^{\text{fold}}_{-J,k}$, $0 \le k \le 2^J - 1$ are independent. It follows that the "folded" structures inherit all the properties (nesting of the spaces, biorthogonality, basis properties,$\cdots$) from the unfolded originals. The filter coefficients corresponding to these folded biorthogonal bases are likewise obtained by folding at the edges corresponding to $x = 0, 1$; if $\psi, \tilde{\psi}, \phi, \tilde{\phi}$ are compactly supported, then only the filter coefficients near the borders will be affected. Examples are given in Cohen, Daubechies, and Vial (1993). Because analyzing f on $[0, 1]$ with these folded biorthogonal wavelets amounts to the same as extending f to all of $\mathbb{R}$ by reflections and analyzing this extension with the original biorthogonal wavelets, we cannot hope, however, to characterize Hölder spaces on $[0, 1]$ beyond Hölder exponent 1 with this technique. This is progress with respect to what periodized wavelets can do, but it is less performant than the orthonormal wavelet bases on $[0, 1]$. For more details, see Cohen, Daubechies, and Vial (1993).

Notes.

1. One example is the following. Let Γ be the hexagonal lattice, $\Gamma = \{n_1 e_1 + n_2 e_2;\ n_1, n_2 \in \mathbb{Z}\}$, where $e_1 = (1, 0)$, $e_2 = (1/2, \sqrt{3}/2)$; Γ defines a partition of $\mathbb{R}^2$ into equilateral triangles. Define V_0 to be the space of continuous functions in $L^2(\mathbb{R})$ that are piecewise affine on these triangles. The orthonormal basis for this multiresolution analysis is constructed in Jaffard (1989). Biorthogonal bases of compactly supported wavelets with this hexagonal symmetry are constructed in Cohen and Schlenker (1993).

2. The one-dimensional conditions of §5.1 can also be cast in a matrix form: in that case $\hat{\psi}(\xi) = m_1(\xi/2)\ \hat{\phi}(\xi/2)$, and the conditions are $|m_0(\xi)|^2 + |m_0(\xi + \pi)|^2 = 1$, $|m_1(\xi)|^2 + |m_1(\xi + \pi)|^2 = 1$, $m_0(\xi)\ \overline{m_1(\xi)} + m_0(\xi + \pi)\ \overline{m_1(\xi + \pi)} = 0$, ensuring the orthonormality of the $\{\phi_{0,n};\ n \in \mathbb{Z}\}$, of the $\{\psi_{0,n};\ n \in \mathbb{Z}\}$ and the orthogonality of these two sets of vectors, respectively. But these conditions are equivalent to the requirement that the matrix $\begin{pmatrix} m_0(\xi) & m_1(\xi) \\ m_0(\xi + \pi) & m_1(\xi + \pi) \end{pmatrix}$ is unitary.

3. If one prefers to index the entries of U with the numbers $1, \cdots, 2^n$ rather than entries of $\{0, 1\}^n$, then one can renumber $\mathrm{s} \in \{0, 1\}^n$ by defining $\sigma = 1 + \sum_{j=1}^n s_j 2^{j-1} \in \{1, \cdots, 2^n\}$.

4. At present, I know of no explicit scheme that provides an infinite family of m_0, for dilation factor 3, with regularity growing proportionally to the filter support width.

5. The same can be done for dilation factor 2, where the factor matrices are even simpler. The basic idea is that if $|m_0(\xi)|^2 + |m_0(\xi+\pi)|^2 = 1$, then, for any $\gamma \in \mathbb{R}$, $n \in \mathbb{Z}$, $m_0^{\#}(\xi) = (1+\gamma^2)^{-1/2}\left[m_0(\xi) + \gamma e^{-i(2n+1)\xi}\,\overline{m_0(\xi+\pi)}\right]$ will also satisfy $|m_0^{\#}(\xi)|^2 + |m_0^{\#}(\xi+\pi)|^2 = 1$. If $m_0(\xi) = \sum_{n=0}^{2N+1} \alpha_n e^{-in\xi}$, then it is convenient to choose $n = N+1$, leading to $m_0^{\#}(\xi) = \sum_{n=0}^{2N+3} \alpha_n^{\#} e^{-in\xi}$. All this can be rewritten in matrix form:
$$\begin{pmatrix} m_0^{\#}(\xi) \\ e^{-i(2N+3)\xi}\,\overline{m_0^{\#}(\xi+\pi)} \end{pmatrix} = (1+\gamma^2)^{-1/2} \begin{pmatrix} 1 & \gamma e^{-2i\xi} \\ -\gamma & e^{-2i\xi} \end{pmatrix} \begin{pmatrix} m_0(\xi) \\ e^{-i(2N+1)\xi}\,\overline{m_0(\xi+\pi)} \end{pmatrix}.$$
The whole operation raises the degree of m_0 by 2. One can, moreover, prove (Vaidyanathan and Hoang (1988)) that any trigonometric polynomial m_0 satisfying $|m_0(\xi)|^2 + |m_0(\xi+\pi)|^2 = 1$ can be obtained by letting products of such γ-matrices act on two-tap filters. We have not used this method in our constructions in §6.4 because it does not preserve the divisibility of m_0 by $(1+e^{-i\xi})$: imposing that the final m_0 is divisible by $(1+e^{-i\xi})$ leads to highly nonlinear constraints on the parameters γ_j. The method has the advantage, however, that it leads to easy implementations of the filters (using the γ_j directly), and that roundoff errors on the γ_j do not mar the exact reconstruction property.

   Anyway, a similar matrix technique can be used for more than two channels, as shown in Doğanata, Vaidyanathan, and Nguyen (1988) or, with more practical matrix factors, in Vaidyanathan et al. (1989). These matrix factorization techniques go back to the circuit theory work of Belevitch (1968).

6. This method for determining m_1, m_2 was indicated to me by W. Lawton and R. Gopinath (personal communication, 1990).

7. Such a one-dimensional filter would not be useful in practice, however!

8. This has been observed by many authors. The oldest reference seems to be McClellan (1973). One can also replace the one-dimensional $\cos\xi$ by $(\alpha\cos\xi + (1-\alpha)\cos\zeta)/2$ for $\alpha \in \mathbb{R}$ arbitrary, but there seems little point in not making the symmetric choice $\alpha = \frac{1}{2}$.

9. One can argue that some of the higher-dimensional schemes discussed in §10.3 correspond to noninteger dilations. In two dimensions, for example, the matrices $D_1 = \begin{pmatrix} 1 & 1 \\ 1 & -1 \end{pmatrix}$ and $D_2 = \begin{pmatrix} 1 & -1 \\ 1 & 1 \end{pmatrix}$ both satisfy $D^8 = 16\,\mathrm{Id}$,

so that a single dilation can be viewed as a dilation by $\sqrt{2}$ (combined with a rotation and/or a reflection).

10. If one also considers orthonormal wavelet bases not resulting from multiresolution analysis, then it is not known whether irrational dilation factors are allowed or not.

11. For large ℓ, the concentration of $\hat{\psi}_{\ell,\epsilon_1,\cdots,\epsilon_\ell}$ is not as good as this discussion suggests, however; see Coifman, Meyer, and Wickerhauser (1992). This is already noticeable on Figure 10.8, where $\hat{\psi}^1, \hat{\psi}^2$ have "sidelobes."

12. In image analysis one often extends f beyond the boundaries of the image by reflection; this extension is continuous, but its derivative still has a "jump." We come back to this at the end of §10.7.

13. Some of these assertions are highly non-trivial! A sizeable part of Meyer (1992) is devoted to their proof; simpler proofs have recently been found by Lemarié and Malgouyres (1991).

14. We have

$$\begin{aligned}
&\int_0^1 dx\ f^{\mathrm{fold}}(x)\ \overline{g^{\mathrm{fold}}(x)} \\
&\quad = \sum_{\ell,\ell'} \int_0^1 dx\ \Big[f(x+2\ell)\overline{g(x+2\ell')} + f(x+2\ell)\overline{g(2\ell'-x)} \\
&\qquad\qquad + f(2\ell-x)\overline{g(x+2\ell')} + f(2\ell-x)\overline{g(2\ell'-x)} \Big] \\
&\quad = \sum_{\ell,m} \int_{2\ell}^{2\ell+1} dx\ f(x)\overline{g(x+2m)} + \sum_{\ell,n} \int_{2\ell}^{2\ell+1} dx\ f(x)\overline{g(2n-x)} \\
&\qquad + \sum_{\ell,m'} \int_{2\ell-1}^{2\ell} dy\ f(y)\overline{g(2m'-y)} + \sum_{\ell,n'} \int_{2\ell-1}^{2\ell} dy\ f(y)\overline{g(y+2n')} \\
&\quad = \int_{-\infty}^{\infty} dx\ f(x) \sum_m \overline{g(x+2m)} + \int_{-\infty}^{\infty} dx\ f(x) \sum_m \overline{g(2m-x)} \\
&\quad = \int_{-\infty}^{\infty} dx\ f(x)\overline{g^{\mathrm{fold}}(x)}\ .
\end{aligned}$$

References

Published papers and papers that are scheduled for publication are labelled by the year in which they appeared or will appear; other preprints are labelled by the year in which they were written. This may play havoc with the chronological ordering, but then, due to differences in publication speed of different journals, so may official publication dates.

M. ANTONINI, M. BARLAUD, AND P. MATHIEU (1991), *Image coding using lattice vector quantization of wavelet coefficients*, Proc. IEEE Internat. Conf. Acoust. Signal Speech Process., pp. 2273–2276.

M. ANTONINI, M. BARLAUD, P. MATHIEU, AND I. DAUBECHIES (1992), *Image coding using wavelet transforms*, IEEE Trans. Image Process., 1, pp. 205–220.

F. ARGOUL, A. ARNÉODO, J. ELEZGARAY, G. GRASSEAU, AND R. MURENZI (1989), *Wavelet transform of two-dimensional fractal aggregates*, Phys. Lett. A, 135, pp. 327–336.

F. ARGOUL, A. ARNÉODO, G. GRASSEAU, Y. GAGNE, E. J. HOPFINGER, AND U. FRISCH (1989), *Wavelet analysis of turbulence reveals the multifractal nature of the Richardson cascade*, Nature, 338, pp. 51–53.

A. ARNÉODO, F. ARGOUL, J. ELEZGARAY, AND G. GRASSEAU (1988), *Wavelet transform analysis of fractals: Application to nonequilibrium phase transitions*, in Nonlinear Dynamics, G. Turchetti, ed., World Scientific, Singapore, p. 130.

E. W. ASLAKSEN AND J. R. KLAUDER (1968), *Unitary representations of the affine group*, J. Math. Phys., 9, pp. 206–211; see also *Continuous representation theory using the affine group*, J. Math. Phys., 10 (1969), pp. 2267–2275.

P. AUSCHER (1989), *Ondelettes fractales et applications*, Ph.D. Thesis, Université Paris, Dauphine, Paris, France.

——— (1990), *Symmetry properties for Wilson bases and new examples with compact support*, preprint, Université de Rennes, France, in Wavelets: Mathematics and Applications, J. Benedetto and M. Frazier, eds., CRC Press, to appear.

——— (1992), *Wavelet bases for $L^2(\mathbb{R})$, with rational dilation factor*, in Ruskai et al. (1992), pp. 439–452.

P. AUSCHER, G. WEISS, AND M.V. WICKERHAUSER (1992), *Local sine and cosine bases of Coifman and Meyer and the construction of smooth wavelets*, in Chui (1992b).

H. BACRY, A. GROSSMANN, AND J. ZAK (1975), *Proof of the completeness of lattice states in the kq-representation*, Phys. Rev., B12, pp. 1118–1120.

R. BALIAN (1981), *Un principe d'incertitude fort en théorie du signal ou en mécanique quantique*, C. R. Acad. Sci. Paris, 292, Série 2.

V. BARGMANN (1961), *On a Hilbert space of analytic functions and an associated integral transform, I*, Comm. Pure Appl. Math, 14, pp. 187–214.

V. BARGMANN, P. BUTERA, L. GIRARDELLO, AND J. R. KLAUDER (1971), *On the completeness of coherent states*, Rep. Math. Phys., 2, pp. 221–228.

M. J. BASTIAANS (1980), *Gabor's signal expansion and degrees of freedom of a signal*, Proc. IEEE, 68, pp. 538–539.

——— (1981), *A sampling theorem for the complex spectrogram and Gabor's expansion of a signal in Gaussian elementary signals*, Optical Engrg., 20, pp. 594–598.

G. BATTLE (1987), *A block spin construction of ondelettes. Part I: Lemarié functions*, Comm. Math. Phys., 110, pp. 601–615.

——— (1988), *Heisenberg proof of the Balian-Low theorem*, Lett. Math. Phys., 15, pp. 175–177.

——— (1989), *Phase space localization theorem for ondelettes*, J. Math. Phys., 30, pp. 2195–2196.

——— (1992), *Wavelets, a renormalization group point of view*, in Ruskai et al. (1992), pp. 323–350.

V. BELEVITCH (1968), *Classical Network Synthesis*, Holden Day, San Francisco.

M. A. BERGER (1992), *Random affine iterated function systems: Curve generation and wavelets*, SIAM Review, 34, pp. 361–385.

J. BERTRAND AND P. BERTRAND (1989), *Time-frequency representations of broad-band signals,* pp. 164–171 in Combes, Grossmann, and Tchamitchian (1989).

G. BEYLKIN, R. COIFMAN, AND V. ROKHLIN (1991), *Fast wavelet transforms and numerical algorithms*, Comm. Pure Appl. Math., 44, pp. 141–183.

B. BOASHASH (1990), *Time-frequency signal analysis*, in Advances in Spectrum Analysis and Array Processing, S. Haykin, ed., Prentice-Hall, Englewood Cliffs, NJ, pp. 418–517 .

J. BOURGAIN (1988), *A remark on the uncertainty principle for Hilbertian basis*, J. Funct. Anal., 79, pp. 136–143.

P. BURT AND E. ADELSON (1983), *The Laplacian pyramid as a compact image code*, IEEE Trans. Comm., 31, pp. 482–540.

A. P. CALDERÓN (1964), *Intermediate spaces and interpolation, the complex method*, Stud. Math., 24, pp. 113–190.

A. S. CAVARETTA, W. DAHMEN, AND C. MICCHELLI (1991), *Stationary subdivision*, Mem. Amer. Math. Soc., 93, pp. 1–186.

C. K. CHUI (1992), *On cardinal spline wavelets*, in Ruskai et al. (1992), pp. 419–438.

——— (1992b), *An Introduction to Wavelets*, Academic Press, New York.

——— (1992c), (ed.), *Wavelets: A Tutorial in Theory and Applications*, Academic Press, New York.

C. K. CHUI AND X. SHI (1993), *Inequalities of Littlewood–Paley type for frames and wavelets*, SIAM J. Math. Anal., 24, pp. 263–277.

C. K. CHUI AND J. Z. WANG (1991), *A cardinal spline approach to wavelets*, Proc. Amer. Math. Soc., 113, pp. 785–793, and *On compactly supported spline wavelets and a duality principle*, Trans. Amer. Math. Soc., to appear.

A. COHEN (1990), *Ondelettes, analyses multirésolutions et filtres miroir en quadrature*, Ann. Inst. H. Poincaré, Anal. non linéaire, 7, pp. 439–459.

——— (1990b), *Ondelettes, analyses multirésolutions et traitement numérique du signal*, Ph.D. Thesis, Université Paris, Dauphine.

A. COHEN AND J. P. CONZE (1992), *Régularité des bases d'ondelettes et mesures ergodiques*, Rev. Math. Iberoamer., 8, pp. 351–366.

A. COHEN AND I. DAUBECHIES (1992), *A stability criterion for biorthogonal wavelet bases and their related subband coding schemes*, Duke Math. J., 68, pp. 313–335.

——— (1993a), *Orthonormal bases of compactly supported wavelets III: Better frequency localization*, SIAM J. Math. Anal., 24, pp. 520–527.

——— (1993b), *Non-separable bidimensional wavelet bases*, Rev. Math. Iberoamer., 9, pp. 51–137.

A. COHEN, I. DAUBECHIES, AND J. C. FEAUVEAU (1992), *Biorthogonal bases of compactly supported wavelets*, Comm. Pure Appl. Math., 45, pp. 485–500.

A. COHEN, I. DAUBECHIES, AND P. VIAL (1993), *Wavelets and fast wavelet transform on the interval*, Applied and Computational Harmonic Analysis, to appear.

A. COHEN AND J. JOHNSTON (1992), *Joint optimization of wavelet and impulse response constraints for biorthogonal filter pairs with exact reconstruction*, AT&T Bell Laboratories, unpublished.

A. COHEN AND J. M. SCHLENKER (1993), *Compactly supported wavelets with hexagonal symmetry*, Constr. Approx., 9, pp. 209–236.

R. R. COIFMAN AND Y. MEYER (1991), *Remarques sur l'analyse de Fourier à fenêtre*, C. R. Acad. Sci. Paris I, 312, pp. 259–261.

R. COIFMAN, Y. MEYER, AND M. V. WICKERHAUSER (1992), *Wavelet analysis and signal processing*, in Ruskai et al. (1992), pp. 153–178; and *Size properties of wavelet packets*, in Ruskai et al. (1992), pp. 453–470.

R. R. COIFMAN AND R. ROCHBERG (1980), *Representation theorems for holomorphic and harmonic functions in L^p*, Astérisque, 77, pp. 11–66.

R. COIFMAN AND M. V. WICKERHAUSER (1992), *Entropy-based algorithms for best basis selection*, IEEE Trans. Inform. Theory, 38, pp. 713–718.

J. M. COMBES, A. GROSSMANN, AND PH. TCHAMITCHIAN (1989), eds., *Wavelets-Time-Frequency Methods and Phase Space*, Proceedings of the Int. Conf., Marseille, Dec. 1987, Springer-Verlag, Berlin.

J. P. CONZE (1991), *Sur le calcul de la norme de Sobolev des fonctions d'échelles*, preprint, Dept. of Math., Université de Rennes, France.

J. P. Conze and A. Raugi (1990), *Fonction harmonique pour un opérateur de transition et application*, Bull. Soc. Math. France, 118, pp. 273–310.

I. Daubechies (1988), *Time-frequency localization operators: a geometric phase space approach,* IEEE Trans. Inform. Theory, 34, pp. 605–612.

——— (1988b), *Orthonormal bases of compactly supported wavelets*, Comm. Pure Appl. Math., 41, pp. 909–996.

——— (1990), *The wavelet transform, time-frequency localization and signal analysis*, IEEE Trans. Inform. Theory, 36, pp. 961–1005.

——— (1993), *Orthonormal bases of compactly supported wavelets II. Variations on a theme*, SIAM J. Math. Anal., 24, pp. 499–519.

I. Daubechies and A. Grossmann (1988), *Frames of entire functions in the Bargmann space*, Comm. Pure Appl. Math., 41, pp. 151–164.

I. Daubechies and A. J. E. M. Janssen (1993), *Two theorems on lattice expansions*, IEEE Trans. Inform. Theory, 39, pp. 3–6.

I. Daubechies and J. Klauder (1985), *Quantum mechanical path integrals with Wiener measures for all polynomial Hamiltonians II*, J. Math. Phys., 26, pp. 2239–2256.

I. Daubechies and J. Lagarias (1991), *Two-scale difference equations I. Existence and global regularity of solutions*, SIAM J. Math. Anal., 22, pp. 1388–1410.

——— (1992), *Two-scale difference equations II. Local regularity, infinite products of matrices and fractals*, SIAM J. Math. Anal., 23, pp. 1031–1079.

I. Daubechies and T. Paul (1987), *Wavelets — some applications*, in Proceedings of the International Conference on Mathematical Physics, M. Mebkkout and R. Sénéor, eds., World Scientific, Singapore, pp. 675–686.

——— (1988), *Time-frequency localization operators: A geometric phase space approach II. The use of dilations and translations*, Inverse Prob., 4, pp. 661–680.

I. Daubechies, A. Grossmann, and Y. Meyer (1986), *Painless nonorthogonal expansions*, J. Math. Phys., 27, pp. 1271–1283.

I. Daubechies, S. Jaffard, and J. L. Journé (1991), *A simple Wilson orthonormal basis with exponential decay*, SIAM J. Math. Anal., 22, pp. 554–572.

I. Daubechies, J. Klauder, and T. Paul (1987), *Wiener measures for path integrals with affine kinematic variables*, J. Math. Phys., 28, pp. 85–102.

N. Delprat, B. Escudié, P. Guillemain, R. Kronland-Martinet, Ph. Tchamitchian, and B. Torrésani (1992), *Asymptotic wavelet and Gabor analysis: extraction of instantaneous frequencies*, IEEE Trans. Inform. Theory, 38, pp. 644–664.

G. Deslauriers and S. Dubuc (1987), *Interpolation dyadique*, in Fractals, dimensions non entières et applications, G. Cherbit, ed., Masson, Paris, pp. 44–55.

——— (1989), *Symmetric iterative interpolation*, Constr. Approx., 5, pp. 49–68.

Z. Doğanata, P. P. Vaidyanathan, and T. Q. Nguyen (1988), *General synthesis procedures for FIR lossless transfer matrices, for perfect reconstruction multirate filter bank applications*, IEEE Trans. Acoust. Signal Speech Process., 36, pp. 1561–1574.

S. Dubuc (1986), *Interpolation through an iterative scheme*, J. Math. Anal. Appl., 114, pp. 185–204.

R. J. Duffin and A. C. Schaeffer (1952), *A class of nonharmonic Fourier series*, Trans. Amer. Math. Soc., 72, pp. 341–366.

P. Dutilleux (1989), *An implementation of the 'algorithme à trous' to compute the wavelet transform*, pp. 298–304 in Combes, Grossmann, and Tchamitchian (1989).

N. Dyn and D. Levin (1990) *Interpolating subdivision schemes for the generation of curves and surfaces*, in Multivariate Interpolation and Approximation, W. Haussman and K. Jeller, eds., Birkhauser, Basel, pp. 91–106.

N. Dyn, A. Gregory, and D. Levin (1987), *A 4-point interpolatory subdivision scheme for curve design*, Comput. Aided Geom. Des., 4, pp. 257–268.

T. Eirola (1992), *Sobolev characterization of solutions of dilation equations*, SIAM J. Math. Anal., 23, pp. 1015–1030.

D. Esteban and C. Galand (1977), *Application of quadrature mirror filters to split-band voice coding schemes*, Proc. IEEE Int. Conf. Acoust. Signal Speech Process., Hartford, Connecticut, pp. 191–195.

G. Evangelista (1992), *Wavelet transforms and wave digital filters*, pp. 396–407 in Meyer (1992b).

J. C. Feauveau (1990), *Analyse multirésolution par ondelettes non orthogonales et bancs de filtres numériques*, Ph.D. Thesis, Université de Paris Sud, Paris, France.

C. Fefferman and R. de la Llave (1986), *Relativistic stability of matter*, Rev. Math. Iberoamer., 2, pp. 119–213.

G. Fix and G. Strang (1969), *Fourier analysis of the finite element method in Ritz–Galerkin theory*, Stud. Appl. Math., 48, pp. 265–273.

P. Flandrin (1989), *Some aspects of non-stationary signal processing with emphasis on time-frequency and time-scale methods*, in Wavelets, J. M. Combes, A. Grossmann, and Ph. Tchamitchian, eds., Springer-Verlag, Berlin, pp. 68–98.

M. Frazier and B. Jawerth (1988), *The ϕ-transform and applications to distribution spaces*, in Function Spaces and Application, M. Cwikel et al., eds., Lecture Notes in Mathematics 1302, Springer-Verlag, Berlin, pp. 233–246; see also *A discrete transform and decompositions of distribution spaces*, J. Funct. Anal., 93 (1990), pp. 34–170.

M. Frazier, B. Jawerth, and G. Weiss (1991), *Littlewood–Paley theory and the study of function spaces*, CBMS – Conference Lecture Notes 79, American Mathematical Society, Providence, RI.

D. Gabor (1946), *Theory of communication*, J. Inst. Electr. Engrg., London, 93 (III), pp. 429–457.

F. Gori and G. Guattari (1985), *Signal restoration for linear systems with weighted inputs. Singular value analysis for two cases of low-pass filtering*, Inverse Probl., 1, pp. 67–85.

G. K. Gröchenig (1991), *Describing functions: atomic decompositions versus frames*, Monatsh. Math., 112, pp. 1–42.

K. Gröchenig (1987), *Analyse multi-échelle et bases d'ondelettes*, Comptes Rendus Acad. Sci. Paris, 305, Série I, pp. 13–17.

K. Gröchenig and W. R. Madych (1992), *Multiresolution analysis, Haar bases and self-similar tilings of* $\mathbb{R}^n$, IEEE Trans. Inform. Theory, 38, pp. 556–568.

A. Grossmann and J. Morlet (1984), *Decomposition of Hardy functions into square integrable wavelets of constant shape*, SIAM J. Math. Anal., 15, pp. 723–736.

A. Grossmann, J. Morlet, and T. Paul (1985), *Transforms associated to square integrable group representations, I. General results*, J. Math. Phys., 27, pp. 2473–2479.

——— (1986), *Transforms associated to square integrable group representations, II. Examples*, Ann. Inst. H. Poincaré, 45, pp. 293–309.

A. Grossmann, M. Holschneider, R. Kronland-Martinet, and J. Morlet (1987), *Detection of abrupt changes in sound signals with the help of wavelet transforms*, in Inverse Problems: An Interdisciplinary Study; Advances in Electronics and Electron Physics, Supplement 19, Academic Press, New York, pp. 298–306.

A. Grossmann, R. Kronland-Martinet, and J. Morlet (1989), *Reading and understanding continuous wavelet transforms*, in Wavelets, J. M. Combes, A. Grossmann, and Ph. Tchamitchian, eds., Springer-Verlag, Berlin, pp. 2–20.

A. Haar (1910), *Zur Theorie der orthogonalen Funktionen-Systeme*, Math. Ann., 69, pp. 331–371.

C. Heil and D. Walnut (1989), *Continuous and discrete wavelet transforms*, SIAM Rev., 31, pp. 628–666.

O. Herrmann (1971), *On the approximation problem in nonrecursive digital filter design*, IEEE Trans. Circuit Theory, CT-18, pp. 411–413.

M. Holschneider and Ph. Tchamitchian (1990), *Régularité locale de la fonction 'non-différentiable' de Riemann*, pp. 102–124 in Lemarié (1990).

M. Holschneider, R. Kronland-Martinet, J. Morlet, and Ph. Tchamitchian (1989), *A real-time algorithm for signal analysis with the help of the wavelet transform*, pp. 286–297 in Combes, Grossmann, and Tchamitchian (1989).

S. Jaffard (1989), *Construction et propriétés des bases d'ondelettes. Remarques sur la contrôlabilité exacte*, Ph.D. Thesis, Ecole Polytechnique, Palaiseau, France.

——— (1989b), *Exposants de Hölder en des points donnés et coéfficients d'ondelettes*, C. R. Acad. Sci. Paris, 308, Série 1, pp. 79–81.

I. M. James (1991), *Organizing a conference*, Math. Intelligencer, 13, pp. 49–51.

C. P. Janse and A. Kaiser (1983), *Time-frequency distributions of loud-speakers: the application of the Wigner distribution*, J. Audio Engrg. Soc., 37, pp. 198–223.

A. J. E. M. Janssen (1981), *Gabor representation of generalized functions*, J. Math. Appl., 80, pp. 377–394.

——— (1984), *Gabor representation and Wigner distribution of signals*, Proc. IEEE, pp. 41.B.2.1–41.B.2.4.

——— (1988), *The Zak transform: a signal transform for sampled time-continuous signals*, Phillips J. Res., 43, pp. 23–69.

——— (1992), *The Smith-Barnwell condition and non-negative scaling functions*, IEEE Trans. Inform. Theory, 38, pp. 884–885.

H. E. Jensen, T. Hoholdt, and J. Justesen (1988), *Double series representation of bounded signals*, IEEE Trans. Inform. Theory, 34, pp. 613–624.

M. Kac (1959), *Statistical independence in probability, analysis and number theory,* no. 12 in the Carus mathematical monographs, Mathematical Association of America.

G. Kaiser (1990), *Quantum Physics, Relativity and Complex Spacetime: Towards a New Synthesis*, North-Holland, Amsterdam.

J. R. Klauder (1966), *Improved version of the optical equivalence theorem*, Phys. Rev. Lett., 16, pp. 534–536; this topic is also discussed in Chapter 8 of J. R. Klauder and E. C. G. Sudarshan (1968).

J. R. Klauder and B.-S. Skagerstam (1985), *Coherent States*, World Scientific, Singapore.

J. R. Klauder and E. C. G. Sudarshan (1968), *Fundamentals of Quantum Optics*, W. A. Benjamin, New York.

J. Kovačević and M. Vetterli (1993), *Perfect reconstruction filter banks with rational sampling rates*, IEEE Trans. Signal Process., 41, pp. 2047–2066.

——— (1992), *Nonseparable multidimensional perfect reconstruction filter banks and wavelet bases for* $\mathbb{R}^n$, IEEE Trans. Inform. Theory, 38, pp. 533–555.

R. Kronland-Martinet, J. Morlet, and A. Grossmann (1987), *Analysis of sound patterns through wavelet transforms*, Internat. J. Pattern Recognition and Artificial Intelligence, 1, pp. 273–301.

E. Laeng (1990), *Nouvelles bases orthonormées de* L^2, C. R. Acad. Sci. Paris, 311, Série 1, pp. 677–680.

H. Landau (1967), *Necessary density conditions for sampling and interpolation of certain entire functions*, Acta Math., 117, pp. 37–52.

——— (1993), *On the density of phase space functions*, IEEE Trans. Inform. Theory, 39, pp. 1152–1156.

H. J. Landau and H. O. Pollak (1961), *Prolate spheroidal wave functions, Fourier analysis and uncertainty, II*, Bell Systems Tech. J., 40, pp. 65–84.

——— (1962), *Prolate spheroidal wave functions, Fourier analysis and uncertainty, III*, Bell Systems Tech. J., 41, pp. 1295–1336.

W. Lawton (1990), *Tight frames of compactly supported wavelets*, J. Math. Phys., 31, pp. 1898–1901.

——— (1991), *Necessary and sufficient conditions for constructing orthonormal wavelet bases*, J. Math. Phys., 32, pp. 57–61.

W. M. Lawton and H. L. Resnikoff (1991), *Multidimensional wavelet bases*, submitted to SIAM J. Math. Anal.

P. G. Lemarié (1988), *Une nouvelle base d'ondelettes de* $L^2(\mathbb{R}^n)$, J. de Math. Pures et Appl., 67, pp. 227–236.

——— (1990), ed., *Les ondelettes en 1989*, Lecture Notes in Mathematics no. 1438, Springer-Verlag, Berlin.

P. G. Lemarié (1991), *La propriété de support minimal dans les analyses multirésolution*, Comptes Rendus de l'Acad. Sci. Paris, 312, pp. 773–776.

P. G. Lemarié and G. Malgouyres (1992), in Meyer (1992b).

P. G. Lemarié and G. Malgouyres (1991), *Support des fonctions de base dans une analyse multirésolution*, Comptes Rendus de l'Acad. Sci. Paris I, 313, pp. 377–380.

E. Lieb (1981), *Thomas-Fermi theory and related theories of atoms and molecules*, Rev. Mod. Phys., 53, pp. 603–641.

F. Low (1985), *Complete sets of wave packets*, in A Passion for Physics – Essays in Honor of Geoffrey Chew, World Scientific, Singapore, pp. 17–22.

Yu. Lyubarskii (1992), *Frames in the Bargmann space of entire functions*, in Entire and subharmonic functions, Vol. 11 of the series Advances in Soviet Mathematics, B. Ya. Levin, ed., Springer-Verlag, Berlin, pp. 167–180.

S. Mallat (1989), *Multiresolution approximation and wavelets*, Trans. Amer. Math. Soc., 315, pp. 69–88.

——— (1989b), *A theory for multiresolution signal decomposition: the wavelet representation*, IEEE Trans. PAMI, 11, pp. 674–693.

——— (1989c), *Multifrequency channel decompositions of images and wavelet models*, IEEE Trans. Acoust. Signal Speech Process., 37, pp. 2091–2110.

——— (1991), *Zero-crossings of a wavelet transform*, IEEE Trans. Inform. Theory, 37, pp. 1019–1033.

S. Mallat and S. Zhong (1992), *Characterization of signals from multiscale edges*, Computer Science Tech. Report, New York University, IEEE Trans. PAMI, to appear.

S. Mallat and W. L. Hwang (1992), *Singularity detection and processing with wavelets*, IEEE Trans. Inform. Theory, 38, pp. 617–643.

H. Malvar (1990), *Lapped transforms for efficient transform/subband coding*, IEEE Trans. Acoust. Signal Speech Process., 38, pp. 969–978.

J. McClellan (1973), *The design of two-dimensional filters by transformations*, in Seventh Annual Princeton Conference on ISS, Princeton University Press, Princeton, NJ, pp. 247–251.

Y. Meyer (1985), *Principe d'incertitude, bases hilbertiennes et algèbres d'opérateurs*, Séminaire Bourbaki, 1985–1986, no. 662.

——— (1986), *Ondelettes, fonctions splines et analyses graduées*, Lectures given at the University of Torino, Italy.

——— (1990), *Ondelettes et opérateurs, I: Ondelettes, II: Opérateurs de Calderón-Zygmund, III: Opérateurs multilinéaires*, Hermann, Paris. An English translation will be published by the Cambridge University Press in 1992.

——— (1992), *Ondelettes sur l'intervalle*, Rev. Math. Iberoamer., 7, pp. 115–133.

——— (1992b) (ed.), *Wavelets and applications*, Proceedings of the International Conference on Wavelets, May 1989, Marseille, France; Masson, Paris.

C. A. MICCHELLI (1991), *Using the refinement equation for the construction of pre-wavelets*, Numer. Algorithms, 1, pp. 75–116.

C. A. MICCHELLI AND H. PRAUTZSCH (1989), *Uniform refinement of curves*, Linear Algebra Appl., 114/115, pp. 841–870.

F. MINTZER (1985), *Filters for distortion-free two-band multirate filter banks*, IEEE Trans. Acoust. Speech Signal Process., 33, pp. 626–630.

J. MORLET (1983), *Sampling theory and wave propagation*, in NATO ASI Series, Vol. 1, Issues in Acoustic signal/Image processing and recognition, C. H. Chen, ed., Springer-Verlag, Berlin, pp. 233–261

J. MORLET, G. ARENS, I. FOURGEAU, AND D. GIARD (1982), *Wave propagation and sampling theory*, Geophysics, 47, pp. 203–236.

J. MUNCH (1992), *Noise reduction in tight Weyl-Heisenberg frames*, IEEE Trans. Inform. Theory, 38, pp. 608–616.

R. MURENZI (1989), *Wavelet transforms associated to the n-dimensional Euclidean group with dilations: signals in more than one dimension*, in Wavelets, J. M. Combes, A. Grossmann, and Ph. Tchamitchian, eds., Springer-Verlag, Berlin, pp. 239–246; see also *Ondelettes multidimensionelles et application à l'analyse d'images*, Ph.D. Thesis (1990), Université Catholique de Louvain, Belgium.

T. PAUL (1985), *Ondelettes et mécanique quantique*, Ph.D. Thesis, Université de Marseille, France; see also Paul and Seip (1991).

T. PAUL AND K. SEIP (1992), *Wavelets and quantum mechanics*, in Ruskai et al. (1992), pp. 303–322.

A. M. PERELOMOV (1971), *On the completeness of a system of coherent states*, Teor. Mat. Fiz., 6, pp. 213–224.

G. POLYA AND G. SZEGÖ (1971), *Aufgaben und Lehrsätze aus der Analysis*, Vol. II, Springer-Verlag, Berlin.

M. PORAT AND Y. Y. ZEEVI (1988), *The generalized Gabor scheme of image representation in biological and machine vision*, IEEE Trans. Pattern Anal. Mach. Intell., 10, pp. 452–468.

M. RIEFFEL (1981), *Von Neumann algebras associated with pairs of lattices in Lie groups*, Math. Ann., 257, pp. 403–413.

O. RIOUL (1992), *Simple regularity criteria for subdivision schemes*, SIAM J. Math. Anal., 23, pp. 1544–1576.

M. B. RUSKAI, G. BEYLKIN, R. COIFMAN, I. DAUBECHIES, S. MALLAT, Y. MEYER, AND L. RAPHAEL (1992), eds., *Wavelets and their Applications*, Jones and Bartlett, Boston.

K. SEIP (1991), *Reproducing formulas and double orthogonality in Bargmann and Bergman spaces*, SIAM J. Math. Anal., 22, pp. 856–876.

K. SEIP AND R. WALLSTÉN (1990), *Sampling and interpolation in the Bargmann–Fock space*, preprint, Mittag-Leffler Institute.

M. J. Shensa (1991), *The discrete wavelet transform: wedding the 'à trous' and Mallat's algorithms*, preprint, Naval Ocean Systems Center, San Diego, IEEE Trans. Signal Process., to appear.

D. Slepian (1976), *On bandwidth*, Proc. IEEE, 64, pp. 292–300.

——— (1983), *Some comments on Fourier analysis, uncertainty and modeling*, SIAM Rev., 25, pp. 379–393.

D. Slepian and H. O. Pollak (1961), *Prolate spheroidal wave functions, Fourier analysis and uncertainty, I*, Bell Systems Tech. J., 40, pp. 43–64.

M. J. T. Smith and T. P. Barnwell III (1986), *Exact reconstruction techniques for tree-structured subband coders*, IEEE Trans. Acoust. Signal Speech Process., 34, pp. 434–441; the basic results were already presented at the IEEE Internat. Conf. Acoust. Signal Speech Process., March 1984, San Diego.

E. Stein (1970), *Singular integrals and differentiability properties of functions*, Princeton University Press.

E. Stein and G. Weiss (1971), *Introduction to Fourier Analysis on Euclidean Spaces*, Princeton University Press, Princeton.

J. O. Stromberg (1982), *A modified Franklin system and higher order spline systems on* $\mathbb{R}^n$ *as unconditional bases for Hardy spaces*, Conf. in honor of A. Zygmund, Vol. II, W. Beckner et al., ed., Wadsworth math. series, pp. 475–493,

D. J. Sullivan, J. J. Rehr, J. W. Wilkins, and K. G. Wilson (1987), *Phase space wannier functions in electronic structure calculations*, preprint, Cornell University.

Ph. Tchamitchian (1987), *Biorthogonalité et théorie des opérateurs*, Rev. Math. Iberoamer., 3, pp. 163–189.

B. Torrésani (1991), *Wavelet analysis of asymptotic signals: Ridge and skeleton of the transform*, in Meyer (1992b); see also Tchamitchian and Torrésani's paper in Ruskai et al. (1992), pp. 123–152.

P. P. Vaidyanathan (1987), *Theory and design of M-channel maximally decimated quadrature mirror filters with arbitrary M, having the perfect reconstruction property*, IEEE Trans. Acoust. Signal Speech Process., 35, pp. 476–492.

——— (1992), *Multirate Systems and Filter Banks*, Prentice-Hall, Englewood Cliffs, NJ.

P. P. Vaidyanathan and P.-Q. Hoang (1988), *Lattice structures for optimal design and robust implementation of two-channel perfect-reconstruction QMF banks*, IEEE Trans. Acoust. Signal Speech Process., 36, pp. 81–94.

P. P. Vaidyanathan, T. Q. Nguyen, Z. Doğanata, and T. Saramaki (1989), *Improved technique for design of perfect reconstruction FIR QMF banks with lossless polyphase matrices*, IEEE Trans. Acoust. Signal Speech Process., 37, pp. 1042–1056.

M. Vetterli (1984), *Multidimensional subband coding: some theory and algorithms*, Signal Process., 6, pp. 97–112.

——— (1986), *Filter banks allowing perfect reconstruction*, Signal Process., 10, pp. 219–244; these results were already presented as *Splitting a signal into subsampled channels allowing perfect reconstruction*, IASTED Conf. on Applied Signal Processing and Digital Filters, June 1985, Paris.

M. VETTERLI AND C. HERLEY (1992), *Wavelets and filter banks: Theory and design*, IEEE Trans. Signal Process., 40, pp. 2207–2232.

M. VETTERLI, J. KOVAČEVIĆ, AND D. LEGALL (1990), *Perfect reconstruction filter banks for HDTV representation and coding*, Image Comm., 2, pp. 349–364.

P. VIAL (1992), *Construction de bases orthonormales de* $\mathbb{R}^n$, preprint, Centre de Physique Théorique, CNRS, Luminy–Marseille, France.

L. F. VILLEMOES (1992), *Energy moments in time and frequency for two-scale difference equation solutions and wavelets*, SIAM J. Math. Anal, 23, pp. 1519–1543.

H. VOLKNER (1992), *On the regularity of wavelets*, IEEE Trans. Inform. Theory, 38, pp. 872–876.

M. V. WICKERHAUSER (1990), *Picture compression by best-basis sub-band coding*, preprint, Yale University.

——— (1992), *Acoustic signal processing with wavelet packets*, in Chui (1992c), pp. 679–700.

K. G. WILSON (1987), *Generalized Wannier Functions*, preprint, Cornell University.

A. WITKIN (1983), *Scale space filtering*, in Proc. Internat. Joint Conf. Artificial Intelligence.

R. M. YOUNG (1980), *An Introduction to Nonharmonic Fourier Series*, Academic Press, New York.

A. ZYGMUND (1968), *Trigonometric Series*, 2nd ed., Cambridge University Press, Cambridge.

Subject Index

Author Index